GUIDE TO MODERN METHODS OF INSTRUMENTAL ANALYSIS

Guide to Modern Methods of Instrumental Analysis

Edited by

T. H. GOUW
Chevron Research Company
Richmond, California

Wiley-Interscience, a Division of John Wiley & Sons, Inc.
New York • London • Sydney • Toronto

Library of Congress Catalog Card Number: 79-171913

ISBN 0-471-31925

Printed in the United States of America.

10 9 8 7 6 5 4 3

Contributors

Edward M. Barrall II, IBM Research Laboratory, San Jose, California

Manfred J. R. Cantow, IBM Corporation, Systems Development Division, San Jose, California

Norman B. Colthup, American Cyanamid Company, Stamford, Connecticut

Robert A. Flath, Western Utilization Research and Development Division, Agricultural Research Service, U.S. Department of Agriculture, Albany, California

D. D. Gilbert, Northern Arizona University, Flagstaff, Arizona

T. H. Gouw, Chevron Research Company, Richmond Laboratory, Richmond, California

Curtis R. Hare, University of Miami, Coral Gables, Florida

R. E. Jentoft, Chevron Research Company, Richmond Laboratory, Richmond, California

Julian F. Johnson, Department of Chemistry and Institute of Materials Science, University of Connecticut, Storrs, Connecticut

Charles P. Poole, Jr., Department of Physics, University of South Carolina, Columbia, South Carolina

Dallas L. Rabenstein, Department of Chemistry, University of Alberta, Edmonton, Alberta, Canada

Victor W. Rodwell, Department of Biochemistry, Purdue University, Lafayette, Indiana

David J. Shapiro, Department of Pharmacology, Stanford University School of Medicine, Stanford, California

Dennis H. Smith, University of Bristol, School of Chemistry, Bristol, England

Preface

Introductions seem to be as important to a book as credits are to a motion picture. They are usually as exciting as a lecture on statistical thermodynamics to students of social welfare. On the other hand the preface appears to be quite indispensable to critics and to graduate students in library science who still consider it their duty to peruse this article of faith. Actually, the introduction is a convenient place for the many loose items that are of possible interest but look odd when scattered haphazardly throughout the body of the text.

This book is intended as an advanced guide for the scientific investigator who wishes to acquire additional knowledge about the more widely used methods of instrumental analysis. It is therefore of greater value to those scientists who are already knowledgeable in one or more of the subjects discussed here. Although it is not an exhaustive survey, the book has a sufficient degree of comprehensiveness and depth to give the reader the most important information without requiring him to delve in the more specialized treatises. This volume should also be very helpful to the research worker who is seeking the solution to a question and who wishes to review the available techniques for possible applicability to his problem. All those who wish to have an overview of instrumental analysis can profit from an advanced textbook on the comparative merits of the most prevalent forms of instrumental analysis. The largest applicability of this book, however, is still for the average scientist who is regularly confronted with problems in analytical chemistry.

The scientific investigator has several courses of action at his disposal. He can choose from a bewildering array of methods, techniques, and commercially available equipment. But he may be unaware of the possible alternatives. Many of the instrumental methods now in routine use were not available when he graduated 20 years ago. Others were present only as a curiosity in some distant laboratory. A majority of the techniques in current use have made giant advances of which he may not be aware.

In some cases the scientist's choice of method is obvious. The boiling range of a chemical product intermediate has since time immemorial been determined in his organization by an ASTM D86 distillation. This test method was prescribed decades ago by intelligent beings who in the meantime have been rewarded for their acuity by being elevated to the presidency or related positions in the organization. So the ASTM D86 test method is again requested.

The younger and more adventurous scientist, unencumbered by tradition, may observe in his quest for an improved method that dissent and confusion are not confined to the college campus he has just left, but also enter his endeavors to establish the best way of solving his problem. He will observe, for example, that to communicate fruitfully with the actual operators of these instrumental techniques he must have some knowledge of and familiarity with the technical lore and terminology. When discussing equipment with a commercial purveyor, he will perceive that the expressed quality and capabilities of the instruments in which he is interested tend to be inversely proportional to the amount of his knowledge of the technique. And I can assure you that large numbers of instruments are acquired more on the basis of the availability of excess funds (usually at the end of the fiscal year) and the rhetorical eloquence of the salesman than as a result of a knowledgeable discussion about actual requirements.

In the absence of a patient, accessible, and erudite research scientist with a good knowledge of most of the available instrumental techniques, the investigator often has no other choice but to consult a book in which the most commonly used methods and techniques are described side by side. Several good publications are available. Some are relatively elementary texts describing an integrated approach. Others are quite detailed and advanced and describe only the techniques themselves.

This book is an outgrowth of a series of lectures that I have been organizing for some time for the University of California Extension in Berkeley. The treatment is at an intermediate level. Emphasis is placed on the instrumental aspects and the comparative status of the techniques in relation to one another—an important factor for reasons already cited and especially because of the enormous increases in capabilities that can be attained by the judicious combination of two or more techniques. The length of each chapter and the depth of the treatment is equivalent to what can be conveyed to a group of intelligent chemists in a three-hour lecture. Although not so exhaustive as that in a specialized text, this treatment is sufficient to give the reader the necessary background to discuss fruitfully the more esoteric details of a technique with the expert in the field.

The proliferation of techniques in the past decade makes it an arbitrary and hazardous venture to choose the 10 to 15 "best" topics to be included in this volume. It is easy to decide which subjects not to include—these are the techniques that are still in the experimental stage, require a large number of k$'s to acquire and operate, or both. All in all, a number of techniques that could justifiably be considered for inclusion had to be left out because of lack of space. I will consider a second volume to include a number of these topics if *you* will cooperate by purchasing a copy of this book, instead of borrowing one from the library. This will help to create the impression for the publishers that this publication is a success, which will be of immense value for the underwriting of the subsequent volumes.

I have mentioned the important new trend within the past few years of integrating two or more techniques. Many combinations are noteworthy for their capabilities in solving problems far beyond those of each component technique. The best-known "marriage" is that of the gas chromatograph to the mass spectrometer.

Better run a GC-MS on that batch right away... before NASA finds out!'

Universal applicability is not always the result or the objective when combining two techniques. A highly successful union with a very specific application is found in the use of an atomic absorption spectrometer as the detector for chromatographic effluents in the analysis of lead alkyls in gasoline. This trend to integrate makes it necessary for an expert in one field of analysis to be more aware of the possibilities in adjacent fields. This knowledge allows him to profit from any possible extension of his capabilities by a judicious union of techniques.

A publication of this scope cannot be compiled without the assistance and advice of many friends and colleagues who have gently and patiently introduced me in the mysteries of their particular field of endeavor and have helped me to review the manuscripts for this volume. Especial thanks are due to Drs. J. Q. Adams, W. Gaffield, E. J. Gallegos, J. W. Green, E. Heftmann, R. E. Jentoft, D. R. Rhodes, J. R. Scherer, L. H. Smithson, R. M. Teeter, and Mr. J. J. Windle.

I am also grateful to the management of Chevron Research Company for allowing me to edit this volume and to my patient and understanding wife, who has been invaluable in helping me prepare this publication.

T. H. GOUW

Richmond, California
October 1971

Contents

GUIDE TO MODERN METHODS OF INSTRUMENTAL ANALYSIS

JULIAN F. JOHNSON

Department of Chemistry and
Institute of Materials Science
University of Connecticut, Storrs

I. Gas Chromatography

I. INTRODUCTION

Gas-liquid chromatography has produced sweeping changes in many fields of research since its introduction in 1952. Basically, it is a separation process capable of extremely high resolving power. Thus it provides a means of completely describing very complex mixtures. These cover a wide molecular-weight range from hydrogen isotopes through waxes and steroids. Its direct applicability is limited to compounds which can be volatilized without decomposition, such as thermally stable, nonionic compounds with a maximum molecular weight of around 400–500. Even then, there are probably more than 150,000 known compounds which can be classified in this category, especially if one would also include those heat-sensitive compounds, from which stable, volatile derivatives can be made.

This powerful analytical method has contributed significantly to advances in many fields, for example, catalyst research, biochemical studies, and flavor analyses. Scaled up in size, gas chromatography also offers a method for preparing very pure compounds. Physicochemical properties may be determined from gas-chromatography measurements. It is rapidly becoming an important process control method widely used in chemical plants and refineries.

Clearly an area of such magnitude cannot be covered in a brief amount of space. Nor can any adequate sampling of the more than ten thousand original publications be included in the bibliography. This chapter is intended to acquaint rapidly the research worker with no or very little experience in the field with the general nature of gas chromatography, to discuss the instrumentation and operating variables, and to outline a few of the myriad applications. The number of references is necessarily brief. A list of books is included and a number of review articles are cited.

1. Outline of the Technique

Chromatography, in general, refers to the physical methods of separation based on the distribution of components between two phases. One phase is stationary, usually with a relatively large surface area; and the other is a fluid that moves in contact with the stationary phase.

The stationary phase may be either a solid or a liquid. If it is a liquid, it is usually supported and kept stationary on an inert bed of solid. In gas-liquid chromatography, for example, crushed brick is a common support. If the stationary phase is a solid and the mobile phase is a gas, then the process is called gas-solid chromatography. If the stationary phase is a liquid and the mobile phase is a gas, it is called gas-liquid chromatography. Similarly, there are other combinations using liquids as the mobile phase.

Table 1.1 summarizes this classification of chromatographic systems. These definitions are not always clear cut. For example, gas-solid chromatography is sometimes carried out with the addition of a small amount of liquid to the solid substrate. The stationary phase is then somewhat intermediate between an active solid and a liquid.

Table 1.1. Classification of Chromatographic Systems

Name	Stationary Phase	Mobile Phase
Gas-liquid chromatography	Liquid	Gas
Gas-solid chromatography	Solid	Gas
Liquid-liquid chromatography	Liquid	Liquid
Liquid-solid chromatography	Solid	Liquid

Hereafter gas chromatography will be used to designate all chromatographic methods in which the moving phase is a gas. Gas-liquid chromatography will describe methods in which the stationary phase is a liquid distributed on a solid support. Gas-solid chromatography will refer to the techniques where separation is achieved by selective adsorption on the surface of an active solid.

Figure 1.1 shows a schematic diagram of a typical gas chromatograph. The mobile phase or carrier gas is usually obtained from a cylinder under pressure. The carrier gases most frequently used are hydrogen, helium, argon, and nitrogen. The gas passes through a flow regulator for adjustment of flow rate. Next the carrier gas goes into a sample injector. Here the sample, initially either as a gas or as a liquid, is introduced into the carrier gas stream. This sample injector is heated in order to ensure rapid vaporization of the liquid samples. Sample sizes are normally small, ranging from a few milligrams down to a microgram or less. The vaporized sample, mixed with carrier gas, now enters the column. The column may be either thermostated or temperature programmed. In gas-liquid chromatography the stationary phase is a liquid, for example, a mineral oil, adsorbed on a solid support. For gas-solid chromatography the stationary phase is an active solid, such as powdered charcoal or silica gel. Column diameters are usually of the order of $\frac{1}{8}$ to $\frac{1}{4}$ in. for packed columns with lengths ranging from a few inches to 50 ft or more. As the vapor passes down the column, it is distributed between the carrier gas or mobile phase and the stationary phase. Some of the components will spend more time in or on the stationary phase than others. Therefore they will travel down the column at different rates. At the end of the column components of the sample, now assumed to

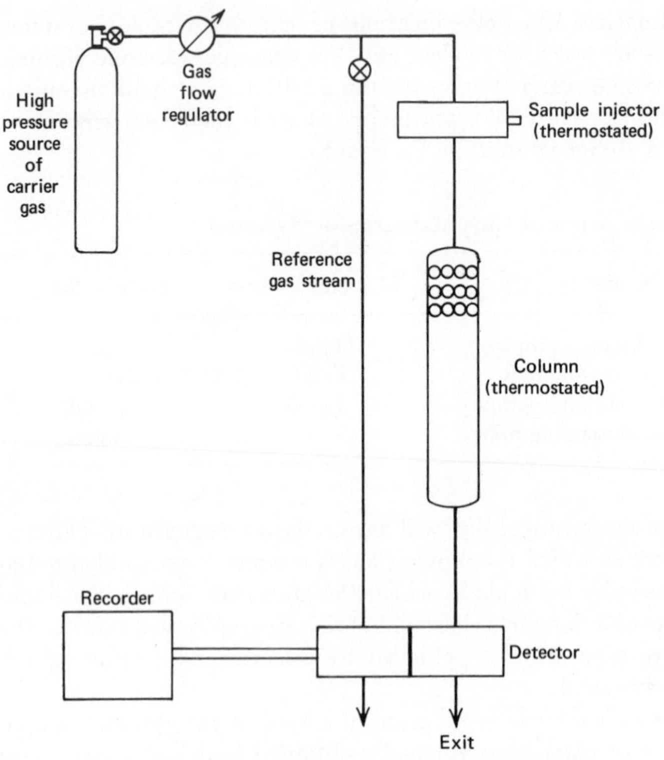

Figure 1.1 Schematic diagram of a gas chromatograph.

be separated in the carrier gas, enter a detector. The detector measures the change of composition of the carrier gas mixed with the components. In Figure 1.1 a common type of differential detector is shown. In one side is pure carrier gas and in the other, the carrier gas mixed with the separated components of the sample. With suitable electronic devices this rapid concentration change can be recorded.

A chromatogram is a plot of the detector signal versus time. Examples are shown in Figures 1.4, 1.8, and 1.16. The times required for components to travel through the column are compared with those for known compounds and serve as a means of identification. The area of the peak is related to the amount of each component present. Thus gas chromatography provides for both qualitative and quantitative analysis.

2. Brief History

In 1941, A. J. P. Martin and R. L. M. Synge published a classic paper that introduced liquid-liquid chromatography (37). This very important chromatographic discovery led to the award of the Nobel Prize in 1952. In the paper they also discussed the use of gas-liquid chromatography, outlined the necessary instrumentation, and speculated on the advantages of the method. No one experimentally tried gas chromatography until 1952 when A. T. James and A. J. P. Martin reported the successful separation and analysis of methyl esters of fatty acids (24). This article contained essentially all of the elements of the apparatus and techniques so widely used today. Interesting accounts of the development of gas chromatography have been given by Martin (36) and by James (23).

II. CHOICE OF OPERATING CONDITIONS

1. Definition of Column Efficiency and Resolution

A basic question in chromatography is the length of column required to make a separation of the two compounds. To find this, it is necessary to define column efficiency and separating power. Figure 1.2 gives a greatly simplified model of how a gas-liquid chromatography column works. This is not intended to represent the actual separation mechanism. Let the column arbitrarily be divided into a number of theoretical plates. For

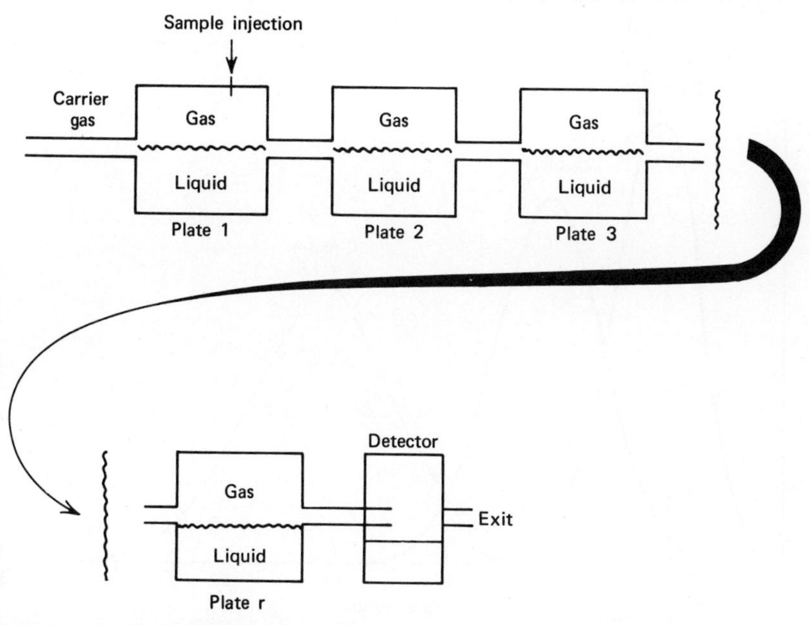

Figure 1.2 Simplified plate model of gas chromatography.

simplicity these plates are assumed to consist of a volume of gas in equilibrium with a volume of liquid; the solid support is ignored. If a sample is now introduced into the isolated first plate, the sample will be distributed between the gas and liquid phase which can be described by a partition coefficient, K, defined by Equation 1.1.

$$K = \frac{(\text{weight of solute in stationary phase})/(\text{volume stationary phase})}{(\text{weight of solute in mobile phase})/(\text{volume mobile phase})} \tag{1.1}$$

The units of K are dimensionless. It is assumed that the partition coefficient is independent of concentration which, at the low concentrations encountered in gas chromatography, is commonly correct. It is further assumed that the partition coefficient of one compound is not changed by the presence of others. Again, with the dilute solutions encountered, this is frequently correct. Now, let the gas from the last plate be instantaneously moved into the detector, the gas from plate 2 be moved to plate 3, and so on, and the gas be replaced on plate 1 from the carrier gas source. The compound then reequilibrates in each isolated plate, another transfer of the gas from plate to plate takes place, and so on, This is an approximation to what is occurring in a gas chromatographic column.

In Figure 1.3 curves representing compounds with two different K values distributed over 25 plates are plotted. The peak maxima are found on

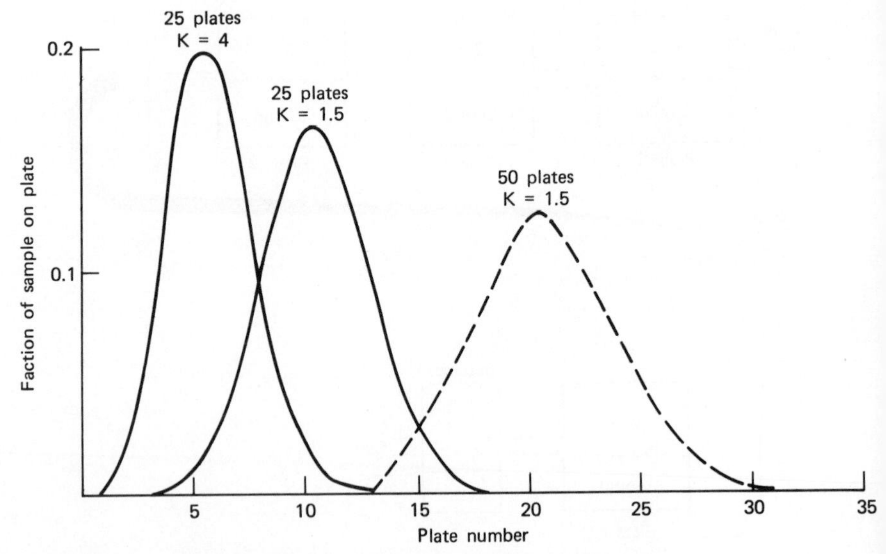

Figure 1.3. Chromatogram assuming model in Figure 1.2.

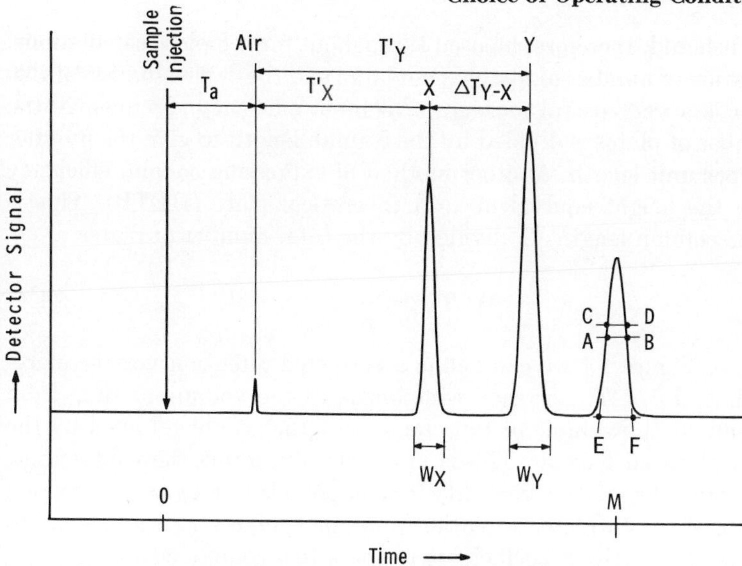

Figure 1.4. Method of calculating column resolution.

different plates illustrating that separation is beginning to take place. Also in Figure 1.3 a plot for the smaller K value shows the broadening of the peak as it continues through the column. From this model it is possible to derive an expression to permit the calculation of the total number of plates, n, in the column. This is done by injecting a small amount of a compound into the column and measuring certain parameters of the resulting peak as shown in Figure 1.4. Here OM is the distance, or time, from the injection of the sample to the maximum of the peak and AB is the width of the peak at one half the peak height. The total number of theoretical plates is then calculated from the formula

$$n = 5.54 \left(\frac{OM}{AB}\right)^2 \tag{1.2}$$

An alternate and equivalent method of calculating n is by drawing in the two tangents to the peak at the inflection points C and D and measuring the distance they intercept on the baseline EF. Then

$$n = 16 \left(\frac{OM}{EF}\right)^2 \tag{1.3}$$

Because of the approximations involved in the derivation of Equations 1.2 and 1.3 the value of n will change somewhat depending on the compound selected. In comparing the effect of any variable on efficiency the same compound should, therefore, be used throughout for the plate calculations. This variation of number of plates is not large and the determination of this parameter is a very useful measure of column efficiency. Frequently the total number of plates is divided by the column length to give the number of plates per unit length. Another method of expressing column efficiency is to give the height equivalent to a theoretical plate (HETP). This is simply the column length, L, divided by the total number of plates,

$$\text{HETP} = \frac{L}{n} \tag{1.4}$$

Referring to Figure 1.4 we can define a corrected retention volume or retention time, V' or T', for each component as the volume or time from injection minus the volume or time for a peak that is not retained by the stationary phase such as air. This is in effect subtracting the void volume of the column. The relative volatility for compounds x and y, α_{xy}, is defined as the ratio of the corrected retention times or volumes. It is also equal to the ratio of the partition coefficients of these two compounds.

$$\alpha_{xy} = \frac{T_y'}{T_x'} = \frac{V_y'}{V_x'} = \frac{K_y}{K_x} \tag{1.5}$$

The degree of separation depends on the distance ΔT_{yx}. If the curves are Gaussian in shape, practically complete separation may be considered to occur when

$$\Delta T_{yx} = T_y' - T_x' = 1.5\,\frac{W_x + W_y}{2} \tag{1.6}$$

Then the number of plates required to separate the two components, n, is given by

$$n = 36\left(\frac{\alpha + T_a/T_x'}{\alpha - 1}\right)^2 \tag{1.7}$$

For packed columns T_x' is often much greater than T_a (see Figure 1.4) so Equation 1.7 becomes

$$n = 36\left(\frac{\alpha}{\alpha - 1}\right)^2 \tag{1.8}$$

It is of interest to see how the number of plates required for a given separation changes with α as shown in Table 1.2.

A major advantage of gas chromatography as a separation method is that there are two routes to improve separation: (1) by increasing the column efficiency, in other words by increasing n; and (2) by increasing α by finding a more selective stationary phase.

Table 1.2. Plates Required for Separation as a Function of Relative Volatility

Relative Volatility	Plates Required	Relative Volatility	Plates Required
10	44	1.10	4,360
5	56	1.07	8,400
2	144	1.05	15,900
1.5	324	1.015	164,000
1.25	900	1.010	367,000

2. Selection of Carrier Gas and Flow Rate

The choice of carrier gas is usually based on detector response, convenience, and column efficiency. The more commonly used gases are helium, hydrogen, nitrogen, argon, and carbon dioxide. While the nature of the carrier gas does influence column efficiency this is not usually a major consideration. Effects on detector sensitivity, considered in a later section, and convenience in terms of availability, cost, and safety, are usually the determining factors.

Flow rate markedly influences column efficiency (see Figure 1.5). Here the HETP for a typical column is plotted against gas velocity which is proportional to flow rate. At low flow rates the HETP is high, that is, the efficiency is low. As the flow rate increases the HETP decreases, passes through a minimum and then increases slowly. Clearly operation at the flow velocity where the HETP is a minimum would give the most efficient column.

The slope of the curve in Figure 1.5 can be described by a simplified equation of the type

$$\frac{L}{n} = H = A + \frac{B}{\bar{u}} + C\bar{u} \tag{1.9}$$

where L is the column length, n the total number of plates, $\bar{u}$ is the time-averaged mean carrier gas velocity, and A, B, and C are coefficients. Equation 1.10 gives a convenient way of measuring $\bar{u}$

$$\bar{u} = \frac{L}{T_{\text{air}}} \tag{1.10}$$

Extensive theoretical and experimental studies have been carried out to elucidate the mechanism associated with each of the coefficients A, B,

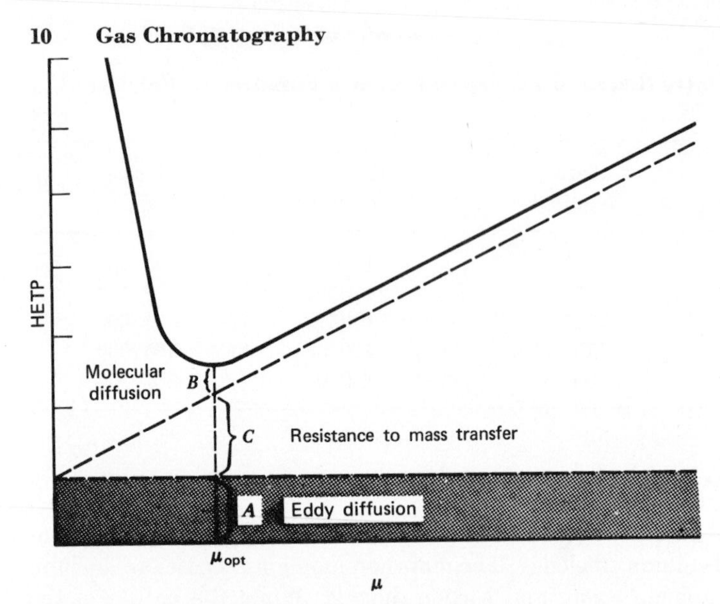

Figure 1.5. Plot of HETP versus gas velocity.

and C. These are not discussed here except in general terms. The multiple path effect is the major contributor to A, molecular diffusion to B, and resistance to mass transfer to C. A somewhat more detailed description of these terms is given in Chapter 2, Section II.

Maintenance of a given flow rate implies that a pressure differential exists between the inlet of the column, p_i, and the outlet of the column, p_o. The carrier gas is compressible and the effect of compressibility on retention times must be considered. To remove this effect a pressure-corrected retention, V_x^0, may be calculated from Equation 1.11

$$V_x^0 = jV_x' \qquad \text{where} \qquad j = \frac{3}{2}\left(\frac{(p_i/p_o)^2 - 1}{(p_i/p_o)^3 - 1}\right) \tag{1.11}$$

The use of Equation 1.11 permits transferring retention data between columns although it should be noted that use of relative retention volumes does not require the pressure correction.

Equation 1.11 shows that the pressure drop associated with the column can materially effect retention volumes. Consider a column packed in such a manner that with the outlet at atmospheric pressure, $p_o = 1$, an inlet pressure of $p_i = 5$ atm is required for a flow rate of 20 ml/min. For this column $j = 0.290$. Another column with the same packing might be more loosely packed and have a flow rate of 20 ml/min for $p_i = 3$ atm, $p_o = 1$ atm. This would yield a value for j of 0.461. Thus the observed corrected retention volume on the first named column would be 1.6 times that of the second column.

The flow of gases through porous media can be described by Equation 1.12:

$$\Delta p = \frac{\bar{u}\eta L}{C'} \qquad (1.12)$$

where Δp is the pressure drop across the column, C' is the specific permeability coefficient, L is the column length, $\bar{u}$ is the average linear flow velocity, and η is the viscosity of the carrier gas.

The flow of gas must be regulated and maintained constant for reproducible results. Commonly a reducing valve is used on the gas cylinder followed by a flow regulator or metering valve in the gas chromatograph. This is usually sufficient. More sophisticated controls are available for ultraprecise measurements. Flow rate may be measured before the sample inlet by rotameters. A precise, convenient way to measure the exit flow rate is provided by a simple soap bubble flowmeter. In this device the time required for a specific volume of gas to emerge is measured. The flowmeter is simply a burette with a T replacing the stopcock. A rubber bulb filled with a soap solution is squeezed which sends a bubble into the burette. The time for the bubble to travel a certain distance is then determined.

The purity of the carrier gas is of importance, particularly for trace analysis. Although commercially available gases are available in high purity there are still often a number of impurities present at the level of several or more parts per million. Several methods have been outlined to remove impurities. Low-temperature, solid-adsorbent traps, frequently charcoal or a molecular sieve operating at liquid nitrogen temperatures, are effective for many compounds. Hydrogen and hydrocarbons may be removed by passage through a heated tube containing catalyst. The precautions necessary will depend on the nature of the analysis.

The effects of flow rate and methods of control described in this section have referred to constant flow rates; the technique of flow programming is described in a following section.

3. Selection of Temperature

The temperature at which the column is operated has a major effect on retention volume and resolution. For this section only isothermal operation is considered; a separate section discusses temperature programming.

The retention volume for any component is a function of its vapor pressure above the liquid phase. The relationship between vapor pressure and temperature has, of course, been extensively studied and can be used to derive retention volume–temperature equations. The two most frequently used in gas chromatography are

$$\log V = \frac{b}{T} + d \tag{1.13}$$

or

$$\log V = \frac{a}{t + c} + e \tag{1.14}$$

where T is the absolute temperature, t is temperature in °C, and a, b, c, d, and e are constants. Plots of $\log V$ versus $1/T$ should therefore be linear over an appreciable temperature range. This has been experimentally verified for many compounds on a wide variety of packings. Figure 1.6 shows typical plots. Table 1.3 gives numerical examples of the experimentally determined constants for representative compounds on two packings (4). It should be emphasized that such equations do not cover an unlimited temperature range but are frequently valid over 100–150°C. From Table 1.3 it is apparent that the equations describing the temperature dependence of elution volume vary with compound by compound type and even within a homologous series. Therefore relative retention volumes

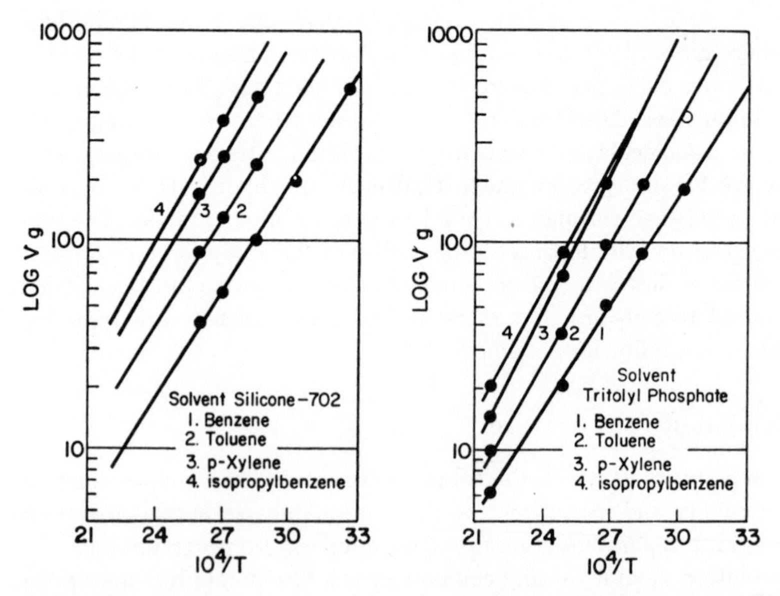

Figure 1.6. Elution volume versus reciprocal absolute temperature (4).

determined at one temperature are not applicable to other temperatures. Values for relative retention times for acetate esters at two temperatures are given in Table 1.4 (4).

Table 1.3. Typical Values for Constants in Equation 1.14 (4)

	Silicone 702 Liquid Phase			Tritolyl Phosphate Liquid Phase		
Compound	*e*	*a*	*c*	*e*	*a*	*c*
Ethyl alcohol	−1.06	540	150	−2.77	452	100
Isopropyl alcohol	−1.13	588	150	−5.85	1810	273
n-Propyl alcohol	−1.09	629	150	−6.07	1985	273
Benzene	+0.11	370	100	−5.88	1930	273
Toluene	−0.20	454	100	−6.13	2135	273

Table 1.4. Relative Retention Times as a Function of Temperature (4) [a]

	Relative Retention Time	
Compound	50°C	98°C
Methyl acetate	0.188	0.286
Ethyl acetate	0.361	0.500
n-Propyl acetate	1.00	1.00
n-Butyl acetate	3.122	1.935
Isopropyl acetate	0.465	0.631
Isobutyl acetate	1.325	1.400
sec-Butyl acetate	1.151	1.251

[a] Liquid phase—tritolyl phosphate.

Clearly compounds emerge more rapidly as the temperature increases. However this is usually accompanied by a loss in resolution. Normally the α value for a given pair of compounds increases as the temperature decreases. This is not always the case but in general it is. Therefore the number of plates required for a separation markedly decreases as the temperature of separation is reduced.

The effect of temperature on the efficiency of the column varies with the type of column. It is normally not a major effect; thus the controlling variables in selecting the temperature are the resolution and elution times.

Although a variety of thermostating methods have been reported, such as liquid baths, heated blocks, resistance heating of columns, and so on, the majority of commercial gas chromatographs use stirred thermostated air baths. Stirring is essential for good temperature regulation. While no further discussion of temperature control will be given it should be pointed out that subambient temperatures are frequently required for separating complex mixtures of low-boiling compounds. Regulation at these temperatures is somewhat more complex but not prohibitively so (8, 14).

4. Selection of Liquid Phase

The liquid phase is a vital part of the chromatographic column but its choice is still to a large extent empirical. Solute–solvent interaction forces are responsible for the selectivity of the liquid phase in separations. These forces include dipole–dipole interactions, dispersion forces, hydrogen bonding, and so on. To compute accurately the effect of the various forces on retention volumes is quite complex and beyond the capabilities of the average gas chromatographer even though promising results have been obtained by a number of specialized workers in the field. With a choice of several hundred liquid phases however, some type of generally descriptive parameter is helpful. One approach to this has been through a simple model of polarity (42). For this model the corrected retention volumes of butadiene and butane are determined for the liquid phase and for squalane and β,β'-oxydipropionitrile. The polarity is calculated from Equation 1.15.

$$P_p = a' \log \frac{V_p \text{ (butadiene)}}{V_p \text{ (butane)}} - \log \frac{V_u \text{ (butadiene)}}{V_u \text{ (butane)}} \tag{1.15}$$

V_p refers to retention volumes on polar phases, V_u on nonpolar phases. Oxydipropionitrile is assigned a polarity of $P = 100$, squalane $P = 0$, and the constant a' is adjusted to give these values. Table 1.5 gives polarities for a number of liquid phases measured in this manner.

This simple model serves to give a numerical index of polarity on which generalizations may be based.

1. Nonpolar solutes are separated in order of their boiling points on nonpolar liquid phases.
2. Polar solutes elute more rapidly than nonpolar solutes of the same boiling point on nonpolar liquid phases.
3. Nonpolar solutes elute more rapidly than polar solutes of the same boiling point on polar liquid phases.

The general concept of polarity is thus useful in selecting general types of liquid phases. For a somewhat more descriptive characterization of the

Table 1.5. Polarity of Liquid Phases (42)

Liquid Phase	Polarity
Squalane	0
Methyl silicone oil	7
Tetraethylsilicane	26
Benzyl ether	44
Diethyl malonate	61
Polyethylene glycol 600	78
Propylene carbonate	83
Oxydipropionitrile	100

liquid substrate several reference compounds can be used to take the polarity of the substances being analyzed into account. The most widely used characterization method appears now to be that introduced by Rohrschneider (43), which measures the difference in retention indices, ΔRI's (see Section IV), of benzene, ethanol, methylethyl ketone, nitromethane and pyridine on the substrate of interest, and on a squalane column under otherwise identical conditions.

In addition to giving the desired separation the liquid phase must have a low vapor pressure at the column temperature and must not irreversibly react with the solute. The requirement for low vapor pressure arises for two reasons. The liquid phase is being removed by the carrier gas at a rate proportional to the vapor pressure. Therefore the useful life of the column will be inversely proportional to the vapor pressure. It is difficult to assign a quantitative value for an upper limit to the vapor pressure that is permissible. An analytical column with a life span of one day might still be acceptable in some circumstances while a column in a process monitor that lasted only a month might well be unsatisfactory.

Another problem associated with liquid packing being removed or "column bleeding" is the effect on detector noise and sensitivity. These effects vary with the type of detector. Differential thermal conductivity detectors are relatively insensitive to the presence of small amounts of liquid phase. Ionization detectors are very sensitive to column bleeding and may become saturated, that is, the constant signal observed is due to the liquid phase present in the carrier gas and not because of the sample.

Some of the upper temperature limits assigned by a number of investigators to various packings are given in Table 1.6. Most of the high-temperature packings are not chemically pure compounds. This is probably the basis for the various "conditioning" procedures. These consist of operating the column at temperatures above the maximum expected operating temperature with a slow stream of carrier gas passing through the column. This removes the more volatile components and the remaining liquid phase may then be more stable. For example, Dow Corning vacuum grease heated at 325–400°C for 4–8 days is stated to be indefinitely stable at 275°C, stable for 2000 hours at 300°C and 200 hours at 325°C.

Table 1.6. Temperature Limits for Liquid Phases

Liquid Phase	Temperature Limits, °C	
	Lower	Upper
Squalane	20	100–125
β,β'-Oxydipropionitrile	0	75
1,2,3-Tris(2-cyanoethoxy)propane	0	150–175
Ucon LB-550-X	0	200
Dimethyl silicone liquid OV-101	0	325–350
Silicone gum SE-30	50	250–350
Diethylene glycol succinate (DEGS)	20	200
Apiezon M	40	275–300
Apiezon L	50	250–300
Carbowax 20M	60	200–250
Dexsil 300 polycarboranesiloxane [a]	50	450

[a] This is the most stable organic liquid phase for high-temperature operation known at this time.

Another approach to obtain stable, high-temperature columns is to use an inorganic- salt eutectic mixture as the substrate. For example, a mixture of sodium, potassium, and lithium nitrates in the ratio of 18.2:54.5:27.3 by weight melts at 150°C and is stable to a least 400°C. Such columns usually offer low efficiencies.

Table 1.6 also shows the lower temperature limits of these liquid phases. Operation below these temperatures results in a sharp reduction of column efficiency and long tailing peaks. This is because the liquid phase has either solidified or has increased so much in viscocity that the resultant increase in mass transfer resistance of the solute in the stationary phase will unfavorably affect the chromatographic process.

The amount of liquid phase present affects both the retention volume and resolution of the column. For a packed column, if the solid support has no effect, the retention volume will be proportional to the amount of liquid phase present. This is usually observed for those cases where the weight of liquid present is in the 5–40% range. At lower levels of liquid the fact that the solid support is usually not completely inert becomes apparent as the adsorptive properties of the solid may influence retention volumes. This is shown in Figure 1.7. Here various amounts of the liquid phase, ethylene glycol adipate polyester (EGA), can lead to reversals in the relative retention times for amino acid derivatives (48). If the effect of the solid support is minimized by deactivation or when an inert support is used, lower liquid phase-to-support ratios often produce more efficient columns. This is discussed more fully in the section on column selection.

5. Sample Size and Injection Temperature

Most gas chromatographs have a sample vaporizer into which the sample is injected. Since the vaporizer temperature can affect the separation efficiency it is usually provided with a separate temperature controller. Normally efficiency increases as the vaporizer temperature increases until the vaporizer is about 50°C higher than the column temperature after which no further change is observed. This can be checked by varying the vaporizer temperature for any given sample although the 50°C differential appears to be a good first guess.

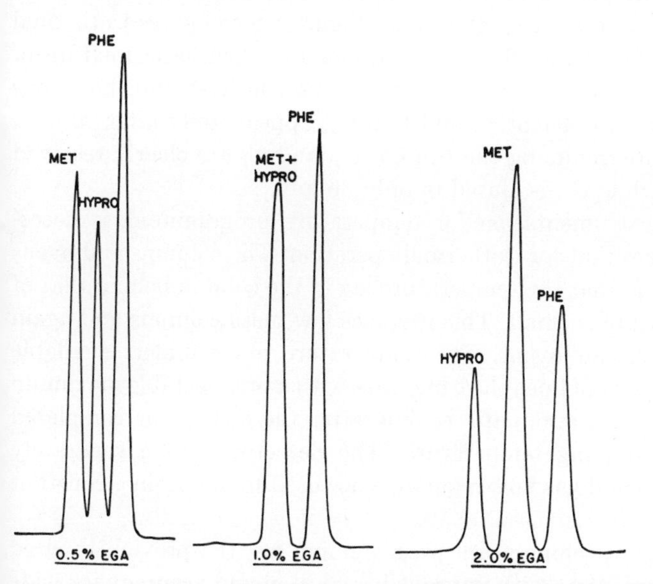

Figure 1.7. Effect of liquid phase support ratio on elution times (48).

Sample size may affect retention times and resolution. The effect on retention times for gas-liquid chromatography is usually negligible although gas-solid chromatography may show appreciable variations. An increase in sample size will cause a decrease in resolution. With the small samples normally used in analytical gas chromatography this effect is usually of minor importance. It becomes a major problem in preparatory-scale chromatography when large throughputs are required. In that case speed is usually sacrificed for resolution (see Figure 2.6).

6. Temperature Programming

If the sample to be chromatographed contains compounds covering a wide range of boiling points and the column is operated isothermally, then the first peaks to emerge will be very narrow and those at the end of the analysis will be very broad. This leads to difficulties in quantitative analysis as it is difficult to determine the area of the very narrow peaks with good precision and the location of the baseline for the broad peaks may introduce significant errors. The resolution for the initially emerging peaks may not be complete due to low α values. Additionally, very long analysis times may be required for the higher-boiling components in the sample.

These difficulties can be minimized by the technique of temperature programming where initially the column temperature is low enough to give good separation and peak shapes and the final temperature is sufficient to elute all compounds in a reasonable time. Figure 1.8 compares isothermal and temperature programmed chromatograms of a synthetic mixture of *n*-hydrocarbons. Note that in the isothermal run the lower-boiling homologs are eluted very close to each other and that C_{15} appears only after 90 min. By temperature programming the run C_6, C_7, and C_8 are clearly resolved from each other while C_{21} is eluted in only 36 min.

The required instrumentation for temperature programming is necessarily different from that for isothermal operation. The columns and ovens must be designed so that the temperature lag in the column behind that of the oven temperature is small. This involves low mass columns and again good stirring in the air ovens. The temperature programmers available usually offer a number of linear heating rates with a preselectable maximum temperature and often automatic cooling after the analysis is completed to a preselected starting temperature. The detector is in a separately thermostated oven and gas flow regulators are used to maintain a constant gas flow rate.

The advantages of temperature programming are the previously cited reduction in analysis time with improved resolution and accuracy for wide

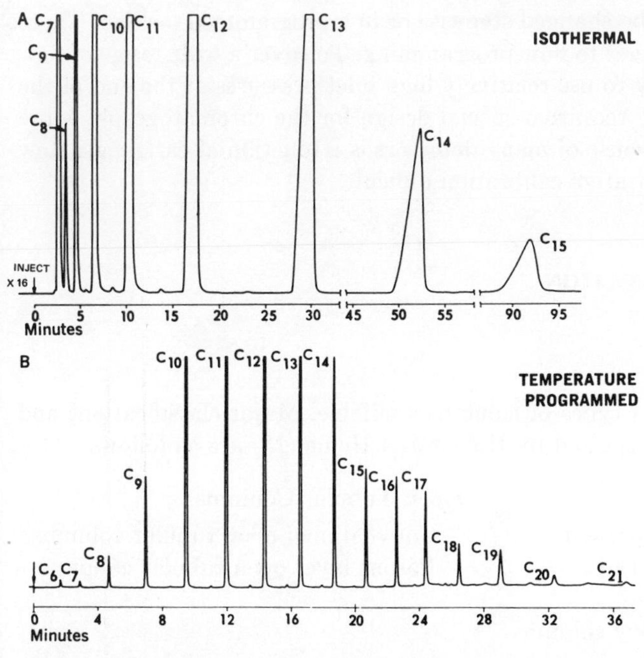

Figure 1.8. Programmed temperature and isothermal chromatograms of *n*-hydrocarbons.

boiling-range mixtures. The disadvantages are that more complex equipment is required and that more careful control of operational variables are required both for qualitative and quantitative analysis.

It should also be emphasized that the liquid phase must be thermally stable over the entire temperature range of operation. If the temperature range is wide this restricts the number of liquid phases available.

7. Flow Programming

Flow programming, that is, increasing the flow rate of the carrier gas during the analysis, may be used to produce the same shorter analysis time as with temperature programming. There are a number of differences between the two techniques. As flow programming operates at lower temperatures the amount of liquid phase removed is less than in temperature programming. The amount of column bleeding increases linearly with carrier gas flow rate but exponentially with increasing temperature. Use of lower temperatures allows for the use of a wider variety of liquid phases than are available with temperature programming.

Flow rates may be changed stepwise or in a programmed fashion. There are two disadvantages to flow programming. To cover a wide range of flow rates it is necessary to use relatively high inlet pressures at the end of the program. This may require a special design for the chromatograph. More important, the response of many detectors is a function of carrier gas flow rate making quantitative calibration difficult.

III. INSTRUMENTATION

1. Columns

There are several types of columns available. Major classifications and nomenclature as suggested by Halasz and Heine (18) are as follows:

Packed columns	Open Tubular Columns
Conventional packed	Conventional open tubular columns
Porous layer bead	Porous layer open tubular columns
Thin layer bead	
Packed capillary columns	

Each column type has advantages and disadvantages. These are discussed briefly.

A. Solid Supports

Common to all packed columns in gas-liquid chromatography is a solid support. There are several desirable features that the ideal solid support should have. It should lead to highly efficient columns, in other words, it should have as many plates per unit length as possible. A low pressure drop prevents excessive inlet pressures and makes possible the use of long columns to increase resolution. The surface area must be sufficient to support the desired amount of stationary phase, and the mechanical strength should be enough to prevent crushing. Normally, it is desirable that the solid support be inert so that it exerts no adsorptive forces on the solute. This statement must be qualified because there are cases where a desired separation is achieved by the combined effect of the stationary phase and an adsorptive solid support. Certainly irreversible adsorption effects and catalyst activity that may change the nature of the solute are undesirable.

A number of materials have been used as solid supports. These include crushed insulating brick, diatomaceous earth, glass beads, sand, and so on. Choice of particle sizes varies considerably but usually a fairly narrow range obtained by sieving is used. Common choices are 30/50, 40/60, and so on, mesh screen fractions. Narrow particle-size distributions produce more efficient columns than those with broad distributions.

Experimentally, activity of the solid support may be observed by the appearance of nonsymmetrical peaks, usually with a trailing edge at the end of the peak. Other experimental observations that show that the solid is exerting adsorptive effects are (1) changes in retention times with both sample size and sequence in which compounds are injected, (2) inversions of retention time as the amount of liquid phase is varied, and (3) irreversible adsorption, in other words, injected compounds may be irreversibly adsorbed and never emerge from the column. Other undesirable effects due to interaction with the solid are catalytic decomposition of the stationary phase and isomerization of the components to be separated. All these effects have been observed on various solid supports. It would appear that the forces involved in the adsorptions due to the supports range from weak van der Waals forces to actual chemisorption. The weaker forces can be neutralized by the use of almost any liquid phase that thoroughly covers the surface. The stronger forces due to hydrogen bonding can be neutralized by the use of polar liquid phases having hydrogen bonding capabilities.

Solid activity may also be eliminated or reduced by chemically removing the hydrogen bonding sites, for example, by treating with hexamethyldisiloxane, adding trace amounts of polar materials to nonpolar liquid phases and the use of a carrier gas saturated with a polar material such as water vapor.

B. Preparation of Conventional Packed Columns

There are a number of commercial sources for packed columns. One method of preparing packed columns will be described briefly. The calculated amount of the selected liquid phase is weighed, diluted with a volatile solvent, and the solid support weighed in to form a slurry. The slurry is gently agitated to evenly distribute the liquid phase with care being taken not to crush the solid support. The volatile solvent is removed by evaporation on a hot plate or, preferably, by a fluidization technique.

A selected length of straight or precoiled column tubing is packed by inserting a small plug of glass wool or a wire screen in one end to retain the packing and then slowly pouring in the packing while gently tapping or vibrating the column to cause even packing. Column diameters usually range from $\frac{1}{8}$ to $\frac{1}{4}$ in. Copper and aluminum have the advantage of being easily coiled to form a small compact unit. Stainless steel may be coiled but with more difficulty; however it is less apt to catalyze decomposition of the compounds to be separated. Glass columns are usually the least reactive and can be used when inertness is necessary. Other very stable materials are Monel, Teflon, and Kel F.

C. *Conventional Packed Column Performance*

Comparing the performance of the various types of columns experimentally is difficult because for each type ten-fold variations in the various parameters may be observed. Table 1.7 lists some characteristics of the various columns. Conventional packed columns are widely used because they are easy to prepare. In addition, because they can handle relatively large sample sizes, insensitive detectors may be used.

Table 1.7. Comparison of Column Performance (15)

Column Type	Minimum Plate Height, cm	Sample Size, μg
Conventional packed	0.05–0.2	10–1000
Porous layer bead	0.05–0.2	1–100
Thin layer bead	0.1 –0.3	1–100
Packed capillary	0.05–0.2	1–50
Conventional open tubular	0.03–0.2	0.1–10
Porous layer open tubular	0.06–0.2	1–50

D. *Porous Layer Bead Columns*

Porous layer bead columns utilize as a solid support a spherical particle with a fluid-impenetrable inner core surrounded by a porous layer with a high surface area. This porous shell is typically 1/10–1/30th of the particle diameter. This packing can thus contain more liquid phase than the thin layer bead column (see below), but not as much as for the completely porous packings. These columns give good efficiencies, relatively short analysis times, and can handle moderately sized samples. The advantages of the porous layer column are more thoroughly discussed in Chapter 2, Section IV.

E. *Thin Layer Bead Columns*

These columns employ solid supports that are impermeable to the liquid phase. Glass beads are used most frequently, although metal helices and springs have also been used. The thin film of liquid packing improves the speed of mass transfer and results in good efficiencies with very short analysis times. Because only very low liquid loadings can be used, the column capacity is very low and only very small samples can therefore be used with these columns.

F. *Packed Capillary Columns*

Packed capillary columns are characterized by small internal diameters, of the order of 0.2–0.5 mm, and large particle-to-column diameter ratios, about 1:3 to 1:5. The use of the large particle diameter of the support in comparison to the internal diameter of the column markedly changes the flow pattern of the carrier gas with respect to that of the conventional packed column. The chief advantage of these columns is the speed of analysis. Separation of methane, ethane, ethylene, and propane in 3 sec has been demonstrated with these columns.

G. *Open Tubular Columns*

Open tubular columns are often referred to as "capillary columns." The use of this type column was predicted by A. J. P. Martin in the early days of gas chromatography and first demonstrated by M. J. E. Golay soon afterwards. As the name implies the column is simply an open tube with a small internal diameter, usually of the order of 0.01–0.02 in. The liquid phase is coated on the inside wall of the tube.

The major advantages of the open tubular column are high resolution and fast analysis times. While Table 1.7 indicates that the minimum attainable plate height is roughly comparable between conventional packed columns and open tubular columns, there are additional aspects to this statement. For open tubular columns this minimum is relatively easy to attain while great care must be exercised to do so with conventional packed columns. Pressure drops in open tubular columns are also very low so that long columns can be used with concomitantly large numbers of plates; 1000 ft is a common length while at least one column a mile long has been reported. This then gives a much larger total number of plates than can be obtained from conventional packed columns. The other major advantage is that the time of analysis is relatively short. Capillary columns can easily generate as much as 100 plates/sec, while packed columns are usually operated at around 10–20 plates/sec.

The disadvantage of the open tubular columns is that only very small samples can be separated so highly sensitive detectors are required. Auxiliary sample injection splitters are generally needed to inject conveniently the small sample required. These columns are also somewhat more difficult to pack than conventional packed columns.

Open tubular columns are usually prepared from stainless steel or glass. Other materials have been used including nylon tubing and copper. A variety of coating procedures has been reported. One widely used method is to force a solution of the liquid phase in a solvent slowly through the tubing. Excess solvent is then evaporated by the carrier gas. The wetting properties of the liquid phase toward the material of the capillary must be suitable and the coating solution should be forced through the column at a low, constant velocity. An apparatus suitable for packing capillary columns has been described by Levy et al. (31). Liquid loadings usually range from 10 to 50 mg of liquid phase per 100 ft of column. This corresponds to a film thickness of about 0.2–1.0 μ.

To inject a small but representative sample into the column may require an indirect split stream sampling system of special design. These involve conventional injection of a normal-size sample into the carrier gas followed by splitting this gas stream into two parts one much smaller than the other. This portion of the sample then enters the open tubular column.

H. Porous Layer Tubular Columns

These columns are also known as support-coated open tubular columns (SCOT columns). They are obtained by a modification of the open tubular column by depositing a thin porous layer inside the column, usually by *in situ* chemical reaction. This porous layer, which is usually around 40–50-μ thick may be used as a solid stationary phase or coated with a liquid phase. This porous layer results in a higher loading capacity of the column while still retaining the low pressure drop and high speed characteristics of conventional open tubular columns. This is especially important for the analysis of rapidly eluting compounds, in other words, compounds with low K values. The higher liquid loading allows for the use of larger samples with less danger of saturating the liquid phase with the solute.

2. Detectors

A. General

A very large number of detectors for gas chromatography have been proposed. A number of these are in common use; others are used only for special purposes, and still others appear to have received little further attention after the initial report. Only the more commonly used detectors are discussed here.

The desirable features of a detector are many. It should operate over a wide range of conditions, temperatures, and flow rates. It should have a linear response, be sensitive, be inexpensive, and so on.

Detector sensitivity may be defined by

$$S = \frac{\Delta R}{\Delta Q} \tag{1.16}$$

Equation 1.16 defines sensitivity, S, as ΔR, the response of the detector, divided by ΔQ, the quantity of sample or change in the concentration of sample. This is a perfectly general expression for any detector. This in itself is not sufficient to define the minimum amount that can detect or be detected. In order to do this, the noise level must be considered. If the noise level, R_n, is the maximum deviation in either direction around the average baseline, a common definition of the minimum concentration change that can be detected is given by

$$Q_{\min} = \frac{2R_n}{S} \tag{1.17}$$

Here $Q_{\min}$ = twice the noise level divided by the sensitivity. In subsequent discussion it will be assumed that the response of the detector is linear over the concentration range used.

Detectors may conveniently be divided into two general classes—nonspecific, that is a detector that responds to any compound emerging from the column, and specific detectors that respond only to a given class or type of compounds. Both have their uses. To analyze mixtures of various types of compounds a nonspecific detector is desirable. To analyze for a specific compound or set of compounds that may be interfered with by others, or to increase the sensitivity specific detectors are often used.

B. *Thermal Conductivity Detectors*

A widely used type of detector is the thermal conductivity detector shown schematically in Figure 1.9. This shows one cell. A second identical cell is used to monitor the effluent from the reference column as shown in Figure 1.1. The cell is thermostated. The resistance elements or filaments may be either resistance wires or thermistors. Consider first a resistance

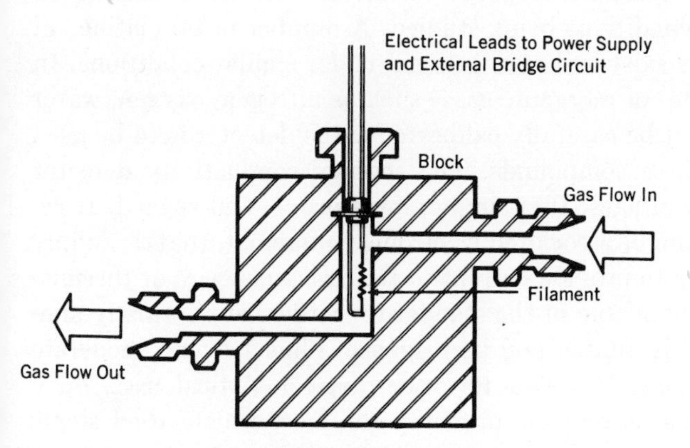

Figure 1.9. Schematic diagram of a thermal conductivity cell.

wire where a given current is passed through both the resistance wires in the two detectors with carrier gas flowing through both sides. The resistance wires will increase in temperature until they reach temperatures at which the rate of heat loss is exactly balanced by the rate of energy input. This is a function of a number of variables but an important one is the thermal conductivity of the flowing gas surrounding the resistance wires. A high-conductivity carrier gas, helium or hydrogen, is normally used.

When an organic vapor emerges from the analytical column and enters the detector it has a lower thermal conductivity than the carrier gas and its mixture has a different thermal conductivity from the pure carrier gas in the reference detector. The resistance wire then assumes a new equilibrium temperature based on the thermal conductivity of the mixture. Usually the two resistance wires are incorporated as two elements of a Wheatstone bridge circuit. The difference in resistance or voltage is amplified by conventional means and recorded to produce a chromatogram.

The choice between resistance wires and thermistors as sensing elements is usually dependent on the desired sensitivity and operating conditions. The change in resistance with temperature is linear for resistance wires and inversely exponential for thermistors. Frequently therefore, the thermistor-type detectors are more sensitive in low-temperature operations, while resistance wires would be the better choice for high-temperature detectors.

The thermal conductivity detector is used almost exclusively for quantitative work with helium or hydrogen as a carrier gas. The large difference between the thermal conductivity of these gases and that of most organic vapors results in a response that is very similar regardless of compound type for many organic compounds. For precise quantitative work however, it is necessary to determine a response factor for each compound to be analyzed under the conditions being studied. A number of tabulations of such response factors exist and may be used under similar conditions. In particular the response of inorganic gases such as nitrogen, oxygen, water vapor, and so on must be carefully calibrated if the detector is to be used quantitatively for these compounds. The thermal conductivity detector has a number of advantages. It is simple, inexpensive, and rugged. It requires only a minimum of associated electronic equipment, that is, simply a stable power supply to provide current to the resistance wire or thermistors. The amplifiers available in the standard recorders are normally adequate. Its sensitivity is limited and its response is linear over a moderate range of concentrations. However for wide-range analytical uses, on a variety of compounds, it remains probably the most widely used single detector.

C. *Ionization Detectors*

There are a number of ionization detectors that operate by measuring the electrical conductivity of a gas. The electrical conductivity is directly proportional to the concentration of the charged particles within the gas. Ionization can be achieved in a number of ways.

For the flame ionization detector (FID) ionization is provided by a flame. Figure 1.10 is a schematic diagram of a FID. In operation the carrier gas from the column containing the organic vapors is continually mixed with hydrogen and air or oxygen and burned at a jet. The flame provides an ionizing mechanism which may be quite complex but at any rate ions are formed. The electrical resistance of the flame is monitored continuously. The resistance decreases when an organic vapor is present to supply a source of ions. The detector responds to almost all compounds; exceptions are a number of low-molecular-weight inorganic gases including air and water. For quantitative measurements a response factor is determined for each compound although tabulated values and calculation methods may be substituted.

In comparison with the thermal conductivity detector the flame ionization detector is considerably more sensitive and has a wider linear range. Operation is relatively simple but the associated electronic equipment is more expensive.

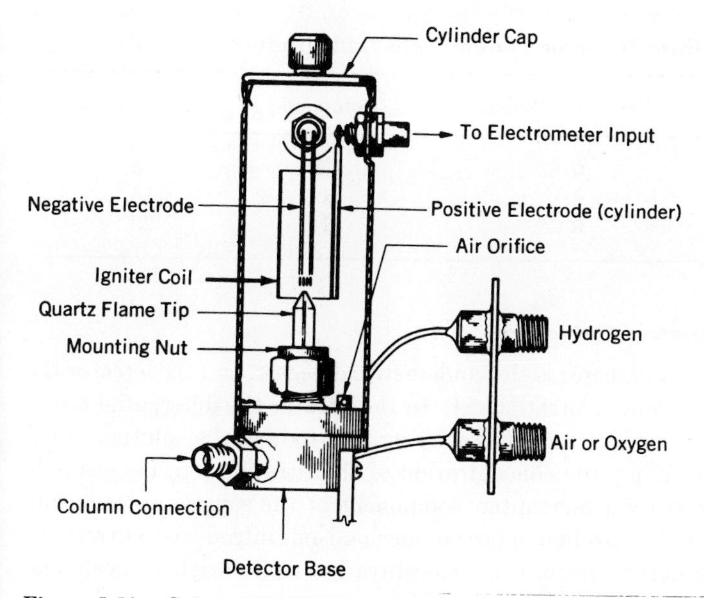

Figure 1.10. Schematic diagram of a flame ionization detector.

The flame ionization detector is nonselective. The same principle may be used to give a selective detector for phosphorus and sulfur in the flame photometric detector. The carrier gas, nitrogen, is mixed with hydrogen and air or oxygen and burned. Organic materials are excited to higher energy states in the reducing atmosphere of the flame. The flame is viewed by a conventional photomultiplier tube using narrow band filters to select a wavelength responsive to the element to be measured. For phosphorus, 526 mμ is used; for sulfur, 394mμ. Response for compounds not containing sulfur, for example, is 5×10^{-3} to 4×10^{-4} less than those with sulfur so the detector is very selective. The same detector may also be operated as an FID providing simultaneous selective–nonselective detection.

Ionization may be achieved in other ways. Figure 1.11 is a schematic diagram of a helium detector. The carrier gas, helium, enters a chamber containing tritium as an ionization source and a voltage gradient of 4000 V/cm. This combination produces metastable helium with an ionization potential of 19.8 eV. The presence of other compounds having a lower ionization potential will allow them to be ionized and a negative signal will result. While in principal this is a nonselective detector it is usually used to measure very small amounts of low-molecular-weight gases. Table 1.8 lists the lower detectable limits for some gases. A similar type detector using argon as a carrier gas has been used but is less sensitive.

Table 1.8. Helium Detector—Lower Detectable limits

Compound	Parts Per Billion	Compound	Parts Per Billion
CO_2	0.8	CH_4	3.5
CO	3	N_2	15
O_2	3	H_2	20

D. Coulometric Detectors

Another type of detector is the coulometric detector. In this detector the carrier gas passes into a titration cell. In the titrator the differential signal between a reference electrode and a sensor electrode in the solution is fed to an amplifier. When the concentration of the titrating ion, for example silver, is reduced by a mercaptan component of the sample entering the cell, the amplifier, through a servo mechanism, introduces current to generate silver ions to restore the concentration to its original level. The

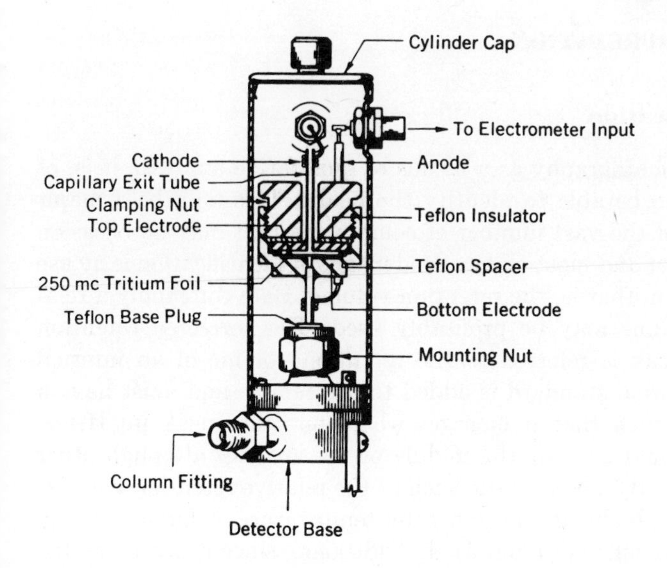

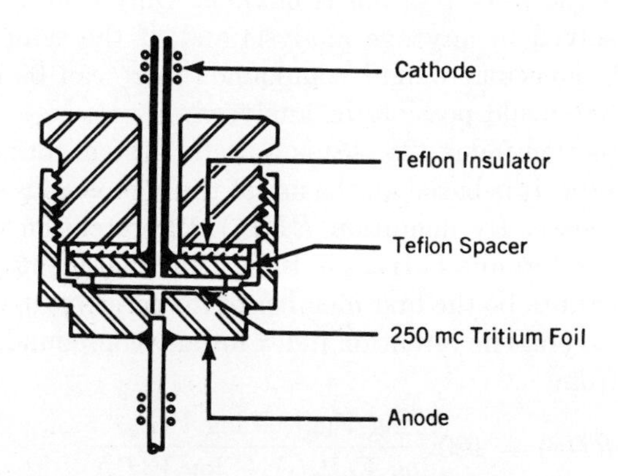

Figure 1.11. Schematic diagram of a helium detector.

current required is displayed on the recorder and is proportional to the amount of the eluting component. An advantage of the coulometric detector is that results are directly quantitative with no calibration required. In practice the coulometric detector is usually employed as a selective detector although by combusting all effluent compounds to carbon dioxide it may be used as a nonselective detector.

IV. DATA INTERPRETATION

1. Peak Identification

To use gas chromatography as a means of qualitative analysis, it is, of course necessary to be able to identify the peaks. This may be a major problem because of the vast number of compounds that may on occasion be present. The first and most widely used means of identification is by use of the peak position, that is, the retention volume. Here commonly a relative retention volume may be profitably used. The corrected retention volume of each peak is referred to the retention volume of an internal standard. The internal standard is added to the sample and must have a retention volume such that it emerges where no other peak interferes. Preferably this would be near the middle of the analysis although other peaks may be used. By using a ratio such as the relative retention volume, the influence of small changes in flow rate, temperature, column packing, and so on, between runs are minimized. Additional, since it is a corrected retention volume, differences in apparatuses are also removed. The limitation on the use of the peak position is obvious. Only a finite number of peaks may be resolved in any one analysis and, if the sample contains moderate- or high-molecular-weight compounds there can be many more isomers present than could possibly be analyzed.

The Kovats retention index (28, 45) is a useful system for standardization of retention data. It is based on the use of the homologous series of the n-alkanes as references. By definition $RI(P_z)$, the retention index of an n-alkane P_z with the formula C_zH_{2z+2} is $100z$. Thus the retention index for hydrogen (considered to be the first member of the series) is 0, $CH_4 = 100$, $C_2H_6 = 200$, and so on. The retention index for any compound, x, is found by interpolation from

$$RI(x) = 100 \frac{\log V(x) - \log V(P_z)}{\log V(P_{z+1}) - \log V(P_z)} \tag{1.18}$$

where $V(P_z) < V(x) < V(P_{z+1})$ and $V(x)$ is the retention volume or retention time of compound x.

The retention volumes for all compounds including the standards must be obtained under the same conditions. The use of the retention index permits formulation of some general relative relationships between classes of compounds to aid in the identification.

A system that is considerably more reliable is to analyze the samples on two different column packings differing markedly in polarity. Here we are trying to take advantage of different α values from the two packings. In this case the relative retention volumes of the measured known compounds are compared on each of the columns with that of the unknown on each of the columns. This requires, of course, that the amount of the unknown, that is, the peak size is such that it may be unequivocally identified on each of the columns even though its retention time may vary widely.

Other subsidiary information is available. When working with homologous series it is well-established that the log of the retention volume versus the carbon number of the homologous series forms a straight line. Thus by interpolation or extrapolation one can predict from running a few members of the homologous series where the others will have their retention volumes. A similar correlation exists over a reasonable boiling point range for the log of the retention volume versus the boiling point. An example is shown in Figure 1.12.

Additional information may be gained by running a nonselective and selective detector in series. For example, the thermal conductivity detector followed by a flame photometric detector would readily distinguish between compounds eluting from the column that did and did not contain sulfur.

A number of chemical modifications of the sample can also be used to aid in identification. These are sometimes referred to as reaction gas chroma-

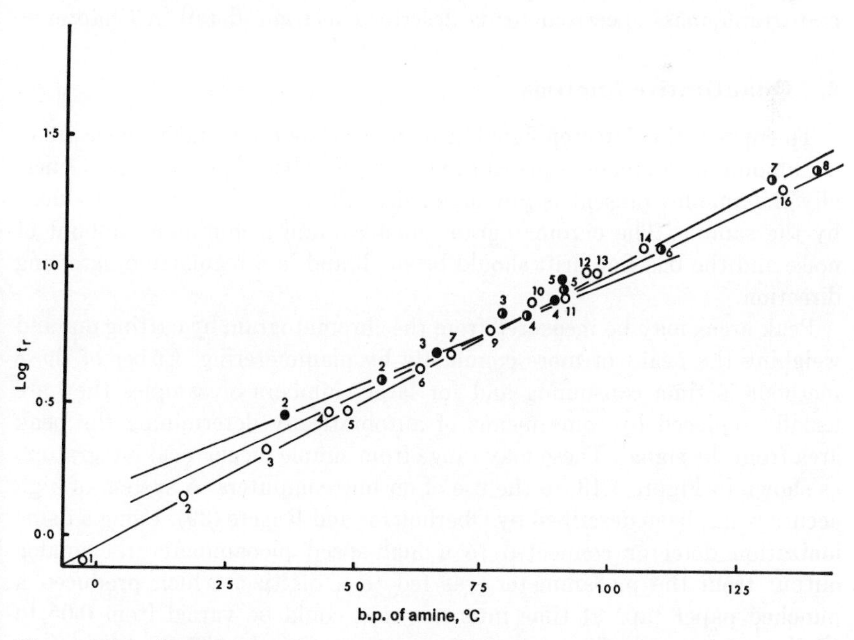

Figure 1.12. Log of retention time versus boiling point of amines (47).
○ : primary amines ◑ : secondary amines ● : tertiary amines

tography. Identification is based on comparing chromatograms before and after treating the sample. Examples are as follows:

Hydrogenation to convert a wide variety of compounds to their corresponding alkanes
Treatments with molecular sieves to remove straight-chain hydrocarbons
Extraction of olefins and aromatics with sulfuric acid
Removal of alcohols with sodium hydroxide
Conversion of alcohols to trimethyl silyl ethers, and
Pyrolysis of the sample under very well controlled conditions

Some of these reactions can be carried out *in situ* in a column prior to the chromatographic column.

Finally the eluting peaks may be collected in suitable traps and identified by any of the myriad standard analytical procedures. Those used have included mass, infrared, ultraviolet, and nuclear magnetic resonance spectrometry, and chemical analysis. These procedures depend on efficient trapping techniques and the ability to analyze very small samples. The problems of trapping can be avoided by coupling the analyzing system to the gas chromatograph. A particularly useful combination, the gas chromatograph–mass spectrometer is described in some detail in Chapter 9.

2. Quantitative Analysis

To convert the detector signal into an accurate quantitative measure of the amount of component present requires a number of precautions. Generally the amount present is proportional to the area of the peak produced by the sample. The chromatogram must contain a minimum amount of noise and the baseline drift should be small and in a regularly progressing direction.

Peak areas may be measured from the chromatogram by cutting out and weighing the peaks or more commonly by planimetering. Either of these methods is time consuming and for large numbers of samples they are usually replaced by some means of automatically determining the peak area from the signal. These may range from simple mechanical integrators, as shown in Figure 1.13, to the use of on-line computers. A system of high accuracy has been described by Oberholtzer and Rogers (39). Using a flame ionization detector connected to a high-speed picoammeter the analog output from the picoammeter was fed to a digitizer which produced a punched paper tape at time intervals that could be varied from 0.05 to 12.8 sec/data point. The data rate was selected so that 20–30 data points were obtained for each chromatographic peak. The data from the punched

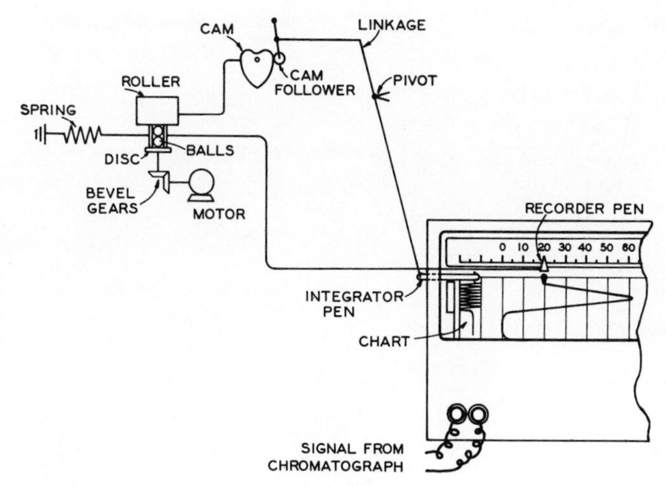

Figure 1.13. Mechanical integrator.

paper tape were analyzed by a high-speed computer. The beginning and end of the chromatogram were determined from the first derivative and a baseline established between these points. The area was then computed. An extensive discussion of the use of digital computers to evaluate chromatograms has been given by Baumann et al. (6).

The area associated with each of two incompletely resolved peaks is difficult to determine precisely. A number of procedures have been suggested. Dropping a perpendicular from the lowest point of the valley between the two peaks is most commonly used (see Figure 1.14), but is known to introduce significant errors. Reconstruction of the two peaks by computer methods is more precise but requires a suitable program and computer to be available (20).

After the area of the peak is known it must be multiplied by a response factor determined for the particular detector and operating conditions. The corrected areas of the peak are then normalized to obtain the percentage of each component. If all components are not eluted an internal standard may be used to obtain a reference area.

3. Combined Gas Chromatograph–Mass Spectrograph

Probably the most powerful analytical instrumentation available today for the analysis of complex mixtures is the combination of a gas chromatograph in series with a mass spectrometer interfaced with a computer for data interpretation. The mass spectrometer can supplement the information

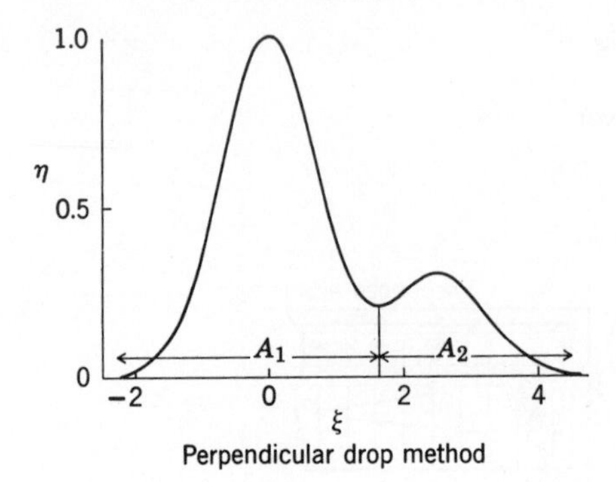

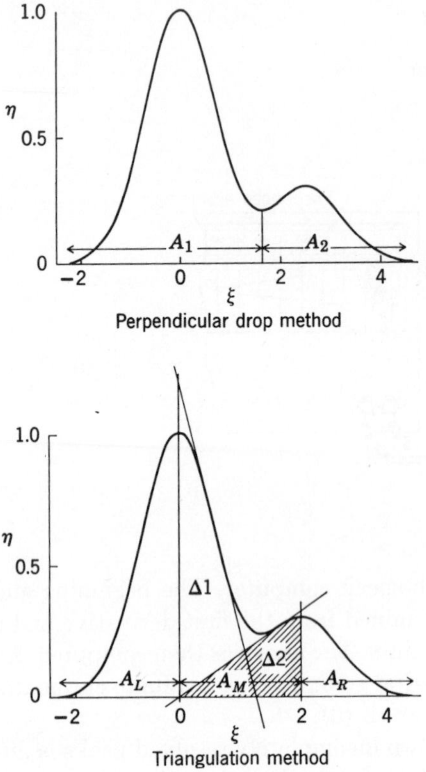

Figure 1.14. Methods for resolving overlapped peaks (54).

from the gas chromatograph in several ways: qualitative identification of a compound that has been separated but whose identity is not known, qualitative identification of a multicomponent mixture emerging as a single peak, and quantitative analysis of multicomponent peaks. Figure 1.15 is a schematic diagram of the combined system. A more detailed description of this technique is given in Chapter 9.

V. SUMMARY

Gas chromatography is an extremely powerful technique to separate, identify, and quantitatively determine the components of a wide variety of natural and synthetic mixtures even though its direct applicability is limited to compounds which can be volatilized without decomposition.

By making stable derivatives or by using very inert supports and glass columns and glass-lined injector systems many heat-sensitive compounds

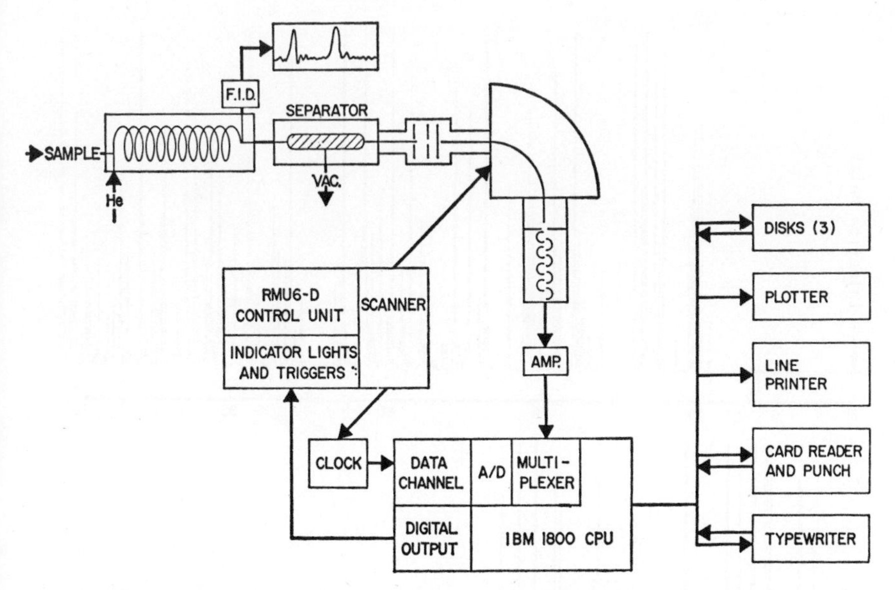

Figure 1.15. Schematic diagram of gas chromatograph-mass spectrometer-computer system (21).

such as those from biological or biochemical origin can also be analyzed by this technique. In many cases however, high-resolution liquid chromatography (see Chapter 2) is preferable for the analysis of this class of compounds.

The cost for the necessary instrumentation need not be excessive. A simple gas chromatograph and recorder can be purchased or built for a few hundred dollars. The more esoteric instruments, featuring multiple detection capabilities, different temperature programming options, a sub-ambient cooling accessory, and solid-state electronics may cost anywhere from \$5000 to \$10,000. Addition of high-quality peripheral equipment, such as digital electronic integrators, automatic sample injection, high-quality, low-response time recorder, may add another \$6000 to \$12,000 to the total costs.

Gas chromatography has been used in highly theoretical studies and in process control applications involving repetitious analysis of a single product stream over many months. Gas chromatography is also a prime candidate for combination with other instrumental techniques. It is the single most useful technique for the rapid analysis of unknown low-molecular-weight mixtures and for the quantitative determination of mixtures where the qualitative composition is known. Whereas other

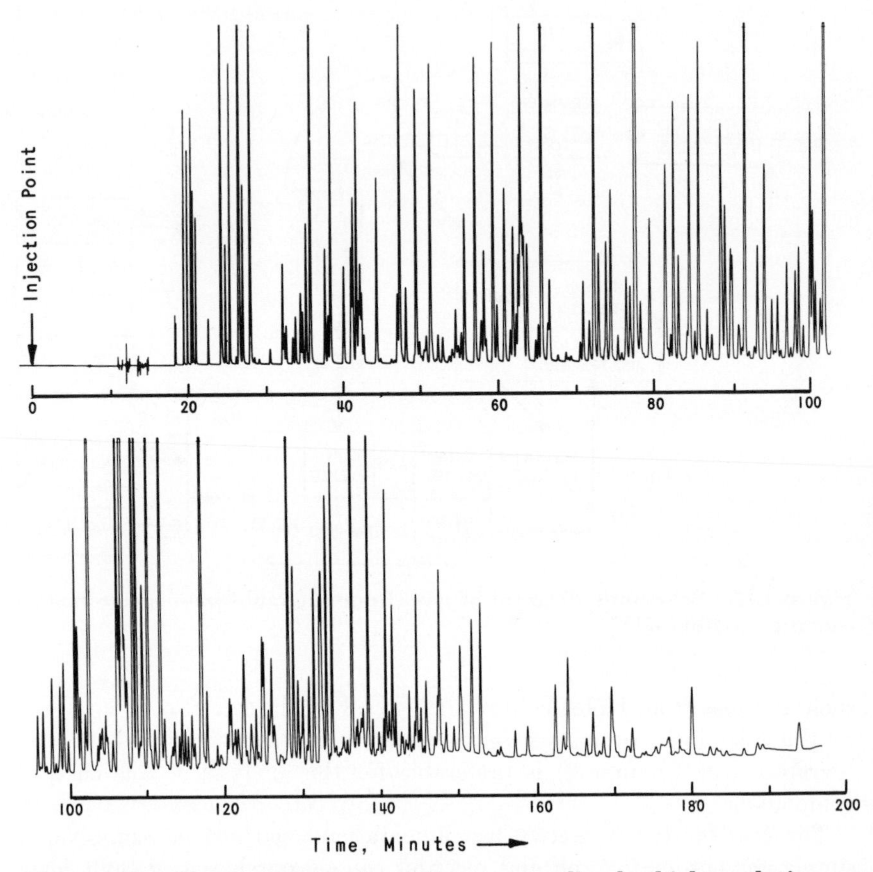

Figure 1.16. Chromatogram of a full-range gasoline by high-resolution gas chromatography. (Courtesy Dr. I. M. Whittemore, Chevron Research Company).

instrumental techniques are only capable of quantitatively analyzing multicomponent mixtures containing not more than a few components, with gas chromatography it is quite possible to analyze mixtures containing hundreds of components. The extremely powerful capabilities of gas chromatography are well illustrated by the chromatogram of a full-range gasoline shown in Figure 1.16. This analysis was carried out on a 1000-ft by 0.02-in. capillary column coated with squalane. A combination of temperature and pressure programming was used. Around 350 compounds can be distinguished in this chromatogram of which around 250 have been tentatively identified. The latter corresponds to about 97% of the gasoline.

BIBLIOGRAPHY

Table 1.9 lists subject headings for a number of review papers. It is hoped that these plus the cited list of books will aid those wishing more detailed information. For the neophyte in the field, the book *Basic Gas Chromatography* by H. M. McNair and E. J. Bonelli, and the programmed text, *A Programmed Introduction to Gas-Liquid Chromatography*, by J. B. Pattison are particularly recommended.

Table 1.9. Subject Index of Review Papers

Subject	Reference
Aroma analysis and applications	51
Atmospheric analysis	3
Automatic chromatographs	49
Automotive exhaust gas analysis	9
Column performance	15
Fatty acids and esters	1
Flame ionization detectors	2
Flow programming	12, 46
Identification of peaks	29, 45
Labeling techniques using GC	11
Lipid analysis	19
Liquid crystals as packings	27
Mass spectrometer plus GC	33, 34, 41
Nonionization detectors	55
Optimum analysis conditions	18
Packed capillary columns	17
Paint industry applications	16
Pesticide analysis	5, 32
Petroleum industry applications	13
Pharmaceutical industry applications	22
Physicochemical measurements	7
Plastic industry applications	38
Preparatory scale GC	44
Process monitoring	35
Pyrolysis plus GC	30 ,40
Radioassay detectors	25
Reaction kinetic studies	53
Steroid analysis	10, 26
Support effects	52
Trace analysis	50

ACKNOWLEDGMENTS

It is a pleasure to acknowledge the courtesy of the American Chemical Society, publishers of *Analytical Chemistry,* for permission to reproduce Figures 1.6, 1.7, 1.12, 1.14, and 1.15, and Varian Aerograph for permission to reproduce Figures 1.5, 1.8, and 1.9–1.11.

REFERENCES

1. R. G. Ackman, *Methods Enzymol.* **14,** 329 (1969).
2. R. G. Ackman, *J. Gas Chromatogr.* **6,** 497 (1968).
3. A. P. Altshuller, *Advan. Chromatogr.* **5,** 229 (1968).
4. D. Ambrose, A. I. M. Keulemans, and J. H. Purnell, *Anal. Chem.* **30,** 1583 (1958).
5. G. H. Bachelder, *Anal. Methods Pestic., Plant Growth Regul., Food Additives* **5,** 491 (1967).
6. F. Baumann, E. Herlicska, and A. C. Brown, *J. Chromatogr. Sci.* **7,** 680 (1969).
7. J. R. Conder, *Advan. Anal. Chem. Instr.* **6,** 209 (1967).
8. W. Dammeyer, *Anal. Chem.* **39,** 1339 (1967).
9. B. Dimitriades, C. F. Ellis, and D. E. Seizinger, *Advan. Chromatogr.* **8,** 327 (1969).
10. K. B. Eik-Nes and E. C. Horning, *Monographs in Endocrinology,* Vol. 2, 1968.
11. H. Elias, *Advan. Chromatogr.* **7,** 243 (1968).
12. L. S. Ettre, L. Mazor, and J. Takacs, *Advan. Chromatogr.* **8,** 271 (1969).
13. D. C. Ford, *Develop. Appl. Spectroscopy* **6,** 373 (1967) (Pub. 1968).
14. T. H. Gouw and I. M. Whittemore, *Chromatographia* **2,** 176 (1969).
15. G. Guiochon, *Advan. Chromatogr.* **8,** 179 (1969).
16. J. Haken, in *Treatise Coatings,* Vol. 2 (Part 1), R. R. Myers, Ed., Dekker, New York, 1969, pp. 191–270.
17. I. Halasz and E. Heine, *Advan. Chromatogr.* **4,** 207 (1967).
18. I. Halasz and E. Heine, *Advan. Anal. Chem. Instr.* **6,** 153 (1967).
19. S. F. Herb, *J. Amer. Chem. Soc.* **45,** 784 (1968).
20. R. A. Hites and K. Biemann, *Anal. Chem.* **42,** 855 (1970).
21. R. A. Hites and K. Biemann, *Anal. Chem.* **40,** 1217 (1968).
22. H. B. Hucker, *Theory Appl. Gas Chromatogr. Ind. Med. Hahnemann Symp.* 281 (1966) (Pub. 1968).
23. A. T. James, in *Gas Chromatography* H. J. Noebles et al., Eds., Academic New York, 1961, p. 247.
24. A. T. James and A. J. P. Martin, *Biochem. J.* **50,** 679 (1952).
25. A. Karmen, *Methods Enzymol.* **14,** 465 (1969).
26. A. Karmen and J. L. Marsh, in *Lipid Chromatogr. Anal.* **2,** 536 (1969); G. V. Marinetti, Ed., Dekker, New York.

27. H. Kelken and E. von Schivizhoffen, *Advan. Chromatogr.* **6,** 247 (1968).
28. E. Kovats, *Helv. Chim. Acta* **41,** 1915 (1958).
29. D. A. Leathard and B. C. Shurlock, *Advan. Anal. Chem. Instr.* **6,** 1 (1967).
30. R. S. Lehrle, *Lab. Pract.* **17,** 696 (1968).
31. L. L. Levy, D. A. Murray, H. D. Gesser, and F. W. Houghen, *Anal. Chem.* **40,** 459 (1968).
32. D. J. Lisk, *Anal. Methods Pestic., Plant Growth Regul., Food Additives* **5,** 363 (1967).
33. A B. Littlewood, *Chromatographia* **1**, 37 (1968).
34. W. H. McFadden, *Advan. Chromatogr.* **4,** 265 (1967).
35. I. G. McWilliam, *Advan. Chromatogr.* **7,** 163 (1968).
36. A. J. P. Martin, in *Gas Chromatography*, V. C. Coates et al., Eds., Academic, New York, 1958, p. 237.
37. A. J. P. Martin and R. L. M. Synge, *Biochem. J.* **35,** 1358 (1941).
38. J. E. Newell, *Theory Appl. Gas Chromatogr. Ind. Med., Hahnemann Symp.* 234 (1966) (Pub. 1968).
39. J. E. Oberholtzer and L. B. Rogers, *Anal. Chem.* **41,** 1234 (1969).
40. S. G. Perry, *Advan. Chromatogr.* **7,** 221 (1968).
41. D. Price, *Chem. Brit.* **4,** 255 (1968).
42. L. Rohrschneider, *Advan. Chromatogr.* **4,** 333 (1967).
43. L. Rohrschneider, *J. Chromatogr.* **22,** 6 (1966).
44. D. T. Sawyer and G. L. Hargrove, *Advan. Anal. Chem. Instr.* **6,** 325 (1967).
45. G. Schomburg, *Advan. Chromatogr.* **6,** 211 (1968).
46. R. P. W. Scott, *Advan. Anal. Chem. Instr.* **6,** 271 (1967).
47. J. R. L. Smith and D. J. Waddington, *Anal. Chem.* **40,** 522 (1968).
48. M. Stefanovic and B. L. Walker, *Anal. Chem.* **39,** 710 (1967).
49. P. B. Stockwell and R. Sawyer, *Lab. Pract.* **19,** 279 (1970).
50. V. Svojanovsky, M. Krejci, K. Tesarik, and J. Janak, *Chromatogr. Rev.* **8,** 90 (1966).
51. R. Teranishi, *Flavour Ind.* **1,** 35 (1970).
52. P. Urone and J. F. Parcher, *Advan. Chromatogr.* **6,** 299 (1968).
53. M. van Swaay, *Advan. Chromatogr.* **8,** 363 (1969).
54. A. W. Westerberg, *Anal. Chem.* **41,** 1770 (1969).
55. J. D. Winefordner and T. H. Glenn, *Advan. Chromatogr.* **5,** 263 (1968).

RECOMMENDED READING

J. B. Pattison, *A Programmed Introduction to Gas-Liquid Chromatography*, Heydon & Son, London, and Sadtler Research Labs., Philadelphia, Pa., 1969.
V. G. Berezkin, *Analytical Reaction Gas Chromatography*, Plenum, New York, 1968.
L. S. Ettre and W. H. McFadden, Eds., *Ancillary Techniques of Gas Chromatography*, Wiley-Interscience, New York, 1969.

H. M. McNair and E. J. Bonelli, *Basic Gas Chromatography*, Varian, California, 1969.

Harry P. Burchfield and E. E. Storrs, *Biochemical Applications of Gas Chromatography*, Academic, New York, 1962.

Herman A. Szymanski, Ed., *Biomedical Applications of Gas Chromatography*, Plenum, New York, 1968.

Malcolm P. Stevens, *Characterization and Analysis of Polymers by Gas Chromatography* (*Techniques and Methods of Polymer Evaluation*, Vol. 3), Dekker, New York, 1969.

Paul G. Jeffery and P. J. Kipping, *Gas Analysis by Gas Chromatography*, Pergamon, London, 1964.

A. V. Kiselev and Ya. I. Yashiu, *Gas Adsorption Chromatography*, Plenum, New York, 1969.

J. H. Knox, *Gas Chromatography* (Methuen) Barnes & Noble, New York, 1962.

Howard Purnell, *Gas Chromatography*, Wiley, New York, 1962.

A. B. Littlewood, *Gas Chromatography*, 2nd ed., Academic, New York, 1970.

O. E. Schupp III, in *The Practice of Gas Chromatography* (Technique of Organic Chemistry, Vol. 13, A. Weissberger, Ed.), Wiley-Interscience, New York, 1968.

A. I. M. Keulemans, *Gas Chromatography*, 2nd ed., Van Nostrand, Princeton, 1959.

R. Porter, Ed., *Gas Chromatography in Biology and Medicine*, Churchill, London, 1969.

F. Polvani, M. Surace, and M. Luisi, Eds., *Gas Chromatographic Determination of Hormonal Steroids*, Academic, New York, 1967.

Herbert H. Wotiz and S. J. Clark, *Gas Chromatography in the Analysis of Steroid Hormones*, Plenum, New York, 1966.

R. Scholler and M. F. Jayle, *Gas Chromatography of Hormonal Steroids*, Gordon and Breach, New York, 1968.

E. C. Horning, Ed., *Gas Chromatography of Lipids* (Progress in Chemistry of Fats and Other Lipids, Vol. 7, Pt. 2), Pergamon, London, 1967.

Ross W. Moshier and R. E. Sievers, *Gas Chromatography of Metal Chelates*, Pergamon, London, 1965.

M. Lipsett, Ed., *Gas Chromatography of Steroids and Biological Fluids*, Plenum, New York, 1965.

S. Dal Nogare and R. S. Juvet, *Gas-liquid Chromatography*, Wiley-Interscience, New York, 1962.

J. K. Grant, Ed., *Gas Liquid Chromatography of Steroids*, Cambridge University Press, Cambridge, 1967.

Rudolph Kaiser, *Gas Phase Chromatography:* Vol. 1, Gas Chromatography; Vol. 2, Capillary Chromatography; Vol. 3, Tables for Gas Chromatography (Butterworth), Plenum, New York, 1963.

K. B. Eik-Nes and E. C. Horning, *Gas Phase Chromatography of Steroids*, Springer, Vienna, 1968.

Austin V. Signeur, *Guide to Gas Chromatography Literature* (2 Vols.), (IFI-Plenum) Plenum, New York, Vol. 1, 1964; Vol. 2, 1967.

T. R. Lynn, C. L. Hoffman, and M. M. Austin, *Guide to Stationary Phases for Gas Chromatography*, 5th ed., Analabs, Hamden, Conn., 1968.

L. S. Ettre, *Open Tubular Columns in Gas Chromatography*, Plenum, New York, 1965.

Leslie S. Ettre and A. Zlatkis, *Practice of Gas Chromatography*, Wiley-Interscience, New York, 1967.

Walter E. Harris and H. Habgood, *Programmed Temperature Gas Chromatography*, Wiley, New York, 1966.

Howard Purnell, *Progress in Gas Chromatography*, Wiley-Interscience, New York, 1968.

A. Anne Patti and Arthur A. Stein, *Steroid Analysis by Gas Liquid Chromatography*, Thomas, Springfield, Illinois, 1964.

Harry S. Kroman and Sheldon R. Bender, Eds., *Theory and Application of Gas Chromatography in Industry and Medicine*, Grune, New York, 1968.

T. H. GOUW AND R. E. JENTOFT

Chevron Research Company
Richmond, California

II. High-Resolution Liquid Chromatography

I. INTRODUCTION

Natural chromatographic processes and the earliest forms of human ingenuity in the application of chromatography as a separation tool were based on the use of a liquid as the percolating agent and a solid adsorbent as the stationary phase. As such, liquid chromatography has a valid claim to being the oldest of all chromatographic processes. It was, however, not until the latter part of the 1960s that a thorough rejuvenation took place, uncovering in this process an extremely powerful technique with a very large potential.

Modern high-resolution liquid chromatography can separate a wide variety of compounds ranging in molecular weight from less than a hundred to upwards of a few million. It is particularly applicable to the fractionation of high-molecular-weight substances, thermally unstable compounds, biologically active material, and products which cannot be volatilized without decomposition. These capabilities are, of course, well known to classical liquid chromatography. Modern techniques, however, will carry out these separations on much smaller sample sizes at efficiencies and speeds several magnitudes larger than previously observed. Table 2.1 summarizes the difference between classical and modern, high-speed, high-resolution liquid chromatography. The major advances in this technique have been attained by the use of high pressures, high flow rates, and especially by the development of highly regular specialty packings with, for example, a solid fluid-impermeable core. Of the approximately 10^6 organic compounds which are currently reasonably well characterized, only about 15% can be volatilized without decomposition. These predominantly lower-molecular-weight compounds should preferentially be analyzed by gas chromatography. The balance, such as amino acids, proteins, synthetic and natural polymers, nucleic acids, steroids, lipids, carbohydrates, vitamins, many synthetic drugs, antibiotics, and other compounds, fall into the domain of liquid chromatography.

Two special forms of liquid chromatography, namely, gel permeation chromatography and ion exchange chromatography, are not discussed in this chapter. Gel permeation chromatography, especially applicable for the analysis of higher-molecular-weight products and for the preliminary separation of many mixtures prior to high-resolution analysis of the desired fractions, is treated in Chapter 4. Ion exchange chromatography will be treated in a subsequent volume.

In the near future one may also expect a new technique, supercritical fluid chromatography, to be quite competitive with a large portion of the

Table 2.1. Comparison of Classical and Modern Liquid Chromatographic Techniques

Parameter	Classical	Modern
Particle size of column packing, μ	75–600	10–50
mesh	30–200	300–400
Packing material	standard grade	specialty packings
Particle-size spread, σ (as % of median)	20–30	<10
Column length, cm	10–100	100–500
Column diameter, cm	2–5	0.1–1.0
Column inlet pressure, atm gage	0.01–1.0	20–300
Column efficiency, theoretical plates	2–50	10^3–10^4
Sample size, g	1–10	10^{-6}–10^{-2}
Separation speed, plates/sec	<0.05	1–20
Analysis time, hr	1–20	0.05–1

field of liquid chromatography. This technique is in many respects quite similar to high-resolution liquid chromatography; and, in general, only small modifications are necessary to allow both techniques to be carried out with the same apparatus.

II. THEORY

1. Differential Migration

In a chromatographic column, a solute molecule migrates at a specific speed. This molecule will continually vacillate between the stationary and the mobile phases. The *fraction* of the time spent in the mobile phase, R, is equal to the probability that the molecule is in that particular phase. Since this molecule can be assumed to be stationary during those periods when it is attached to the substrate, R is also equal to the fraction of the velocity of the mobile phase at which the molecule travels down the column. On a macroscopic scale where a sufficiently large number of solute molecules are involved to justify statistical considerations, R is also equal to the fraction of the solute molecules in the moving phase; $1 - R$ is, thus, the fraction of the solute in the stationary substrate.

The ratio of the concentrations of a solute in these two phases in equilibrium with each other is, at low solute concentrations, a constant which is only dependent on the temperature. In chromatography, this ratio is called the distribution coefficient, partition coefficient, or partition ratio.

$$K = \frac{\text{equilibrium solute concentration in the stationary phase}}{\text{equilibrium solute concentration in the mobile phase}} \tag{2.1}$$

Hence

$$\frac{R}{1 - R} = \frac{1}{K}\frac{V_m}{V_s}$$

or

$$R = \frac{V_m}{V_m + KV_s} \tag{2.2}$$

where V_m and V_s are the volumes of the mobile and stationary phases, respectively.

If t_m is the time for a mobile-phase molecule to travel through the column, the time for a solute molecule to travel the same distance would be

$$t_1 = \frac{t_m}{R_1} = t_m\left(1 + K_1\frac{V_s}{V_m}\right) \tag{2.3}$$

For a solute molecule of a different species, the retention time would be

$$t_2 = t_m\left(1 + K_2\frac{V_s}{V_m}\right) \tag{2.4}$$

A measure of the separation capability of a system is the difference in retention times per unit length. Since for a given mobile phase velocity t_m is proportional to the column length, one can write:

$$\frac{\Delta t}{t_m} = \frac{t_1 - t_2}{t_m} = \frac{V_s}{V_m}(K_1 - K_2) = K_2\frac{V_s}{V_m}(\alpha - 1) \tag{2.5}$$

The basic cornerstone of chromatography is the existence of differences in the speeds of migration. To increase $\Delta t/t_m$, we will have to increase K, V_s/V_m, and/or α, where $\alpha = K_1/K_2$.

A. K

One of the main advantages of liquid chromatography is that the mobile phase allows for an added dimension in the choice of K. In gas chromatography, the concentration of the solute in the mobile phase (hence the K) is chiefly determined by its vapor pressure. The choice of the gas has little influence on the distribution coefficient, and changes in K are usually carried out by controlling the temperature of the column. Because of the much stronger intermolecular interactions in a liquid, on the other hand, this mobile phase can be quite competitive with the stationary substrate for the possession of the solute molecules. The increased solubility of the solute results in a reduction in the observed values of the K's by a factor of around 10^3 compared to gas chromatography.

This is especially critical in the chromatographic analysis of higher-molecular-weight compounds where in *gas* chromatography distribution coefficients of 10^3 or higher are observed. High temperatures can be used to decrease the retention time, but thermal stability of both the solute and the stationary phase may pose some problems. Using *liquid* chromatography, many of these compounds can be chromatographed at room temperatures under conditions where K is in the more practical range of 1–10^2.

The use of a liquid as the mobile phase allows the K to be modified in a range of as much as 10^{10}. Although Equation 2.5 suggests the use of very high K values, for good separations we would, in practice, attempt to work with K values between 1 and 10^2. With a $K < 1$, the components are eluted too fast with too little resolution; with a $K > 100$, excessively long times are necessary for routine separations (36). Equation 2.8 will show, as a matter of fact, that the increase in resolution by increasing the K is small for $K > 10$.

B. V_s/V_m

Equation 2.5 shows that Δt is also increased by increasing the V_s/V_m ratio. In practice, this means that Δt is increased by using higher liquid loading levels in partition chromatography, increasing the active surface of the adsorbent in adsorption chromatography, increasing the charge density in ion exchange chromatography, or increasing the number of the pores in the polymer lattice in gel permeation chromatography. We later note that this increase results in an increase in resistance to mass transfer in the stationary phase. For analytical, high-speed work, this increase in resistance outweighs the advantages of increasing V_s/V_m.

C. α

An important advantage of using a liquid as the mobile phase is the possibility of varying its selectivity to allow many difficult separations to be carried out in a relatively short column. Many workers underestimate these capabilities when choosing their mobile phase. Varying α is, as a matter of fact, the preferred way to increase $\Delta t/t_m$ in Equation 2.5, particularly for liquid-solid chromatography (and liquid-liquid chromatography with bonded substrates). Thoma (53) gives some rules on how to choose the best chromatographic solvent. Section V discusses some further aspects of this problem.

2. Column Efficiencies

If the two solute species to be separated should consist of only one molecule each, a complete separation should always be possible. In practice, even the most sensitive detector would need as many as 10^{10} molecules to give a recognizable signal: the majority of the detectors in use require a sample several magnitudes larger. This detection capability limits the minimum size of the sample. Hence we are faced, in practice, with the separation of very large numbers of molecules so that we must consider relative and not absolute separations.

Solute molecules travel in a zone, and zones have the unfortunate tendency to broaden during the migration process through the column. Zone or band broadening is due to flow velocity inequalities in the packing, longitudinal diffusion, and resistance to mass transfer in both the stationary and mobile phases. With more than one component, the zone shapes at the outlet of the column may have the appearance shown in Figure 1.4.

A measure of the separation between two components is the resolution, R_s, defined as

$$R_s = \frac{2\Delta t}{(W_x + W_y)} \tag{2.6}$$

Hence resolution is a function of both Δt and W_x and W_y, the widths of the peaks. Two compounds with their peak maxima at some distance from each other may not necessarily be adequately resolved if their peak widths are very large. This is the usual case with samples eluted after long periods from classical liquid chromatographic columns.

The peak width for any compound is a function of the length, L, and the efficiency of the column. The latter is usually defined by an efficiency parameter, H, the plate height, or the HETP, the height equivalent to a theoretical plate, according to

$$W = 4\sqrt{LH} \tag{2.7}$$

The peak width can be decreased by reducing L and/or H. Since Δt is directly proportional to L for a given system and mobile phase velocity, it can be seen from Equations 2.5, 2.6, and 2.7 that the resolution is proportional to $\sqrt{L}$. For liquid adsorption chromatography (50), the resolution can be related to the separation parameters by

$$R_s = \tfrac{1}{4}(\alpha - 1)\sqrt{N}\,\frac{\overline{K}}{(1 + \overline{K})} \tag{2.8}$$

where α is a measure of the selectivity of the solvent, N is the number of plates in the column, and $\overline{K}$ is the average of K_1 and K_2. This equation shows the importance of choosing a solvent (combination) which gives a high α for the compounds of interest. This equation also shows that for $\overline{K} > 10$ an increase in K will only result in a relatively small increase in the resolution. For liquid-liquid partition chromatography one should substitute k' for K, where $k' = KV_s/V_m$, the capacity factor.

Up to this point, we have not considered the effect of the flow rate on the separation efficiency. Flow rate, however, is a very important operational parameter to determine the efficiency and speed of the separation. The relation of H to u, the flow rate, as a function of the column parameters can be best seen from one of the most successful rate theories, the one given by van Deemter et al.(6) and subsequently modified by Giddings (9).

$$H = \frac{B}{u} + C_s u + \frac{1}{1/A + 1/C_m u} \tag{2.9}$$

In equation 2.9, the constants A, B, and C have the following significance: $A = 2\lambda\, d_p$. The term A is the contribution to the plate height of the spread in path lengths and the flow velocity inequalities due to the packing structure. This term is also called the "eddy diffusivity". The value of λ ranges from 0.5 to 8 and is smaller for more homogeneous packings; d_p is the particle diameter of the packing. $B = 2\gamma\, D_m$. The term B is the contribution of the longitudinal molecular diffusion to the plate height, γ is a structural constant <1 showing the degree of obstruction to diffusion, and D_m is the molecular diffusion coefficient of the solute in the mobile phase.

$$C_s = Q_1 \frac{(k')}{(1 + k')^2} \frac{d_f^2}{D_s} \tag{2.10}$$

C_s is the resistance to mass transfer in the stationary phase, where k' is KV_s/V_m, the capacity factor; d_f is the film thickness of the stationary phase; and D_s is the molecular diffusion coefficient in the stationary phase.

$$C_m = Q_2 \frac{(k')^2}{(1 + k')^2} \frac{d_p^2}{D_m} \tag{2.11}$$

The term C_m is the resistance to mass transfer in the mobile phase. The terms Q_1 and Q_2 are constants which are characteristic of the system.

In treatises of a more theoretical nature, use is preferentially made of the following dimensionless reduced parameters to allow column performance to be compared on a more fundamental basis (10, 11). The reduced plate height.

$$h = \frac{H}{d_p} \tag{2.12}$$

The reduced mobile phase velocity.

$$\nu = \frac{ud_p}{D_m} \tag{2.13}$$

The reduced mass transfer coefficient in the mobile phase.

$$\omega_m = \frac{C_m D_m}{d_p{}^2} \tag{2.14}$$

By rearranging the expression for A, we find that

$$\lambda = \frac{A}{2d_p}. \tag{2.15}$$

Equation 2.9 can thus be rewritten as

$$h = \frac{2\gamma}{\nu} + \omega_s \nu + \frac{1}{1/2\lambda + 1/\omega_m \nu} \tag{2.16}$$

where ω_s is the mass transfer coefficient in the stationary phase arranged in reduced form and which is equivalent to $C_s D_m / d_p{}^2$. The general form of Equations 2.9 and 2.16 is shown in Figure 1.5. To obtain an idea of the magnitudes involved, some typical values of the properties listed in the previous equations are given in Table 2.2. For comparative purposes, values are given for a gas, liquid, and supercritical fluid. The numbers quoted only give the order of magnitude.

Table 2.2. Physical Properties of a Gas, Liquid, and Supercritical Fluid

Property	Symbol	Units	Gas	Liquid	Supercritical Fluid
Density	ρ	g/ml	10^{-3}	1	0.3
Diffusivity	D	cm²/sec	10^{-1}	5×10^{-6}	10^{-3}
Dynamic viscosity	η	poise (g/cm. sec)	10^{-4}	10^{-2}	10^{-4}

Because D_m is very small in liquids, the first term in Equations 2.9 and 2.16 can usually be ignored in liquid chromatography, especially in modern, high-speed analysis. Mobile-phase velocities in liquid chromatography are much lower than in gas chromatography. In regular packed columns where the flow is laminar and d_p is less than a tenth of the column diameter, peak broadening is mainly due to C_m, the mass transfer resistance in the mobile phase (18).

The use of the reduced parameters allows for a comparison of the relative behavior of a gas and a liquid chromatography column. For the same reduced velocities and particle diameters (Equation 2.16), observed flow rates should be 10^5 times slower in liquid than in gas chromatography. Since the viscosity of a liquid is 10^2 that of a gas, the pressure drop over the column will only be 10^{-3} that of a gas system for a comparable performance of the column.

On the other hand, it is quite feasible to work with high pressure in liquid systems. A liquid is much less compressible than a gas, and safety precautions are more easily carried out. In addition, high pressure on a liquid does not have the deleterious effect on the separation process it has on gas chromatography.

This large reserve in pressure capabilities allows the use of much smaller particles for packing materials. From Equations 2.9 and 2.11, it can be seen that the plate height is reduced by decreasing the particle size. Figure 2.1 shows this effect graphically (50). The available pressure margin can also be used to increase the length of the column. This will proportionally increase the number of plates.

The excess efficiencies generated by these changes can be traded off by operating at higher mobile-phase velocities. The latter is necessary because the minimum value of H occurs at a flow velocity at only around 10^{-2} to

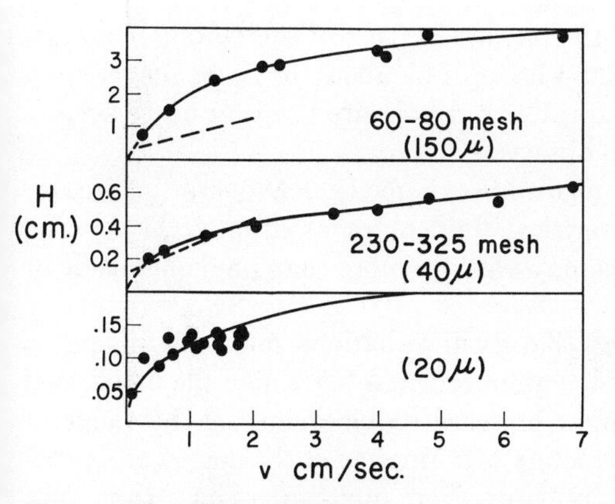

Figure 2.1. Plate height *H* as function of the mobile phase (*n*-pentane) velocity. Columns are 123 × 0.46 cm dry-fill packed with 4% $H_2O \cdot SiO_2$ (Davison Code 62). The solute is dibenzyl (courtesy of the authors and *Journal of Chromatographic Science*).

10^{-3} cm/sec. At these velocities, most separations would take too much time to carry out.

By operating at higher velocities, the plate height is increased. Because of the "coupled" form of the two parameters in the last term in Equations 2.9 and 2.16, one will notice from Figure 1.5 that the increase in plate height is less pronounced than the increase in flow velocity. At higher flow rates, this "coupling" phenomenon tends to reduce the relative rate of increase in plate height even further (12). In this region of interest, Waters et al. (56), observed that for a wide range of liquid chromatographic systems the relation of H to u can be expressed by

$$H = D'u^n \tag{2.17}$$

where D' is a constant for a particular column, depending on the size of the column, the column packing, and the method of column packing. The value of n was found to vary betweeen 0.2–0.6. For $n = 0.3$, for example, an increase of the flow velocity by a factor of 10 will only decrease the efficiency of the column by a factor of 2 per unit length. Hence by using a 10 times higher flow velocity and a two times longer column than necessary for optimum operation, an 80% gain in time can be obtained for the same total number of plates. For Reynolds numbers in excess of 10, the reduced plate height becomes independent of velocity and has a value of about 2 for axial dispersion and about 0.2 for radial dispersion (18).

A further gain in speed can be obtained by increasing the mobile phase velocity to induce turbulence in the column (42). In a packed bed, turbulence starts at around a Reynolds number of 10.

Turbulence makes the radial mass transport in the mobile phase largely convective and independent of the physical nature of the mobile phase. In this case, however, columns would have to be about 10 times longer than used in laminar flow chromatography, and pressure drops of up to several hundred atmospheres would be observed.

Table 2.3 lists the order of magnitude of some typical operating parameters in gas, liquid, and supercritical fluid chromatography. Depending on the conditions, actual values may vary by more than one magnitude or more from the numbers given.

By judicious combination of the given equations one can, if one so wishes, theoretically derive the minimum analysis time, the maximum separation number, the minimum pressure requirement, or the range of the values of the operating parameters as a function of the degree of separation, the particle diameter, the pressure drop, and the analysis time (35). For the intrepid reader interested in a more detailed understanding of the mechanisms and dynamics of liquid chromatography, we refer to two excellent textbooks available (9, 48).

Table 2.3. Typical Values of Operating Parameters in Packed-Bed Analytical Chromatography

Parameter	Symbol	Units	Gas	Liquid	Super-critical Fluid
Particle diameter	d_p	cm	2×10^{-2}	5×10^{-3}	5×10^{-3}
Mobile phase velocity	u	cm/sec	10	0.2	2.5
Modified Reynolds number	N_{Re}	Dimensionless	0.5	0.1	10
Void fraction	ϵ			0.6	
Shape factor	Φ			0.7	
Friction factor	f_c		7	60	0.7
Pressure drop (0.5 × 400-cm column)	ΔP	atm	0.5	10	2
Reduced velocity	ν	Dimensionless	1	200	10

III. INSTRUMENTATION

A schematic diagram of a high-resolution liquid chromatograph is shown in Figure 2.2 The unit consists of a high-pressure pump to force the liquid mobile phase through the column, a presaturator (optional), an injector, the column, and the detection system. The column and the injection system may be enclosed in a thermostated oven. This is necessary if the unit is also going to be used for supercritical fluid chromatography.

As in all chromatographic techniques, the sample is introduced into the system through the injector; the components in the sample are fractionated during their passage through the column. The detector system senses these components as they elute from the column and generates a signal proportional to the amount of the solutes passing through the system. The separate components of the chromatograph will now be discussed.

1. Pump

High pressures of up to at least 100 atm and preferably much higher are needed for high-resolution, high-speed liquid chromatography. The pumps in general use are piston-type pumps or large reservoirs with gas pressure to drive the mobile solvent. It should be noted that these two pump types operate on different principles. With mechanical piston pumps, the "flow rate" is set to the desired value; and the pressure is observed as a dependent variable. With displacement pumps using gas pressure as the driving force,

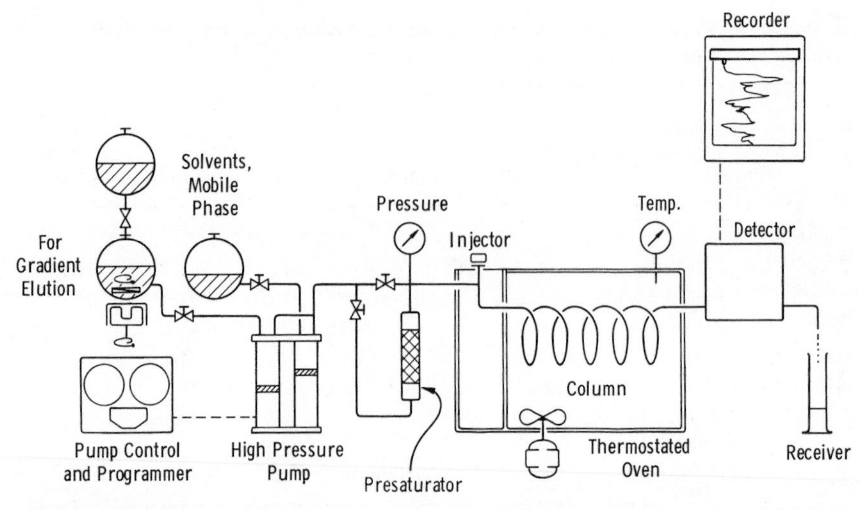

Figure 2.2. Schematic diagram of a high-speed, high-resolution liquid chromatograph.

the inlet pressure is the controlling parameter, and the flow rate is obtained as a derived variable.

Small-volume, piston-type pumps will give pulses which are undesirable because it may affect detector stability, column resolution, and reproducible operation. To smooth out the ripples, one can use a multipiston pump where the action of the individual pistons is spaced at regular intervals of a complete stroke cycle. Ripples can be further decreased by using bellows and restrictors. An elegant method to smooth out pressure fluctuations is to use a several-meters-long 1.5-mm ID nylon tube between the pump and the chromatographic column.

The best pump of this type is the large-volume syringe pump. These are actually large syringes where spring-loaded Teflon seals are used in the plungers to minimize leakage around the pistons at high pressure. The volume of a charge may be several hundred milliliters. By using a gear mechanism to drive the plunger, a constant and reproducible flow is obtained. By the use of stainless steel cylinders, operating pressures of several hundred atmospheres can be obtained.

One of the first pumps specifically designed for high-pressure, high-resolution liquid chromatography used high-pressure nitrogen as the pressure source and mercury as the displacing agent (22). It is especially useful in pressure programming work, which is an important capability if supercritical fluid chromatography is to be carried out in the same unit. Instead of mercury, the barrier can also be a polyethylene film or a metal bellows (39).

For the analysis of mixtures with a wide range of partition coefficients, solvent programming, also called gradient elution (see Section V), is a good technique to improve the analysis. In many cases, this capability has to be borne in mind when choosing a pump. Solvent programming is usually carried out by continuously adding a more polar solvent to the mobile-phase feed reservior, thereby increasing the polarity of the eluant as a function of time. If a single pump only is used, this pump should have a small holdup volume. For high-pressure, pulseless operation, the choice will now be limited to the multiple plunger-type pump. An excellent solution, especially when large-volume displacement pumps are used, is to use separate pumps feeding different solvents or solvent mixtures concurrently into the column and programming the output of each pump. A simple attachment to allow any pump to acquire solvent programming capabilities is shown in Figure 2.3*a*. It consists essentially of a high-pressure, magnetically stirred reservoir placed between the outlet of the pump and the inlet of the column. Before the start of the run, the least polar solvent is introduced into this reservoir. The more polar solvent is used as feed to the pump. The main disadvantages of this setup are the concave form and the relative inflexibility of the solvent gradient. For many problems, it is more advisable to have a linear or a convex gradient, in

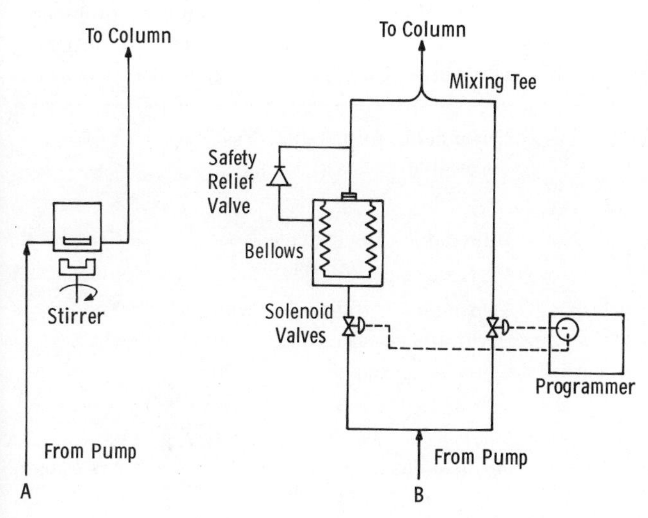

Figure 2.3. Attachments for gradient elution chromatography.

other words, where the polarity would increase slowly in the beginning and much faster later in the run.

Another attachment which is more versatile is shown in Figure 2.3*b*. A *T* is connected to the outlet of the pump with a solenoid valve mounted in each branch. One solvent is introduced into the bellows assembly, while the other solvent is used as feed to the pump. When the left solenoid is activated, the second solvent is pumped into the closed container, forcing out an equivalent amount of the first solvent to the mixing *T*. By programming the time intervals between opening and closing of these two solenoid valves, the desired solvent gradient can be obtained as feed to the column.

A pump which has not received much attention is the large displacement pump based on the pressure-intensifier principle. Very high pressures can be attained with only low primary pressure sources. Table 2.4 lists some comparative features of these pumps.

2. Presaturator

Since a liquid is capable of dissolving approximately 10^4 as much solute as a gas of the same volume, we now have the drawback that the mobile

Table 2.4. Pumps for High-Pressure Liquid Chromatography

Pump Type	Price	Ripples, Pulses	Pressure	Flow	Solvent Programming	Applicable to Supercritical Work
Small-volume piston	Medium	Yes	Dependent variable	Adjustable	Yes	Yes
Pressure vessel	Low to medium	No	Adjustable	Dependent variable	No	Very good
Large-volume syringe	High	No	Dependent variable	Adjustable	Multiple pumps, programmed output	Yes
Pressure intensifier	Medium	No	Adjustable	Dependent variable	No	Very good

phase will also take up substrate material from the column. Depletion of the active component results in a change of characteristics of the packing, usually for the worse. In liquid-liquid chromatography, some of the liquid substrate might be dissolved in the mobile phase; in liquid-solid chromatography, this might be the water which is removed from the adsorbent. This problem can usually be solved by presaturating the solvent with the stationary liquid substrate or with water before it enters the column (23). A presaturator is generally unnecessary with the newer type packings where the substrate is bonded to the stationary phase.

3. Injector

Because of the much greater solvent properties of the mobile phase and because of the high pressures involved, the injector systems used in gas chromatography fail rapidly if used in high-pressure liquid chromatography. This objection can be partly overcome by the use of Teflon-backed septums; but leaching is difficult to obviate, and the unwary chromatographer may find that he has more peaks in his chromatogram than he can account for on the basis of his sample composition. Some of the injectors described use a switching valve, O-ring closures, and the use of plungers to position the sample in the stream. Injector systems can also be used where the inlet pressure is first relieved, the sample deposited on top of the column, and the injector access closed off again with a plug. The loss of resolution is supposedly quite low because of the very low solute diffusion in the mobile liquid phase.

Good design of the injector is important. Smuts et al. (47) believe that the injection system may ultimately be the limiting factor in high-resolution liquid chromatography. A discussion on injector design has been given by Scott et al. (44).

4. Column and Column Packing

The column is undoubtedly the most important section of the unit. Theoretical analysis has shown that much smaller particles should be used than in gas chromatography; d_p in analytical liquid chromatography currently ranges from 50–20 μ (150–400 mesh) and occasionally even to smaller particle sizes. These values indicate the range of particle sizes used; in practice, very narrow mesh-range particles are prescribed to decrease both λ and γ in the constants A and B in Equation 2.9. With the exception of the modern specialty packings, commercially available column packings are often not good enough. Additional grading, preferably by elutriation, is necessary to obtain a narrow mesh-range particle size with negligible amounts of fines. Although column diameters have been described with less than three times the particle diameter, the preferred diameter would be in the $>50\text{-}d_p$ range. The columns in current use generally are in the 0.1–2.0-cm ID range.

Although the column diameter does not occur in the equations presented so far, this parameter is important in the performance of the column. In packed chromatographic columns, a large contribution to the band broadening is due to large-scale unevenness of flow. This becomes worse as the column diameter increases. The effect of uneven flow may be decreased by reducing the diameter of the column or by increasing the mobile-phase velocity. Practical problems, however, may be insurmountable if the column diameter becomes less than 0.05 cm because very small particle sizes and very high pressure drops will have to be used.

Transcolumn velocity differences are smoothed out by increasing the mobile-phase velocity to above approximately 1 cm/sec. Above this value, the contribution of the A term (Equation 2.9) is still large; but it becomes unimportant in relation to the other contributions.

The capacity of an analytical column is obviously very small, and some resolution will have to be sacrificed if larger bore columns are to be used for preparative work. Preparative columns may have a diameter of 2–3 cm. Even then, by using small particle sizes, high inlet pressures, and a good column packing technique, obtained resolutions and speed will be at least a magnitude better than in classical liquid chromatography.

The column length is, of course, determined by the degree of resolution desired since resolution is proportioned to $\sqrt{L}$. The usual columns in practice range from 1–4 m.

Of major importance in obtaining a good column is a good packing procedure. With the very small particle sizes in liquid chromatography, it is more difficult to obtain homogeneous packing than in gas chromatography. Since the influence of good column packing is much more pronounced in liquid than in gas chromatography, many techniques have been proposed to obtain the ideal, efficient, and reproducible column. In practice, the preparation of a good column necessitates a judicious combination of patience, savoir faire, and applied witchcraft.

The method of column packing is dependent on the type, regularity, and the particle size of the packing used (50). In general, a good way appears to be to add small incremental amounts of packing at a time while continuously tapping around the column. A mechanical vibrator will only promote particle size segregation. For gentlemen of leisure and those who have access to free slave labor, packing can also be carried out by adding small amounts of material at a time and *gently* compacting with a rod. The procedure is recommended if very small particle sizes are to be used. Too strong tamping may result in inferior columns because the friability of the packing may also contribute to this phenomenon.

For relatively uniform particles, 50 μ appears to be a natural break point. Particles larger than 50 μ can usually be packed to give good columns; for particles below 50 μ ,special care is necessary for good results.

There is also a correlation between the degree of irregularity of the packing and the difficulty in obtaining a good column. Glass beads and Corasil (see Section IV.1.D), for instance, are readily packed to give efficient, reproducible columns, even for particle diameters as small as 20–40 μ. This is due to the greater density and smoothness of these particles. A highly irregular packing, such as alumina, on the other hand, is difficult to pack well. Here the break point is closer to 100 μ.

Very small particles are almost impossible to pack well in the dry state. The effect of electrostatic forces becomes relatively large in relation to the gravitational forces. In these cases, one may choose to suspend the packing in a liquid of the same density and to pump the slurry into the column to be packed. Although this method is somewhat unreliable, it sometimes gives exceptionally good results.

True bore columns are also believed to be beneficial for maximum column efficiencies. This is especially the case for the smaller diameter columns.

A special method of packing larger columns employs a distributor head for feeding the packing into the column while the latter is being rotated and tapped. Sie et al. (46) found that a five fold increase in efficiency in comparison to standard packing techniques can be obtained by this method. The same authors also advocate a totally different approach which is especially applicable to irregular packings. This packing procedure, called the "wet-fill technique," uses gypsum to consolidate the adsorbent. Packing is carried out as a slurry. Results indicate that the highest gain in efficiency is obtained with the roughest particles, such as alumina and Sil-O-Cel. This advantage is, however, less pronounced for column diameters below 1.0 cm. The presence of $CaSO_4$ might also be disadvantageous in some applications.

The long columns advocated in the preceding paragraphs may create some problems since most laboratories are not shaped like bowling alleys. For small-diameter columns which are coiled to conserve space, especially when the column has to be mounted in a chromatographic oven, Scott et al. (44) found that coiling the packed column can cause a serious increase in HETP because the packing is disturbed during the coiling process. Good results can, as a matter of fact, be obtained by preforming the empty tube before packing. In this case, one may wish to use a column packer which combines mechanical vibration with feeding the packing under pressure. A very tight packing is obtained if one strokes a coiled column along the axis of the coil with a piece of metal. This apparently enhances the natural vibration of the system and aids in the formation of good, reproducible packings.

5. Detector

The absence of a versatile, universal, and relatively inexpensive detector has been one of the main reasons why development in liquid chromatography has not been as rapid as in gas chromatography. The best detectors have only limited applicability, while the so-called universal detectors have one or more of the following disadvantages: they are expensive, are difficult to operate, and have only a limited linearity range. An excellent survey of liquid chromatography detectors has been published by Conlon (5).

A. Ultraviolet Absorption Detector

Although the applicability is limited to compounds which absorb in the UV (ultraviolet) and to mobile phases which are transparent in the wavelength region of interest, the UV absorption detector is widely used because of its ease of operation. Flow-through detectors with a 1-cm path length and cell volumes as low as 10–20 μl are now quite common (8, 21, 23). Almost any UV recording spectrophotometer can be modified to perform as a detector. Relatively inexpensive commercial units are also available from a number of sources. Most of them, however, operate at fixed wavelengths.

Very high sensitivities can be attained with this detection system. Units giving a full-scale detection for 0.01 absorbance at a noise level of 5×10^{-4} absorbance can be obtained commercially. As little as 10^{-10} moles/ml of ribonucleosides or 1×10^{-8} g of diuron (32) can be detected with this detector (19). The UV detector is not sensitive to variations in flow and can be used with gradient elution techniques. When a recording spectrophotometer is used, there is the added capability of arresting the flow momentarily and scanning the UV absorption spectrum of the eluting components to aid in the identification of these solutes.

B. Differential Refractometer

The principle of this detector is based on the measurement of the difference in refractive index between the solution and of the pure solvent. This is usually carried out by measuring the bending of a monochromatic light beam as it passes through wedge-shaped sample cavities. Cell volumes are in the order of 70 μl and higher, and careful control of the sample temperature and the optical system is necessary. These detectors are quite stable and sensitive. The differential refractometer is currently the most widely used detector in gel permeation chromatography.

Another group of differential refractometers is based on the Fresnel effect. These refractometers determine the intensity of reflectance which, in turn, is inversely proportional to the refractive index. Instead of reflectance, the detector can also be constructed to measure transmittance. With this principle, smaller cell volumes and easier temperature control are feasible than with the deflection-type instruments. On the other hand, the Fresnel-type refractometer is more sensitive to bubbles and dirt in the cell. Erratic results can also be obtained if films are formed on the glass prisms.

The index of refraction is temperature dependent. To measure absolute refractive indexes to within 10^{-5} units, the temperature has to be controlled to within 0.01°C. The use of a *differential* refractometer, however, allows the detection of as little as 2×10^{-7} units difference with the same overall quality of the temperature controller. How much material one can detect with this detector is, of course, dependent on the difference in refractive index of the solute and of the solvent. With isooctane as the mobile phase, a deflection of 3×10^{-7} units would correspond roughly to a concentration of 2×10^{-6} g/ml of pseudocumene in the solution. To attain this sensitivity, the *difference* in temperature between the two cells should be less than 0.003°C for aqueous solutions and 0.007°C for organic solvents. This temperature control is not too difficult to achieve since both the sample and the reference cavity are mounted in the same unit. These detectors are difficult to operate in conjunction with solvent programming unless specially developed solvent combinations are used. An interesting example is the use of 30% cyclohexane–70% diethyl ether with a combined refractive index of 1.3752 as the starting mixture to which increasing amounts of a 58.5% tetrahydrofuran–41.5% methanol mixture, with the same combined refractive index as the starting material, is added. A disadvantage of the differential refractometer is its noticeable sensitivity to variations in flow.

C. *Flame Ionization Detector*

A detector system which enjoyed some publicity in the earlier days of high-resolution liquid chromatography, and is now enjoying a renaissance, employs a moving chain (28), wire (57), belt (3), band, or disk in conjunction with a flame ionization detector (FID). A schematic diagram of a moving-wire detector system is shown in Figure 4.4. The multifilament wire first goes through a prepyrolyzer compartment to burn off any residual organic material, such as lubricants, which might be present on the surface. It is subsequently coated with an aliquot of the eluant from the chromatographic column. The solvent is flashed off in the stripping compartment; and the residual solute is pyrolyzed in the pyrolysis oven, which is held at a slightly lower temperature than the prepyrolyzer compartment. All ovens are thermostatically controlled to decrease the effect from environmental variations. The pyrolysis products are swept into the flame ionization detector, and the signal coming from this detector is then processed in the usual manner. This detector is not flow sensitive at all; it can be used with gradient elution techniques, and it is probably the best candidate to be the "universal" detector.

The FID is very sensitive. It has a wide, dynamic range and has a quantitative response. On the other hand, these detectors tend to be very noisy and very difficult to operate. Many of the problems can be traced to the behavior of the column effluent on the wire or band.

The better-quality instruments are generally very bulky and expensive. The sensitivity of the detector can be as low as $3–5 \times 10^{-6}$ g/l. By suitable modification, an improvement by a factor of 100 should be possible (43). A lot of work is currently being done to improve this detector; and hopefully we may expect good, relatively inexpensive detectors on the market soon.

D. Heat of Adsorption Detector

This detector is based on the measurement of the evolution of the heat of adsorption and the heat uptake at desorption as the solutes in the effluent stream come into contact with an adsorbent in which a sensitive thermistor is embedded. There is the necessity for exceptionally good temperature control. The detector is usually operated differentially by mounting an additional thermistor in an adjacent chamber. As little as 10^{-4}°C temperature change can be detected by this system. This corresponds to around 5×10^{-8} g fructose if detected on Amberlite CG-400 or about 5×10^{-5} g proline when detected on Sephadex G-10. Figure 2.4 shows the cross-section of a differential heat of adsorption detector with the adsorption cell mounted on top of the reference cell. The upstream detector cavity is packed with an adsorbent, such as alumina, silica, or porous glass beads. The reference cell (downstream detector) is usually packed with smooth, inactive glass beads to reduce band broadening. The two glass-encased thermistors, bonded into disks made of Teflon, are very closely matched and are positioned in a Wheatstone bridge.

Adsorption detectors are inherently flow sensitive. Later developments include the use of a center disk to decrease the influence of flow variations and a dual-detection system to cancel out shifts in baseline due to changes in flow rate (40). This calorimetric-type detector yields differential curves

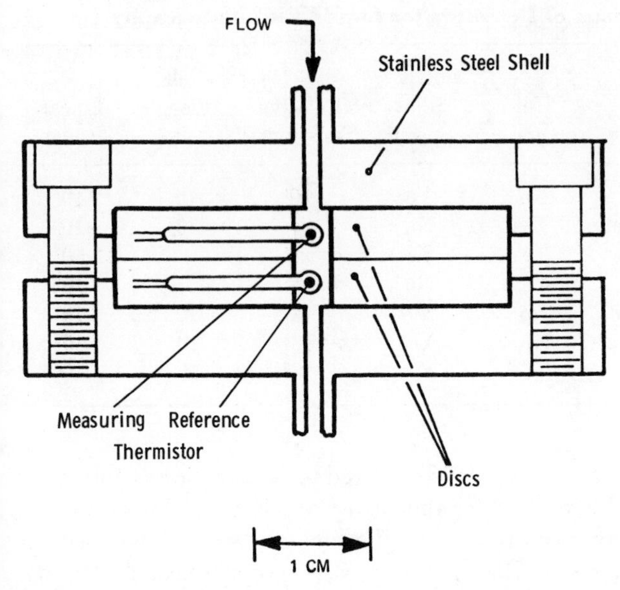

Figure 2.4. Dual-microadsorption detector (courtesy of Varian Aerograph, Walnut Creek, Calif.).

because both the heat of adsorption and the heat of desorption are measured as the solute passes through the adsorbent. Integration of these peaks will result in the usual curves observed with other detectors. Because the detection system is part of a packed column, zero dead volumes are possible. The detector can be used with gradient elution systems, but a shift in baseline will be obtained because of the different heats of adsorption of the components of the mobile phase. The region where the response is linear with solute concentration is also quite narrow. A disadvantage of this detector is that it is not suitable for large solute concentrations.

E. Other Detectors

Other detectors are based on the principle of colorimetry, polarography (25, 29), radioactivity, electrical conductivity, thermal conductivity (41), pH, dielectric constant, and the ultrasonic sound velocity. The colorimetric detector is especially applicable in amino-acid analysis using ninhydrin as the reagent. The electrical conductivity detector is one of the more simple and rugged detectors and should be considered if one is working with ionic species. This detector can be used with both aqueous and nonaqueous systems. A comparison of some of the discussed detectors is given in Table 2.5.

Table 2.5. Comparison of Detectors for Liquid Chromatography

Detector	Universal	Solvent/ Solute Specific	Relative Sensitivity	Cell Size, μl	Linear Range
UV absorption	No	Yes	10^2	10	10^4
Differential refractometer	Yes	No	1	10	10^3
Flame ionization	Yes	Yes	1		10^3
Heat of adsorption	Yes	No	1	1	10^2
Electrical conductivity	No	Yes	1	10	10
Polarography	No	Yes	10^2	10	10^5
Fluorimetry	No	Yes	10^3	10	10^2

Only three of the detectors can be considered to be more or less universal; in other words, they can be used for almost any application. Unfortunately, they are all relatively insensitive. The fluorimeter can only be used for substances which fluoresce. The excitation wavelength is usually 360 mμ. The number of applications is, therefore, limited; but in those cases where this detector can be used, the sensitivity is very high. Quinine sulfate, for example, can easily be detected to the parts per billion level.

IV. COLUMN PACKINGS AND CHROMATOGRAPHIC MODES

1. Liquid–Solid Adsorption Chromatography

An excellent treatise on this technique has recently been published by Snyder (48). Liquid-solid adsorption chromatography is especially applicable for compounds in the 300–1000 molecular weight range. The technique is not suitable for highly polar or ionic water-soluble samples. It is otherwise a very flexible tool which can successfully process anything that can be handled by thin layer chromatography. In liquid-solid chromatography, the migration rate of a solute through a column is dependent on the shape of the adsorption isotherm. The slope of this curve is a function of the type adsorbent and of the polarity of the eluant. The characteristics of the adsorbent can be modified by pretreatment of the packing, for example, by heating, by adding a known percentage of water or salts to the adsorbent, or by deactivation by silanization. In the latter process, the active sites on the surface are covered by alkyl groups. A large variety of adsorbents is available for this technique.

A. *Silica*

This is the most widely used chromatographic adsorbent. It can be successfully used in probably more than 90% of all separations by liquid chromatography. It is a typically polar material, and it is a good first choice as a general-purpose adsorbent. Its activity can be modified by the addition of, for example, 2.5–10% $MgSiO_3$, K_2SiO_3, or sodium tetraborate. The amount of adsorbed water is also somewhat important, but 10% appears to be a good compromise. There is a vast amount of literature on the use of silica gel in adsorption chromatography. One of the important properties of this material is its almost complete inertness towards labile material. It is commercially available in a large range of grades and forms. A special form is the spherical porous silica bead, commercially available under the tradename Porasil. For column chromatography, this material shows a number of improved characteristics over regular silica gel such as greater ease in packing. It can be used as the solid phase in liquid-solid chromatography, as the support in liquid-liquid chromatography, or as a column packing for gel permeation chromatography.

B. *Alumina*

A typical polar adsorbent, it is the second most popular material. Its activity is usually adjusted by the addition of controlled amounts of water. It can also be obtained commercially in a wide variety of grades and forms. It is a good all-purpose adsorbent, although some samples can be modified by the catalytic action of the surface. Highly acidic samples with $pK_a < 5$ tend to be chemisorbed. There is also a vast amount of literature available on this material. Active alumina is especially useful in the separation of mixtures of aromatic hydrocarbons; many isomers can be separated quite readily on this substrate.

C. *Other Less-Used Adsorbents*

These are charcoal, magnesia, florisil (a coprecipitate of silica and magnesia), bentone (derivative of montmorillonite) (7, 13), polycaprolactam, diatomaceous earth, and silver-impregnated adsorbents which are selective for olefinic double bonds (26).

D. *Specialty Packings*

These are manufactured to have precise geometries and characteristics designed to give higher performance than the regular adsorbents. Examples are the porous silica beads mentioned earlier (1); the porous layer beads (37), also called controlled surface porosity packing (30–33); or pellicular sorbents (19–21). These hard, spherical particles pack uniformly without fracture and make possible improved column efficiency.

In using regular adsorbents as column packing, d_f (Equation 2.9), the film thickness of the stationary phase, is equivalent to d_p since the porosity of the adsorbent allows the solute to diffuse through the whole particle. Since C_s is proportional to d_f^2 and since C_s contributes linearly to the plate height with increase in flow velocity, for high-speed liquid chromatography the particle diameter should be kept as small as possible. A reduction of d_f has been attained very neatly without concurrent reduction of d_p by coating a thin porous layer of the adsorbent on a fluid-impermeable glass sphere. These *porous layer beads* typically show a d_f of 0.03–0.1 d_p. These packings are commercially available under the tradenames of Corasil (Waters Associates, Framingham, Massachusetts) and Zipax (E. I. duPont de Nemours, Wilmington, Delaware). Since, for high-velocity work, the main contribution to the plate height would be C_s, the resistance to mass transfer in the stationary phase (the mass transfer resistance in the mobile phase being reduced by increased axial dispersion), the use of these types of supports allows for an increase in mobile-phase velocities from 10^{-1} cm/sec to around 1–10 cm/sec. Under these conditions, the main cause of peak dispersion will be the slow mass transfer in the mobile phase, where coupling is particularly advantageous. In comparison, in gas chromatography the stationary-phase mass transfer dominates even at low reduced velocities.

Another advantage is that the higher degree of shape regularity allows for the preparation of more efficient and reproducible columns. Bed porosities generally range around 0.40. One disadvantage of this packing is its much lower capacity. Much smaller sample sizes and very sensitive detectors are hence necessary.

2. Liquid–Liquid Partition Chromatography

In this technique, a high-molecular-weight liquid which is immiscible with the mobile phase is deposited on a solid surface and used as the stationary phase. Since one can choose from a wide variety of compounds, there is an additional parameter with which to modify α in Equation 2.8. This added selectivity is useful when difficult separations have to be carried out. Many of the liquid substrates used have found prior applicability in gas-liquid chromatography. A liquid substrate which has enjoyed some popularity is 1,2,3-tris(cyanoethoxy)-propane; but substrates like Carbowax, silicone gum rubber, propylene carbonate, β,β'-oxydipropionitrile, and many others have been used. By judicious combination of different liquids, highly selective substrates can be obtained (23).

Porous layer beads also make excellent supports for work in liquid-liquid partition chromatography (37). The relatively open, wide-pored surface layer (shell) has a large area where the liquid substrate can be deposited as a thin layer on the solid, thus making it very accessible to the solute molecules. The depth of the shell is limited; deep pools of liquid substrate which would increase C_s are not present. Porous silica beads can be used similarly but with results between those expected for coated porous layer beads and a coated regular support such as diatomaceous earth.

Another specialty packing support consists of surface-textured beads (27) having no internal porosity. These are relatively spherical. Their surface roughness extends to a depth of about 2–4 μ, and the width of the voids is about 3–5 μ in diameter. When coated with a thin layer of a suitable liquid phase, their low resistance to mass transfer in and out of the voids leads to high efficiency performance. The surface area of these beads is considerably less than that of porous layer beads. Another product with a similar geometry but with a somewhat different surface texture is available from Pittsburgh Corning Corporation. This ceramic support material is marketed under the tradename CeraBeads.

Figure 2.5 shows some comparative data obtained with columns packed with a "controlled surface porosity support," a "textured" glass bead support, and a "Gas Chrom" P "diatomaceous earth support" (30). For average linear carrier velocities in the range of up to 2.0 cm/sec, the 500-Å controlled surface porosity packing shows a leveling off of the plate height at approximately 15 particle diameters. The diatomaceous earth support, on the other hand, shows a continuing sharp increase of H with increase in u. At a mobile-phase velocity of 1.5 cm/sec, the former packing shows an H of around 0.7 mm, while the latter shows a plate height of approximately 2.5 mm. This large difference is, of course, also due to the different levels of liquid loading on the supports. The "textured" glass beads have a very low surface area, in other words, 0.043 m^2/g, as compared to 0.65 m^2/g for the 1000-Å and 0.33 m^2/g for the 500-Å CSP supports. This drastically limits the amount of liquid phase which the support can tolerate before "bridging" with a concurrent increase in C_s takes place (34). At very low speeds, there is, of course, little difference between the different packings. Equilibration distances of 2 d_p's have been reported (23) for flow velocities of 10^{-2} cm/sec. However since the trend in liquid chromatography, as in life in general, is to higher speeds, the concept of an active layer over an inert sphere is one of the major advances in liquid chromatography.

A major problem in this technique is the solubility of the stationary phase in the mobile liquid. In practice, there is no such thing as two completely immiscible liquids. In isothermal operation and in applications where the

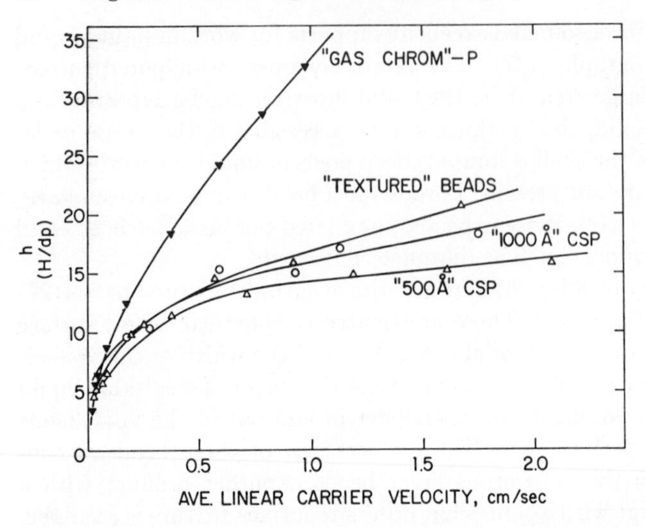

Figure 2.5. Effect of porous layer supports on plate efficiencies as function of the mobile phase velccity (30). Columns are 50 × 0.32–cm internal diameter; the carrier is hexane; the sample is 5 μl of a 5 mg/ml benzyl alcohol solution in hexane; the liquid loading is 0.25% β, β′-oxydipropionitrile on all except on "Gas Chrom" P (4%) support. Particle diameters for the "1000 Å" CSP support is 20–27 μ, for the "textured" glass beads, 105–125 μ, and for "Gas Chrom" P, 53–64 μ (courtesy of the author and *Journal of Chromatographic Science*).

pressure drop over the column is not too large, the use of a presaturator is generally adequate to prevent the mobile phase from stripping the stationary phase from the solid support. In those operations where varying conditions of temperature, eluant composition, and pressure are present or where large pressure drops are used over the column, it is very difficult to maintain reproducible results. The use of a highly polymeric stationary phase reduces the solubility in the mobile phase; but these liquid phases have high viscosities, which result in low solute diffusion and a high C_s. Column performance can be improved by using very thin films, 5–10 monolayers thick; but the danger in this process is the possible presence of bare, active sites on the surface.

One solution to this problem is to use a "brush" packing, a stationary phase where the liquid substrate is chemically bonded to the solid support (15). When 3-hydroxypropionitrile is esterified to Porasil C, packings are obtained for which high separation speeds are possible. In this case, the organic compounds are oriented as bristles on the surface. This substantially increases the accessibility of the stationary liquid phase to the solute molecules, effectively decreasing the C_s term in Equation 2.9. In addition, because the liquid substrate is chemically bonded to the solid support, this substrate will not dissolve in the mobile phase. The polarity of these packings appears to be less a function of the chemical composition of the organic chain but rather of the chain length of these molecules (14).

Although several packings of this type are now commercially available, results show that there are still several problems which have not yet been solved. Long hydrocarbon chains, for example, tend to lie flat on the surface of the support. Halasz claims, however, that this phenomenon has little influence on the speed of separation (14). For the brushes to stand up like bristles, short chains with mutually repelling substituents, for example, nitrile groups, on the chain are necessary. The support itself is porous, and the gain in increased accessibility to the liquid substrate is partly counteracted by the solute retention in the core of the packing. In addition, the hydrolytic and thermal stability of these silicate esters is poor.

A more stable bond is possible if the attachment to the surface is via a silicon–carbon bond (55). Ether-bonded polymeric silicone stationary phases (33) are also found to be much more stable than silicate esters on porous glass (15). These coatings may be prepared with functional groups which range from very polar to nonpolar, yielding widely diverse selectivities. (31)

3. Ion Exchange Chromatography

One of the major applications of liquid chromatography is in ion exchange chromatography, usually in aqueous buffer solutions. A more detailed description of this important technique will be included in the next volume of the series. Resolution can be adjusted by the use of pH and ionic strength gradient elution. Ion-exchange resins are cross-linked polymers to which ionic end groups are attached. The principle of a thin shell of active material on a fluid-impermeable sphere has also been successfully applied to ion exchange resins (19, 20, 30). With this pellicular-type packing, high-speed, high-resolution ion-exchange chromatography can be successfully carried out. An example of this technique is given in Figure 2.11.

4. Gel Permeation Chromatography

This special application of high-resolution liquid chromatography is treated in Chapter 4.

V. OPERATIONAL PARAMETERS

1. Mobile Phase

The choice of the mobile phase is based on the same considerations as in classical liquid chromatography. For liquid-liquid and liquid-solid chromatography, the elutropic series of solvents established by Trappe in 1940 is still a very good guide to obtain the solvents with the correct strength to obtain K values between 1 and 100 for the solute components of interest (54). Besides the solvent strength, however, one should also consider α, the solvent selectivity. The importance of this latter aspect has not been sufficiently recognized by a great majority of the workers in the field. As an example, for methyl-1-naphthyl ketone and 1,5-dinitronaphthalene on alumina with 23% CH_2Cl_2 in pentane as the mobile phase, the observed values for the K's are 5.5 and 5.8, respectively. From these data, we can compute the solvent strength $[=\overline{K}/(\overline{K}+1)]$ to be 0.85 and the selectivity ($\alpha = K_2/K_1$) to be 1.05. For a baseline separation of these two peaks, the necessary efficiency of the column has to be around 17,000 plates. Even under the best circumstances, this necessitates an exceptionally long column. However, when the mobile phase is a mixture of 0.05% CH_3SOCH_3 in pentane, the observed K's are 1.0 and 3.5, respectively. The solvent strength is approximately the same, in other words, 0.70; but the α is now 3.5. In this case, only 70 plates are necessary; and a short column will suffice to yield the desired separation (49).

It is, however, not always easy to predict on theoretical grounds what solvent combination to choose to obtain optimum results. Snyder suggests a corrected polarity for a solvent-solute-substrate system by taking into account the dispersion effects (e.g., from the Hildebrand solubility parameter), polar interactions, and hydrogen bonding interactions, in other words,

$$\delta^2 = \delta^2_{\text{disp}} + \delta^2_{\text{pol}} + \delta_A\delta_H \tag{2.18}$$

where the latter term is broken down into a donor and an acceptor part (49). Preliminary results are very encouraging.

There are also other factors to keep in mind when choosing solvent combinations for liquid-liquid chromatography. Solute adsorption at an interface can modify the expected partition ratio (4, 37). Also, the wettability between very polar and nonpolar phases can reduce the expected efficiency in the case of a bonded-phase liquid substrate (30). This might even be a factor for mechanically held liquid phases.

For ion exchange chromatography, aqueous buffer solutions are generally used as the mobile phase. Both pH and ionic strengths can be adjusted to obtain a mobile phase with the desirable characteristics.

2. Sample Size

The minimum sample size is determined by the detector sensitivity and the minimum amount of a specific component which one wishes to detect. In general, analytical sample sizes should be as small as possible, although below a certain value there is no appreciable increase in resolution. Since large samples lead to overloading and loss of resolution, the maximum sample size in preparative work is determined by the minimum resolution which is still acceptable for the separation. Column overloading ean be recognized because of the change in retention times and by a distortion of the peak shapes.

The relation between sample size, resolution, and speed of analysis is best shown by the triangle in Figure 2.6. An increase in one parameter can only be achieved by a decrease in the values of the two others. The region *A* is where analytical separations are carried out. Sample size is sacrificed for speed and resolution. By decreasing the speed, more resolution can be attained and vice versa. Preparative work is carried out in the region denoted by *B*, where speed is sacrificed for capacity. This triangle also shows that excess resolution can be traded off for a further increase in sample size.

The sample capacity of a column is also determined by the type of packing. The usual sample sizes range from 10^{-2} to 10^{-5} g. This is one to several magnitudes larger than used in gas chromatography. The use of a porous layer packing reduces the capacity because a large fraction of the packing consists of the impermeable glass core. A 1-mm ID column packed with 270–325-mesh ion-exchange resin deposited on glass beads, for example, has a loading capacity of only around 10^{-8} moles.

Liquid chromatography, has, of course, the large advantage that quantitative recovery of the column effluent is very easy. Fractions thus collected can be concentrated for further analysis by UV, IR, laser-Raman, mass spectrometry, and other spectroscopic techniques.

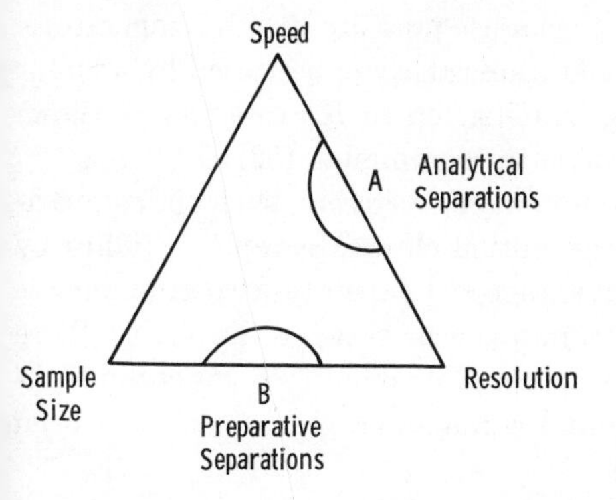

Figure 2.6. Speed, resolution, and sample size.

A little-publicized but important advantage of liquid chromatography is in the area of radiochromatographic separations. Since the fraction of interest can be quantitatively recovered and counted in a low-background counter over longer periods of time, much higher dilution levels can be tolerated. An increase of at least 10^2 in sensitivity should be observed in comparison to radio-gas chromatographic on-line detection techniques, even though gas counters are much more efficient than liquid scintillation counters. Combined with the larger sample size possible, one can hence work with a 10^4 or even much larger radiochemical dilution than in gas chromatography.

This ability to quantitatively recover solutes in undegraded form is also important in biochemical separations, where the fractions have to be used for further studies. Although the *gas* chromatographic separation of many products of biochemical origin have been reported, it is doubtful whether most of these compounds would remain unaltered in their passage through the hot chromatographic system.

3. Recycling

Instead of using very long columns to achieve a large number of plates to separate two or more closely resembling compounds, one can also use a shorter column and recycle the mobile phase. Because of the incompressibility of the liquid, this technique can be carried out with much less loss of resolution in the recycling pumps than would be the case in gas chromatography. This technique has been used in gel permeation chromatography to resolve, among others, the oligomers of a number of polymers (17).

4. Programming Options

Programming allows mixtures of widely varying partition ratios to be eluted in a reasonable time without loss of resolution for the fastest eluting components. In chromatography one can use pressure (flow), temperature, and solvent programming. Band migration rates can be varied by a factor of up to 10^2 by temperature programming, up to 10^2 by pressure (flow) programming, and up to 10^{10} by solvent programming (52).

The effect of temperature programming is, therefore, slight in comparison with the capabilities of gradient elution chromatography. Studies by Scott (45) show that another disadvantage of temperature programming is the large time necessary for equilibrium to occur between the phases. Were it not for the large success of temperature programming in gas chromatography, interest in this option for liquid chromatography would have been virtually nonexistent. The exception is when very large molecules are concerned, such as in gel permeation chromatography. In that case, a combination of solvent and temperature programming can be used to aid the separation because of solubility effects. Recent work with operating temperatures above the normal boiling point of the solvent indicates that in some cases this approach leads to an increased mass transfer for the solute and to more efficient separations (58). Even though temperature programming may have little applicability, thermostatic control of the column is advisable if reproducible results are desired (38).

Pressure programming, on the other hand, shows more promise to decrease analysis time without significant loss in resolution. The range is much more limited than in solvent programming; but for many applications, this technique is quite adequate to yield very useful results. In supercritical fluid chromatography, on the other hand, pressure programming is a very important option (24).

For extremely wide-range mixtures, solvent programming is currently the only solution to obtain good chromatograms. In solvent programming, also called gradient elution, the polarity of the solvent is increased during the run. This technique is comparable in importance to temperature programming in gas chromatography. Studies by Snyder and Saunders on the optimizing of solvent programs have been a valuable contribution to the correct understanding of this technique (52). An example of the capabilities of solvent programming is shown in Figure 2.7. A 17-component mixture ranging from chlorobenzene to 3,4-benzacridine can be separated with good resolution in less than 50 min. Even higher speeds would have been possible if a porous layer packing had been used.

5. Coupled Columns

This technique, which has been overlooked in the recent developments, has outstanding practical advantages for those applications involving repetitive, high-efficiency separations (51). After the sample is introduced into the primary column, selected fractions are further separated on one or more secondary columns. If more than one secondary column is used, they are operated concurrently and parallel to each other. The outlet of these columns goes to a common detector. By judicious choice of varying column lengths and by the use of columns with different packings, the solutes from the different columns can be adjusted to elute at different times. The use of these different columns allows the resolution in those regions of interest to be sharply improved, compounds of minor interest to be eluted in one band, and heavy components to be detected in much

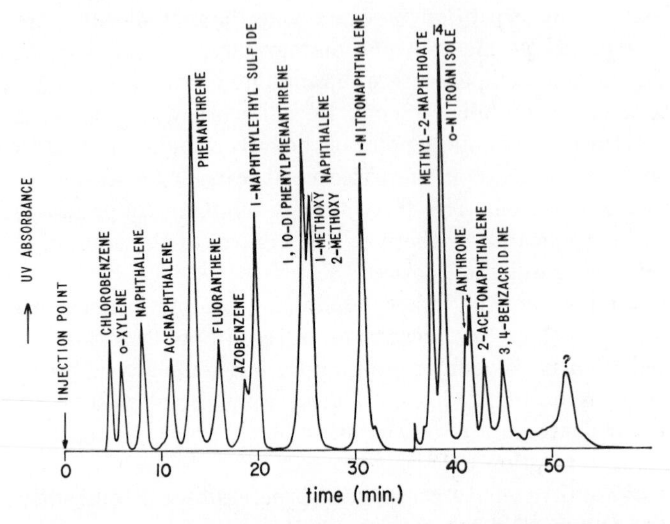

Figure 2.7. Separation of a 17-component synthetic mixture by logarithmic solvent programming (courtesy of Dr. L. R. Snyder, Union Oil Company of California and the *Journal of Chromatographic Science*).

shorter times. Because of the special setup, the technique is mainly applicable to those analyses involving repetitive, high-efficiency separations.

VI. APPLICATIONS

The success of a technique is obviously dependent on its applicability to a wide variety of problems. The following shows some good chromatograms obtained by this technique.

Figure 2.7 shows the separation of 17 components by liquid-solid adsorption chromatography, using a logarithmic gradient elution technique. The eluant consists initially of pentane to which increasing amounts of ether were added to change the polarity of the mobile phase. The column was a 6- by 1200-mm column packed with 20-μ silica and equilibrated with 50% water-saturated solvent. The sample size was 5 mg.

Figure 2.8 is an example of a separation by liquid-liquid partition chromatography, where the stationary liquid phase is chemically bonded via an ether linkage to the surface of a porous layer bead support (33). The mobile phase is 5% $CHCl_3$ in hexane. Even though the stationary phase is quite soluble in $CHCl_3$, no presaturator is necessary because this stationary phase is attached to the solid support.

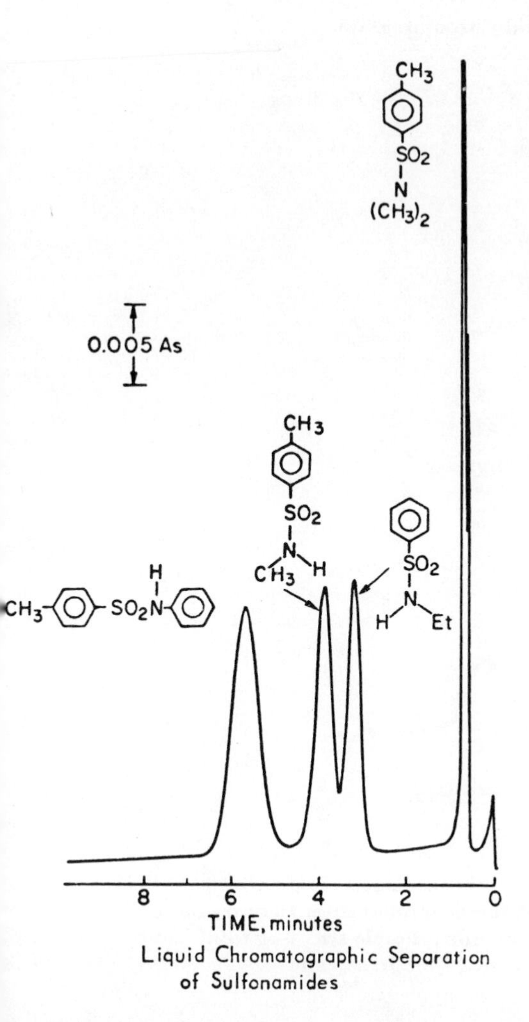

Figure 2.8. Separation of sulfonamides by high-resolution liquid-liquid partition chromatography (33). Mobile phase: 5% $CHCl_3$ in hexane; 0.77% of an aliphatic "ether-bonded" phase on $<37\ \mu$ "Zipax"; flow rate is 2.66 ml/min; column inlet pressure is 860 psi; temperature is 27°C (courtesy of the authors and *Journal of Chromatographic Science*).

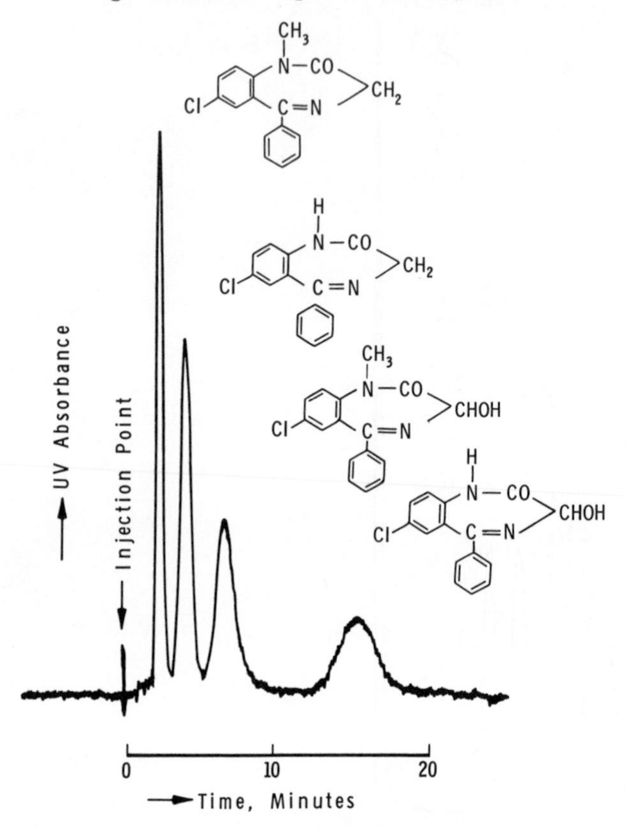

Figure 2.9. Separation of some benzodiazepines on Durapak "OPN" 36–76 μ. Column: 100 × 1-mm ID at room temperature. Carrier; hexane-isopropanol (80–20 v/v); Carrier flow, 1.0 ml/min; sample size, 8 μg total (courtesy of Dr. C. G. Scott, Hoffman-La Roche, Inc., New Jersey and the *Journal of Chromatographic Science*).

Figure 2.9 is another example of high-resolution liquid-liquid partition chromatography. The chromatogram shows the results of the separation of a synthetic mixture of benzodiazepines. The column used is 1-mm ID by 100-cm long packed with 36–75-μ "Durapak" OPN brush packing. The organic phase bonded to the porous glass core is β,β′-oxydipropionitrile. The mobile phase is hexane-isopropanol 4:1. The carrier flow is 1.0 ml/min. The sample size is 8 μg total. Detection was carried out with a UV spectrophotometer at 254 mμ.

Figure 2.10 is another example of rapid liquid-liquid chromatography (2,56). The chromatogram shows the separation of a synthetic mixture of five insecticides on a 50-cm by 2.3-mm column packed with Corasil II. A

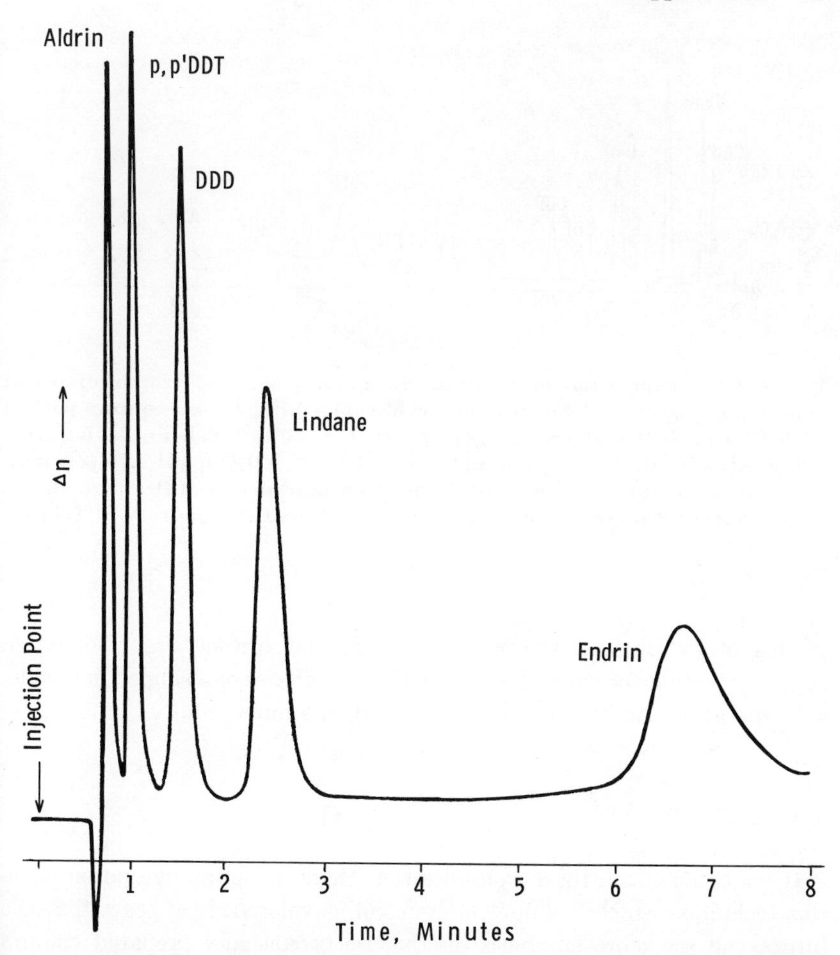

Figure 2.10. Separation of insecticides by rapid liquid chromatography (courtesy of Waters Associates, Framingham, Massachusetts and the *Journal of Chromatographic Science*).

differential refractometer is used to monitor the effluent. The inlet pressure is 280–320 psi. The flow rate of the mobile phase, *n*-hexane, is 1.5 ml/min, corresponding to a flow velocity of around 1.35 cm/sec. Under these conditions, about 5 plates/sec are obtained for *p,p'*DDT ($k' = 0.64$); and about 3 plates/sec are observed for DDD ($k' = 1.32$).

Figure 2.11 shows some results obtained by rapid ion-exchange chromatography. This chromatogram shows the separation of a wide-range mixture on a 30-μ pellicular anion exchange resin. All 12 mono-, di-, and triphos-

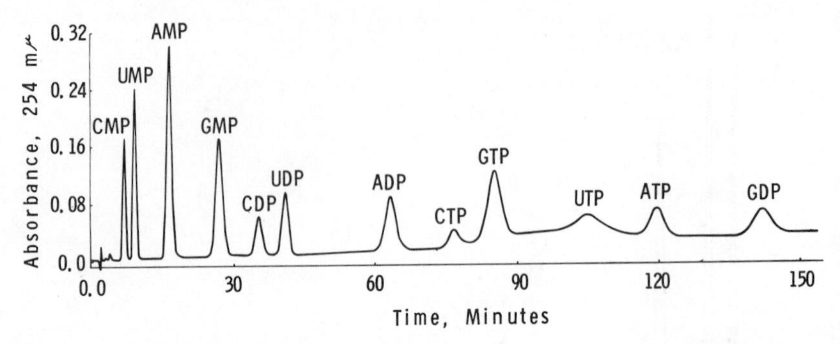

Figure 2.11. Separation of 10 μg of the mono-, di-, and triphosphates of cytidine, guanosine, adenosine, and uridine on a 0.1 × 300-cm column packed with 30 μ pellicular anion exchange resin. T = 80°C. Gradient elution with 1.0 *M* KH_2PO_4 pH 4.2 solution added to a 0.01 *M* KH_2PO_4 pH 3.25 solution. Flow rate 20 ml/hr. Inlet pressure 1800–2200 psig (courtesy of Dr. C. A. Burtis et al., Varian Aerograph, Walnut Creek, Calif. and the *Journal of Clinical Chemistry*).

phates of cytidine, guanosine, adenosine, and uridine are resolved in slightly less than 150 min. To resolve the phosphates of a single nucleoside, the separation can be carried out in less than 4 mins.

VII. FUTURE

It is obvious from these examples that there is still a huge potential in this technique which has not yet been fully explored. It appears that the future will see more emphasis on the use of specially prepared column packings. One direction is to use micron and submicron particles, operating pressures of 300 atm and higher, and still much higher flow velocities. With these changes, analysis speed can be increased to as high as 20–50 plates/sec, which, even for gas chromatographic standards, would be quite fast. As many as 100 peaks could be separated in an hour, which compares favorably with present-day temperature-programmed gas chromatography's capabilities. The use of wide-range solvent programming and further improvements in the reliability and sensitivity of the detectors will improve the scope so much further that one can envision that in a few years this technique will be indispensable for all laboratories.

NOMENCLATURE

A = contribution to the plate height due to "eddy diffusion".
B = contribution to the plate height due to longitudinal molecular diffusion.
C = contribution to the plate height due to resistance to mass transfer.
D = diffusion coefficient or solute diffusivity.
H = plate height.
$HETP$ = height equivalent to a theoretical plate.
K = partition coefficient, distribution ratio, or partition ratio.
$\overline{K}$ = $(K_1 + K_2)/2$.
L = column length.
R = fraction of the velocity of the mobile phase.
R_s = resolution.
V = volume in column.
W = bandwidth.
d = diameter, layer thickness.
h = the reduced plate height.
k' = the capacity factor.
t = retention time.
u = flow rate.
α = K_1/K_2.
γ = obstruction factor for diffusion in the mobile phase.
λ = structural parameter of packing.
ν = reduced mobile phase velocity.
ω = reduced mass transfer coefficient.

Subscripts

m = mobile phase.
s = stationary phase.
i = component i.
p = particle.
f = film.

REFERENCES

1. K. J. Bombaugh, R. N. King, and A. J. Cohen, *J. Chromatogr.* **43,** 332 (1969)
2. K. J. Bombaugh, R. F. Levangie, R. N. King, and L. Abrahams, *J. Chromatogr Sci.*, **8,** 657 (1970).
3. H. Coll, H. W. Johnson, Jr., A. G. Polgar, E. E. Seibert, and F. H. Stross *J. Chromatogr. Sci.* **7,** 30 (1969).
4. J. R. Conder, D. C. Locke, and J. H. Purnell, *J. Phys. Chem.* **73,** 700 (1969)
5. R. D. Conlon, *Adal. Chem.* **41,** 107A, April (1969).
6. J. J. van Deemter, F. Zuiderweg, and A. Klinkenberg, *Chem. Eng. Sci.* **5,** 2 (1956).
7. W. W. Emerson, *Nature* **180,** 48 (1957).
8. H. Felton, *J. Chromatogr. Sci.* **7,** 13 (1969).
9. J. C. Giddings, Dynamics of Chromatography, Part 1, Dekker, New York 1956.
10. J. C. Giddings, *Anal. Chem.* **36,** 1891 (1964); **37,** 60 (1965).
11. J. C. Giddings and R. A. Robinson, *Anal. Chem.* **34,** 885 (1962).
12. J. C. Giddings and P. D. Schettler, *J. Phys. Chem.* **73,** 2577, 2582 (1969).
13. R. E. Grim, *Clay Mineralogy*, McGraw-Hill, New York, 1953.
14. I. Halasz, *1969 Eastern Anal. Symp.*, New York, New York, 1965, pp. 19–21
15. I. Halasz and I. Sebestian, *Angew. Chem., Int. Ed.* **8,** 453 (1969).
16. I. Halasz and P. Walkling, *J. Chromatogr. Sci.* **7,** 129 (1969).
17. W. Heitz and H. Ullner, *Makromolek. Chem.* **120,** 58 (1968).
18. D. S. Horne, J. H. Knox, and L. McLaren, *Sepn. Sci.* **1,** 531 (1966).
19. C. Horvath and S. R. Lipsky, *Anal. Chem.* **41,** 1227 (1969).
20. C. Horvath and S. R. Lipsky, *J. Chromatogr. Sci.* **7,** 109 (1969).
21. C. G. Horvath, B. A. Preiss, and S. R. Lipsky, *Anal. Chem.* **39,** 1422 (1967)
22. R. E. Jentoft and T. H. Gouw, *Anal. Chem.* **38,** 949 (1966).
23. R. E. Jentoft and T. H. Gouw, *Anal. Chem.* **40,** 923, 1787 (1968).
24. R. E. Jentoft and T. H. Gouw, *J. Chromatogr. Sci.* **8,** 138 (1970).
25. P. L. Joynes and R. J. Maggs, *J. Chromatogr. Sci.* **8,** 427 (1970).
26. G. Jurriens, *Riv. Ital. Sostanze Grasse* **42,** 116 (1965).
27. B. L. Karger, K. Conroe, and H. Engelhardt, *J. Chromatogr. Sci.* **8,** 242 (1970)
28. A. Karmen, *Anal. Chem.* **38,** 286 (1966).
29. W. Kemula and D. Sybelska, *Anal. Chim. Acta* **38,** 97 (1967).
30. J. J. Kirkland, *J. Chromatogr. Sci.* **7,** 7, 361 (1969).
31. J. J. Kirkland, *J. Chromatogr. Sci.* **9**, 206 (1971).
32. J. J. Kirkland, *Anal. Chem.* **40,** 391 (1968); **41,** 218 (1969).
33. J. J. Kirkland and J. J. DeStefano, *J. Chromatogr. Sci.* **8,** 309 (1970).
34. J. H. Knox, *Anal. Chem.* **38,** 253 (1966).
35. J. H. Knox and M. Saleem, *J. Chromatogr. Sci.* **7,** 614 (1964).
36. D. C. Locke, *Anal. Chem.* **39,** 921 (1967).
37. R. E. Majors, *J. Chromatogr. Sci.* **8,** 338 (1970).

38. R. J. Maggs, *J. Chromatogr. Sci.* **7,** 145 (1969).
39. S. L. Miller, K. Dus, and J. Kraut, *J. Chromatogr.* **45,** 135 (1969).
40. M. N. Munk and D. N. Raval, *J. Chromatogr. Sci.* **7,** 48 (1969).
41. K. Ohzeki, T. Kambara, and K. Saitoh, *J. Chromatogr.* **38,** 393 (1968).
42. V. Pretorius and T. W. Smuts, *Anal. Chem.* **38,** 274 (1966).
43. R. P. W. Scott, *J. Chromatogr. Sci.* **7,** 21A, August (1969).
44. R. P. W. Scott, D. W. J. Blackburn, and T. Wilkins, *J. Gas Chromatogr.* **5,** 183 (1967).
45. R. P. W. Scott and J. G. Lawrence, *J. Chromatogr. Sci.* **7,** 65 (1969).
46. S. T. Sie and N. van der Hoed, *J. Chromatogr. Sci.* **7,** 257 (1969).
47. T. W. Smuts, D. J. Solms, F. A. van Niekerk, and V. Pretorius, *J. Chromatogr. Sci.* **7,** 24 (1969).
48. L. R. Snyder, Principles of Adsorption Chromatography, Dekker, New York, 1968.
49. L. R. Snyder, private communication, July 1970.
50. L. R. Snyder, *J. Chromatogr. Sci.* **7,** 352 (1969).
51. L. R. Snyder, *J. Chromatogr. Sci.* **8,** 692 (1970).
52. L.R. Snyder and D. L. Saunders, *J. Chromatogr. Sci.* **7,** 195 (1969).
53. J. A. Thoma, *Anal. Chem.* **37,** 500 (1965).
54. W. Trappe, *Biochem. Z.* **305,** 150 (1940).
55. K. Unger, K. Berger, and E. Gallei, *Kolloid U. Z. Poly.* **234,** 1108 (1969).
56. J. L. Waters, J. N. Little, and D. F. Horgan, *J. Chromatogr. Sci.* **7,** 293 (1969).
57. T. E. Young and R. J. Maggs, *Anal. Chim. Acta* **38,** 105 (1967).
58. J. N. Little and W. N. Pauplis, private communication.

BIBLIOGRAPHY

J. J. Kirkland, Ed., *Modern Practice of Liquid Chromatography,* Wiley, New York, 1971.

L. R. Snyder, *Principles of Adsorption Chromatography,* Dekker, New York, 1968.

Nina Hadden, F. Baumann, F. McDonald, M. Munk, R. Stevenson, D. Gere, and F. Zamaroni, *Basic Liquid Chromatography,* Varian Aerograph, Walnut Creek, Calif., 1971.

DAVID J. SHAPIRO*
VICTOR W. RODWELL

Department of Biochemistry
Purdue University, Lafayette, Indiana

III. Thin Layer and Paper Chromatography

---

* Predoctoral fellow of the National Institutes of Health. Present address, Department of Pharmacology, Stanford University School of Medicine, Stanford, California.

I. INTRODUCTION

The wide appeal of paper chromatography (PC) and of thin layer chromatography (TLC) reflects both their simplicity and their applicability to isolation, identification and, on occasion, quantitation of essentially all low-molecular-weight organic and many inorganic compounds. PC and TLC, which evolved from column chromatographic techniques first introduced over 60 years ago, are of comparatively recent vintage. In comparison to other techniques described in this volume they are, however, of long standing. Key developments in PC and TLC were as follows:

1903–1906. Resolution of plant pigments by column chromatography and introduction of the term "chromatographic analysis" by Michael Tswett (8).

1941–1944. Development of partition chromatography on paper by Martin and Synge (5) and its application to amino acid separations by Consden, Gordon, and Martin (1).

1956. Promulgation by Egon Stahl of quality standards for TLC adsorbents and development of equipment for reproducible TLC (7).

The origins of both PC and TLC extend back to the 19th century. A form of circular PC was developed by Friedrich Runge (1795–1867), a dye chemist. His suggestion that inclusions be added to the paper support to improve separations forecasted the future development of TLC. In 1861 Christian Schönbein originated "capillary rise analysis," a form of ascending PC exploited in the latter part of 19th Century by Friedrich Goppelsröder. Unlike Twsett, none of these early chromatographers used solvent to develop "chromatograms," and most lacked appreciation of the role of the adsorbent in effecting separations. Many years later Ismailov and Schraiber utilized "thin" (2-mm) layers of alumina on glass plates for chromatographic separations—an early form of TLC. The applications of both PC and TLC have grown prodigiously in recent years. Over 20,000 publications using these techniques have appeared since 1945. PC and TLC remain living, expanding techniques.

II. APPLICATIONS

Applications of PC and TLC may be considered as basically qualitative, quantitative, or preparative in nature. The lines of demarcation are, however, not clear-cut, and a semiquantitative result frequently may be achieved without recourse to strict quantitative techniques. PC and TLC are excellent for trace component analysis. Examples of their application include detection of a by-product in a synthetic process or determination of the presence of a radioimpurity in a radioisotopic biological (quality control); identification of a by-product, an unusual urinary amino acid or a new natural product; monitoring the time course of an organic synthesis, an enzyme-catalyzed reaction or the decay of a short-lived intermediate; removal of a trace impurity and/or its isolation for further study and identification. Many chromatographic separations contain elements of several of the above applications.

Although gas chromatography (GC) may prove to be more convenient than PC or TLC, especially in quantitative analysis, there are other advantages of PC and TLC over gas chromatography:

1. PC and TLC are applicable to the analysis of thermally labile compounds.
2. Quantitative recovery is often easier in PC and in TLC.
3. In PC and in TLC the mobile phase plays an important role in the separation process. The ease with which the mobile phase can be varied is an important reason why these techniques are so versatile.
4. Two-dimensional PC and TLC can yield data and separations which are not feasible in a single chromatographic run.
5. The cost outlay for PC and TLC is relatively low.
6. Several samples and reference compounds can be compared simultaneously on a single chromatogram.
7. Color tests can be used to supplement R_f data in identification work.

Gas chromatography can be regarded as a technique complementary to TLC and PC. Another important alternative technique is high-speed, high-resolution liquid chromatography (see Chapter 2). It is certain, however, that TLC and PC will not be replaced by this new technique. The advantages listed above under items 4–7 also apply to TLC and PC over high-speed, high-resolution liquid chromatography.

Although PC and TLC are often mentioned interchangeably in this chapter one would, in most cases, prefer to use TLC if a choice were available. TLC is more sensitive and gives sharper zones, hence better resolutions. In addition, corrosive reagents cannot be used in paper chromatography.

III. TERMINOLOGY

1. R_f Value

Both in PC and TLC, the distance a sample travels during chromatography is expressed in one of two ways:

1. As the ratio of the distance (D) travelled by the sample to the distance travelled by the mobile phase, measured from the point of sample application. This ratio is termed R_f.

$$R_f = \frac{D_{\text{sample}}}{D_{\text{solvent}}} \tag{3.1}$$

2. As the mobility relative to that of a standard compound on the same chromatogram.

$$R_s = \frac{D_{\text{sample}}}{D_{\text{standard}}} \tag{3.2}$$

The name of the standard employed is usually given as a subscript, for example, R_{alanine}, $R_{\text{nitrophenol}}$, and so on.

2. Partition Chromatography

In normal partition chromatography a less-polar mobile phase passes over an immiscible more polar phase adsorbed on a support which is preferably inert. For reversed-phase partition chromatography the polarities of the mobile and stationary phases are reversed. A more polar phase moves over a stationary less polar phase. In both cases components of the mixture to be separated are partitioned or distributed between the phases in accordance with their solubilities in each phase. The partition coefficient (K) for a specific component is the ratio of the concentrations present in each phase:

$$K = \frac{C_1}{C_2} \tag{3.3}$$

Repeated partition of the sample between the stationary and mobile phases permits the separation of compounds differing only slightly in partition coefficient. Chapter 1, Section II and Chapter 2, Section II treat in more detail the degree of resolution of two compounds as a function of the ratio of their partition coefficients and of the efficiency of the chromatographic system.

3. Adsorption Chromatography

Whereas in partition chromatography the ideal support or adsorbent is inert, in adsorption chromatography it plays an active role in the chromatographic separation. Sample and solvent compete for active sites on the adsorbent. Sample mobility thus depends not only on the relative affinities of the different components in the samples, but is also a function of the degree of competition between sample and solvent for the available adsorbent sites. The ratio of the mole fraction of sample adsorbed (C_A) to the mole fraction in the solvent (C_S) defines the adsorption coefficient, K'.

$$K' = \frac{C_A}{C_S} \tag{3.4}$$

A difference in the K' of two compounds permits their separation by adsorption chromatography.

In what follows, the term "support" will denote the inert material supporting the thin layer of adsorbent in TLC. The adsorbents most commonly used in adsorption TLC are silica gel and alumina. Silica gel, the most widely used adsorbent, functions chiefly by solvent and sample interactions with its hydroxyl groups. Polar solvents, such as water, form hydrogen bonds with these hydroxyl groups, reducing the number of active adsorbent sites available for the sample.

IV. EQUIPMENT

1. Adsorbent Supports

In PC, the paper sheet is both adsorbent and support. Paper is a very weak adsorbent, at best. For TLC, thin layers of adsorbent are bonded to a rigid or semirigid sheet support, generally a flexible synthetic polymer, such as polyethylene terephthalate, glass, or metal foil. Glass is most often used in hand-made layers. Although most glass supports are regular $\frac{1}{8}$-in.-thick square or rectangular plates measuring 2, 4 or 8 by 8 in., thinner glass plates may be used. The substitution of other materials for glass in commercial precoated layers results in part from greater ease in shipping.

2. Adsorbents

In PC the adsorbent is a fibrous cellulose sheet specially manufactured for this use. Two widely used, general-purpose papers are Whatman #1 (lower capacity) and Whatman #3 MM (higher capacity).

A wide variety of TLC adsorbents alone (Table 3.1, Appendix Section 1), or precoated on glass, plastic or metal foil (Appendix Section 2) are available commercially. Although some adsorbents (e.g., powdered cellulose) form stable layers and bond well to glass, most require addition of a binder (e.g., $CaSO_4$). Commercial precoated layers are generally of highly uniform quality. The adsorbent is less prone to flake off than that of most self-prepared plates. Layer stability is due to the use of polyvinyl pyrrolidone or similar organic binders which results in strong bonding of the layer to the support. Precoated plates are also available with nonorganic binders. The uniformity and reproducibility of the commercial product is difficult to duplicate under most laboratory conditions. We therefore recommend use of commercially prepared plates, particularly for the beginner who may lack the skill for producing homemade plates of comparable quality. For quantitative applications, precoated plates are generally superior to those made by anyone but an expert.

Table 3.1. Representative Adsorbents for TLC

Adsorbent	Particularly Applicable to Chromatography of	Basis of Separation
Silica gel	Practically all classes	Adsorption, partition, ion exchange
Alumina	Bases, steroids	Adsorption, partition, ion exchange
Cellulose powder	Polar compounds including amino acids and nucleotides	Partition
Kieselguhr	Sugars	Partition
Polyamides	Anthocyanins, flavones	Adsorption
DEAE,[a] TEAE-cellulose,[b] CM-cellulose[c]	Nucleotides, phospholipids, glycolipids, and pigments	Ion exchange, partition

[a] DEAE, diethylaminoethyl.
[b] TEAE, triethylaminoethyl.
[c] CM, carboxymethyl.

3. Chromatographic Chambers

There are two basic styles of chromatographic chambers, each available in a variety of shapes, sizes and materials. Most familiar is the large-volume chamber (Plate 3.1) which is suitable for most TLC and some PC

Plate 3.1. Large-volume chambers measuring approximately 12 × 12 × 4 in. (photographs courtesy of Brinkman Instruments, Westbury, N. Y.).

applications. A wide-mouth jar suffices for small plates. The sides of the chamber frequently are lined with filter paper to facilitate solvent equilibration with the chamber atmosphere. The "small volume," "sandwich," or "S" chamber (Plate 3.2) is designed primarily for TLC, but may also be used for small-scale PC. The difference between these two chamber designs is basic. Large-volume chambers permit chromatography in an atmosphere partially or totally saturated with solvent vapors. In "S" chambers, the solvent composition of the atmosphere above the layer may be quite different from that of the bulk solvent. This can have profound effects on the chromatographic behavior of the system. This subject is discussed in greater detail in Section VIII.

V. ADSORBENT AND SOLVENT SELECTION

Under optimum conditions, as many as 20 compounds may be completely separated by a single PC or TLC development. Mixtures containing larger numbers of components must be subjected to a preliminary fractionation.

Although cellulose paper may be coated with a less-polar liquid for reversed-phase partition chromatography, this technique is now largely obsolete. A certain degree of variability can be attained by varying the more polar stationary phase and the less polar mobile solvent. While in PC one selects only a solvent, in TLC one employs an adsorbent-solvent pair which together form the chromatographic system.

1. Adsorbent Selection

Partition chromatography, either on PC or TLC, is best suited to the separation of more polar compounds or those compounds which are closely related and which show only small differences in polarity. Adsorption chromatography by TLC is indicated for the more lipophilic compounds. Although a wide variety of adsorbents may be used for TLC, either silica gel or powdered cellulose is suitable for most separations. While cellulose is most frequently used for partition chromatography, silica gel is suitable for use with both adsorption and partition chromatographic separations. Other adsorbents used include alumina, ion exchangers, polyamides, and other nonionic polymers.

When dry, silica gel is an extremely active adsorbent, in other words, it has a large number of adsorbent sites which interact strongly with solutes and solvents. If exposed to air, silica adsorbs moisture and becomes less active. Activated silica gel may be better for separations of molecules capable of fairly weak interactions with its adsorbent sites. Deactivated

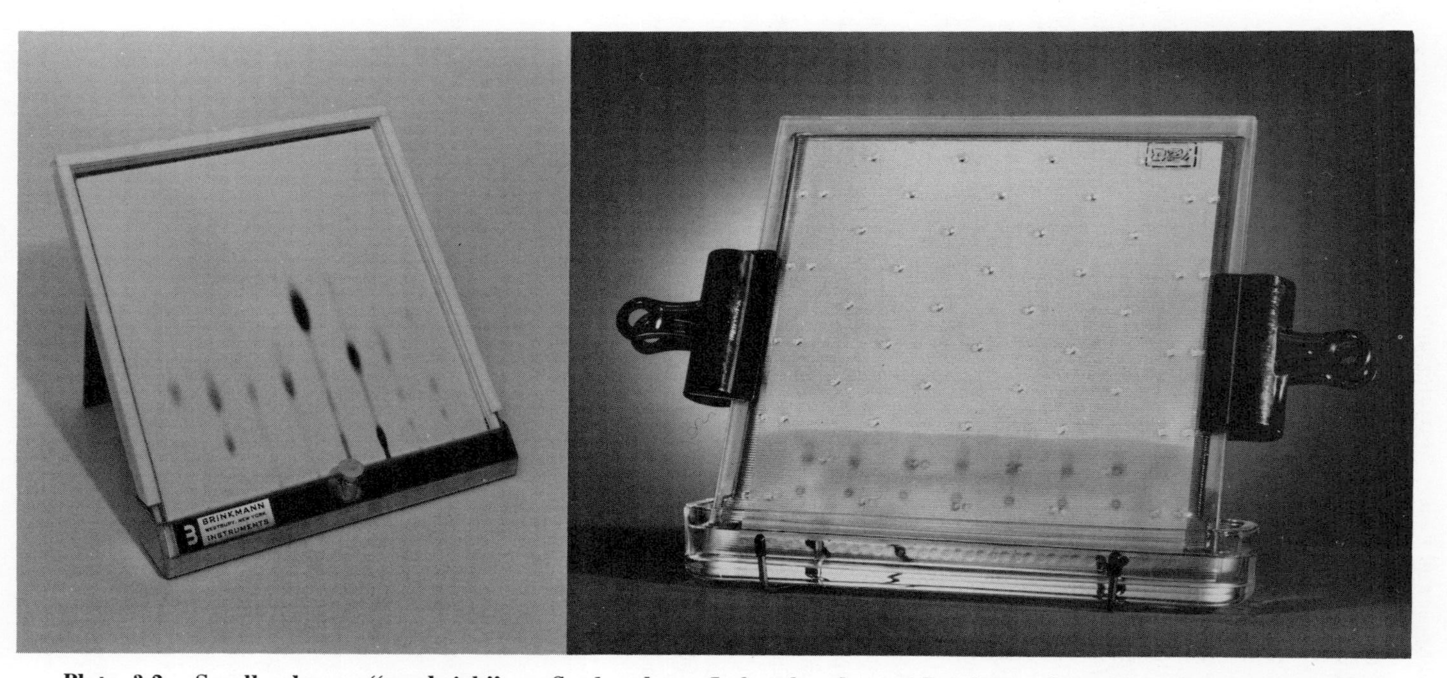

Plate 3.2. Small-volume, "sandwich" or S chambers. Left: chamber of Brinkman design for adsorbents on glass, paper, plastic or metal foil supports (photograph courtesy of Brinkman Instruments, Westbury, N. Y.). Right: chamber of Eastman design for adsorbents on supports other than glass (photograph courtesy of Eastman Kodak Company, Rochester, N. Y.).

silica gel, on the other hand, is preferable for molecules with functional groups which interact strongly with the adsorbent. Cellulose, a relatively inactive adsorbent for most compounds, is best used for highly polar molecules which interact strongly with adsorbent sites.

2. Solvent Selection

If the identity of the sample is known, the sections on chromatographic data of the *Journal of Chromatography* should be consulted. The *Bibliography of Paper and Thin Layer Chromatography 1961–1965* contains extensive lists of compounds which have been successfully chromatographed. This bibliography also gives complete listings under classes of compounds. Separations achieved by PC generally can be duplicated by cellulose TLC.

The worker will in most cases have at least a general idea of the nature of the compounds to be separated. The solvents commonly used for that class of compounds should be tested (see Section XIII). Since many solvents or solvent combinations may effect the desired separation, the first system that gives adequate separation should be employed and the search should be terminated at that point.

If little or no information is available, solvents may be selected by simple experimentation. A rapid method of solvent selection is to apply a series of sample spots on the adsorbent. Solvents to be tested are then applied to the center of the spots. In this way, a miniature circular chromatogram is produced. Solvents which give a series of bands with an approximate R_f of 0.5 are further investigated. A better method is to spot the sample on coated microscope slides or strips of adsorbent coated on plastic. These miniature chromatograms can be developed in small jars containing different solvents. The representative eluotropic series below provides a guide to solvent selection.

An indication of the polarity of a solvent can also be obtained from the dielectric constant, ϵ.

Solvents Listed in Order of Increasing Polarity

Solvent	ϵ_{20}	ϵ_{25}
n-Hexane	1.89	
Cyclohexane	2.02	
Carbon tetrachloride	2.24	
Benzene	2.29	
Toluene	2.44(ϵ_0)	
Trichloroethylene	3.4(ϵ_{16})	
Diethyl ether	4.34	
Chloroform	4.81	

Solvent	ϵ_{20}	ϵ_{25}
Ethyl acetate		6.02
n-Butanol	17.8	17.1
n-Propanol		20.1
Acetone		20.7
Ethanol		24.3
Methanol		32.6
Water	80.4	78.5

If a pure solvent is not satisfactory, solvents of suitable polarity can be obtained by trying out mixtures in various proportions of solvents which are listed adjacent to each other on the list given above. Generally a solvent or solvent mixture which gives an R_f of 0.2–0.8 for the sample should be selected.

VI. PREPARATION OF TLC LAYERS

1. The Support

Glass plates are first washed with an abrasive cleaner (e.g., scouring powder), then rinsed thoroughly with water, and finally flushed with acetone. Fingerprints are removed with alcohol after the plates are on the spreading board.

2. Spreading the Thin Layer

To coat glass plates with powdered cellulose, mix 15 g cellulose with 90 ml water. Blend until smooth, then stir gently to remove air bubbles—a prime cause of poor plates. Although manufacturers generally specify other proportions for silica gel plates, 30 g silica gel G (with $CaSO_4$) may be mixed with about 20 ml absolute ethanol and 40 ml water. A smooth slurry is prepared in a mortar and plates are spread immediately. The $CaSO_4$ sets or hardens rapidly; it is therefore necessary to carry out these operations swiftly.

Moderately satisfactory TLC layers may be produced with equipment available in most laboratories. Clean microscope slides may be dipped vertically into a thick slurry of adsorbent (e.g., silica gel plus $CaSO_4$) in water or alcohol. To avoid coating both sides of the slide with adsorbent, two slides are held together, dipped simultaneously in the slurry, and then separated. The excess adsorbent is drained off and the slides are placed on a flat surface to dry.

More uniform layers are obtained by the use of mechanical spreading devices. These consist of a plate holder and an adsorbent reservoir and are of two general types. Either the plates remain stationary and the spreader moves, or the plates are passed under a reservoir of adsorbent. A simple spreader of the latter type may be constructed for less than $10 (9) Commercial spreaders generally embody a device to hold the plates firmly in place and a flat gate which can be adjusted to the desired height above the plates. A variety of equipment, varying in cost and complexity, is available (See Appendix). Most spreaders permit the layer thickness to be varied, but layers spread at a wet thickness of 250 μ suffice for most purposes. When spreading plates, a smooth continuous motion is desirable. Stopping or changes in speed result in an irregular layer thickness. This can be observed as variations in the surface consistency.

Plates are allowed to air-dry on the support, then transferred to a drying rack. Silica or cellulose plates may be air-dried or dried in vertical position (to permit free flow of solvent vapors) for 10 min at 100°C. For best results it is imperative that drying be carried out in a clean, dust-free atmosphere. Otherwise, spurious peaks may be observed on the TLC plates after development.

3. Activation

Activation—removal of water or other polar solvents from the adsorbent—profoundly affects development by adsorption chromatography. Silica or alumina plates may be activated by heating in a vertical position for 1 hr at 100–110°C. Although activated plates may be stored in a cabinet containing a dessicant, it is preferable to reactivate the adsorbent just prior to use. It is not necessary to heat cellulose layers.

VII. SAMPLE APPLICATION

For TLC the sample is dissolved in a small volume of a volatile solvent (e.g., ethanol, benzene, or ether). In adsorption chromatography (e.g., silica gel TLC) the least polar solvent which will dissolve the sample should be used. Polar solvents inactivate the adsorbent at the point of sample application and may adversely affect separations. Using a micropipette or capillary tube, samples are applied in small spots about 2.5 cm from one edge of the plate and at least 1 cm apart.

If more than 5 μl are to be applied, spot the sample repetitively, 5 μl at a time. Remove the solvent between applications by directing a stream of warm or cold air from a blower or hair dryer. Cold air is advisable if the sample is thermally labile. Repeated application is easier with abrasion-resistant, commercial TLC plates. Large samples are applied as a line or as a continuous streak either by hand or by using a commercial applicator. Lines consisting of a series of spots frequently are too wide and uneven for optimum separations.

Since PC is used primarily for partition chromatography in polar solvents, the sample may be applied in water or alcohol. The high capacity for aqueous samples is a primary advantage of PC. By repetitive application as much as 1 ml may be applied to Whatman #3 MM or similar papers. If possible, aqueous samples should be desalted prior to application.

Chromatographic standards are applied beside the sample. Literature R_f values, unless qualified by a detailed description of conditions, have only limited value in predicting sample R_f or identity.

VIII. DEVELOPING THE CHROMATOGRAM

Once the samples and standards have been applied, the chromatogram is ready to be developed. Commercial sandwich or S chambers designed for use with adsorbents bonded to thin sheets of plastic or metal foil are used without solvent equilibration of the chamber atmosphere. The adsorbent layer is placed between the glass plates and the trough is filled with solvent.

1. Large-Volume Chamber Development

It is customary to partially saturate the atmosphere of large-volume chambers with solvent vapors prior to use. Lining the chamber with filter paper shortens the time required to achieve equilibration. This varies widely with solvent composition. It is doubtful whether true vapor equilibrium is achieved and maintained, for uncovering the tank to insert the chromatogram will disrupt the equilibration process.

Thorough equilibration may be undesirable where solvent components can react with one another, for example, when *n*-butanol plus glacial acetic acid are used. For reproducible results in TLC the prime factor is consistency. Select a standard time to equilibrate the chamber (e.g., 30 min) and observe this period in subsequent runs.

Whereas single-component solvents may be reused until exhausted, the process of developing a chromatogram changes the composition of multicomponent solvents. The adsorbent generally binds one component more strongly than another, reducing its concentration both in the advancing solvent front (see 2, below) and in the solvent reservoir. Although this is a minor factor with large volumes of solvent containing approximately equal volumes of components, a component present at the 1–5% level may be totally removed from the solvent by selective binding to the adsorbent. For this reason, and because chemical reactions between components may change the solvent composition, reuse of solvents should be discouraged.

2. Adsorption of Solvent Vapors by the Adsorbent Layer

Although the area of a TLC plate ahead of the advancing front appears to be dry, the adsorptive properties of the layer ensure that it contains solvent vapor adsorbed from the chamber atmosphere. The quantities adsorbed depend on the vapor composition, the adsorbent properties of the layer, and the time of exposure. The quantity of solvent vapor taken up is large; as much as 3.6 mmoles of acetone or 6.7 mmoles of chloroform can be sorbed on a typical 8 × 8 in. plate. Even after 20 min, adsorption is often incomplete (10). In a typical open chamber the times required for silica gel to reach half-saturation are about 1.5 and 5 min for acetone and chloroform, respectively (10).

Large-volume chamber development is usually carried out without prior equilibration of the layer with the chamber atmosphere. The advancing solvent front thus reaches areas progressively richer in solvent vapors as development proceeds. With a binary or ternary solvent mixture, the composition of the adsorbed vapors ahead of the solvent will also vary with the time of development. This behavior is not necessarily deleterious to the separation process. As a matter of fact, some of the factors pointed out here are partially responsible for the remarkable resolving power of TLC.

Considerations such as the above, while useful to explain apparently anomalous behavior of chromatograms developed under differing conditions, pose few practical problems for the casual user. If reproducible results are desired, it is necessary for the experimenter to maintain rigidly standardized procedures.

3. Comparison of Large-Volume and S Chambers

The R_f values in multicomponent solvent mixtures may vary widely from large-volume to S chambers. This is illustrated in Table 3.2 for the chromatography of four hypnotics. Although solvent I separates the hypnotics in either chamber, the R_f values differ significantly. Even more striking differences in mobility between the two chambers are observed using solvent II. These differences result from the interplay of two factors: solvent demixing and differential adsorption of solvent vapors from the chamber atmosphere (10).

Table 3.2. Influence of Chamber Geometry on R_f

	$R_f \times 100$[a]			
	Saturated Large-Volume Chamber		S Chamber	
Hypnotic	I	II	I	II
Heptobarbital	11	19	28	100
Cyclobarbital	23	45	43	100
Pentobarbital	29	59	48	100
Pentothiobarbital	60	62	78	100

[a] Hypnotics were applied to activated silica gel G plates and chromatographed in chloroform–acetone (90:1) (I) or chloroform–isopropanol–25% NH_4OH (9:9:2) (II). Development was carried out both in a large-volume, vapor-saturated chamber and in a small-volume S chamber (10).

In both chambers solvent demixing occurs as the solvent ascends the chromatographic plate. With solvent I, the leading edge is essentially pure chloroform since acetone is a minor component and is more strongly adsorbed. With solvent II, the ammonia is strongly bound by the silanol groups of the silica gel and is completely removed from the solvent front. Solvent demixing alone does not determine the composition of the solvent front. In a large-volume chamber the adsorbent contains significant quantities of acetone (solvent I) or ammonia (solvent II). In an S chamber, the vapors adsorbed ahead of the solvent front are derived principally from the front itself. As noted above, this front is free of acetone (solvent I) or ammonia (solvent II). In an S chamber, little or no diffusion of ammonia vapor can occur and the hypnotics migrate as free acids with the solvent front. In the large-volume chamber, adsorption of ammonia vapor occurs over the entire plate and the acids migrate as their ammonium salts. Some of the shortcomings of S chambers can be overcome, for example, by impregnating the plate with ammonia vapor prior to development. In many instances, however, S chambers produce decidedly superior separations.

It should be clear that data obtained using S chambers cannot be applied to large volume chambers, and vice versa. In S chambers, as in column chromatography, vapor diffusion and impregnation are difficult. The adsorption of vapors on the chromatographic plate can have a pronounced beneficial influence on the separation process. Hence TLC in large-volume chambers may produce separations superior to those obtained on columns.

IX. PREPARATIVE-SCALE PC AND TLC

In the course of examining a reaction mixture, a supposedly pure compound, or an extract of biological material by PC or TLC, additional components are frequently detected. Preparative-scale PC or TLC offers a simple way to isolate such components for structural determination, analysis, or use in a synthetic or biosynthetic process. Another application of preparative-scale TLC is the preliminary fractionation of compounds according to solubility class prior to further fractionation by GLC or TLC. The term "preparative-scale," as applied to PC or TLC, means chromatography of 0.1 mg to about 1 g of material. If larger quantities are required, other techniques, such as column chromatography, should be used.

In preparative-scale PC thick, high-capacity paper, such as Whatman #3 MM is used. The sample is repeatedly streaked across the entire width of the paper. After development, the sample band is cut out and eluted by descending development with a suitable solvent.

Although large-scale plates and development tanks are available, preparative-scale TLC may be accomplished using several ordinary 8×8 in. plates. Up to six plates may be developed simultaneously in a standard large-volume chamber. Following chromatography, the sample band is scraped or aspirated from the plate, the sample is dissolved, and the adsorbent is removed by filtration or centrifugation. As much as 0.2 g of material, applied as a streak across the entire plate, may be separated on 0.25-mm-thick plates. Thicker layers (0.25–2.0 mm) may be used for larger quantities, but these are more difficult to prepare and tend to develop fissures which adversely affect resolution. It is difficult to achieve uniform solvent migration in layers more than one mm thick.

Less solvent than normal should be used in preparing the slurry for thicker preparative layers. Good quality preparative-scale TLC plates are commercially available. Preparative or analytical work on greater than gram quantities usually requires column chromatography. TLC and PC are most profitably applied to samples in the 10^{-1} 10^{3}mg range.

X. VISUALIZATION

Unless the compounds chromatographed are colored, physical or chemical means must be used to reveal their presence on the chromatogram. Although detection techniques and reagents may be classified as either general or specific, the distinction is not clear-cut. A more useful distinction is between those visualization techniques which destroy all or part of the sample and those which do not. Only selected examples are discussed.

1. Nondestructure Visualization

A. Water

Chromatograms lightly sprayed with water frequently reveal hydrophobic compounds as optically dense "waxy" areas. The technique is restricted primarily to TLC and hydrophobic compounds.

B. Viewing Under Ultraviolet Light

Inspection of a chromatogram under UV light of suitable wavelength in a darkened box or room reveals most aromatic and many nonaromatic compounds as dark spots against a light background. (Beware of possible retinal damage when using UV light.) Some compounds may exhibit fluorescence. The dark or "quenched" regions where UV chromophores are present are easier to see if the chromatogram is first treated with a phosphor (e.g., 2,7-dichlorofluorescein). Dark areas then appear against a brilliantly lit background. Adsorbents and precoated layers are both available with phosphor added. These additives generally do not affect development, and layers with phosphors are suitable for most applications. Paper chromatograms must be sprayed with phosphor, but it is easier to use fluorescent-cellulose TLC. Although most compounds are relatively stable to UV light, others (e.g., steroids and vitamins) can be destroyed when UV visualization techniques are used.

C. Iodine

Chromatograms are either sprayed with a 1% methanolic I_2 solution or placed in a chromatographic chamber to which several crystals of I_2 have been added some time previously. The latter method is most often used. Compounds are visualized as brown areas which progressively darken with continuing exposure to the I_2 vapors. Eventually the entire chromatogram may darken if it is not removed from contact with the iodine. Compounds with double bonds or other sources of π electrons stain extremely rapidly. Eventually, however, almost all organic compounds react. The organic binders used in some commercial TLC plates react strongly with I_2. When the chromatogram is removed from the chamber it must be marked or photographed, since the colors fade with time. Although one would expect the visualization process to be reversible and the test to be nondestructive, this is not always true.

2. Destructive Visualization Techniques

A. Oxidation by H_2SO_4

H_2SO_4 or various combinations of H_2SO_4 plus dichromate are widely used as destructive visualization agents. The chromatogram is sprayed,

Table 3.3. Representative Specific Detection Reagents

Reagent	Specificity	Composition	Color Development
1. Ninhydrin	All three reagents react most rapidly with α-amino acids. Peptides and amines react more slowly. Reagents B and C give characteristic colors with many amino acids. The colors formed with C are stable.	A. 1.0–2.5% Ninhydrin in acetone or n-butanol + 3% acetic acid. B. Reagent A plus 1% collidine added just prior to use. C. Mix, in order, 100 mg cadmium acetate, 10 ml H_2O, 5 ml glacial acetic acid, 100 ml acetone, and 1 g ninhydrin.	Leave at room temperature or heat briefly at 80–110°C
2. Bromcresol green	Carboxylic or other acids as yellow spots on a green background.	To 3% bromcresol green in methanol add NaOH or HCl until green.	None
3. Bromthymol blue	Lipids visualized as yellow spots on a blue background.	0.04% in 0.01 N NaOH.	None
4. Antimony trichloride	Steroids, glycosides, vitamin A, some lipids.	A. $SbCl_3$–glacial acetic acid (1:1) (w/v). B. Saturated solution of $SbCl_3$ in ethanol. Discard after use.	95–100°C for 5–10 min. Observe under UV light.
5. Anisaldehyde	Reagent A gives specific colors with many sugars. Reagent B gives blue, red, or green colors with phenols, terpenes, steroids, or sugars.	A. 0.5 ml anisaldehyde, 9 ml 95% ethanol, plus a few drops of acetic acid. B. 5 ml anisaldehyde, 50 ml glacial acetic acid, plus 1 ml conc. H_2SO_4.	90–110°C, 5–10 min

6. Phosphomolybdic acid	Blue spots on a yellow background with bile acids, cholesterol esters, glycerides, and unsaturated fatty acids.	5–10% Phosphomolybdic acid in 95% ethanol.	100–150°C, 1–10 min
7. Chlorosulfonic acid	Fluorescence with triterpenes and steroids.	$ClSO_3H$–glacial acetic acid (1:2).	130°C, 5 min. View under UV light (365 nm).
8. Acidic vanillin	Blue-green spots with higher alcohols, and ketones (A) or with steroids (B).	A. 3 g vanillin, 100 ml ethanol plus 5 ml conc. H_2SO_4. B. 1% Vanillin in 50% H_3PO_4.	120°C, 10–20 min
9. Chloroplatinic acid	Characteristic colors with alkaloids.	3 ml 10% chloroplatinic acid, 97 ml H_2O, 100 ml 0.1 *N* KI. Store in brown bottle.	None
10. $HgNO_3$	Black or white colors on a grey background with barbiturates.	1% $HgNO_3$.	None
11. Diazotized sulfanilic acid	Yellow-orange with phenols and phenolic natural products.	0.1 g diazonium salt in 20 ml 10% NaOH.	None
12. $SbCl_5$	Terpenes, essential oils, resins.	$SbCl_5$–CCl_4 (1:4) (w/v).	120°C, 5–10 min
13. Bromcresol purple	Yellow spots on purple background with halide ions.	1% Bromcresol purple in ethanol plus dilute NaOH until the color just starts to change.	None
14. Dichlorophenol-indophenol	Inorganic ions; vitamin C.	2 g 2,6-Dichlorophenolindophenol (Na salt) in 100 ml 95% ethanol. Add 3 g $AgNO_3$. Filter. Discard after use.	None

streaked, or dipped in reagents such as 5–50% methanolic H_2SO_4, then heated for 5–30 min at 100–120°C. Ultraviolet fluorescence or colors produced depend on the reagent used, the duration of heating and the nature of the sample. Eventually all colors darken to a gray-black hue. If sample destruction presents no problem, the general applicability of this technique for nonvolatile organic compounds offers great advantages. Plates with organic binders are not suitable for this visualization technique.

B. Other Techniques

Under this heading we group all chemical reagents giving color reactions with specific functional groups or compounds. Some selected examples are given in Table 3.3. PC reagents may be used for TLC, usually with a considerable increase in sensitivity.

C. Preparative Chromatography

Since visualization of an entire preparative chromatogram with a destructive reagent is a self-defeating process, one may cover the central portion before applying the reagent. Colors then form only at the edges of the chromatogram. If the solvent front is straight, the desired band may readily be located and either cut out (in PC) or scraped off (in TLC). This process allows the bulk of the sample to be recovered intact.

3. Radioactive Compounds

Isotope detection methods are several orders of magnitude more sensitive than chemical methods. Since radioactive scanning does not alter the chemical composition of the sample, additional operations can be carried out on the chromatogram. One may carry out chemical detection methods or recover the sample. The greatest utility of these detection methods, however, lies in their ability to permit ready quantitation of compounds on chromatograms.

Several special techniques are available, each of which utilizes the radioactivity of the sample. Each radioactive spot should contain a minimum of 1000 counts/min if ^{14}C is used, or 10,000 counts/min if the isotopic label is ^{3}H. Paper strips may be analyzed by a strip scanner which gives quantitative information on the radioactivity in each region. TLC layers on plastic or aluminum foil supports may be treated similarly. Strip scanners specifically for TLC plates are commercially available. (see Appendix Section 4). The tritium counting efficiency of most strip scanners is quite low.

An alternative method of detecting radioactive compounds is radioautography (autoradiography). The chromatogram is pressed tightly against a sheet of x-ray film with a glass plate and kept in the dark for a period of time. Exposure time depends both on film sensitivity and the number of disintegrations per minute per unit area. Development of ^{14}C compounds usually takes several days. For 10,000 counts/min of ^{14}C, a readily visible spot is obtained after two days.

The low energy and short range of the β-radiation of tritium necessitates special techniques. The chromatogram can be coated directly with a photographic emulsion, then stored in the dark for several days or weeks prior to development. A simpler method is to convert the emissions to light quanta and use these to expose the film. This conversion is achieved by incorporating anthracene or other scintillators into the adsorbent or by spraying a solution of these compounds on the developed plates (4). These products are commercially available.

XI. REPRODUCIBILITY OF R_f VALUES

1. Physical State of the Adsorbent Layer

This can vary widely from one preparation to another. If formulations are changed, profound effects on specific separations will often be observed. Where possible, a single batch of adsorbent should be used for comparative work. Formulations in commercially prepared plates are usually quite consistent. Large orders for commercial plates are generally filled with material from the same batch.

A major factor in adsorption chromatography is the degree of hydration of the adsorbent. Although polar solvents are removed from silica layers by drying at 100–110°C, silica takes up water rapidly and should, therefore, be used soon after activation. Water uptake can occur from several sources. The local cooling caused by evaporation of the spotting solvent can condense moisture from the air at the point of sample application, as may blowing on the spots to speed evaporation. To circumvent this problem, one can use a slow stream of warm air or N_2. It should always be borne in mind that oxygen, warm air, or sunlight can cause decomposition of sensitive compounds spread over a large area.

2. Temperature

Changes in temperature alter partition coefficients. The longer times required for development of PC make it especially vulnerable to changes in ambient temperature. Many workers use a controlled-temperature room for PC. TLC on silica gel is less sensitive to small temperature changes.

Even then, one should take reasonable precautions to avoid large temperature fluctuations in the working environment. Tanks for development, for example, should not be placed in air currents or with one edge against a wall, because the air temperature next to a wall differs from that of the room as a whole.

3. Solvent Purity

Solvents should, when possible, be of reagent grade. Many solvents may require prior purification. Contaminants may have especially deleterious effects in adsorption chromatography. Traces of water adversely affect development in nonpolar solvents, and the 0.5–1.0% ethanol present in chloroform as a preservative can produce major changes in its properties as a chromatographic solvent. Gross impurities in the solvent or adsorbent may be observed as a dark band near the solvent front of a developed chromatogram.

4. Solvent Demixing

During the development process, multicomponent solvents penetrating the adsorbent undergo partial demixing. The advancing front deposits solvent components on the adsorbent. In this process the composition of the solvent front is altered and trace components may even be entirely removed. To test for demixing, a mixture of a polar and a nonpolar compound is spotted repeatedly at various heights diagonally across a TLC plate and developed with the solvent mixture of interest. The resulting R_f values reflect the solvent composition at various heights on the plate.

5. Evaporation

In development with multicomponent solvent systems in incompletely saturated chambers the more volatile solvent will evaporate faster. The rate of evaporation decreases toward the middle of the plate, increasing the R_f of many compounds. The "edge effect" can be largely prevented by proper chamber equilibration. Tanks may be equilibrated with the dry chromatogram in place. More solvent may then be added through a hole in the cover to initiate the development. The cover must make a vapor-tight seal with the chamber.

XII. QUANTITATION

Quantitation of samples on chromatograms generally requires analysis by physical or chemical methods such as those discussed elsewhere in this book. Considerable progress has been made in adapting instruments and techniques to yield more accurate results. It is certain that this progress will continue, and the use of TLC for quantitative analysis will become increasingly commonplace. At present, it is not too difficult to obtain results within a few percent of the true value. Under rigorous conditions, extreme care, and the use of internal standards, results reliable to within less than 1% may be obtained. The average laboratory would, however, probably do well to look for other techniques when results of this precision are desired.

Quantitative analysis may be performed following elution or *in situ*, in other words, without removal of the sample from the layer. Both techniques are widely used. The latter method offers the advantages of speed and ease in analysis, but requires costly instruments and is generally less accurate. The first is slower, but requires little additional equipment.

1. *In Situ* Quantitation

A. Photodensitometry

Reflected or transmitted light is used for quantitation. Absorption of light of a specific wavelength by the sample permits direct analysis. Alternatively, the chromatogram may be first treated chemically to produce a color. Amino acids, for example, may be converted to colored complexes by treatment with ninhydrin, and lipids may be charred with H_2SO_4 under carefully controlled conditions.

A variety of instruments designed for TLC densitometry are available (Appendix Section 3). These consist of a plate holder and drive mechanism which passes the plate under a narrow slit illuminated by the light source. Reflected or transmitted light is sensed by a phototube circuit and the amplified response is fed to a strip chart recorder. Since irregularities in layer thickness may be a significant source of error, some instruments incorporate double-beam optics. The recorder then displays the difference between the transmission or reflectance properties of the sample area and that of an adjacent blank area of adsorbent.

Difficulties in photodensitometric quantitation more often reflect shortcomings of the TLC system itself than of the instrument. Irregular, streaked, or overloaded chromatograms cannot be expected to yield satisfactory results. For superior results the spots should be circular. These spots are most easily obtained in a chamber saturated with solvent. If these spots are not round, a different adsorbent-solvent system should be used. Imperfections in the adsorbent layer and variations in color development conditions are other potential sources of trouble. Extreme care and the inclusion of appropriate standards are essential for accurate quantitative analysis.

B. *Fluorimetry*

Fluorescent materials absorb UV light and emit light of longer (frequently visible) wavelength. Fluorimetry may thus be used like photodensitometry to quantitate samples on chromatograms. Instruments designed for this purpose consist of a UV source, a monochromator to select the desired wavelength for irradiation, a sample holder, a second monochromator to select the desired wavelength for detection, and a phototube–amplifier–output assembly. The fluorimeter is mounted on a plate tracking device and irradiates and records the intensity of the light emitted by samples on the plate. This technique has, for example, been successfully applied to the quantitation of pregnanediol and of fluorescent derivatives of amino acids (3).

C. *Radioisotopes*

The widespread use of radioisotopes in conjunction with PC and TLC has stimulated the development of instruments designed to detect radioactive substances on chromatograms. The available instruments are in most cases adaptable to both PC and TLC. For the former, 4 π geometry may be achieved, but for TLC 2 π geometry must be used. In most instruments the plate is mounted on a movable carriage with the adsorbent layer upward. The radiation detectors are generally gas-flow proportional counters with diaphragms positioned close to the adsorbent layer. Windowless operation, mandatory for detection of ^{3}H, is also possible in some instruments. The detector output is amplified and the analog output signal is delivered to a suitable strip chart recorder (11).

Although sensitivities vary with layer thickness and with instrument design, the following activity levels are generally detectable: 10^{-3} μCi of ^{14}C, ^{32}P, or of other hard β-emitters, and 10^{-2} μCi of ^{3}H. Depending on the adsorbent layer thickness, the counting efficiency varies from 10–25% for ^{14}C, 45–55% for ^{32}P and 0.1-3% for ^{3}H.

2. Elution Prior to Quantitation

Color development may either precede or follow removal of the sample from the chromatogram. The former is obviously advantageous for accurate location of the sample; in many cases this capability is a major factor for accurate quantitation (2). In other cases, however, superior results are achieved by elution prior to chemical modification of the sample.

Samples on PC are cut out; the paper is then snipped into pieces and the compound is dissolved in an appropriate solvent in a test tube. The sample can also be eluted by descending development. The latter technique is

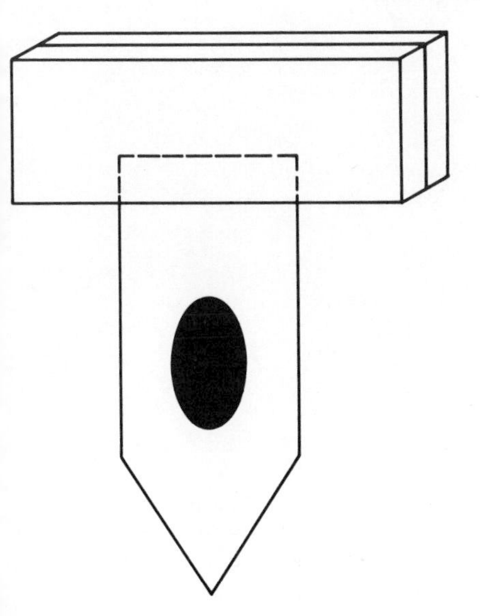

Figure 3.1. Sample recovery in paper chromatography. One edge of the paper is clamped between microscope slides.

preferable if one wishes to recover the sample in a small volume. A simple method to carry out this recovery is to cut out an area of paper around the spot. This piece of paper is then pointed and placed with part of one edge between two microscope slides (Figure 3.1).

The slides are placed in a trough of elution solvent and a few drops of effluent are collected (Figure 3.2).

If the paper contains substances that are eluted with the sample this may interfere with accurate quantitation. In this case, the papers should be washed by development with the eluting solvent and thoroughly dried just prior to use.

In TLC the adsorbent containing the spot is either scraped off or aspirated into a small vacuum flask. The sample is extracted with an appropriate solvent and the adsorbent is subsequently removed by filtration or by centrifugation. The sample may then be assayed. Certain sensitive compounds may break down if polar solvents are used to remove them from silica or other highly active adsorbents.

Elution of the sample followed by scintillation counting is highly accurate and can be readily applied to the quantitation of small sample spots in TLC. Samples should be eluted in a small volume to permit scintillation counting without concentrating the sample. If the location of the sample is known, the paper can be cut up, placed in a scintillation vial and counted directly in a dioxane liquid scintillation counting solution. One milliliter of water in the dioxane may improve counting efficiency by solubilizing the

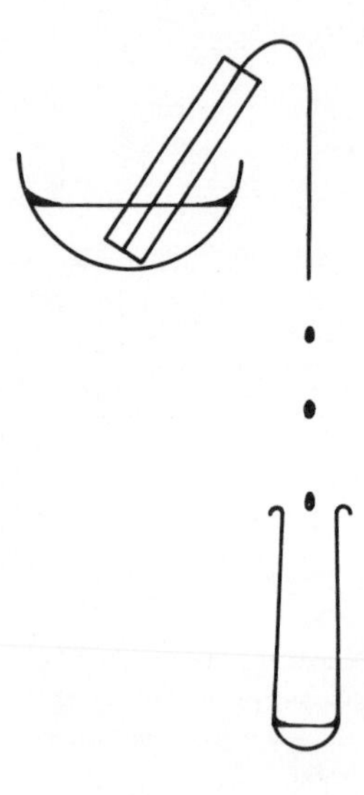

Figure 3.2. Sample recovery in paper chromatography. The sample spot is eluted off the paper.

sample. Samples aspirated from TLC plates do not have to be separated from the adsorbent prior to counting. Both sample and adsorbent (preferably without fluorescent indicator) are added to the dioxane scintillation counting solution. With care, quantitation by scintillation counting may exceed 2% accuracy (6).

XIII. SEPARATIONS OF CLASSES OF COMPOUNDS

We do not discuss all separations in detail. The literature is—to say the least—extensive, and numerous books and monographs provide detailed guides to specific separations. What follows is a summary of some of the solvents, adsorbents, and techniques which have found general and widespread acceptance. The section is intended as an introduction to the subject and as a point of departure for further reading.

1. Amino Acids and Peptides

The α-amino acids present in proteins were the first class of closely related compounds successfully separated by PC. The adsorbent (filter paper), solvents, and detection reagents developed in 1945–1955 are still in favor today and are also applicable, with minor modification, to the separation of other amino acids and of oligopeptides.

A. Adsorbents

Cellulose, either as filter paper sheets (PC) or as microcrystalline layers (TLC), remains a popular support for these highly polar compounds. TLC on silica gel or on mixtures of cellulose and silica gel are useful in specific instances; activation of layers is unnecessary in this case. Both one- and two-dimensional development are employed, the latter for mixtures of more than 5 or 6 components. Most separations are improved and development times are in all cases substantially decreased by substitution of cellulose TLC for PC.

B. Solvents

The partition chromatographic solvent mixtures developed for PC are used without modification for cellulose TLC and in many instances for silica gel TLC also. As with all separations based on partition chromatography, development is slow (2–24 hr). Two excellent pairs of general solvents are (1) *n*-butanol–glacial acetic acid–water (12:3:5) (v/v) and phenol–water (3:1) (v/v) and (2) *n*-butanol–acetone–ammonia–water (10:10:5:2) (v/v) followed by isopropanol-formic acid–water (20:1:5) (v/v). The *n*-butanol–glacial acetic acid–water system must be freshly prepared prior to use. For two-dimensional development these solvents are used in the order listed. Both large-volume and S chambers give satisfactory results.

In many cases the best separations are achieved by combining electrophoretic and chromatographic techniques. This is especially true for the separation of peptides resulting from the partial hydrolysis of proteins. Electrophoresis at a pH of 6.5 followed by descending PC in *n*-butanol–acetic acid-water is widely used as a "fingerprint" technique for peptide "mapping."

C. Detection Reagents

Although a few specific reagents are available for certain α-amino acids, ninhydrin in either acetone or butanol (Table 3.3) is the most widely used detection reagent. Color development proceeds slowly at room temperature but is accelerated by heating at 80–100°C for about 5 min. Most amines react to produce purple colors. Because α-amino acids react most rapidly with this reagent, it is possible to distinguish α- from other amino acids and peptides by delaying the heating process until all α-amino acids have reacted. If a trace of γ-collidine is added to the ninhydrin reagent just prior to use, many amino acids give characteristic blue, brown, yellow, or red colors.

D. References

V. M. Ingram, *Nature* **178,** 792 (1956).
A. L. Levy and D. Chung, *Anal. Chem.* **25,** 396 (1953).
G. Pataki, *J. Chromatogr.* **17,** 580 (1965).
G. Pataki, *Techniques of Thin Layer Chromatography in Amino Acid and Peptide Chemistry*, Ann Arbor Science Publishers, Ann Arbor, 1968.
G. Pataki, E. Baumann, U. P. Geiger, P. Jenkins, and U. Kupper, in *Chromatography*, E. Heftmann, Ed., Reinhold, New York, 1967, p. 373.
D. T. N. Pillay and R. Mehdi, *J. Chromatogr.* **47,** 119 (1970).
W. J. Ritschard, *ibid.* **16,** 327 (1964).
I. Smith, *Chromatographic Techniques*, Interscience, New York, 1958, p. 59.

2. Carbohydrates

As with amino acids, the polar character of carbohydrates and of phosphorylated sugars found in biological cells dictates the use of partition chromatographic techniques.

A. Adsorbents and Solvents

PC and TLC on silica gel or on buffered kieselguhr are widely used for carbohydrates (Table 3.4). Mixed adsorbents such as silica gel plus starch or aluminum oxide and adsorbents such as polyamides or polycarbonates also find some applications in this field. The most useful solvents contain water, a low-molecular-weight alcohol, ketone or ester, and a base such as ammonia or pyridine, but acidic or neutral solvents are also used.

Table 3.4. TLC of Simple Sugars on Buffered Kieselguhr[a]

Sugar	$R_f \times 100$	Color with Anisaldehyde–Sulfuric Acid
Lactose	4	Green
Maltose	6	Violet
Sucrose	8	Violet
Glucose	17	Blue
Galactose	18	Gray-green
Mannose	23	Green
Fructose	25	Violet
Ribose	49	Blue

[a] Chromatography in ethyl acetate–isopropanol–water (650:234:117). From E. Stahl and U. Kaltenbach, *J. Chromatogr.* **5**, 351 (1961).

B. *Detection Reagents*

The most commonly used reagents contain H_2SO_4 or oxidizing agents that change color when reduced. Some carbohydrates are reducing agents; others may be degraded to reducing compounds by periodate oxidation.

C. *References*

J. P. Marais, *J. Chromatogr.* **27,** 321 (1967).

F. Percheron, in *Chromatography,* E. Heftmann, Ed., Reinhold, New York, 1967, p. 573.

E. Stahl and U. Kaltenbach, *J. Chromatogr.* **5,** 351 (1961).

D. W. Vomhof and T. C. Tucker, *J. Chromatogr.* **17,** 300 (1965).

3. Lipids

The lipids constitute a large and heterogeneous class of compounds of substantial industrial and biological interest. One subclass, the steroids, is discussed separately below. Both TLC and GLC are used extensively in lipid analysis and complement each other. Analysis of a lipid mixture may involve a preliminary fractionation by solvent partition, followed by separation into classes by TLC, and then into individual components by TLC and/or GLC. A complete lipid analysis may involve a combination of TLC, PC, GC, Sephadex column chromatography, DEAE cellulose, silicic acid, Florisil, silicic acid–silicate and/or TEAE-cellulose column chromatography.

A. *Adsorbents and Solvents*

Adsorption chromatography on silica gel layers has superseded reversed-phase PC techniques. Layers are generally activated prior to use. Solvents for separation into lipid classes usually contain benzene or petroleum ether and diethyl ether as major components; glacial acetic acid is sometimes also present. Improved class separations may be achieved by partial development in one of the above solvents, followed by development with a solvent containing the same components in different proportions. Table 3.5 lists a number of solvent systems which are used in the chromatographic analysis of various lipid classes.

B. *Adsorption Chromatography*

Hydrocarbons may be fractionated in petroleum ether–diethyl ether mixtures. Resolution of linear hydrocarbons with functional groups at the α or β position is improved by the addition of traces of acetic acid to the solvents. Methyl esters of fatty acids, which have to be fractionated by TLC prior to analysis by GLC, may also be chromatographed in the above solvents.

Table 3.5. Solvents for Class Separations of Lipids by TLC on Silica Gel

Lipid Class Separated	Solvent
Waxes, oils, hydrocarbons	Petroleum ether (b.r. 60–70°C)–diethyl ether (19:1
Plant and animal fats	Petroleum ether (b.r. 60–70°C)–diethyl ether-glacial acetic acid (90:10:1 or 80:10:1)
Fatty acids and alcohols	Petroleum ether (b.r. 60–70°C)–diethyl ether-glacial acetic acid (35:15:1)
Phospholipids	Chloroform–methanol–water (65:25:4 or 13:6:1)

C. *Reversed-Phase Partition Chromatography*

Silica gel plates impregnated with hydrocarbons may be used to separat members of a homologous series of fatty acids by partition chromatography Coating agents include paraffin oil, silicone oil, and undecane. Activate silica gel plates are carefully immersed in a 5–15% solution of the hydro carbons in a volatile organic solvent. Paraffin- and silicone oil-coated plate may be stored, but undecane-coated plates must be used soon after prep aration. Polar solvents, such as glacial acetic acid–acetonitrile (1:1) o glacial acetic acid–water (9:1 or 1:1) are used for development. In partitio chromatography, the R_f value of a compound is inversely proportional t its chain length. Functional groups and double bonds increase the polarit of the sample and decrease the observed R_f values.

D. *Inclusions*

Silica gel plates impregnated with $AgNO_3$ are used to separate fatty acid and glycerides differing in the number of double bonds. The silver com plexes with the π electrons, retarding trienes more than dienes, and mon oenes more than saturated fatty acids. Silica layers are prepared in a solven containing 5–10% $AgNO_3$ and stored in the dark. Solvents used includ benzene–ether (4:1) or chloroform containing either 0.5% acetic acid o 1–2% ethanol.

E. *Visualization*

2,7-Dichlorofluorescein, rhodamine, and bromthymol blue are used i addition to acid charring.

F. *References*

C. B. Barrett, M. S. J. Dallas, and F. B. Padley, *J. Amer. Oil Chem. Soc.* **40**, 58 (1963).

H. K. Mangold and R. Kammereck, *J. Amer. Oil Chem. Soc.* **39**, 201 (1962).

H. P. Kaufmann and Z. Makus, *Fette, Seifen, Anstrichmittel* **62**, 153 and 1014 (1960); H. P. Kaufmann, Z. Makus, and B. Das, *ibid.* **63**, 807 (1961).
D. C. Malins and H. K. Mangold, *J. Amer. Oil Chem. Soc.* **37**, 576 (1960).
L. J. Morris and B. W. Nichols, in *Chromatography*, E. Heftmann, Ed., Reinhold, New York, 1967, p. 466.
C. G. Tedeschi, Ed., *Neuropathology: Methods and Diagnosis*, Little-Brown, 1970.
H. Wagner, L. Hörhammer, and P. Wolff, *Biochem. Z.* **334**, 175 (1961).

4. Steroids

The steroids form a diverse group of compounds whose common feature is the steroid (cyclopentanoperhydrophenanthrene) nucleus. The structural similarity of many steroids, coupled with their biological and clinical importance, account for the extensive literature on their chromatographic behavior.

The major steroid classes in order of decreasing polarity are: bile acids, estrogens, corticosteroids, and sterols. Steroids present in extracts of biological fluids or tissues generally are first fractionated by TLC to separate them into polarity classes. These are then further resolved by TLC on the basis of minor structural differences within a polarity class. GLC provides another powerful tool for resolution and quantitation of steroids.

A. Methods and Techniques

The complex mixtures of steroids present in blood and urine are difficult to fractionate in a single PC or TLC development. These extracts contain compounds of highly similar as well as widely differing polarity. Separation of steroids by classes, followed by elution and rechromatography, are common. A number of steroids break down on prolonged exposure to silica. Sunlight, water, and oxygen may accelerate this process.

For PC according to Zaffaroni, filter paper is impregnated with a stationary phase, such as cellosolve or glycerol. The samples are spotted, and the chromatogram is then developed. According to Bush, the samples are spotted first and the filter paper is then equilibrated with solvent vapors prior to development. Although excellent separations may be achieved by these methods, TLC on silica gel is far simpler and faster.

Reagents 4–8 in Table 3.3 are especially useful in steroid detection. Many steroids fluoresce strongly after treatment with sulfuric acid.

B. Bile Acids

Bile acids are C_{24} hydroxy acids with additional oxygen functions at C_6, C_7, and/or C_{12}. The carboxyl group may form an amide linkage with glycine or taurine. Anisaldehyde and phosphomolybdic acid are good detection reagents. Many supposedly pure commercial bile acid preparations are grossly contaminated with other bile acids and are unsuitable for use as chromatographic standards. Table 3.6 gives the mobilities of the common bile acids in a number of solvent systems.

Table 3.6. TLC of Bile Acids on Silica Gel

	Mobility, Expressed as $R_f \times 100$, in Solvent[a]			Mobility, Expressed as $R_D \times 100$, in Solvent[a]		
Bile Acid	I	II	III	IV	V	VI
Deoxycholic	60	26	59	100	100	100
Chenodeoxycholic	55	21	53	91	88	100
Cholic	30	7	16	76	22	17
Hyodeoxycholic	43	14	39	67	50	63
Lithocholic	82	47	82	128	160	254

[a] Mobilities are expressed either as $R_f \times 100$ or as $R_D \times 100$ (the mobility relative to that of deoxycholic acid). Solvents are: 1, diethyl ether–petroleum ether–methanol–glacial acetic acid (70:30:8:1); II, isooctane–diisopropyl ether–glacial acetic acid (2:1:1); III, carbon tetrachloride–diisopropyl ether–isoamyl acetate–*n*-propanol–benzene–glacial acetic acid (4:6:8:2:2:1); IV, trimethylpentane–isopropanol–glacial acetic acid (30:10:1); V, trimethylpentane–ethyl acetate–glacial acetic acid (5:5:1); VI, benzene–dioxane–glacial acetic acid (75:20:2) Solvents I and II are from D. Kritchevsky, D. S. Martak, and G. H. Rothblat *Anal. Biochem.* **5**, 388 (1963); solvent III from A. F. Hofmann, *J. Lipid Res.* **3** 127 (1962); and Solvents IV–VI from P. Eneroth, *J. Lipid Res.* **4**, 11 (1963).

C. *Steroid Hormones*

The estrogens are 3-hydroxysteroids with an aromatic A ring. Their ring electrons and keto and hydroxy groups interact strongly with silica gel, and moderately polar solvents are employed as the mobile phase (Table 3.7).

The corticosteroids are α,β-unsaturated ketones. Their low concentrations in biological materials, their structural similarity, and the large number of biosynthetic intermediates and breakdown products make separations extremely difficult. The relationship between functional groups and steroid mobilities is illustrated in Table 3.8. Another widely used separatory technique for steroid hormones is gas-liquid chromatography.

Table 3.7. TLC of Estrogens on Silica Gel

Estrogen	$R_f \times 100$ in Solvent[a]		
	I	II	III
Estrone	72	72	59
17 β-Estradiol	61	64	41
6-Ketoestrone	60	60	33
16 α-Hydroxyestrone	57	60	32
16–Epiestriol	43	40	18
Estriol	29	30	6

[a] I = ethyl acetate–cyclohexane–ethanol (9:9:2); II = ethyl acetate–*n*-hexane saturated with water–ethanol (16:3:1); III = ethyl acetate–cyclohexane (1:1). From B. P. Lisboa and E. Diczfalusy, *Acta Endocrinol.* **40**, 60 (1962).

Table 3.8. TLC of Steroids on Silica Gel

Steroid	Number of Functional Groups			$R_f \times 100$ in Solvent[a]					
	—OH	=O	C=C	I	II	III	IV	V	VI
Cholesterol	1	0	1	38	67	52	75	73	85
Estrone	1	1	3	33	63	35	83	65	83
Estradiol	2	0	3	28	47	3	64	56	76
Testosterone	1	1	1	26	48	26	53	66	78
Androstenedione	0	2	0	37	65	57	68	70	85
Pregnenolone	1	1	1	26	58	38	62	66	83
Cortisone	2	3	1	0	20	0	14	23	71
Cortisol	3	2	1	0	6	0	12	8	70

[a] Solvents I and II are chloroform–acetone either, 19:1 (I) or 4:1 (II). Solvents III and IV are benzene–ether, either 2:3 (III) or 1:9 (IV); V is cyclohexane–chloroform–glacial acetic acid (7:2:1); VI is benzene–acetone (1:1).

D. Sterols

The sterols are low in polarity, and the methods described for fatty acids are also applicable to this class of compounds.

E. *References*

I. E. Bush, *Biochem. J.* **50,** 370 (1952).

E. Heftmann and H. H. Wotiz, in *Chromatography*, E. Heftmann, Ed., Reinhold, New York, 1967, p. 539.

A. F. Hofmann, in *New Biochemical Separations*, A. T. James and L. J. Morris, Eds., Van Nostrand, New York, 1964, p. 261.

R. Neher, *Steroid Chromatography*, Elsevier, Amsterdam, 1964.

R. Tschesche, G. Wulff, and K. H. Richert, in *New Biochemical Separations*, A. T. James and L. J. Morris, Eds., Van Nostrand, New York, 1964, p. 197.

A. Zaffaroni, R. B. Burton, and E. H. Keutmann, *Science* **111,** 6 (1950).

5. Nucleosides and Nucleotides

Purine and pyrimidine ribo- or deoxyribonucleosides and ribo- or deoxyribonucleotides are most often encountered as products of chemical or enzymic hydrolysis of nucleic acids. Various mono-, di- and triphosphorylated nucleosides also occur as normal constituents of cells and body fluids. This class of compounds is highly charged and hence very water soluble. Electrophoresis can therefore be used as a useful adjunct to chromatography. Another very successful method is ion-exchange chromatography on columns.

A. *Adsorbents and Solvents*

Due to their high water solubility, solutions of nucleosides and nucleotides isolated from tissues frequently contain large quantities of inorganic salts which may adversely affect chromatography. The high capacity of thick filter paper for aqueous samples and its relative insensitivity to high salt concentrations therefore may make PC the method of choice in many cases. Solvents useful for PC include: 2-propanol–15 *N* NH_4OH–water (7:1:2), isobutyric acid–15 *N* NH_4OH–water (66:1:33), saturated aqueous $(NH_4)_2SO_4$ solution–isopropanol–1 *M* ammonium acetate (40:1:9).

If salts are not present in high concentrations, excellent separations are possible by TLC on cellulose or anion-exchange layers. For cellulose, *n*-butanol–acetone–glacial acetic acid–15 *N* NH_4OH–water (7:5:3:3:2) or *t*-amyl alcohol–formic acid–water (3:2:1) are suitable solvents.

Although cation exchangers give better resolution than cellulose, the preparation, storage, and handling of diethylaminoethyl (DEAE) and polyethyleneimine (PEI) cellulose layers present some difficulties. On DEAE-cellulose, highly negatively charged nucleotides, such as adenosine triphosphate are strongly retarded and have low R_f values. Lithium or sodium chloride solutions, ranging from 0.1 to 2 *M* or 0.02 to 0.04 *N* HCl are used to resolve nucleotides on PEI- or DEAE-cellulose layers.

B. Detection

These aromatic compounds quench strongly when illuminated with 260-nm UV light and appear as dark areas. Use of cellulose which contains a phosphor improves the sensitivity of detection substantially.

C. References

PC and cellulose TLC.

R. Markham and J. D. Smith, *Biochem. J.* **49,** 401 (1951).
A. M. Michelson, *J. Chem. Soc.* **1959,** 1371.
K. Randerath and H. Struck, *J. Chromatogr.* **6,** 365 (1961).
K. Randerath, *Biochem. Biophys. Res. Commun.* **6,** 452 (1961/1962).
G. W. Rushizky and H. A. Sober, *J. Biol. Chem.* **237,** 2883 (1962).
G. Pataki, *Adv. Chromatogr.* **7,** 47 (1968).

Ion exchange TLC.

P. Grippo, M. Iaccarino, M. Rossi, and E. Scarano, *Biochim. Biophys. Acta* **95,** 1 (1965).
E. Randerath and K. Randerath, *J. Chromatogr.* **16,** 111 and 126 (1964); *Anal. Biochem.* **12,** 83 (1965).

6. Alkaloids

Alkaloids comprise a complex and heterogeneous class of compounds that are distinguished by their basicity. Most alkaloids possess tertiary or quaternary nitrogen atoms, a significant feature for their separation by PC or TLC. The polar character of many alkaloids permits the use of partition PC for their separation.

A. Adsorbents and Solvents

For the resolution of tobacco alkaloids as their salts, PC on paper buffered with acetate at pH 5.6 is used with *t*-amyl alcohol or similar solvents as the mobile phase. Partition PC with formamide as the stationary phase and benzene or chloroform as the mobile phase may be used to chromatograph ergot alkaloids.

Table 3.9 lists adsorbent–solvent combinations useful for the separation of alkaloid classes by TLC. Reviews on the chromatography of alkaloids have been published by Waldi and by Macek.

Table 3.9. TLC of Alkaloids

Alkaloid Class	Adsorbent	Solvent
Belladonna	Silica gel	Dimethylformamide–diethylamine–ethanol–ethyl acetate (1:1:6:12)
Ergot	Silica gel	$CHCl_3$–ethanol (5:1) or (10:1); C_6H_6–heptane–$CHCl_3$ (6:5:3)
Morphine	Silica gel	C_6H_6–methanol (4:1)
Opium	Silica gel	$CHCl_3$–methanol (9:1); C_6H_6–ethanol (4:1); xylene–methylethyl ketone–methanol–diethylamine (20:20:3:1)
Purine	Silica gel	$CHCl_3$–96% ethanol (5:1)
Rauwolfia	Cellulose impregnated with formamide	Heptane–methyl ethyl ketone (1:1) (NH_3 atmosphere); heptane–methyl ethyl ketone (4:1)
	Silica gel	$CHCl_3$–diethylamine (9:1), $(C_2H_5)_2NH$, CH_3OH, C_2H_5OH, etc.

B. References

D. Waldi, K. Schnackerz, and F. Munter, *J. Chromatogr.* **6,** 61 (1961).

D. Waldi, in *New Biochemical Separations,* A. T. James and L. J. Morris, Eds., Van Nostrand, London, 1964, p. 157.

K. Macek, in *Chromatography,* E. Heftmann, Ed., Reinhold, New York, 1967, p. 606.

7. Pesticides

Public interest in the biosphere has focused attention on contamination of living organisms by persistent insecticides. TLC provides a rapid and sensitive technique for separating and identifying pesticide residues in foods and animal tissues. For the analysis of halogenated pesticides, GLC with an electron-capture detector is probably the best technique.

Large amounts of lipid material present in biological extracts may seriously affect chromatography and necessitate preliminary purification. Sample preparation generally entails homogenization followed by hexane extraction from acetonitrile or dimethylsulfoxide. This purification process may be followed by column chromatography on Florisil or carbon–cellulose columns. Occasionally, samples are first oxidized or saponified, and the insecticide breakdown products are then chromatographed and identified.

A. Halogenated Insecticides

Halogenated insecticides have been successfully chromatographed on both silica gel and alumina. Table 3.10 lists solvents and R_f values for several representative systems. Mendoza et al. have employed double development with pentane followed by pentane–acetone (19:1) to separate chlorinated insecticides. High-speed liquid column chromatography has also been successfully used for the separation of these insecticides (Waters et al.).

Table 3.10. TLC of Chlorinated Insecticides

	$R_f \times 100$ in Indicated Solvent on Silica Gel[a]					$R_f \times 100$ in Indicated Solvent on Alumina[a]	
Insecticide	I	II	III	IV	V	I	II
Aldrin	70	67	69	64	67	95	82
DDE *p,p′*	65	65	62	66	62	95	78
DDT *o,p*	50	59	58	56	57	89	73
DDT *p,p′*	42	57	54	55	54	89	69
Dieldrin	12	65	48	34	49	37	52
Endrin	13	49	52	37	54	51	61
Heptachlor	58	49	62	63	65	95	78
Methoxychlor		65	36	23	32		
TDE *p,p′*	25	52	46	45		71	57

[a] Solvents are: I, hexane; II, petroleum ether (b.r. 40–60°C)–liquid paraffin–dioxane (94:5:1); III, cyclohexane–liquid paraffin–dioxane (7:2:1); IV, hexane–ether (9:1); V, hexane–acetone (9:1). Solvents I–III are from D. C. Abbot, H. Egan, and J. Thompson, *J. Chromatogr.* **16**, 481 (1964); and solvents IV and V from K. C. Walker and M. Beroza, *J. Ass. Offic. Agr. Chem.* **46**, 250 (1963).

B. Organophosphorus Insecticides

These generally lipophilic compounds fall into four classes: phosphorothioates, phosphorothionates, phosphorothiolates, and phosphates. They are occasionally esterified prior to chromatography. Silica gel TLC and solvents of relatively low polarity can be used for all classes except for the phosphates. Table 3.11 lists the solvents and R_f values for some widely used organophosphorus insecticides. Detection sensitivity is in the range of 0.01–3 μg.

Table 3.11. TLC of Organophosphorous Insecticides on Silica Gel

Insecticide	$R_f \times 100$ in Solvent[a]							
	I	II	III	IV	V	VI	VII	VIII
Phosphorothioates:								
Malathion	63	73	74	44	35	54	27	100
Ethion	80	87	80	74	57	88	72	100
Phorate	75	86	78	72	60	100	93	100
Dimethoate	11	13	28	12	7	11	2	56
Phosphorothionates:								
Chlorothion	0	0	75	23,61	15			
Bayer 25141	18	20	39	18	7	16	0	58
Methylparathion	70	81	75	61	34	60	45	100
Parathion	73	84	76	67	44	83	68	100
Diazinon	0	0	0,54	28,35	28,35	93	22	100
Demeton	33,38	41,86	54,81	35,60	31,61	95	57	100
Phosphates:								
Phosphamidon	9	9	24	7	4	9	0	40

[a] Solvents are: I, chloroform–ether (9:1); II, chloroform–ethyl acetate (9:1); III, benzene–methanol (9:1); IV, benzene–glacial acetic acid (9:1); V, hexane–acetone (4:1); VI, cyclohexane–acetone–chloroform (14:4:1); VII, 2,2,4-trimethylpentane–acetone–chloroform (14:5:1); VIII, acetone–diisopropyl ether–cyclohexane (2:2:1). Some compounds gave two spots. Solvents I–V are from K. C. Walker and M. Beroza, *J. Ass. Offic. Agr. Chem.* **46**, 250 (1963); and solvents VI–VIII from M. E. Getz and H. G. Wheeler, *J. Ass. Offic. Agr. Chem.* **51**, 1101 (1968).

C. *References*

D. C. Abbott, H. Egan, and J. Thomson, *J. Chromatogr.* **16,** 481 (1964).

D. C. Abbott and J. Thomson, *Residue Rev.* **11,** 1 (1965).

M. Beroza, K. R. Hill, and K. H. Norris, *Anal. Chem.* **40,** 1608 (1968).

M. E. Getz and H. G. Wheeler, *J. Ass. Offic. Agr. Chem.* **51,** 1101 (1968).

C. E. Mendoza, P. J. Wales, H. A. McLeod, and W. P. McKinley, *J. Ass. Offic. Agr. Chem.* **51,** 1095 (1968).

E. Stahl, *Arch. Pharm.* **293,** 531 (1960).

K. C. Walker and M. Beroza, *J. Ass. Offic. Agr. Chem.* **46,** 250 (1963).

J. L. Waters, J. N. Little, and D. F. Horgan, *J. Chromatogr. Sci.* **7,** 293 (1969).

J. R. Wessel, *J. Ass. Offic. Agr. Chem.* **50,** 430 (1967).

8. Vitamins

The vitamins include both polar (B vitamins, vitamin C) and nonpolar representatives (A, D, E, K, Q). The nonpolar or fat-soluble vitamins are best chromatographed in low-polarity solvents on silica gel or alumina, while the polar, water–soluble group can be analyzed in highly polar solvents on silica gel.

A. Fat-Soluble Vitamins

Vitamin A derivatives can be separated in single-component solvents such as benzene or chloroform. Vitamin D derivatives may be chromatographed in hexane, in acetone–hexane–methanol (15:135:13), hexane–ethyl acetate (9:1) or hexane–acetone (9:1). Reversed-phase chromatography on paraffin-impregnated silica gel with acetone–water (4:1) as the mobile phase separates vitamins D_2 and D_3. Tocopherols may be separated by chloroform, benzene, or benzene–methanol (49:1) on silica gel. Stowe used petroleum ether (b.p. 60°C)–diisopropyl ether–acetone–diethyl ether– glacial acetic acid (85:12:4:1:1) to separate β- and γ- in the presence of α- and δ-tocopherols. Vitamin K derivatives may be chromatographed on silica gel in cyclohexane–diethyl ether (4:1).

B. Water-Soluble Vitamins

The water-soluble vitamins are chromatographed with polar solvents. Some representative separations are shown in Table 3.12. Vitamin B_1 derivatives have been separated on silica with pyridine–glacial acetic acid–water (10:1:4) at pH 5.8 and B_6 derivatives can be resolved on silica in acetone–dioxane–25% NH_4OH (9:9:5).

C. Precautions

Lipid-soluble vitamins are extremely sensitive to atmospheric oxidation and decompose rapidly on dry, activated silica gel layers. Ultraviolet light should be excluded. Vitamin B_{12} derivatives are decomposed even by ordinary light, and chromatography must be carried out in dimly lit rooms.

D. Visualization

A variety of specific color reagents are used to quantitate individual vitamins. Microbiological assays are also used, as are UV quenching and H_2SO_4 charring. The sensitivity of UV detection ranges from 0.01 μg for riboflavin to 3 μg for nicotinic acid. Biotin does not absorb UV light.

Table 3.12. TLC of B Vitamins on Silica Gel

Vitamin	$R_f \times 100$ in Solvents[a] I	II
Folic acid	0	0
B_1 (thiamine)	5	0
B_{12} (cyanocobalamin)	22	0
B_2 (riboflavin)	42	35
B_6 (pyridoxine)	51	15
Nicotinamide	50	65
Pantothenic acid	60	89
Nicotinic acid	76	75
Biotin	70	80
C (ascorbic acid)	97	30

[a] I, water; II, glacial acetic acid–acetone–methanol–benzene (1:1:4:14). From H. Gänshirt and A. Malzacher, *Naturwissens.* **47**, 279 (1960). Vitamin C is included for comparison.

E. References

R. L. Blakley, *The Biochemistry of Folic Acid and Related Pteridines*, Wiley-Interscience, New York, 1969, p. 97.

G. Katsui, in *Chromatography*, E. Heftmann, Ed., Reinhold, New York, 1967, p. 699.

H. K. Mangold and D. C. Malins, *J. Amer. Oil Chem. Soc.* **37,** 383 (1960).

C. K. Parekh and R. H. Wasserman, *J. Chromatogr.* **17,** 261 (1965).

H. D. Stowe, *Arch. Biochem. Biophys.* **103,** 42 (1963).

9. Inorganic Ions

In addition to their well-known use for organic compounds, PC and TLC provide a useful adjunct to standard methods of qualitative analysis for inorganic anions and cations. Early work on PC separation of inorganic compounds has been reviewed by Lederer.

A. Anions

Anions may be detected by a solution of one part 10% aqueous $AgNO_3$ and 5 parts 0.2% sodium fluorescinate in absolute alcohol.

1. Halides. Halides migrate on silica gel with an R_f directly proportional to their molecular weight. Acetone–*n*-butanol–15 *N* NH_4OH–water (13:4:2:1) as solvent readily separates F^- Cl^- Br^- and I^-. Their R_f values increase in the order listed. An ammoniacal solution of 0.1% bromcresol purple in alcohol may be used for visualization.

2. Phosphate. Phosphate ions are separable in binder-free silica gel by methanol–15 N NH_4OH–10% trichloroacetic acid–water (10:3:1:6). The R_f values increase in the order: pyrophosphate < orthophosphate < phosphite < hypophosphite.

3. Other. Useful general adsorbent–solvent combinations for other anions are: silica gel with acetone–water (10:1), methanol–*n*-butanol–water (3:1:1), or *n*-butanol saturated with 2 N HNO_3.

B. Cations

A good general cation detection reagent is 0.2% tetrahydroxyquinone in ethanol, followed by exposure to ammonia vapor. Even binder-free silica gel contains many anions and cations, especially Fe^{3+}. These may be removed prior to chromatography by development in methanol–12 N HCl (9:1), which moves most ions to the solvent front.

1. Alkali Metals. These separate well on silica gel in ethanol–glacial acetic acid (19:1), giving R_f values of 0.98 (Li^+), 0.89 (Na^+), 0.60 (K^+), and 0.53 (Rb^+). Violuric acid or tetracyanoquinodimethanide is used for visualization.

2. Alkali Earth Metals. Chromatography by PC or cellulose TLC in methanol–12 N HCl–water (8:1:1) gives R_f values of : 0.0 (Ra^{2+}), 0.19 (Ba^{2+}), 0.42 (Sr^{2+}), 0.51 (Ca^{2+}), 0.70 (Mg^{2+}), and 0.85 (Be^{2+}). For visualization 1% 8-hydroxyquinoline in methanol is sprayed on the plate followed by exposure to ammonia vapors and inspection under UV light. Other useful solvents include 95% ethanol–glacial acetic acid (19:1), and *t*-butanol–2 N HCl (19:1).

3. Sulfide Group. Moller and Zeller (1964) have described the PC of cations in *n*-butanol–5 N HCl–acetone–acetyl acetone (125:25:150:1). R_f values are: 0.2 (Ni^{2+}), 0.12 (Co^{2+}), 0.21 (Mo^{2+}), 0.28 (Pb^{2+}), 0.48 (Cu^{2+}), 0.70 (As^{5+}), 0.87 (Bi^{2+} and Sb^{5+}), 0.94 (Hg^{2+}), 0.95 (Zn^{2+}), and 0.97 (Sn^{2+}). The sulfide group may also be resolved with PC or cellulose TLC by *n*-butanol saturated with 3 N HCl. Mobilities, in increasing order, are: Ag^+, Cu^{2+}, Pb^{2+}, Sb^{5+}, Bi^{3+}, As^{5+}, Cd^{2+}, Sn^{4+}, and Hg^{2+}. The ammonium sulfide cations may be resolved by PC in glacial acetic acid pyridine–12 N HCl (4:3:1). Mobilities, in increasing order are: Cr^{3+}, Al^{3+}, Ni^{2+}, Mn^{2+}, Co^{2+}, Zn^{2+}, and Fe^{3+}.

C. *References*

L. F. Druding, *Anal. Chem.* **35,** 1582 (1963).
M. Lederer, *Chromatogr. Rev.* **3,** 134 (1961).
H. G. Möller and N. Zeller, *J. Chromatogr.* **14,** 560 (1964).
G. Nickless and F. H. Pollard, in *Chromatography*, E. Heftmann, Ed., Reinhold, New York, 1967, p. 735.
E. Pfeil, A. Friedrich, and Th. Wachsmann, *Z. Anal. Chem.* **158,** 429 (1957).
W. Schneider and B. Patel, *Arch. Pharm.* **297,** 97 (1964).

APPENDIX 3.1

Sources of TLC Equipment and Supplies

Applied Science Labs., Inc., P.O. Box 440, State College, Pa. 16801.
Brinkman Instruments, Inc., Cantiague Rd., Westbury, N. Y. 11590.
Camag, Inc., 2855 South 163rd St., New Berlin, Wis. 53151.
Kensington Scientific Corp., 1399 64th St., Emeryville, Calif. 94608.
Kontes Glass Company, Spruce St., Vineland, N. J. 08360.
Quickfit, 7 Just Rd., Fairfield, N. J. 07006.
Rodder Instrument, 775 Sunshine Dr., Los Altos, Calif. 94022.
Shandon Scientific Company, Inc., 515 Broad St., Sewickley, Pa. 15143.
Supelco, Inc., Supelco Park, Bellefonte, Pa. 16823.
Ultra-Violet Products, Inc., 5114 Walnut Grove Ave., San Gabriel, Calif. 91778.

APPENDIX 3.2

Manufacturers of Precoated Plates for TLC

Analtech, Inc., Blue Hen Industrial Park, Newark, Del. 19711.
Applied Science Labs, Inc., P.O. Box 440, State College, Pa. 16801.
J. T. Baker Chemical Company, 222 Red School Lane, Phillipsburg, N. J. 08865.
Brinkman Instruments, Inc., Cantiague Rd., Westbury, N. Y. 11590.
Corning Glass Works, Laboratory Products Dept., Houghton Pk, Corning, N. Y. 14830.
Eastman Kodak Company, 343 State St., Rochester, N. Y. 14650.
Gelman Instrument Company, 600 S. Wagner Rd., Ann Arbor, Mich. 48106.
Kensington Scientific Corp., 1399 64th St., Emeryville, Calif. 94608.
Mallinckrodt Chemical Works, 2nd and Mallinckrodt Sts., St. Louis, Mo. 63160.
Schwarz–Mann, Mountain View Ave., Orangeburg, N. Y. 10962.
E. Merck, A. G., Darmstadt, Germany.
Quantum Industries, 341 Kaplan Dr., Fairfield, N. J. 07006.
Schleicher and Schuell, Inc., 543 Washington St., Keene, N. H. 03431.

APPENDIX 3.3

Manufacturers of Densitometers

American Instrument Company, Inc., 8030 Georgia Ave., Silver Spring, Md. 20910.
Baird-Atomic, Inc., 125 Middlesex Tnpk., Bedford, Mass. 01730.

Farrand Optical Company, Inc., 117 Wall St., Valhalla, N. Y. 10595.
Nester/Faust Mfg. Corp., 2401 Ogletown Rd., P.O. Box 565, Newark, Del. 19711.
Photovolt Corp., 1115 Broadway, New York, N. Y. 10010.
Schoeffel Instrument Corp., 24 Booker St., Westwood, N. J. 07675.
G. K. Turner Associates, 2524 Pulgas Ave., Palo Alto, Calif. 94303.
Varian Aerograph, 2700 Mitchell Dr., Walnut Creek, Calif. 94598.
Carl Zeiss, Inc., 444 Fifth Ave., New York, N. Y. 10018.

APPENDIX 3.4

Manufacturers of Radiochromatogram Scanners

Baird-Atomic, Inc., 125 Middlesex Turnpike, Bedford, Mass. 01730.
Camag, Inc., 2855 South 163rd St., New Berlin, Wis. 53151.
Forro Scientific Company, 833 Lincoln St., Evanston, Ill. 60201.
Infotronics Corp., 1062 Linda Vista Way, Mt. View, Calif. 94040.
Kahl Scientific Instrument Corp., P. O. Box 1166, El Cajon, Calif. 92022
Nuclear-Chicago, 2000 Nuclear Drive, Des Plaines, Ill. 60018.
Nuclear Equipment Chemical Corp., 165 Marine St., Farmingdale, N. Y. 11735.
Packard Instrument Company, Inc., 2200 Warrenville Rd., Downers Grove, Ill. 60515.
Tracerlab Division of ICN, 630 20th St., Oakland, Calif. 94612.
Varian Aerograph, 2700 Mitchell Drive, Walnut Creek, Calif. 94598.

REFERENCES

1. R. Consden, A. H. Gordon, and A. J. P. Martin, *Biochem. J.* **38,** 224 (1944).
2. K. V. Giri, A. N. Radhakrishnan, and C. S. Vaidyanathan, *Anal. Chem.* **24,** 1677 (1952).
3. D. Jänchen and G. Pataki, *J. Chromatogr.* **33,** 391 (1968).
4. U. Lüthi and P. G. Waser, *Nature* **205,** 1190 (1965).
5. A. J. P. Martin and R. L. M. Synge, *Biochem. J.* **35,** 91 (1941).
6. D. J. Shapiro, R. L. Imblum, and V. W. Rodwell, *Anal. Biochem.* **31,** 383 (1969).
7. E. Stahl, *Pharmazie* **11,** 633 (1956).
8. M. Tswett, *Ber. Bot. Ges.* **24,** 316 (1906).
9. R. L. Whistler, M. Lamchen, and R. M. Rowell, *J. Chem. Educ.* **43,** 28 (1966).
10. R. A. de Zeeuw, *Anal. Chem.* **40,** 915 (1968).
11. R. Stevenson, *Amer. Lab.*, May 1970, p. 49.

RECOMMENDED READING

K. Macek, in *Chromatography*, E. Heftmann, Ed., Reinhold, New York, 1967, p. 139.

E. Stahl, *Thin Layer Chromatography—a Laboratory Handbook*, 2nd ed., Springer Verlag, New York, 1969.

MANFRED J. R. CANTOW
IBM Corporation
Systems Development Division
San Jose, California

JULIAN F. JOHNSON
Department of Chemistry and
Institute of Materials Science
University of Connecticut, Storrs

IV. Gel Permeation Chromatography

I. INTRODUCTION

Gel permeation chromatography (GPC) is a special form of liquid chromatography. Elution is carried out on a rigid, porous support, and separation is based on molecular size and shape. The smaller the hydrodynamic volume, the deeper a particle penetrates into the support. Consequently the largest particles are eluted first. The method was originally used to separate biopolymers in aqueous solution on cross-linked dextrane gels. The introduction of solvent-resistant support materials made it possible to fractionate synthetic polymers. Suitable products are now available to carry out separations of components in any molecular weight range. However most of the work is presently done on systems in the molecular weight range of a few thousand to a million. The technique is applicable to a large variety of problems, for example, group separation of high-molecular-weight natural products, prefractionation of wide-range mixtures prior to separation by other liquid chromatographic methods, indirect molecular weight determination, and analytical and preparative fractionation of polymers. These latter applications to polymers are presently furthest developed with respect to theory and instrumentation. Instead of attempting to cover the whole field we will deal primarily with this aspect of gel permeation chromatography to describe and elucidate the many and varied applications and problems associated with this technique. This approach allows us to treat in detail some of the more interesting topics of this subject. Other applications of gel permeation chromatography are listed in Table 4.4.

Because GPC is a special form of liquid chromatography, many of the general theoretical and operational parameters are also treated in Chapter 2. The components in the instrumentation involved are also quite comparable to those used in high-resolution liquid chromatography. The interested reader is therefore referred to Chapter 2 for more information on these aspects of the technique.

1. Outline of Method

Polymers as a class of materials are of ever increasing importance. One of the characteristics of polymers is that they usually have a distribution of molecular weights. This distribution may vary over wide limits and most of the physical properties of the polymer, including the "average" molecular weight, will depend on the shape of the molecular weight distribution. Therefore the determination of the molecular weight distribution is of major importance in characterizing a polymer.

Fundamentally, molecular weight distributions are measured by separating the polymer into a number of fractions of different average molecular weights. Enough fractions must be obtained so that each will have a reasonably narrow molecular weight range. Once the amount and molecular weight of each fraction are determined, a molecular weight distribution curve for the polymer can be constructed. Numerous experimental techniques for the fractionation of polymers have been devised. These are the subject of a recent book (121).

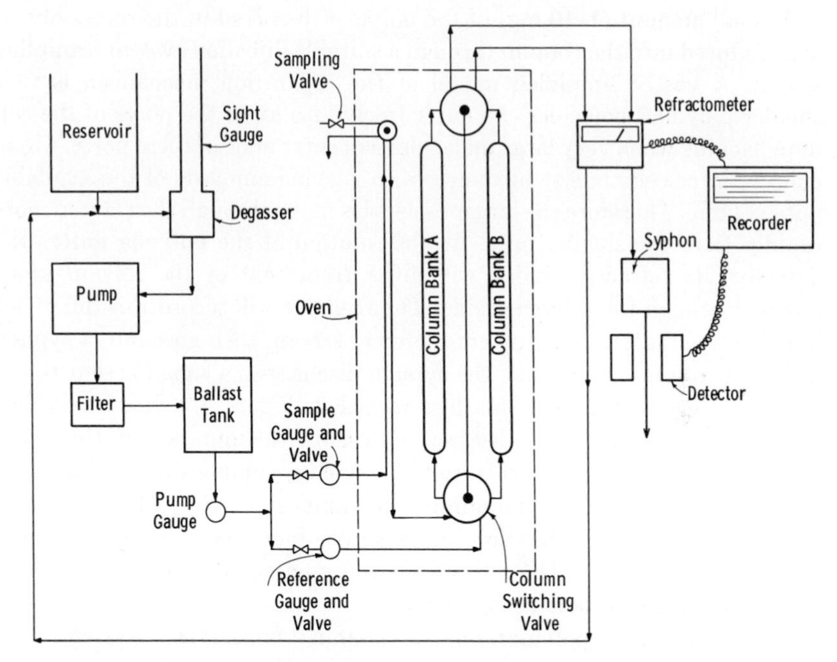

Figure 4.1. Schematic diagram of gel permeation chromatograph (11).

Gel permeation chromatography is a recently developed method of fractionation. It has the major advantages of providing good fractionation in relatively short times; the data on the fractions can be collected automatically. Because of these advantages it is now the most widely used method for the separation and isolation of high-molecular-weight fractions and for determining polymer molecular weight distributions.

Figure 4.1 is a schematic diagram of a gel permeation chromatograph (11). In the design shown, a differential refractometer is used as the detector. There are two banks of columns, one of which serves as a reference for the solvent only; the polymer is fractionated on the other column bank. The usual column diameter is 0.5–2.5 cm with lengths varying from 1 to 10 m. The column packing is a porous material containing pores of various diameters. Some of the materials used are cross-linked polystyrene gels and porous glass. The columns are mounted in a thermostated oven. Solvent is continually circulated through the columns by a constant volume pump; a degasser is used to remove dissolved gases from this mobile phase prior to introduction into the columns. Typical solvent flow rates are 0.5–2.5 ml/min.

A small amount (1–10 mg) of the polymer dissolved in the same solvent is introduced into the column through a suitable injection system (sampling valve). A vastly simplified model of the separation mechanism is that smaller polymer molecules can enter freely into all of the pores of the column packing while very large molecules can enter none of these pores. Molecules of intermediate size have access to varying amounts of the available pore volume. Therefore the larger molecules move through the column more rapidly than the smaller ones. As the solution of the polymer enters the detector, its refractive index will differ from that of the solvent alone passing through the reference side. The recorder will record this difference in refractive indices. The solvent–polymer stream then goes into a syphon of known volume. Each time the syphon discharges, a signal is sent to the recorder where a reference mark is recorded. Figure 4.2 shows a typical chromatogram (14). The abscissa represents the number of times the syphon has discharged; in this case the syphon volume was 5 ml and the elution volume is simply the number of counts times five. The ordinate is proportional to the difference in refractive index between solvent and solvent plus polymer. The polymer is represented by curve A, the low-molecular-weight impurities by curve B.

For a specific polymer–solvent–column–temperature system, the molecular size of the emerging components is a function of the amount of solvent eluted at that point. Thus if fractions with narrow molecular

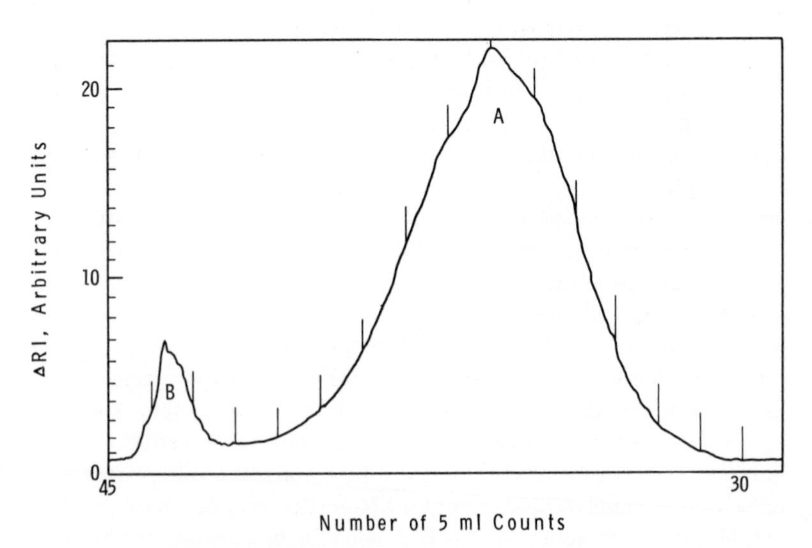

Figure 4.2. Typical chromatogram (14).

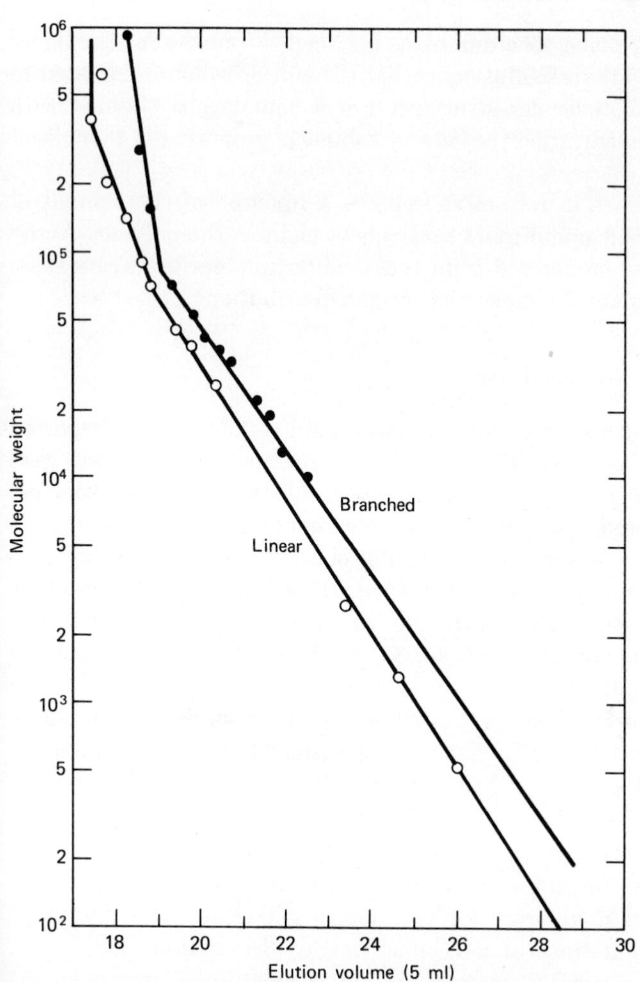

Figure 4.3. Calibration curve for polyethylene (133).

weight distributions are chromatographed and the average molecular weights of the fractions are determined by some independent method, a calibration curve of molecular weight versus elution volume can be established. Figure 4.3 shows a typical plot for polyethylene (133). It was found that the logarithm of molecular weight versus elution volume gives a straight line over an appreciable molecular weight range. The abrupt change in slope at the high-molecular-weight range indicates that all molecules above this molecular weight are excluded and therefore no fractionation is taking place. The difference between the curves for the linear and branched polyethylene illustrates that the elution volume is related to molecular size and molecular structure. For a homologous polymer with no difference in configuration the elution volume is proportional to molecular weight.

Since the difference in refractive index is a function of the amount of polymer present, the amount and molecular weight of the polymer at any elution volume can be derived from these calibration curves. From these data one can calculate the molecular weight distributions.

2. History and Nomenclature

Separations by gel permeation chromatography (GPC), were reported at least as early as 1950. The majority of the materials studied were naturally occurring high-molecular-weight substances. The principal column packing used was a dextran gel (trade name, Sephadex) and the solvents were water or aqueous buffer solutions. Excellent separations on a wide variety of materials were achieved. This work with hydrophilic materials will not be considered in this chapter, because excellent books (41, 121) and review articles (37, 42, 83, 84, 116, 138, 148) have treated the subject in detail.

Synthetic polymers which were soluble in organic solvents were not so readily fractionated by GPC. Early work included that of Boldingh on fatty acids (18), Vaughan on polystyrene (151), Brewer on polyisobutene (27), and others (26, 152).

In 1962, Moore reported on two major improvements in GPC: a method for producing cross-linked gels with a wide range of pore sizes, and the use of a differential refractometer for continuous detection. This permitted automated determinations of molecular weight distributions (104, 105). Shortly thereafter, a GPC instrument embodying these improvements became commercially available (Waters Associates, Inc., Framingham, Mass.) and many workers soon reported on extensive applications of the technique.

The terms used to designate this separation method are far from uniform. The following designations and possibly others have been used: gel permeation chromatography (105); gel chromatography (41); gel filtration (123); exclusion chromatography (118); molecular sieve chromatography (5, 122); restricted diffusion chromatography (145); and molecular sieve filtration (69). The authors have chosen to use gel permeation chromatography in this article simply because it appears currently to be the most widely used designation.

II. CHOICE OF INSTRUMENTATION

1. Detectors

Detectors used for continuous monitoring of column effluent include differential refractometers, IR and UV spectrometers, flame ionization detectors, conductivity monitors, and recording colorimeters. Some of these detectors are also discussed in Chapter 2, Section III.5.

The differential refractometer is probably the most widely used detector in gel permeation chromatography. It is sensitive with a detection limit of about 1×10^{-6} refractive index units. The high sensitivity is necessary because under normal operating conditions the polymer elutes from the column as a solution of only a few milligrams in 20–100 ml of solvent. Its use as a quantitative detector depends on the proportionality between the change in concentration, Δc, and the change in refractive index, Δn,

$$\Delta c \approx \Delta n \tag{4.1}$$

which is generally valid for the low concentration ranges observed in GPC. It is important to minimize the volume of the detector cell to reduce dispersion and to increase sensitivity. By reducing the cell volume from 0.070 ml to 0.01 ml, Billmeyer and Kelley (13) demonstrated a threefold increase in sensitivity and a 50% reduction in zone broadening.

Determination of polymer concentrations using a differential refractometer assumes that a signal will be observed which is proportional to the amount of polymer in the effluent, regardless of the molecular weight of the eluting polymer. The assumption that the refractive index of a polymer is independent of molecular weight is valid for high-molecular-weight compounds, but there may be small systematic variations at the other end of the spectrum (10, 86). The magnitude of this variation and the molecular weight limit below which this variation becomes significant depend upon the polymer–solvent system and the wavelength of light employed in the refractometer. As a rough approximation, the response of the detector as a function of molecular weight should be experimentally evaluated for molecular weights below 3000–5000.

Infrared spectrometers have in certain cases advantages over differential refractometers as detectors for GPC. At the high temperatures necessary for maintaining polyethylene in solution, Ross and Casto (130) found an IR detector somewhat more sensitive than a differential refractometer.

There appears to be no reported systematic study to explore a further increase in sensitivity by the use of double-beam instruments or to compare quantitatively the sensitivity of the two detectors under a variety of conditions.

Another important use of the IR detector is to monitor not only the effluent concentration but also to determine its composition. For example, in copolymer analysis Rodriguez et al. (147) used the band at 1731 cm^{-1} to determine methylmethacrylate, and the band at 698 cm^{-1} for measuring styrene. He also compared these amounts with an internal standard, in other words, phenyl benzoate. Although methylmethacrylate and styrene were determined in separate chromatographic runs, a single fractionation would have sufficed. The mobile phase can be interrupted at fixed intervals while a complete spectrum is being recorded. It has been shown that little loss in resolution occurs when flow is interrupted for periods ranging up to several hours. A very useful example of this technique is the determination of the identity of low-molecular-weight additives in polymers. A wide variety of compounds are used as antioxidants, inhibitors, and so on, and their analysis is often time-consuming. Since in GPC these compounds frequently emerge as separate components, the use of an IR spectrometer would facilitate their identification.

Based on the same considerations as discussed in Chapter 2, Section II, longer and smaller columns should increase resolution in GPC. However a more sensitive detector would be required for the smaller samples. A number of so-called "belt," "chain," or "wire" detectors have been devised for the purpose. A particularly sensitive design is due to Johnson, Seibert, and Stross (82). Figure 4.4 is a schematic diagram of this detector. Liquid emerging from the column is deposited on the moving belt and the solvent is removed by flash evaporation. The polymer residue enters a higher temperature zone where the polymer is pyrolyzed into low-molecular-weight fragments. These fragments are swept into a sensitive flame ionization detector. Results from the belt detector are shown in Figure 4.5. Amounts as small as 0.1 μg may be detected. This detector, being rather complex, is not as simple to operate as those previously described. Other continuous detectors have been used for specialized systems. For inorganic electrolytes simple conductivity detectors may be used. An example is that reported by Saunders and Pecsok (135). Polymers that can form colored complexes may be measured with a recording colorimeter; the work of Kondo et al. (86) on polyether polyols is an example of this technique.

The detectors described above are all of the continuous type. Fractions from the GPC may also be evaporated and weighed. Batch characterization can also be carried out on the separate fractions by polarimetry (98), UV spectroscopy (154), and other methods of analysis.

2. Column Packings and Column Geometry

GPC columns are usually constructed of stainless steel. For low pressure work, glass columns may be used.

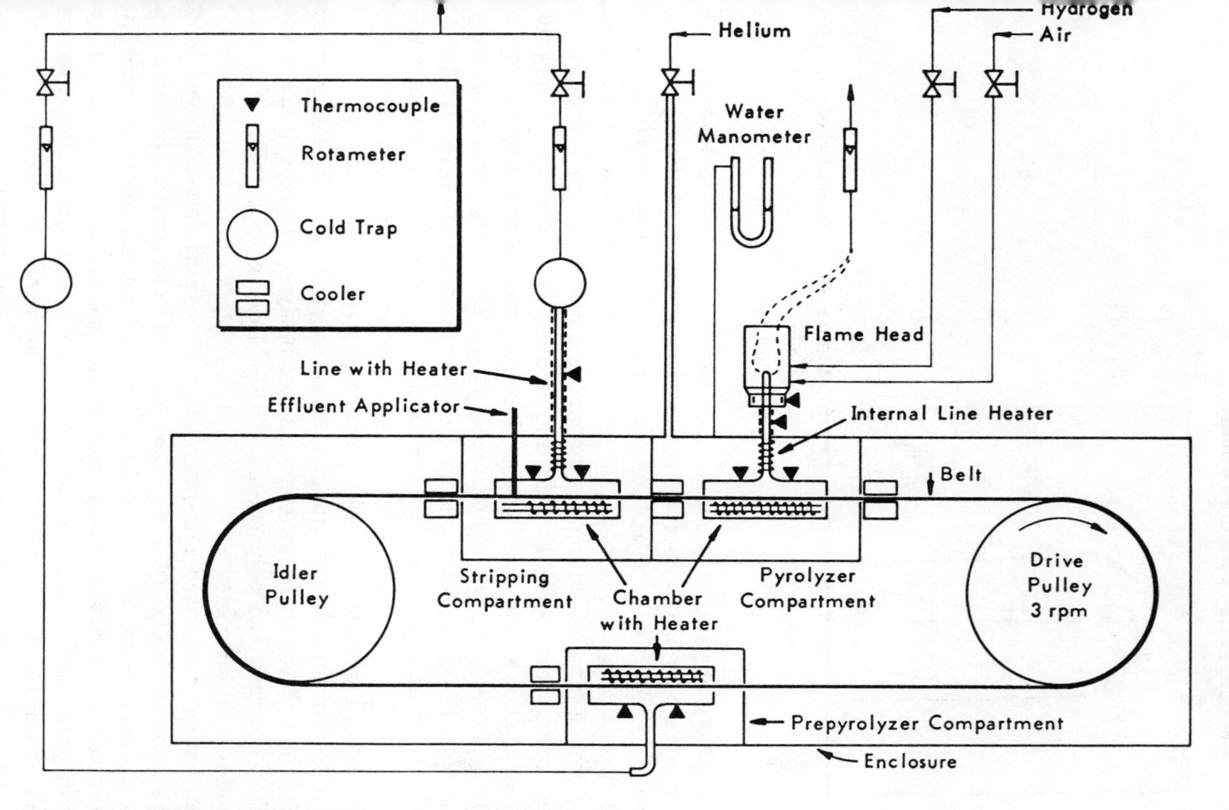

Figure 4.4. Schematic diagram of belt detector (82).

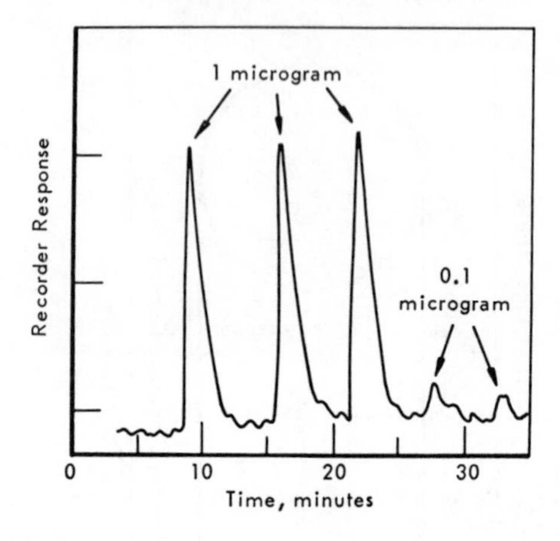

Figure 4.5. Response of belt detector (82).

A variety of packing materials have been used for GPC. Probably the most widely used packing is cross-linked polystyrene. Early applications (152) were not always successful, particularly for lower-molecular-weight polymers. The work of Moore (105) showed that the amount and nature of the diluent present during the polymerization of the gel markedly influenced the pore sizes of the resulting product. This capability to vary the pore size allows these gels to be used for the analysis of a large variety of polymers over a wide molecular weight range. These gels are commercially available. Detailed procedures for their preparation have been given (4).

Other organic gels that have been used are chlorobutyl rubber cross-linked by zinc oxide (25), copolymers of methylmethacrylate and ethylene glycoldimethacrylate (43) and polymethylmethacrylate and polybutylmethacrylate cross-linked with glycoldimethacrylate (70, 72). For use with water or buffered solutions, the most commonly employed gels are cross-linked dextrans and polyacrylamide (51). Ion-exchange resins (88) and cellulose (98) have been used.

Although cross-linked gels, principally polystyrene, have achieved good separations on a variety of polymers, they do have some limitations. The upper temperature limit is about 150°C and column life is appreciably shortened at elevated temperatures. If the gels are allowed to dry, they disintegrate, so they must be kept in a solvent at all times. This is inconvenient and the possibility of loosing a column exists if air is accidentally admitted. The gel columns have a relatively high pressure drop. An important disadvantage is the low mechanical strength. The packing compresses under pressure. After changing from one pressure or temperature to another set of conditions, a return to the initial conditions frequently requires recalibration (24).

To obtain better mechanical stability, a number of workers have successfully used silica gel with a variety of pore and mesh sizes (94, 95, 152). Porous silica beads with uniform spherical particle sizes and an extended range of pore sizes were developed by de Vries (153) and are commercially available. The silica beads have low flow resistance and give good separations.

Another mechanically stable packing is porous glass. Vaughan (152) suggested its use. Haller developed a method of preparing a uniform pore size glass with variable pore size (65) and good results have been obtained using this packing (106). Other porous glasses with a wider pore size distribution have also given successful separations (5, 19, 30, 130). Both the silica gel and porous glass packings may exhibit adsorptive effects. This undesirable feature can be removed by blocking the active sites, for example, by treatment with hexamethyldisilazane. While no extensive quantitative studies are available, the cross-linked gels appear to be somewhat more efficient than columns with other packing material. For columns with rigid particles this lower efficiency can be offset by the use of longer columns. This is quite feasible because of the much lower pressure drop of this packing. For theoretical studies the rigid packings have the advantage that their surface area, pore volume, and pore size distributions may be determined from nitrogen adsorption and desorption isotherms and by well-known methods such as electron microscopy and mercury penetration as a function of pressure. This is not usually possible with the gels, although a technique for studying concentration gradients in gels exists (64).

The effect of particle size and particle size distribution has not been extensively evaluated. For particles with the same average diameter, de Vries et al. found an increase in efficiency with a decrease in the width of distribution (153). However this study also showed an increase in efficiency with an increase in particle size contrary to the results of others (53). Kelley and Billmeyer (85), in a study on the dispersion effects in GPC, used nonporous glass beads and found the values given in Table 4.1 for cyclohexane at a flow rate of 1 ml/min. It appears that the effect of particle diameter may be due to several mechanisms which cannot be easily distinguished.

Variations of column effectiveness as a function of the column geometry have been the subject of few studies. If the column is uniformly packed, the column efficiency should be a linear function of column length. For columns 2–6 m in length, this has been experimentally verified (153). Bombaugh et al. (22) used a 49-m-long column and found a total of 180,000 theoretical plates. This would indicate an almost proportional increase from values for short columns with similar packing. Relatively short (1–1½ m) sections of columns are connected in series with low-volume tubing to

Table 4.1. Nonporous Glass Bead Column

Average Particle Diameter, μ	Plates/ft
385	150
163	475
137	650
115	1400
68	900

obtain the desired length. For gels that may compact under pressure this approach has the advantage of minimizing the pressure drop. The work of Bombaugh et al. (22) indicates that for carefully prepared columns the reduction in efficiency because of the coupling of shorter columns is small. Billmeyer and Kelley (13) found a small loss in efficiency due to the end fittings connecting the columns and Gamble et al. (55) reported that poor connections may result in appreciable peak shape distortions.

Column efficiency should increase as the ratio of column diameter to particle diameter decreases. Some quantitative data support this (53, 58, 153), but the effect of wide variations in this ratio have not been studied.

3. Pumping and Injection Systems

Typical solvent flow rates are 0.5–5 ml/min. corresponding to pressure drops of 250 psi or higher for long columns of polymer gel (22). An adjustable constant volume pump covering these ranges is required. In addition, pressure fluctuations must be minimized because these can cause noise in the detector. Many positive displacement pumps with adjustable stroke are available in this pressure and volume range. Usually a surge tank is connected in series to reduce pressure oscillations.

A small effective design has been described by Ross and Casto (130) and is shown in Figure 4.6. This consists of a stainless steel bellows pressurized on one side with gas to absorb the pressure fluctuations.

A pulseless pump for use in chromatography employs mercury as a movable barrier between the mobile-phase solvent and the driving phase. The latter may be either a compressed gas or an oil pumped in under pressure. The dual chamber construction allows one chamber to be filled while the other is pumping, providing for uninterrupted flow (80). Such pumps give very constant flow rates over a wide range of pressures but they are complex and involve relatively large quantities of mercury.

A simple inexpensive pump that can be quite effective consists of a polyethylene wash bottle in a container that can be pressurized. When

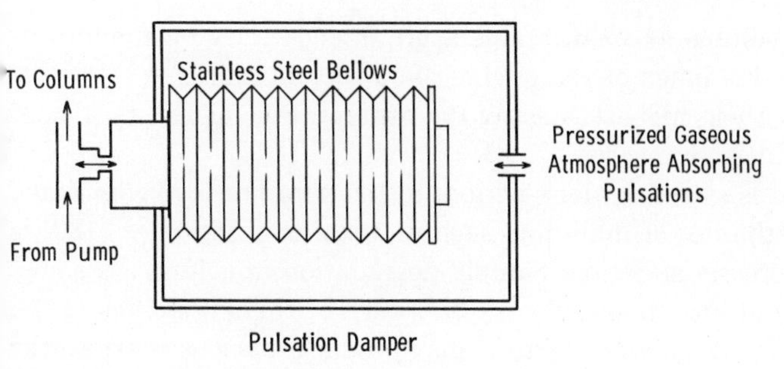

Figure 4.6. Pulsation damper (130).

used in series with an empirically selected short section of capillary tubing, constant flow rates are observed over about two thirds of the volume of the wash bottle. This is usually more than required for a single GPC separation.

A method of extending the effective column length is to operate the pump in recycle. Here the exit from the column returns to the inlet side of the pump (22). In this case, the smallest available pump volume is desirable to reduce the loss in resolution due to mixing in the pump.

Multiport valves and direct-on-column sample injection with a hypodermic syringe are the two principal means of introducing the sample in the system. The multiport valves contain a sample loop which is filled with a solution of the sample and then placed in series with the column. The volume of the loop is usually adjustable. In general, smaller sample volumes produce less zone broadening (35). The multiport valve gives reproducible sample volumes and is easy to incorporate into an automatic sampling system. In the usual design some sample is lost in filling the loop; this problem is avoided by using direct injection with a hypodermic syringe.

Since many gel permeation chromatographs can operate unattended after sample injection, it is desirable to have some method of introducing samples automatically. A number of systems have been described; one specifically designed for use with GPC is available from Waters Associates, Framingham, Mass. (Automatic Sample Injection System).

III. CHOICE OF OPERATIONAL VARIABLES

1. Solvent

The choice of solvent is based on sample solubility and detector response. Operation near room temperature produces the least experimental difficulties and the lowest detector noise level. Preferably then, a solvent in which the polymer is soluble at room temperature is chosen. A maximum difference in refractive index of the poylmer and solvent will give refractive index detectors their highest sensitivity. For infrared detectors a nonabsorbing solvent is necessary.

As the solvent is stored for long periods in the pump reservoir, nitrogen blanketing and the use of inhibitors such as Ionol and Santonex (130) is advisable to decrease oxidation. Sample degradation in solution is sometimes possible and one should always be aware of this possibility (137). Dissolved gases in the polymer solution may produce spurious peaks on the chromatogram (47). These low-molecular-weight peaks normally do not interfere with data interpretation, but, if they do, the solvent used to dissolve the polymer may be degassed prior to use. For some theoretical studies, interpretation of results can be simplified by operation with a theta solvent (106). Table 4.2 summarizes information on several mobile phases used in gel permeation chromatography.

Table 4.2. Solvents for Gel Permeation Chromatography

Solvent	Boiling Point, °C	Refractive Index	Operating Temperature, °C	Soluble Polymers
Tetrahydrofuran	65	$1.404^{25°}$	25–50	Asphalt, cellulose acetate and nitrate, epoxy resins, polycaprolactam, polyethers, polyglocol, polystyrene, polyurethane, polyvinyl acetate, polyvinyl chloride, silicones
Dimethylformamide	135	$1.42^{25°}$	60–80	Acrylics, ethyl cellulose, polycarbonate, polyvinyl alcohol, polyvinyl butyral
Toluene	110	$1.49^{25°}$	80	Neoprene, polyvinyl formal
Trichlorobenzene	213	$1.517^{25°}$	135	Ethylene-propylene rubber, Hycar, polybutadiene, polyisobutylene, polyethylene, polypropylene
m-Cresol	202		130	Mylar, Nylon and other polyamides, polyesters, gelatin

2. Temperature

Many polymers, for example the polyolefins, are not soluble at room temperature and operation at elevated temperatures is required. The elution volume is a function of the temperature, so it is necessary to prepare calibration curves at the operating temperature (33). Changes in elution volume as a function of the temperature may be due to several factors. The hydrodynamic volume, on which the elution volume depends, is temperature dependent. In the case of gels, increasing temperature will cause expansion of the gels, thus decreasing the interstitial volume and possibly changing the pore size distribution. In addition, changes in solvent viscosity and polymer diffusion coefficients may affect the elution volume. As cited in the section on choice of solvent, the temperature to give theta conditions is sometimes selected to simplify theoretical studies. An interesting observation is that by operating at temperatures above the atmospheric boiling point of the solvent an increased mass transfer of the solute is observed which can lead to more efficient separations (162).

3. Sample Size

Sample size in gel permeation chromatography is a variable of considerable importance, as it can change elution volume, resolution, and detector response. The change in elution volume with sample size has been the subject of a number of investigations (23, 35, 44, 111, 129, 136). Although the results are not in complete agreement, they may be summarized qualitatively as follows. The elution volume increases with increasing sample size and with increasing molecular weight. If a single calibration curve is used, the computed molecular weight will therefore decrease as the sample size is increased. The change in elution volume is more pronounced for narrow molecular weight distributions than for broad distributions. If the detector response is sufficiently high, the easiest approach to obviate this problem is to operate at sample sizes small enough that this effect is negligible. In many cases this is possible with samples in the 2–5 mg range. The validity of the approach can readily be checked by comparing two chromatograms obtained with different sample sizes. If there are no differences, obviously there is no effect. Construction of the calibration curve (see also Section IV.1) can also frequenly be done at low enough concentrations so that there is no change in elution volume. If this is not possible, Boni et al. (23) have recommended extrapolating the calibration curve to zero concentration. The use of an extrapolated curve with larger size broad distribution samples is justified on the basis that the concentration at any point in this sample is much lower than in the narrow calibration standards.

4. Flow Rate

The effect of solvent flow rate on elution volume has been studied extensively with results that are not always in agreement with each other. The majority of the earlier workers concluded that, within the usual flow rate ranges of 0.1–2 ml/min, the elution volume was independent of flow rate (1, 25, 94, 106, 108). Most of this work was carried out on narrow distribution polymers. For broad-molecular-weight distribution polymers, the volume to the maximum in the chromatogram has been reported to be independent of the flow rate, although the shape of the chromatogram changes, presumably due to changes in column resolution (1, 103, 158).

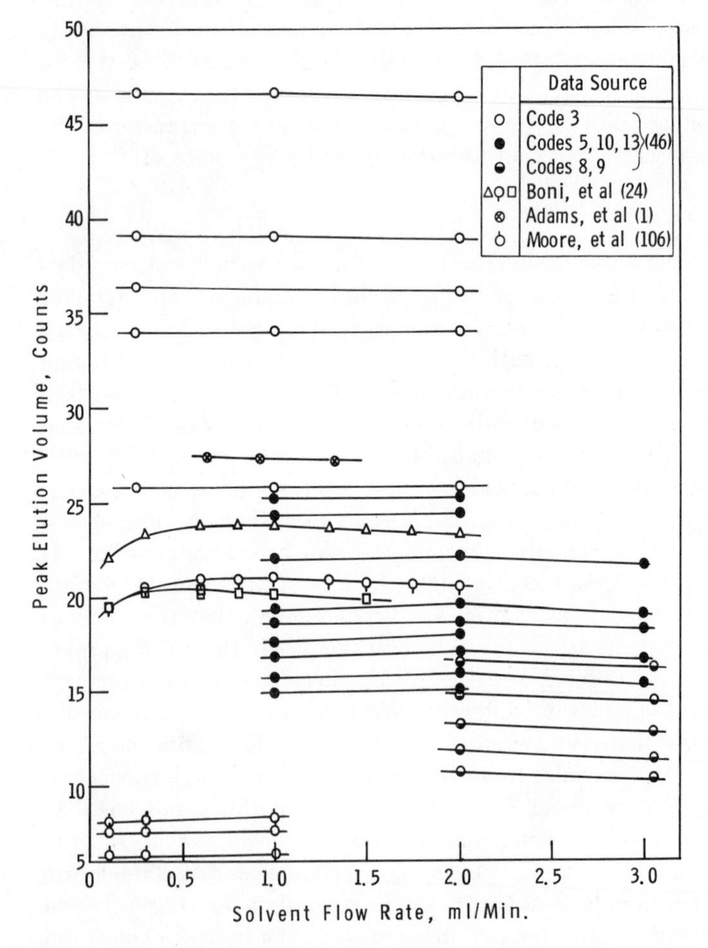

Figure 4.7. Elution volume versus flow rate (44).

However careful studies by Smith and Kollmansberger (142) showed
hat the elution volumes increased as the flow rate decreased, while Boni
t al. (24) reported a maximum in the plot of elution volume versus flow
ate. Bombaugh and Levangie found no difference, within experimental
rror, between the elution volumes observed over a wide range of flow
ates (21). Figure 4.7 plots the data of a number of investigators (44).

One phenomenon that may explain some of the conflicting data is the
erformance of the syphon used in many chromatographs to measure
luant volume. Yau et al. (160) found that the volume discharged by the
yphon varied with flow rate. The variation is due to the continued flow of
luant into the syphon during the finite discharge period. Another source is
variable discharge volume due to evaporation. This may be largely elimi-
ated by saturating the vapor space around the syphon with solvent vapor.

Other possible changes in elution volume due to viscosity differences,
iscosity profile, pressure drop effects, and nonequilibrium diffusion will
equire further studies for complete understanding (44). At present, the
nost practical approach appears to be to calibrate at the flow rate at
vhich the GPC is to be operated.

Column efficiency increases as flow rate decreases. Typical experimental
esults are shown in Figure 4.8 where the height equivalent to a theoretical
late, HETP, is plotted as a function of flow rate. The increase in efficiency
or decrease in HETP) has been treated theoretically (58) and extensively
erified experimentally (1, 25, 26, 44, 75, 94, 106, 108, 142, 153). The re-
uction of flow rate to improve resolution must also be considered in view
f the increased analysis time.

V. EVALUATION OF DATA

. Calibration

A widely used procedure in calibrating a gel permeation chromatograph
s to plot the elution volume, V_e, as a function of log molecular weight. Fre-
uently, for a specific polymer–solvent–temperature system this results in
straight line over an extended span of molecular weights that may be
epresented by

$$V_e = -B \log M + C \tag{4.2}$$

This is especially convenient for computer calculations. The plot is prepared
y determining the elution volumes for a series of narrow molecular weight
istribution polymers of known molecular weight. Figure 4.9 is an example
f a calibration plot for anionically polymerized polystyrene fractions

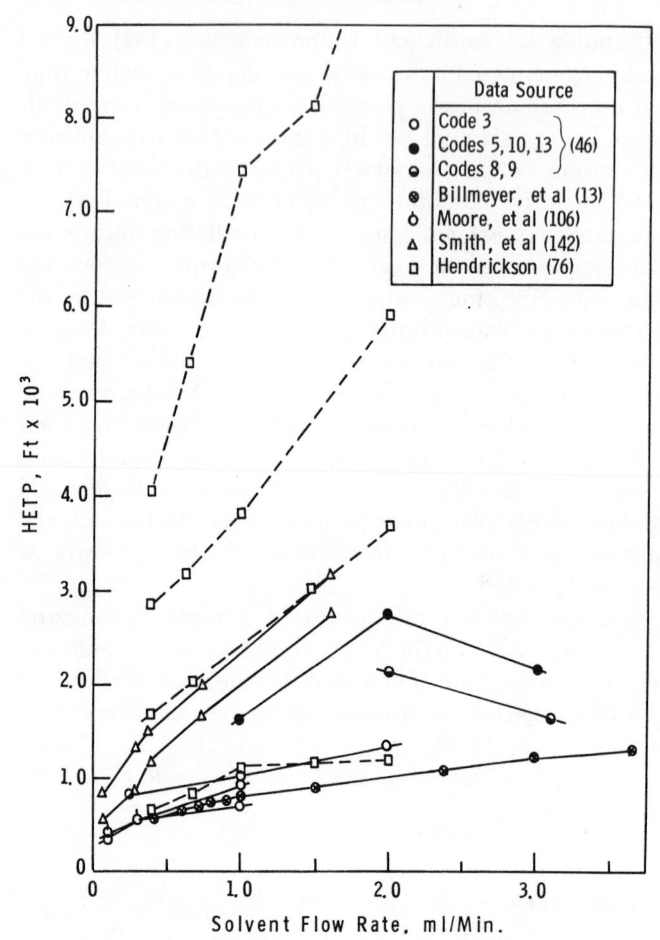

Figure 4.8. Column efficiency versus flow rate (44).

(146), including an extrapolation to zero concentration as discussed in previous section (24). The fractions must be narrow distributions becaus the molecular weight corresponding to the center of the chromatogram i higher than the number average weight but lower than the weight averag molecular weight. This type of calibration has been successfully employe for a variety of polymers (5, 29, 157). Other properties such as the mola volume have been used in calibration plots, particularly for low-molecula weight materials, but they offer no particular advantages (49, 74, 76) ove the described approaches.

In order to avoid the necessity for narrow distribution fractions whic

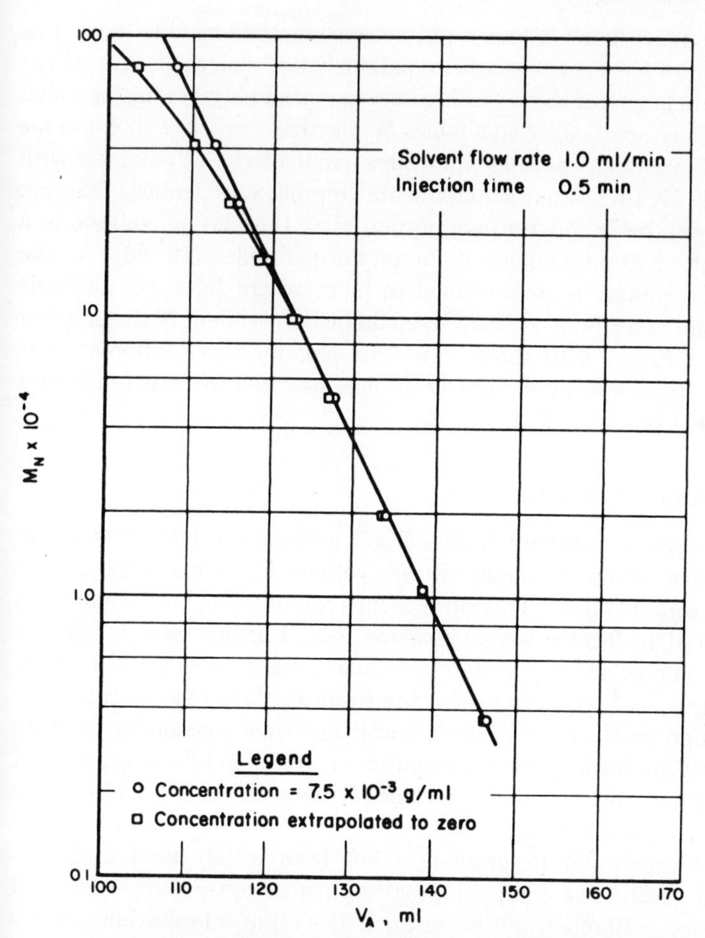

Figure 4.9. Calibration plot (24).

are not available for many polymers, attempts to use known broad distribution polymers to prepare calibration curves have been made (32). Recently a convenient computer technique has been devised which rapidly determines calibration curves from broad molecular weight distribution polymers with good accuracy (7).

What is important is that the calibration be established using the polymer that is to be chromatographed. It would clearly be advantageous to be able to calibrate with one polymer and by some method be able to use this calibration curve for any other type of polymer. Early methods related the projected chain length of the repeating unit to that of polystyrene for which narrow fractions are readily available. While this worked well for some polymers this approach has failed for others, particularly for polymers with branching (56, 57, 107). A more fundamental approach by Grubisic, Rempp, and Benoit was based on the assumption that the elution volume is a function of the hydrodynamic volume of the molecules (12, 63). As the hydrodynamic volume is proportional to $[\eta]M$, where $[\eta]$ is the intrinsic viscosity, a plot of log $[\eta]M$ versus V_e should be independent of the polymer type (38, 87). Figure 4.10 shows this type plot for three different type polystyrenes. Its applicability has been demonstrated for many different polymers (24, 142).

2. Calculation of Results

It is necessary to compute from the gel permeation data, that is the polymer concentration, c, versus elution volume V_e, a molecular weight curve versus concentration. This curve is then converted to both cumulative and differential molecular weight distributions. Further results that are convenient to know are the various molecular weight averages and ratios. In addition it is also desirable to plot the resulting data to save time. The required computations are straightforward, but time consuming, and are therefore best performed by a computer (1, 107). While a number of shortcut methods for approximate values have been reported (16), they have found comparatively little use.

One typical computer program that has been widely used is due to Pickett et al. (120). The complete program instructions, source deck and object deck are available from the authors. The input information consists of a table of elution volumes versus recorder deflection, sample concentration, recorder sensitivity, and the exponent in an applicable Mark-Houwink molecular weight–intrinsic viscosity equation. From this, the program computes the various weight averages, inhomogeneities, and differential and cumulative weight distributions.

The precision on repetitive runs on the same type of polymer has seldom been evaluated. This program provides two features of interest in this respect; the reduced specific area, A, is the total area divided by the sample weight times the response factor of the particular solvent–polymer system. This serves as a quality control reference, because for any given polymer solvent system, A should be a constant. In practice A has been found to remain constant within $\pm 5\%$ or less. In addition the program will compute viscosity average molecular weight, providing an applicable viscosity

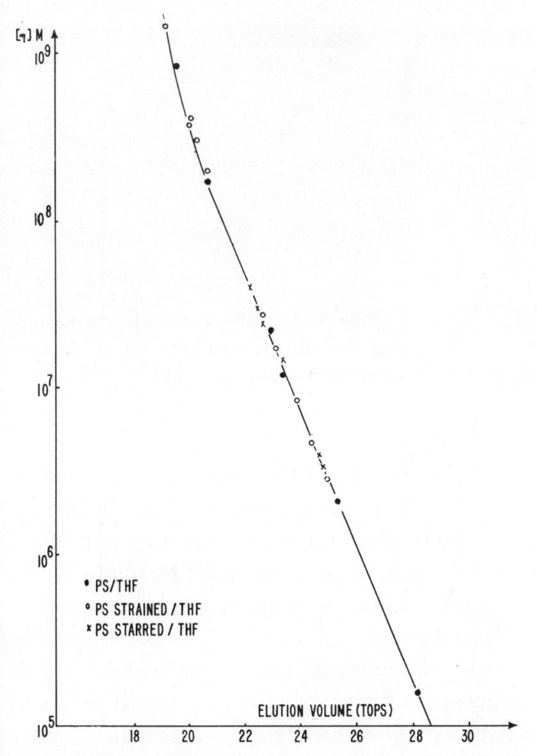

Figure 4.10. Hydrodynamic radius versus elution volume (12).

molecular weight equation is available. Thus one can in a separate experiment readily determine $\bar{M}_v$, the viscosity average molecular weight, and compare this independently determined value with that determined from the GPC. This is most frequently done by a single or multipoint solution viscosity measurement. The availability of high-speed dynamic osmometers provides another method for rapidly cross-checking results.

3. Correction for Dispersion

Figure 4.11 shows the effect of dispersion on peak shape. Part *A* shows the idealized peak shape for a monodispersed compound, while *B* shows representative behavior of a monodispersed compound on a real column. This peak broadening may be reduced by increasing the efficiency of the column. In the computational procedure for determining molecular weight distribution, an alternative approach is to remove this broadening effect mathematically. A number of methods to do this have been reported. The

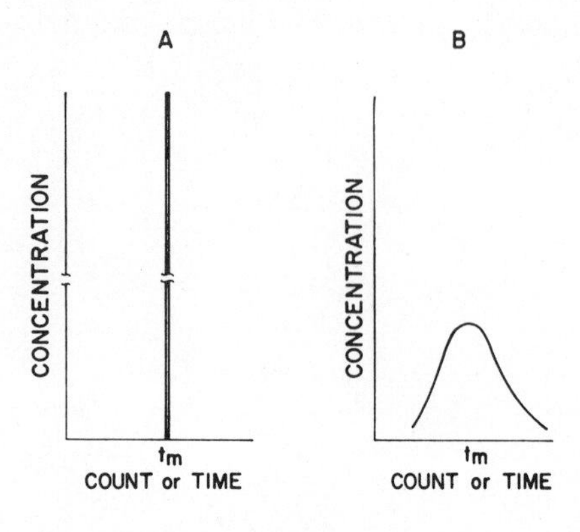

Figure 4.11. Effect of zone broadening on chromatogram of a monodisperse substance (124).

variations deal with the choice of the model and the methods of evaluating the dispersion (66, 67, 150, 77, 114, 141, 119). Duerksen and Hamielec (46) concluded in an extensive study that none of these methods were totally satisfactory but that the best was the one proposed by Tung (150), which assumes a Gaussian distribution for a single molecular species and an average dispersion width for all species.

Recently, Balke and Hamielec (8) have proposed a method for interpreting chromatograms which accounts for skewing and symmetrical axial dispersion. The program is particularly good for high flow rate experiments, in other words, for analysis with low resolution and high molecular weights. So far this program has not been evaluated for nonlinear calibrations or multimodal distributions but it appears to be suitable for such cases. Therefore, it would seem that currently this is the most satisfactory method available to correct for dispersion. It remains a matter of philosophical choice as to whether to select a high-resolution column with its attendant long analysis times but reduction in the size of the correction due to dispersion, or to use the mathematical correction of dispersion on very rapid runs where low efficiency of the column may be expected. Further work in this field will doubtless improve both methods.

4. Accuracy and Precision

Studies of accuracy of gel permeation chromatography have been carried out in one of two ways. The first is to compare the various molecular weight averages: $\bar{M}_n$, $\bar{M}_w$, $\bar{M}_z$, and so on, as determined by GPC with those determined by absolute methods. The second technique is to compare the integral or differential molecular weight distributions measured by GPC and by some other method. Both of these have some uncertainties.

Absolute molecular weight determinations are seldom more precise than a few percent. In the case of osmometry, for example, the results may be affected significantly by the presence of small amounts of low-molecular-weight material. In gel permeation, on the other hand, this low-molecular-weight material would be readily separated out from the polymer and would not be used in the computation of the number average molecular weight. Similarly when comparing distribution curves determined by two different methods it is sometimes implied that one method is absolutely correct. Again, this cannot be established with any certainty. Statistical studies of precision can be made more readily although in general not a great deal of work has been done in this area.

Muller and Alexander (109) show a reproducibility of area versus elution volume of the order of 2%. This is in accord with general experience. Gamble et al. (56), in replica analyses of butyl rubber, showed a 2-σ precision for the weight average and number average molecular weight of about 10%.

One assumption inherent in several of the detector methods must be examined with some care. This is the assumption that the response of the detector is proportional to concentration only and is independent of the molecular weight of the polymer. For example, for differential refractometer detectors, this assumption is correct at high molecular weight but is not exact at low molecular weights. For polystyrene solutions in toluene there is a small but regular increase in refractive index difference with increasing molecular weights up to a molecular weight of the order of 10,000 or higher. Similarly, Kondo et al. (86) found changes at lower molecular weights. Thus for highest precision at low molecular weights it is necessary to determine the detector response as a function of molecular weight and to include this response factor in the calculation procedure. This is easily done. In most cases for polymers of medium-to-high molecular weight this will be unimportant, but it should be considered (40). A number of studies have shown good agreement between values of $\bar{M}_w$ and $\bar{M}_n$ obtained from GPC and from absolute methods (1, 6). Some typical examples for polyglycols are given in Table 4.3. The more detailed comparison of distributions determined by different procedures has been extensively reported. A number of early studies (48) between GPC and precipitation fractionation appeared to show major differences between the two methods. Now most of these seem to have been due to incorrect calibration procedures for the GPC calculations.

Extensive comparisons of molecular weight distribution from GPC have been made with those determined from solvent precipitation (28), turbidimetric titration (54), solvent coacervation (97), and elution chromatography (3, 34, 155). These various techniques when compared with GPC

Table 4.3. Comparison of Molecular Weights of Polyglycols Determined by GPC and by Absolute Methods (6)

Gel Permeation Chromatography			Absolute Methods		
$\bar{M}_w$	$\bar{M}_n$	$\bar{M}_w/\bar{M}_n$	$\bar{M}_w$	$\bar{M}_n$	$\bar{M}_w/\bar{M}_n$
400	370	1.08	490	420	1.16
740	690	1.07	760	700	1.09
980	880	1.11	1010	920	1.10
1430	1300	1.10	1270	1130	1.12
2200	2130	1.03	2100	1950	1.08

have shown good agreement. There appears to be no reason to suspect that gel permeation chromatography is less accurate than any of the other techniques for determining molecular weight distribution. This implies of course that concentration effects are avoided and calibration procedures are carefully performed.

V. APPLICATIONS

1. Specific Polymer Systems

Table 4.4 lists GPC studies on specific polymer systems as well as those areas where GPC has found application.

2. Preparatory Scale

The use of larger-scale columns and more automation to permit the collection of sizable fractions of narrow molecular weight distribution polymer is an obvious extension of analytical gel permeation chromatography. Column scale-up has principally been attempted in the direction of making larger-diameter columns. A commercial apparatus is available that automatically injects a preselected amount of sample and then takes fractions on either a time or volume basis. At the end of the cycle, the procedure automatically starts again, with the fractions on the repetitive run being collected in one container for each fraction. Thus on a time basis, relatively large amounts of polymer may be fractionated.

One particular advantage of large-scale fractionation in GPC is that the recycling technique can be used. In this procedure, one portion of the fractionated sample is isolated and continuously recycled through a column.

Table 4.4. Gel Permeation Chromatographic Applications

Application	Reference
Adsorption studies	52
Alkyl resins	95
Antioxidants	68
Asphaltenes	3, 4
Asphalts	128, 143
Blending distributions	68, 141
Butadiene (poly)	1, 136
Butadiene, branched (poly)	156, 158
Butadiene–acrylic acid–acrylonitrile copolymer	93
Butyl rubbers	56
Cellulose esters	28, 99
Cellulose trinitrates	103, 109, 137
Chloroprene	97
Coal tars	46, 78, 113
Coating industry applications	11
Computer program for blending control	141
Copolymer	73
Crude oil	39
Crystals, chain length studies	17
Defined distributions, preparation of	140
Degradation of polymers	124, 125, 139
Desalting	53, 123
Dextrans	62
Dimethylsiloxanes (poly)	93
Elastomers	100
Epichlorohydrin-bisphenol	48
Epoxy resins	49
Etherpolyols	86
Ethylene (poly)	97, 111, 130, 133
Ethylene, branched (poly)	133
Ethylene glycols (poly)	6
Ethylene-propylene copolymers	54
Fractionation, use of GPC to study	117
Inorganic salts, separation and diffusion coefficients	135
Interaction in solutions, GPC to study	61
Irradiation studies	96
Isobutene (poly)	26

(Continued)

Table 4.4. (*Continued*)

Application	Reference
Kinetic studies	101
Lipids	112
Low-molecular-weight compounds	25, 47, 74, 98, 142
Mechanical degradation studies	42, 126
Mechanism of polymerization	115
Metal protein complexes	50
Methylene (poly)	2
Mineral oil separation from polymer	27
Oligophenylenes	70, 72
Oligourethanes	70
Peptides	134
Pesticides	131
Phenolic resins	57
Polymerization reactor control	68
Polynuclear aromatic hydrocarbons	47
Processability of polymers	127
Process control	161
Propellant analysis	93
Propylene (poly)	119
Propylene glycol (poly)	6, 107
Methylmethacrylate (poly)	155
Proteins	51, 134
Radiolabeled polymers	29
Radiolysis studies	132
Reactor design	45
Reactor kinetics	66
Silicones	129
Surfactants	149
Styrene (poly)	43, 97
Tetrahydrofuran-polypropylene oxide copolymer	14
Textiles	137
Vinyl alcohol (poly)	20
Vinyl chloride (poly)	97
Water soluble polymers	110
Wood industry	109

As the degree of fractionation increases with column length, this approach permits the achievement of almost any desired degree of fractionation. This high resolution is attained at the expense of discarding the upper and lower end of the fraction.

VI. THEORY

1. Mechanism of Separation

The mechanism by which separation takes place in gel permeation chromatography is not as yet completely understood. The mechanisms proposed to date will be briefly reviewed.

A. Steric Exclusion

The original view, and a widely expressed one, is that the mechanism of separation is due to steric exclusion. That is, different amounts of pore volume are available for different sizes of molecules (83, 121). This mechanism inherently assumes that the separation is in diffusion equilibrium. This requires that the residence time of the solute zone is much larger than the time required for a solute molecule to diffuse in and out of a pore. Calculations on simple molecules have been made that support this view (37, 42).

A number of studies of the relationship between elution volume and various molecular parameters have been made. For example, Porath (122) used a model where the pores were assumed to have hollow cone-like shapes. From this he derived a relationship between the molecular weight and distribution coefficient.

Squire (144) used a different model including various shapes for the spaces in the particles. Laurent and Killander (92) used a network of rods as a model. Many other relationships and models have been suggested (51, 70, 72). In general, these are applicable only for a specific class of compounds. It would appear that the relationship between elution volume and hydrodynamic radius (see section IV.1) is the most satisfactory one for a wide variety of polymers.

Attempts have been made to calculate the available pore volume from molecular dimensions (9, 31). Casassa (36) concluded, based on a model that includes the assumption of uniform void surfaces, that the distribution of the polymer between the outside phase and inside the voids is controlled by the loss of conformational entropy due to the transfer of a chain into the space within a void. De Vries (94) has derived relationships between elution volumes and the pore size distributions of the gel. He concluded that the maximum resolving power for a column of given length occurs when the pore size distribution of the gel covers the range of molecular sizes in the sample but does not extend beyond this range. Experimental studies agree with this conclusion (71).

B. Restricted Diffusion

Another mechanism that may be responsible for the separation process is that of restricted diffusion, in other words, the depth of penetration of the molecule into the pore is governed by its diffusion coefficient which in turn is related to its molecular size (116). One difficulty with this mechanism is that it implies that elution volume should depend on flow rate and frequently such dependence is not observed (see Section III.4). Yau and co-workers (157, 159) however have derived an expression relating diffusion to elution volume and have found good agreement with experimental results.

Haller (65) prepared glasses with uniform pore size distributions. He found separations as good as those achieved with broad pore size distribution materials, casting doubt on how steric exclusion could function as a separating mechanism. He measured diffusion rates of various materials into the glass and compared these with experimental results based on separation and found the agrrement with the restricted diffusion theory to be poor.

Other approaches to defining the separating mechanism consider solution theory (79), or interfacial tension (90), or are based on comparison with semi-permeable membranes (102) and electrical analogues (91).

C. Statistical Theory

Extensive development of statistical theory for chromatography has been made by Giddings and co-workers, and others. This is sometimes referred to as theory for zone broadening, peak spreading, or peak dispersion. Specific applications of this theory for the case of gel permeation chromatography have been reported in the literature (58, 60, 130).

ACKNOWLEDGMENTS

It is a pleasure to acknowledge the courtesy of the following editors and publishers for permission to reproduce the following figures:

Interscience Publishers, a division of John Wiley & Sons, Inc., publishers of *Journal of Polymer Science*, for Figures 4.2, 4.3, 4.6, 4.9, and 4.11 and the *Journal of Applied Polymer Science*, for Figures 4.7 and 4.8.

The American Chemical Society, publishers of *Analytical Chemistry*, for Figures 4.4 and 4.5.

Waters Associates, for Figure 4.10.

Journal of Paint Technology, Figure 4.1.

REFERENCES

1. H. E. Adams, K. Farhat, and B. L. Johnson, *Ind. Eng. Chem. Prod. Res. Develop.* **5,** 126 (1966).
2. T. Arakawa and B. Wunderlich, *J. Polymer Sci.* **C16** (2), 653 (1967).
3. K. H. Altgelt, *J. Appl. Polymer Sci.* **9,** 3389 (1965).
4. K. H. Altgelt, *Makromol. Chem.* **88,** 75 (1965).
5. D. M. W. Anderson and J. F. Stoddart, *Anal. Chim. Acta* **34,** 401 (1966).
6. M. D. Baijal and L. P. Blanchard, *J. Appl. Polymer Sci.* **12,** 169 (1968).
7. S. T. Balke, A. W. Hamielec, and B. P. LeClair, *Ind. Eng. Chem. Prod. Res. Develop.* **8** (1), 54 (1969).
8. S. T. Balke and A. E. Hamielec, *J. Appl. Polymer Sci.* **13,** 1381 (1969).
9. E. M. Barrall II and J. Cain, *J. Polymer Sci.* **C21**, 253 (1968).
10. E. M. Barrall II, M. J. R. Cantow, and J. F. Johnson, *J. Appl. Polymer Sci.* **12,** 1373 (1968).
11. R. L. Bartosiewics, *J. Paint Technol.* **39,** 28 (1967).
12. H. Benoit, Z. Grubisic, P. Rempp, D. Decker, and J. G. Zilliox, *Preprints*, Third International Seminar on Gel Permeation Chromatography, Geneva, May 1966.
13. F. W. Billmeyer, Jr., and R. N. Kelley, *J. Chromatogr.* **34,** 322 (1968).
14. L. P. Blanchard and M. D. Baijal, *J. Polymer Sci.* **A1** (5), 2045 (1967).
15. D. D. Bly, *J. Polymer Sci.* **A2** (6), 2085 (1968).
16. D. D. Bly, *J. Polymer Sci.* **C21**, 13 (1968).
17. D. J. Blundell, A. Keller, I. M. Ward, and I. J. Grant, *J. Polymer Sci.* **B4,** 781 (1966).
18. J. Boldingh, *Rec. Trav. Chim.* **69,** 247 (1950).
19. K. J. Bombaugh, W. A. Dark, and R. N. King, *J. Polymer Sci.* **C21,** 131 (1968).
20. K. J. Bombaugh, W. A. Dark, and J. A. Little, *Anal. Chem.* **41** (10,) 1337 (1969).
21. K. J. Bombaugh and R. F. Levangie, *Anal. Chem.* **41** (10), 1357 (1969).
22. K. J. Bombaugh, W. A. Dark, and R. F. Levangie, *J. Chromatogr. Sci.* **7,** 42 (1969).
23. K. A. Boni, G. A. Sliemers, and P. B. Stickney, *J. Polymer Sci.* **A2,** 1567 (1968).
24. K. A. Boni, F. A. Sliemers, and P. B. Stickney, *J. Polymer Sci.* **A2,** 1579 (1968).
25. P. I. Brewer, *Polymer* **6,** 603 (1965).
26. P. I. Brewer, *J. Inst. Petrol.* **48,** 277 (1962).

27. P. I. Brewer, *Nature* **188,** 934 (1960).

28. R. J. Brewer, L. J. Tanghe, S. Bailay, and J. T. Burr, *J. Polymer Sci.* **A1** (6), 1697 (1968).

29. K. E. Brierly-Jones, J. M. Patel, and F. W. Peaker, *Preprints,* Third International Seminar on Gel Permeation Chromatography, Geneva, May 1966.

30. M. J. R. Cantow and J. F. Johnson, *J. Appl. Polymer Sci.* **11,** 1851 (1967).

31. M. J. R. Cantow and J. F. Johnson, *J. Polymer Sci.* **A1** (5), 2835 (1967).

32. M. J. R. Cantow, R. S. Porter, and J. F. Johnson, *J. Polymer Sci.* **A1** (5), 1391 (1967).

33. M. J. R. Cantow, R. S. Porter, and J. F. Johnson, *J. Polymer Sci.* **A1** (5), 987 (1967).

34. M. J. R. Cantow, R. S. Porter, and J. F. Johnson, *J. Polymer Sci.* **C16,** 13 (1967).

35. M. J. R. Cantow, R. S. Porter, and J. F. Johnson, *J. Polymer Sci.* **B4,** 707 (1966).

36. E. F. Casassa, *J. Polymer Sci.* **B5,** 773 (1967).

37. J. Cazes, *J. Chem. Educ.* **43,** A567, A625 (1966).

38. H. Coll and L. R. Prusinowski, *J. Polymer Sci.* **B5,** 1153 (1967).

39. H. J. Coleman, D. E. Hirsch, and J. E. Dooley, *Anal. Chem.* **41,** 800 (1969).

40. P. Crouzet, F. Fine, and P. Mangin, *J. Appl. Polymer Sci.* **13,** 205 (1969).

41. H. Determann, *Gel Chromatography,* Springer-Verlag, New York, 1968.

42. H. Determann, *Angew. Chem. Intern. Ed.* (*Engl.*) **3,** 608 (1964).

43. H. Determann, G. Luben, and T. Wieland, *Makromol. Chem.* **73,** 168 (1968).

44. J. H. Duerksen and A. E. Hamielec, *J. Appl. Polymer Sci.* **12,** 2225 (1968).

45. J. H. Duerksen and A. E. Hamielec, *J. Polymer Sci.* **C25,** 155 (1968).

46. J. H. Duerksen and A. E. Hamielec, *J. Polymer Sci.* **C21,** 83 (1968).

47. T. Edstrom and B. A. Petro, *J. Polymer Sci.* **C21,** 171 (1968).

48. G. D. Edwards, *J. Appl. Polymer Sci.* **9,** 3845 (1965).

49. G. D. Edwards and Q. Y. Ng, *J. Polymer Sci.* **C21,** 105 (1968).

50. D. J. R. Evans and K. Fritze, *Anal. Chim. Acta* **44,** 1 (1969).

51. J. S. Fawcett and C. J. O. R. Morris, *Separ. Sci.* **1,** 9 (1966).

52. R. E. Felter and E. S. Moyer, *J. Polymer Sci.* **7,** 529 (1969).

53. P. Flodin, *J. Chromatogr.* **5,** 103 (1961).

54. L. W. Gamble, presented at the International Symposium on Microchemical Techniques, Pennsylvania State University, August 1965.

55. L. W. Gamble, E. A. McCracken, and J. T. Wade, *Preprints,* Third International Seminar on Gel Permeation Chromatography, Geneva, May 1966.

56. L. W. Gamble, L. Westerman, and E. A. Knipp, *Rubber Chem. Technol.* **38,** 823 (1965).

57. J. J. Gardikes and F. M. Konrad, *Amer. Chem. Soc. Div. Org. Coatings Plastics Chem. Preprints* **26** (1), 131 (1966).

58. J. C. Giddings and K. L. Mallik, *Anal. Chem.* **38,** 997 (1966).

59. J. C. Giddings, *Anal. Chem.* **39,** 1027 (1967).

60. J. C. Giddings, *Anal. Chem.* **40** (14), 2143 (1968).
61. G. A. Gilbert, *Anal. Chim. Acta* **38,** 275 (1967).
62. K. I. Granath and P. Flodin, *Makromol. Chem.* **48,** 160 (1961).
63. Z. Grubisic, P. Rempp, and H. Benoit, *J. Polymer Sci.* **B5,** 753 (1967).
64. E. F. Gurnee, *J. Polymer Sci.* **A2** (5), 799 (1967).
65. W. Haller, *J. Chromatogr.* **32,** 676 (1968).
66. A. E. Hamielec, J. W. Hodgins, and K. Tebbens, *A.I.Ch.E.J.* **13,** 1087 (1967).
67. A. E. Hamielec and W. H. Ray, *J. Appl. Polymer Sci.* **13,** 1319 (1969).
68. D. J. Harmon, *J. Appl. Polymer Sci.* **11,** 1333 (1967).
69. E. Heftmann, *Chromatography*, Reinhold, New York, 1967.
70. W. Heitz, K. L. Platt, H. Ullner, and H. Winau, *Makromol. Chem.* **102,** 63 (1967).
71. W. Heitz, B. Bomer, and H. Ullner, *Makromol. Chem.* **121,** 102 (1969).
72. W. Heitz, H. Ullner, and H. Hocker, *Makromol. Chem.* **98,** 42 (1966).
73. J. Heller, J. F. Schimscheimer, R. A. Pasternak, C. B. Kingsley, and J. Moacanin, *J. Polymer Sci.* **7,** 73 (1969).
74. J. G. Hendrickson, *Anal. Chem.* **40,** 49 (1968).
75. J. G. Hendrickson, *J. Polymer Sci.* **A2** (6), 1903 (1968).
76. J. G. Hendrickson, *Proceedings*, Fourth International Symposium on GPC, Miami Beach (1967).
77. M. Hess and R. F. Kratz, *J. Polymer Sci.* **A2** (4), 731 (1966).
78. B. C. B. Hsiek, R. E. Wood, L. L. Anderson, and G. R. Hill, *Anal. Chem.* **41,** 1066 (1969).
79. M. L. Huggins, *Amer. Chem. Soc. Div. Polymer Chem. Preprints* **8,** 439 (1967).
80. R. E. Jentoft and T. H. Gouw, *Anal. Chem.* **38,** 949 (1966).
81. H. W. Johnson, Jr., V. A. Campanile, and H. A. LeFebre, *Anal. Chem.* **39,** 32 (1967).
82. H. W. Johnson, Jr., E. E. Seibert, and F. H. Stross, *Anal. Chem.* **40,** 403 (1968).
83. J. F. Johnson, R. S. Porter, and M. J. R. Cantow, *Encyclopedia of Polymer Science and Technology*, N. M. Bikales, Ed., Vol. 7, Interscience, New York, 1967 (pp. 231–260).
84. J. F. Johnson, R. S. Porter, and M. J. R. Cantow, *Rev. Macromol. Chem.* **1,** 393 (1966).
85. R. N. Kelly and F. W. Billmeyer, Jr., *Anal. Chem.* **41,** 874 (1969).
86. K. Kondo, M. Hori, M. Hattori, *Bunseki Kagaku* **16,** 414 (1967).
87. W. R. Krigbaum and D. K. Carpenter, *J. Phys. Chem.* **59,** 1166 (1955).
88. K. A. Kun and R. J. Kunin, *J. Polymer Sci.* **C16,** 1457 (1967).
89. J. Kwok, L. R. Snyder, and J. C. Sternberg, *Anal. Chem.* **40,** 118 (1968).
90. G. Langhammer and L. Nestler, *Makromol. Chem.* **88,** 179 (1965).
91. T. C. Laurent and E. P. Laurent, *J. Chromatogr.* **16,** 89 (1964).
92. T. C. Laurent and J. Killander, *J. Chromatogr.* **14,** 317 (1964).
93. R. D. Law, *J. Polymer Sci.* **C21,** 225 (1968).
94. M. Le Page, R. Beau, and A. J. de Vries, *J. Polymer Sci.* **C21,** 119 (1968).

95. D. G. Lesnini, *J. Paint Technol.* **38** (500), 498 (1966).

96. B. J. Lyons, and A. S. Fox, *J. Polymer Sci.* **C21,** 159 (1968).

97. D. MacCallum, *Makromol. Chem.* **100,** 117 (1967).

98. L. F. Martin and S. P. Rowland, *J. Polymer Sci.* **A1** (5), 2563 (1967).

99. L. F. Martin, F. A. Blouin, N. R. Bertoniere, and S. P. Rowland, *J. Technic. Assoc. Pulp Paper Ind.* **52** (4), 708 (1969).

100. R. D. Mate and H. S. Lundstrom, *J. Polymer Sci.* **C21,** 317 (1968).

101. J. A. May, Jr., and W. B. Smith, *J. Phys. Chem.* **72,** 216 (1968).

102. G. Meyerhoff and V. Stannett, *J. Polymer Sci.* **C23,** 277 (1968).

103. G. Meyerhoff and S. Jovanovic, *J. Polymer Sci.* **B5,** 495 (1967).

104. J. C. Moore, *Chem. Eng. News*, pp. 4–5, October 18, 1965.

105. J. C. Moore, *J. Polymer Sci.* **A2,** 835 (1964).

106. J. C. Moore and M. C. Arrington, *Preprints*, Third International Seminar on Gel Permeation Chromatography, Geneva, May 1966.

107. J. C. Moore and J. G. Hendrickson, *J. Polymer Sci.* **C8,** 233 (1965).

108. L. D. Moore and J. I. Adcock, "Characterization of Macromolecular Structure," *Publ. No. 1573*, National Academy of Sciences, Washington, D. C., 1968.

109. T. Muller and W. J. Alexander, *J. Polymer Sci.* **C21,** 283 (1965).

110. S. Moroi and K. Hosoi, *J. Appl. Polymer Sci.* **11,** 2331 (1967).

111. N. Nakajima, *J. Polymer Sci.* **C21,** 153 (1968).

112. E. Nystrom and J. Sjovall, *Anal. Biochem.* **12** (2), 235 (1965).

113. H. H. Oelert, *Erdöl und Kohle-Erdgas-Petrochemie* **22,** 19 (1969).

114. H. W. Osterhoudt and L. N. Ray, Jr., *J. Polymer Sci.* **C21,** 5 (1968).

115. W. A. Pavelich and R. A. Livigni, *J. Polymer Sci.* **C21,** 215 (1968).

116. R. L. Pecsok and D. Saunders, *Separ. Sci.* **1,** 613 (1966).

117. H. L. Pedersen, *J. Polymer Sci.* **B5,** 239 (1967).

118. K. O. Pedersen, *Arch. Biochem. Biophys. Supp.* **1,** 57 (1962).

119. H. E. Pickett, M. J. R. Cantow, and J. F. Johnson, *J. Polymer Sci.* **C21,** 67 (1968).

120. H. E. Pickett, M. J. R. Cantow, and J. F. Johnson, *J. Appl. Polymer Sci.* **10,** 917 (1966).

121. M. J. R. Cantow, Ed., *Polymer Fractionation*, Academic Press, New York, 1967.

122. J. Porath, *Pure Appl. Chem.* **6,** 233 (1963).

123. J. Porath and P. Flodin, *Nature* **183,** 1657 (1969).

124. R. S. Porter, M. J. R. Cantow, and J. F. Johnson, *J. Appl. Polymer Sci.* **11,** 335 (1967).

125. R. S. Porter, M. J. R. Cantow, and J. F. Johnson, *Polymer* **8,** 87 (1967).

126. R. S. Porter, M. J. R. Cantow, and J. F. Johnson, *J. Polymer Sci.* **C16,** 1 (1967).

127. J. R. Purson, Jr. and R. D. Mate, *Preprints*, Third International Seminar on Gel Permeation Chromatography, Geneva, May 1966.

128. W. B. Richman, "Molecular Weight Distribution of Asphalts," paper available from Waters Associates, Inc., Framingham, Massachusetts.

129. F. Rodriguez, R. A. Kulakowski, and O. K. Clark, *Ind. Eng. Chem. Prod. Res. Develop.* **5,** 121 (1966).
130. J. H. Ross and M. E. Casto, *J. Polymer Sci.* **C21,** 143 (1968).
131. J. H. Ruzicka, J. Thomson, B. B. Wheals, and N. F. Wood, *J. Chromatogr.* **34,** 14 (1968).
132. R. Salovey, W. E. Falconer, and M. Y. Hellman, *J. Polymer Sci.* **C21,** 183 (1968).
133. R. Salovey and M. Y. Hellman, *J. Polymer Sci.* **A2** (5), 333 (1967).
134. P. M. Sanfelippo and J. G. Surak, *J. Chromatogr.* **13,** 148 (1964).
135. D. Saunders and R. L. Pecsok, *Anal. Chem.* **40,** 44 (1968).
136. R. M. Screaton and R. W. Seeman, *J. Polymer Sci.* **C21,** 297 (1968).
137. L. Segal, *J. Polymer Sci.* **C21,** 267 (1968).
138. J. Seidl, J. Malinsky, K. Duvek, and W. Heitz, *Fortschr. Hochpolymer Forsch.* **5,** 113 (1967).
139. M. T. Shaw and F. Rodriguez, *J. Appl. Polymer Sci.* **11,** 991 (1967).
140. J. H. Short, R. P. Zelinski, and P. V. McKinney, *Preprints,* Tokyo IUPAC Meeting, 1966, VI-119.
141. W. N. Smith, *J. Appl. Polymer Sci.* **11,** 639 (1967).
142. W. B. Smith and A. Kollmansberger, *J. Phys. Chem.* **69,** 4157 (1965).
143. L. R. Snyder, *Anal. Chem.* **41,** 1223 (1969).
144. P. G. Squire, *Arch. Biochem. Biophys.* **107,** 471 (1964).
145. R. L. Steere and G. K. Ackers, *Nature* **196,** 475 (1962).
146. M. Szivare, *Makromol. Chem.* **35,** 132 (1960).
147. S. L. Terry and F. Rodriguez, *J. Polymer Sci.* **C21,** 191 (1968).
148. A. Tiselius, J. Porath, and P. A. Albertsson, *Science* **141,** 13 (1963).
149. F. Tokiwa, K. Ohki, and I. Kokubo, *Bull. Chem. Soc. Japan* **41,** 2845 (1968); **42,** 575 (1969).
150. L. H. Tung, *J. Appl. Polymer Sci.* **10,** 375, 1271 (1966); **13,** 775 (1969).
151. M. F. Vaughan, Brit. Pat. 1,000,185, August 4, 1965.
152. M. F. Vaughan, *Nature* **195,** 801 (1962).
153. A. J. de Vries, M. Lepage, R. Beau, and C. L. Guillemin, *Anal. Chem.* **39,** 935 (1967).
154. M. Werner, *J. Chromatogr.* **25,** 63 (1966).
155. S. Yamada, S. Kitahara, Y. Hattori, and Y. Konakahara, *Kobunshi Kagaku* **24,** 97 (1967).
156. S. Yamada, S. Imai, and S. Kitahara, *Kobunshi Kagaku* **23,** 577 (1966).
157. W. W. Yau and C. P. Malone, *J. Polymer Sci.* **B5,** 663 (1967).
158. S. Yamada, S. Imai, and S. Kitahara, *Kobunshi Kagaku* **23,** 400 (1966).
159. W. W. Yau, and S. W. Fleming, *J. Appl. Polymer Sci.* **12,** 2111 (1968).
160. W. W. Yau, H. L. Suchan, and C. P. Malone, *J. Polymer Sci.* **A2** (6), 1349 (1968).
161. R. J. Zeman and F. P. Adams, *Preprints, Paper 40A,* 61st National Meeting Amer. Inst. Chem. Eng., Houston, Texas, 1967.
162. J. N. Little and W. J. Pauplis, private communication, October 1970.

CURTIS R. HARE
University of Miami
Coral Gables, Florida

V. Visible and Ultraviolet Spectroscopy

I. INTRODUCTION

Visible and ultraviolet spectroscopy are widely used techniques for the qualitative and quantitative analysis of organic and inorganic systems (1–11). Many compounds or their derivatives have spectra in the visible and ultraviolet region and are, therefore, amenable to analysis by these spectroscopic techniques. The widespread use of these methods is due to many factors, probably the most important of which is the availability of simple-to-use instruments at relatively low cost. The analytical methods developed are generally sensitive and selective, and the results are easy to interpret.

Spectroscopic techniques are based on the measurement of absorption or emission of electromagnetic radiation. In visible or UV spectroscopy the absorption or emission process is a result of an *electronic* rearrangement in atoms or molecules. In the absorption process the atom or molecule is first in a ground electronic state; on absorbing the incident radiation, the system rises to a higher energy electronic excited state. The emission is the reverse of the absorption process; that is, a molecule already in an excited state reverts to the ground state by emission of energy in the form of electromagnetic radiation. In each case, the energy absorbed or emitted is proportional to the frequency of the electromagnetic radiation; that is, it obeys the Planck condition

$$E = h\nu \tag{5.1}$$

in which E is the energy of a photon (quantity of energy absorbed or emitted by each molecule), h is the proportionality constant (6.62×10^{-27} (erg)(sec)/molecule), and ν is the frequency of the radiation in sec^{-1} (or in the presently accepted unit, hertz, Hz). Only the absorption process will be considered in this chapter.

The frequency of the radiation is seldom measured. One would generally determine the wavelength, λ, which is simply the speed of light, C, divided by the frequency

$$\lambda = \frac{C}{\nu} \tag{5.2}$$

The dimension of the wavelength is length; the nanometer (nm $= 10^{-9}$ m) is the unit presently used in visible and UV light spectroscopy.* The visible region is from 750 to 400 nm, which corresponds to the region from red to violet light (Table 5.1). The UV region is between 400 and 200 nm. Oxygen and water absorb beyond 200 nm; this region is called the far ultraviolet. Since the energy is proportional to the frequency, another convenient unit sometimes used in spectroscopy is the wavenumber (cm^{-1}), ω. This unit is also proportional to the energy.

$$\frac{\nu}{C} = \frac{1}{\lambda} = \omega \tag{5.3}$$

In wavenumbers the visible region ranges from 13,300 to 25,000 cm^{-1}; the UV region is from 25,000 to 50,000 cm^{-1} (Table 5.1).

The intensity of absorption depends on the concentration of the absorbing species c, and the path length traversed by the radiation l. The molar extinction coefficient, $\mathcal{E}$ is defined from Beer-Lambert's Law as

$$\mathcal{E} = \frac{1}{cl}\log_{10}\frac{I_0}{I} \quad \text{or} \quad A = \mathcal{E}cl \tag{5.4}$$

* The nanometer is equal to the millimicron (mμ), that is, 10^{-7} cm. The millimicron is generally used in the older literature. The Angstrom unit (1 Å $= 10^{-8}$ cm $= 10$ nm) is also used in some applications.

Table 5.1. Correlation of Frequency, Wavelength, and Wavenumber of Electromagnetic Radiation in the Visible Region

Color	Frequency, Hz	Wavelength, nm	Wavenumber, cm^{-1}
Red	4.8×10^{14}	630	16,000
Orange	5.1×10^{14}	590	17,000
Yellow	5.4×10^{14}	560	18,000
Green	5.9×10^{14}	510	19,500
Blue	6.3×10^{14}	480	21,000
Indigo	6.7×10^{14}	450	22,000
Violet	7.1×10^{14}	420	24,000
Ultraviolet	8.6×10^{14}	350	29,000

and is a measure of the efficiency with which the molecules absorb. The quantity A, which is equal to the log of the ratio of the intensity of the incident (I_0) and transmitted (I) radiation, is called the *absorbance* or *optical density*. Most spectrophotometers record the spectrum as the absorbance versus wavelength; older instruments may record percent transmittance, instead of absorbance.

The extinction coefficient at the wavelength of maximum absorbance is the quantity which is considered in quantitative analysis. Once determined for a given compound under a given set of reproducible conditions, it may be used to calculate the concentrations of the compound in unknown systems. It is important to emphasize that the conditions must be nearly the same in comparing known and unknown samples, since the extinction coefficient is generally solvent- and temperature-dependent and is sometimes even dependent on concentration (deviations from Beer-Lambert Law through molecular interactions). However under ambient conditions and concentrations less than 0.5 M, only the solvent dependencies are important.

The larger the extinction coefficient of a compound, the more sensitive will be the method for the detection. In the visible region in which inorganic complexes and organic dyes absorb, the extinction coefficients of the complexes are usually between 10 and 100. Dyes, on the other hand, show values of ε between 10,000 and 50,000. Thus the limit of detection may be around 10^{-2} M for inorganic complexes, while for the dyes it may be possible to estimate concentrations lower than 10^{-5} M. The extinction coefficients of most compounds (inorganic and organic) are generally larger in the UV region than in the visible region. In UV spectroscopy the limits

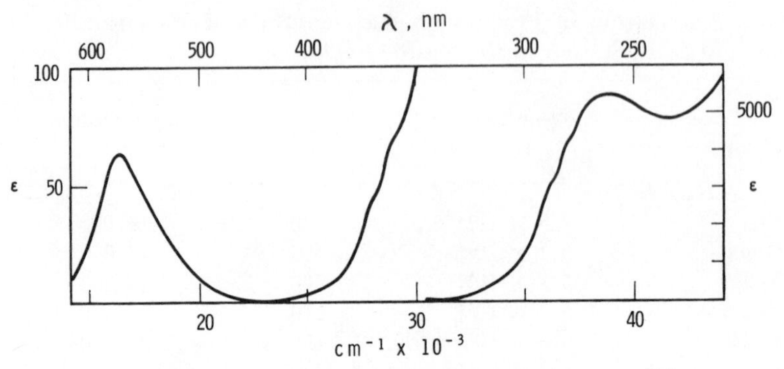

Figure 5.1. The spectrum of bis(phenylglycinato) copper(II).

of estimation are therefore lower and the methods developed are more sensitive.

The spectrum of bis(phenylglycinato)copper(II) (Figure 5.1) is typical of a visible and ultraviolet spectrum. The extinction coefficient in the visible region (left portion of curve) is between 10 and 100; the ε in the ultraviolet, on the other hand, shows values between 1000 and 5000. The band at 16,000 cm^{-1} is a ligand field transition (d–d) of the metal ion, while the bands in the UV are characteristic of a bound phenyl group (36,000 cm^{-1}) and a carboxyl group metal ion charge transfer band (39,000 cm^{-1}).

II. ELECTRONIC SPECTRA

A detailed sojourn in quantitative molecular orbital theory is not required in order to understand the basic origin of electronic spectra of molecules. Some knowledge of the qualitative aspects of molecular orbital theory however can result in a greater appreciation of the nature of the electronic rearrangements that arise from absorption of radiation.

1. Transitions and Intensity

Let us first consider a molecule in its normal, ground state G which has a total energy $E(G)$ and is described by a molecular wave function $\Psi(G)$. On absorption (Figure 5.2) of visible or ultraviolet radiation the molecule undergoes a transition to the excited state (higher energy) S which has an energy $E(S) = E(G) + \Delta E$ where ΔE is the energy of the absorbed radiation ($\Delta E = h\nu$). In order for the transition to occur with a large intensity

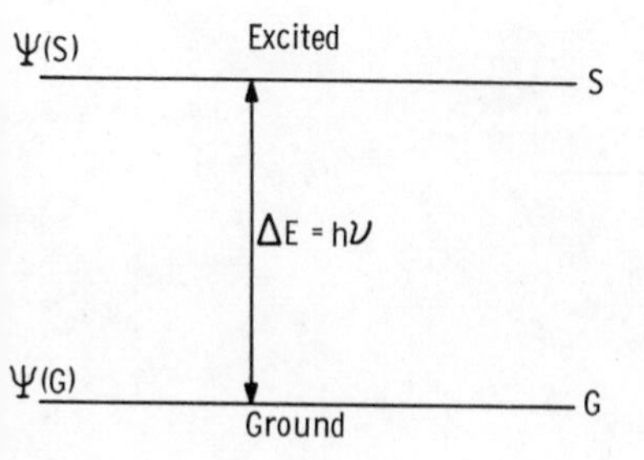

Figure 5.2. Transition from ground to excited state which corresponds to the energy ΔE.

or probability (P), the two states G and S must be coupled by the electric dipole (er) of the light.* The probability is given by the coupling expression

$$P = \int \Psi(S)\ (er)\ \Psi(G)/dT \tag{5.5}$$

in which er is the electric dipole operator and the integration is carried out to cover the overall space.

One would expect the spectrum of the transition between these two states to appear as a sharp line (Figure 5.3), and the width of the line to be given by the Heisenberg uncertainty principle. In molecular spectra, however, particularly those obtained in solution, sharp lines as shown in Figure 5.3 are never observed. The reason is that although the energy of visible and UV radiation is greater than the energy of molecular vibrations and rotations, there is a coupling of these molecular motions with the electronic rearrangements. This situation is best illustrated by looking at the potential energy versus internuclear distance diagrams for the states G and S of

* Electromagnetic radiation is composed of two components; an oscillating electric dipole and a magnetic dipole. In electronic spectra, the electric dipole contributions to the intensity are usually very much greater than contributions from the magnetic dipole.

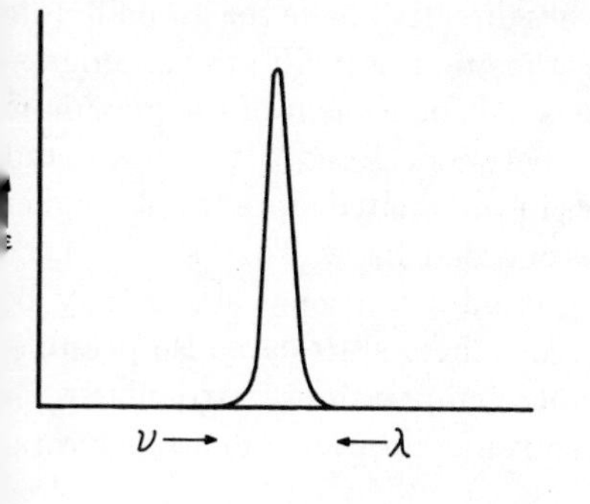

Figure 5.3. The sharp line spectrum expected in the absence of molecular vibrations and rotations.

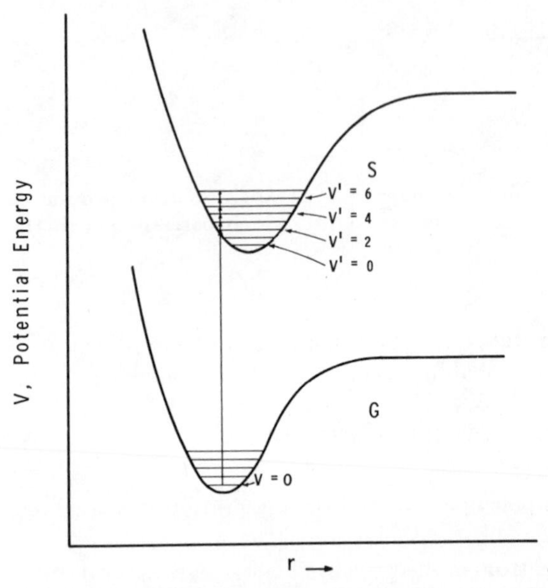

Figure 5.4. The vertical transition from the ground (*G*) vibrationless state ($V = 0$) to the several vibrational states ($V' =$ N) of the excited electronic state (*S*).

the molecule (Figure 5.4). At low temperatures the molecule will be in its ground electronic state (*G*). It is also most probably in the ground vibrational state ($V = 0$). When the electron jumps from the ground state to the excited state, the transition takes place so fast that the nuclei can be considered to be at rest. There is therefore no change in *r*, the internuclear distance. Usually the minima in the potential energy surfaces of excited states are at greater distances than those of the ground state. When the electron undergoes a transition in the vertical direction, it may be accommodated in numerous excited vibrational states (V'). The most probable state is the V' vibration which lies directly above the ground-state equilibrium position. The result of these transitions is a vibrational progression, as illustrated in Figure 5.5*a*. There is a displacement of the maximum of the band from the energy difference between *G* and *S* by an amount corresponding to the vibrational energies in the excited state (5). At higher temperatures the spectra will appear as a broad band. This is due to the fact that at high temperatures other ground vibrational states may be populated by electrons and transitions from these states are also possible. Broad bands are also observed at ambient temperatures, particularly for spectra obtained from solutions where solvent vibrations, molecular rota-

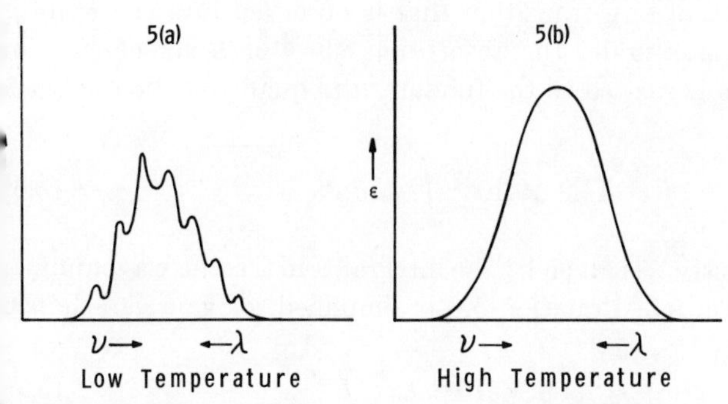

Figure 5.5. Spectrum at low temperatures showing a vibrational progression in the excited state (*a*) and the structureless, broad band at high temperatures (*b*).

tions, and molecule–solvent vibrations all contribute to the band broadening. Here the band envelope arises from the overlap of many gaussians corresponding to the many possible vibrational and rotational states.

The previous discussion applies to the very intense, or "allowed," transitions. Numerous transitions are observed in molecules that are strictly forbidden by the selection rules of spectroscopy. Allowed transitions are those in which (1) the change in angular momentum $\Delta L = 0$ or ± 1, (2) there is no change in the spin orientations of the electrons, and (3) the product of the group representations of the two states and the electric dipole vector are totally symmetric.

The first rule usually is not a problem since most states are within one unit of angular momentum of one another. The second is the spin-forbidden rule, and states which do involve a change in spin orientation of the electrons are observed to have low intensities. These states are also at lower energy than the spin-allowed since energy is lost in reorienting spins. The last rule can best be treated by group theory (5). For symmetric ground states however, it is necessary that the components of the electric dipole vector and the excited state belong to the same representation. Transitions which do not obey these rules are called forbidden, and are usually low in intensity. They arise because the selection rules are not rigorously obeyed by the molecule. That is, the ground state of the molecule may undergo slight static or vibrational distortions which void or relax the rules. The static contributions to the intensity will be temperature-independent, while the vibrational or vibronic contributions are temperature-dependent (25). The term vibronic is used to indicate a coupling of vibrational and electronic motions.

The total intensity of any transition that is observed in a spectrum is proportional to the area under the absorption curve of $\mathcal{E}$ versus cm^{-1} or $\mathcal{E}$ versus ν. This area is related to the fundamental quantity, the oscillator strength f by

$$f = 4.32 \times 10^{-9} \int \mathcal{E}\omega d\omega \tag{5.6}$$

where the area in question is given by the integral, and ω is the wavenumber described in Equation 5.3. Equation 5.6 is simplified for gaussian-shaped bands to

$$f = 4.60 \times 10^{-9}\ \mathcal{E}_{max}\ \delta \tag{5.7}$$

where δ is the width of the band (in cm^{-1}) measured at half height.

2. Organic Spectra

A. Nonaromatic Compounds

The bonding of organic compounds involves the interaction of the atomic s and p electrons to produce molecular orbitals (Figure 5.6). The molecular electronic states arise from the various population distributions of the molecular orbitals. The overlap of the atomic orbitals along the bonding axis from the s or p_σ orbitals gives rise to σ bonding and σ^* antibonding states. The π bonding and π^* antibonding orbitals are formed by the p_π atomic orbitals and are distinguished by a single nodal plane containing the bonding axis. Atomic orbitals which do not interact produce nonbonding (n) molecular orbitals. In most organic compounds the molecular orbitals (Figure 5.6) are filled completely to the nonbonding level and the higher energy antibonding levels are unoccupied. The lowest energy electronic transitions will be from the nonbonding level to the π^* antibonding level ($n\rightarrow\pi^*$). The energy of these transitions varies considerably from compound to compound and may appear between 600 and 250 nm (Table 5.2). The next highest energy transitions are either the $n\rightarrow\sigma^*$ or the $\pi\rightarrow\pi^*$ transitions. They are observed at the limit of the UV region at wavelengths less than about 250 nm. The highest energy transitions are the $\sigma\rightarrow\sigma^*$ transitions which usually appear in the vacuum UV region. An example of a $\sigma\rightarrow\sigma^*$ transition is observed in the saturated hydrocarbons which have absorption maxima at about 130 nm. The intensity of the transitions increases with increasing energy; that is, the shorter wavelength transitions are more intense. The energies of the transitions are all sensitive to the substituents near the chromophore. This is particularly true if the chromophore may enter into conjugation (π delocalization of the electrons) with the substituents.

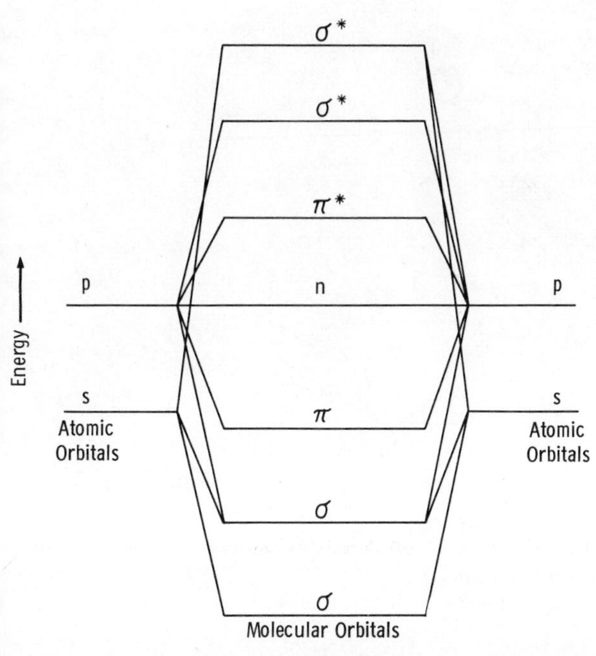

Figure 5.6. A molecular orbital diagram for nonaromatic chromophores. The levels up to *n* are filled with electrons and the transitions take place from these filled levels to the unoccupied antibonding levels.

Table 5.2. Wavelength of Maximum Absorption of $n \rightarrow \pi^*$ Transitions of Selected Chromophores

Chromophore	$n \rightarrow \pi^*$, nm
>C=O	280
>C=N—	300
—N=N—	350
>C=S	500
>N=O	650

The carbonyl group, which is probably the best understood of all chromophores, is desci ibed by the molecular orbital diagram given in Figure 5.6. The electronic arrangements of the various states are illustrated in Figure 5.7. The $n \rightarrow \pi^*$ transition is symmetry-forbidden and appears as a weak band except in asymmetric cases where static distortions allow the transi-

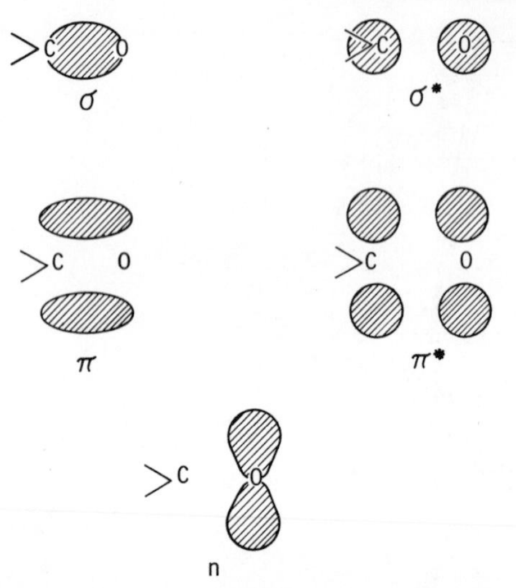

Figure 5.7. An illustration of the electron densities associated with the molecular orbitals of the carbonyl group.

tion to take place. The spectral data for a series of acetyl compounds are given in Table 5.3. Here the bands are shifted with little change in intensity further to the UV. This shift is probably due to a repulsion of the π-level electrons by the lone electron pairs of the substituents. The $n \rightarrow \pi^*$ transitions are also sensitive to the solvent in which the spectrum is measured. The more polar the solvent (electron pair rich), the further the maximum in the band will be shifted toward the UV. This is a higher energy, so-called blue, or *hypsochromic*, shift. The $\pi \rightarrow \pi^*$ transition of an isolated carbonyl occurs at about 170 nm in the vacuum UV. However in carbonyls which are conjugated with olefinic or other carbonyl groups, the levels are reduced in energy and the transitions appear in the UV. In crotonaldehyde, for example, the $\pi \rightarrow \pi^*$ transition is at 214 nm (ε = 15,000) and the $n \rightarrow \pi^*$ transitions are at 340 nm (ε = 25). In quinone, which is highly conjugated, the $\pi \rightarrow \pi^*$ transition is at 245 nm and the $n \rightarrow \pi^*$ is at 435 nm. The $\pi \rightarrow \pi^*$ transitions also undergo a shift in spectrum with change in solvent polarity but the effect is opposite to that of the $n \rightarrow \pi^*$ transitions. That is, the more polar the solvent, the more the band is shifted to lower energy (longer wavelength). This is called a red, or *bathochromic*, shift.

The spectra of conjugated dienes such as 1,3-butadiene also have bands in the UV region. In hexane solution the band of butadiene is at 217 nm (ε = 21,000). Substituents on butadiene or cyclization to form cyclic dienes tend to shift the bands to lower energy (bathochromic shift). Cyclic

Table 5.3. Substituent Dependence of Carbonyl $n \rightarrow \pi^*$ Transition in Acetyl Compounds (CH_3—C(=O)—X)

X	λ_{max}, nm	ε	Solvent
—H	290	17	Isooctane
—CH_3	279	13	Isooctane
—CH_2—CH_3	279	16	Isooctane
—Cl	235	53	Hexane
—O—C(=O)—CH_3	225	47	Isooctane
—NH_2	214		Water
—O—CH_2—CH_3	204	41	Ethanol
—OH	207	69	Petroleum ether

polyenes such as the steroids can have intense $\pi \rightarrow \pi^*$ absorptions in the near ultraviolet.

B. Aromatic Compounds

The spectrum of benzene is composed of two bands in the UV region at 256 nm (ε = 200) and 204 nm (ε = 8000) and one in the far UV at 184 nm (ε = 60,000). The two transitions in the UV are symmetry-forbidden but their intensities are large due to static and vibronic distortions. The spectra are also characterized by vibrational structure in nonpolar solvents. The spectra of a series of monosubstituted phenyl compounds are given in Table 5.4. The substituents tend to shift the bands to lower energy (bathochromic shift) with a concomitant increase in intensity. The fine structure present in benzene is also less distinct in its substituted compounds. These alterations in the benzene spectrum are due to the overlap of the substituents with the π orbitals of the ring. Those substituents which may interact strongly with the π orbitals by resonance or delocalization of the π clouds are characterized by low-energy shifts of the bands. At the same time, these bands also become more intense. A few examples of compounds showing these effects are given in Table 5.5.

In going to the more complicated acenes such as naphthalene, phenanthrene, and anthracene, the bands are also red-shifted and intensified.

Table 5.4. Dependence of Aromatic Absorption Bands of Selected Phenyl Compounds (C_6H_5—X)

	First Band		Second Band		
X	nm	ε	nm	ε	Solvent
—H	256	180	204	7,900	Hexane
—Cl	265	240	210	7,600	Ethanol
—SH	269	700	236	10,000	Hexane
—O—CH_3	269	1,500	217	6,400	Methanol
—OH	270	1,450	210	6,200	Water
—O^-	287	2,600	235	9,400	Alkaline H_2O
—NH_2	280	1,430	230	8,600	Water
—NH_3^+	254	160	203	7,500	Acidic H_2O

Table 5.5. Aromatic Transitions of Conjugated Chromophores

	First Band		Second Band	
Compound	nm	ε	nm	ε
Benzene	256	180	204	7,900
Styrene	282	450	244	12,000
Nitrobenzene	280	1,000	252	10,000
Benzoic acid	270	800	230	10,000
Biphenyl	Masked		246	20,000
Phenylcyanide	271	1,000	224	13,000
p-Nitrophenol	Masked		318	10,000
o-Nitroaniline	412	4,500	283	5,400
p-Nitroaniline	Masked		381	13,500

Other heterocyclic compounds such as furan, pyrrole, pyridine, thiophene, and their substituted compounds have spectra which are similar to those of benzene and its derivatives. For further practical details of spectra of organic compounds, the reader is referred to the Sadtler index (20), or to the text by Silverstein and Bassler (22). Other books which go further into the theory of organic spectra are given in the references (12–19, 21–24).

C. *Charge-Transfer Bands*

The reaction of an electron pair donor with an electron pair acceptor to form a molecular complex is quite analogous to the Lewis concept of acid–base reactions. Many materials acting as donors or acceptors may interact to form complexes, but even in the best substantiated cases this interaction is generally very weak. The molecular complexes which are formed are called charge-transfer complexes because a charge is donated to the acceptor. The detection of charge-transfer complexes is practically always carried out by UV spectroscopy. The technique consists of mixing a donor and acceptor in an inert solvent and comparing the spectrum of the mixture with those of the pure donor and acceptor in that solvent. The presence of a new band which is dependent on donor and acceptor concentrations is indicative of interaction or complex formation.

In the simplest terms one can think of a donor orbital interacting with an acceptor orbital to form molecular orbitals of the complex (Figure 5.8). The spectral transition (charge-transfer band) is from the donor-like orbital to the acceptor-like orbital of the complex. The energy of the charge-transfer band ΔE is related to the ionization potential of the donor I, the electron affinity A of the acceptor, and the Coulombic energy of the electron C.

$$\Delta E = I - A - C \tag{5.8}$$

In order for the charge-transfer band to be observable, it should appear at a higher wavelength than the bands associated with transitions in either the donor or acceptor levels. The charge-transfer bands are usually broad and quite intense, which aids in their detection even when the equilibrium constants for the formation of the complexes are small.

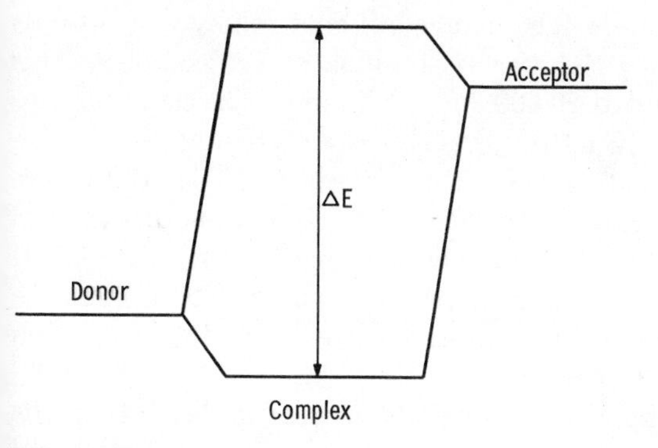

Figure 5.8. Combination of donor and acceptor states to give complex (charge transfer) states.

The best known acceptor molecule is iodine (I_2), which forms complexes with amines, alcohols, benzene, and even hydrocarbons (very weak). Of these complexes, the most well-known is the benzene–iodine complex which has a characteristic absorption at 290 nm. Other common acceptors are chloranil and trinitrobenzene. Another example of charge-transfer complexes is the complex resulting from DDT (the pesticide) and nerve tissue. This will be illustrated in a subsequent section. The electronic spectra book by Murrell (18) has an excellent discussion on charge-transfer bands.

3. Transition Metal Complexes

A. Ligand Field Theory

It is known from atomic spectra that in free space each of the five d orbitals of a transition metal ion has the same energy (are degenerate). These five orbitals are designated in Cartesian coordinates as d_{z^2}, $d_{x^2-y^2}$, d_{xy}, d_{xz}, and d_{yz}. The interaction of the ion with ligands to form a complex with octahedral geometry increases the energy of the system and splits the orbitals into two sets. The d_{z^2} and $d_{x^2-y^2}$ orbitals form one set (e_g) which is directed toward the ligands. Their interaction with the ligands is very large which results in a shift to higher energy. The set of orbitals denoted by d_{xy}, d_{xz}, d_{yz} (t_{2g}) lies inbetween the ligands and their interaction brings about a shift to lower energy (Figure 5.9). The separation between the two sets is the energy Δ or 10 Dq. The term Dq is the crystal field parameter which has been developed from a point-charge model of the metal–ligand interaction. In this model, Dq depends on the charge on the ligand, the average radius of the d orbital, and the metal–ligand distance. The parameter is seldom evaluated from theory, but rather is evaluated from the energy of the first band in the spectrum of transition metal complexes.

The complex ion $[Ti(H_2O)_6]^{3+}$ is a system where one d electron (Figure 5.9) occupies the five d orbitals. The octahedral field splits these orbitals and the electron populates the lower t_{2g} orbitals of the complex. This results in an absorption band at 20,000 cm^{-1} (500) nm). The same description applies to the complex $[Cu(H_2O)_6]^{2+}$. This is a system of nine d electrons; it has one electron void, or hole, in the d orbitals. The orbitals are split as before by the octahedral ligand field. The population distribution of the t_{2g} and e_g orbitals is such that the hole is present in the e_g orbital. If the electron is excited from the lower t_{2g} orbital to the e_g orbital, then the hole appears in the t_{2g} orbitals. This process corresponds to a transition which occurs at 12,500 cm^{-1} (800 nm).

The extent of the splitting of the orbitals (Δ or 10 Dq) depends on the type and strength of the metal–ligand interaction. The various ligands are usually classified by their splitting ability in terms of the value of Dq

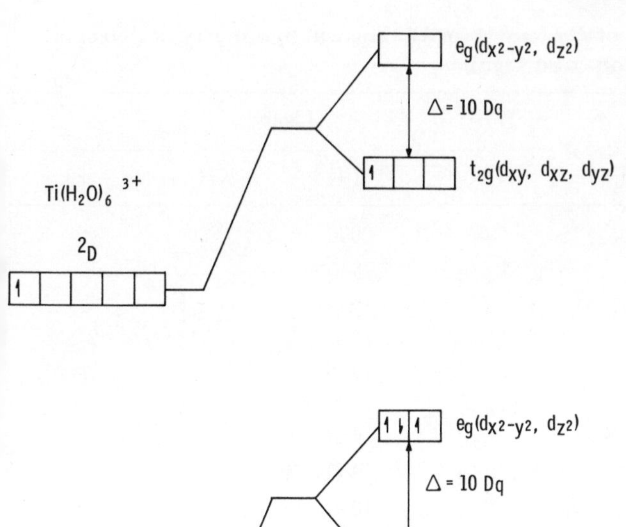

Figure 5.9. The splitting of the *d* orbitals by the octahedral field of the ligands into the orbitals t_{2g} and e_g. The electronic transitions take place from lower to higher orbitals.

(Table 5.6), since 10 *Dq* roughly corresponds to the first absorption band (lowest in energy) of the complex. For a given ligand the value of *Dq* is larger for a higher charge on the metal ion. Also, the better the ligand is as a donor, the larger will be the value of *Dq*. For a given metal ion the various ligands have been classified in a spectrochemical series which is in order of greater (higher energy) splitting of the orbitals; that is, in order of the ability of the ligands to shift the bands toward the ultraviolet (33). A partial listing is $I^- < Br^- < Cl^- < S^{2-} < F^- < OH^- <$ oxalate $< H_2O <$ $NCS^- <$ glycinate $<$ pyridine $< NH_3 <$ ethylenediamine $< NO_2^- \sim$ bipyridyl $<< CN^- < CO$. In order of the donor atoms the listing is as follows: $I < Br < Cl < F < O < N \ll C$.

Ligand field theory may also be applied to complexes other than those with the more prevalent octahedral geometry. The results are simplest however for cubic and tetrahedral complexes since they may be related to the octahedral splittings. This occurs because all three geometries can be derived from a cubic arrangement. The tetrahedral splittings of the orbitals

Table 5.6. Values of *Dq* (cm^{-1}) for Octahedral Symmetry of Selected Transition Metal Ions and Ligands

Configuration	Ion	Ligand			
		Cl^-	H_2O	NH_3	*en*
d^1	Ti^{3+}	1550	2030		
d^2	V^{3+}	1300	1840		
d^3	V^{2+}		1235		
	Cr^{3+}	1318	1750	2150	2188
d^4	Cr^{2+}	(1300) [a]	1410		1800
	Mn^{3+}	(2000)	2100		
d^5	Mn^{2+}	750	850		1010
	Fe^{3+}	(1100)	1430		
d^6	Fe^{2+}		(−1040)		
	Co^{3+}		(−1820)	−2290	−2320
	Rh^{3+}	−2032	−2700	−3410	−3460
	Ir^{3+}	−2500			−4140
d^7	Co^{2+}		−950	−1050	(−1100)
d^8	Ni^{2+}	−750	−860	−1080	−1120
d^9	Cu^{2+}		−1250	(−1500)	(−1640)

[a] The values given in parentheses are less certain.
en = ethylenediamine

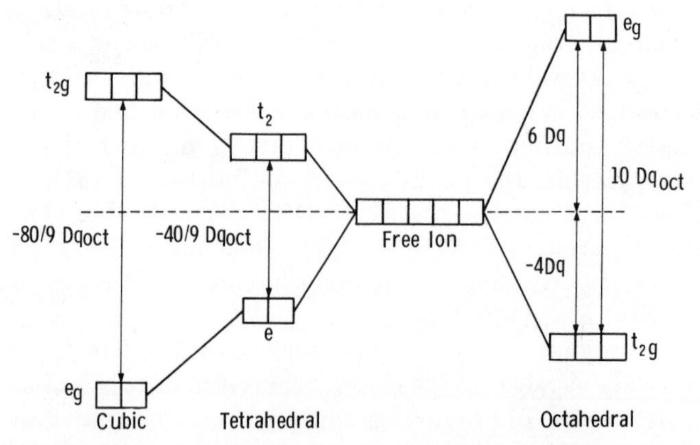

Figure 5.10. Splittings of the *d* orbitals in octahedral, cubic, and tetrahedral fields in terms of the octahedral parameter Dq_{oct}.

(Figure 5.10) are opposite those of the octahedral. For a given ligand the tetrahedral splitting is approximately half (4/9) as much as observed for the octahedral case. The cubic splittings are in the same direction as the tetrahedral but twice as large. Thus the values of Dq obtained from the spectra of octahedral complexes may be used as first approximations for complexes of nonoctahedral geometry.

Another approximation which is very useful for the determination of the configuration of transition metal ion complexes is the rule of average environment. This states that the ligand field about the metal ion is approximately equivalent to the average of all the ligands. Thus in the fictitious complex

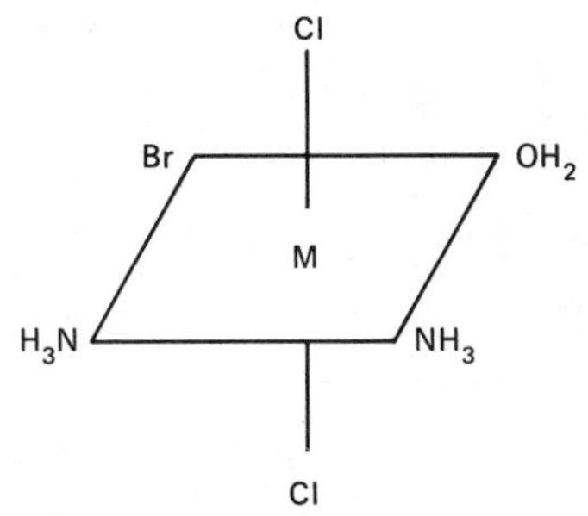

the approximate value of Dq for the complex is

$$Dq\ (\text{complex}) = \frac{2Dq(\text{Cl}^-) + 2Dq(\text{NH}_3) + Dq(\text{H}_2\text{O}) + Dq(\text{Br}^-)}{6} \tag{5.9}$$

The Dq values of the ligands are obtained from investigations of the homoligand octahedral complexes of the metal ion. This is a very powerful tool for the determination of binding sites and the geometry of complexes, since the observed spectrum may be compared with that anticipated for each possibility.

The spectra of d^1 and d^9 octahedral or tetrahedral complexes are characterized by one ligand field band which is defined above as 10 Dq. The following generalizations are for octahedral complexes for a given number number of d electrons (d_n). In the case of d^2 and d^8, there are three bands possible and the low-energy band is 10 Dq for d^8 and approximately so for d^2. For d^3 and d^7 three bands are also possible, though usually only two are observed. The low-energy band is 10 Dq for d^3 and approximately so for d^7. The spectra of d^4, d^5, and d^6 are complicated and depend on the metal–ligand interaction because of changes in the ground state properties. The d^4 complexes are usually weak and the low-energy band (only one of ligand field origin) is 10 Dq. The d^6 complexes are usually in strong environments and two bands are predicted, and the low-energy band approximates 10

Dq. The spectra of d^5 complexes are usually void of intense ligand field bands and for weak ligands no band has a simple dependence on 10 *Dq*. For a given d^n the number of bands for tetrahedral complexes is the same as for octahedral systems, and the first band is close to 10 *Dq*. The several bands which occur in the d^2 through d^8 complexes arise from excited states which are generated by interelectronic repulsions. The theory of the behavior of these bands is treated in several books (26, 28–34).

The spectra of a series of Ni(II) complexes (Figure 5.11) illustrates the relationship between the color and spectrum of a complex and the shift of the bands according to the spectrochemical series. The $[Ni(H_2O)_6]^{2+}$

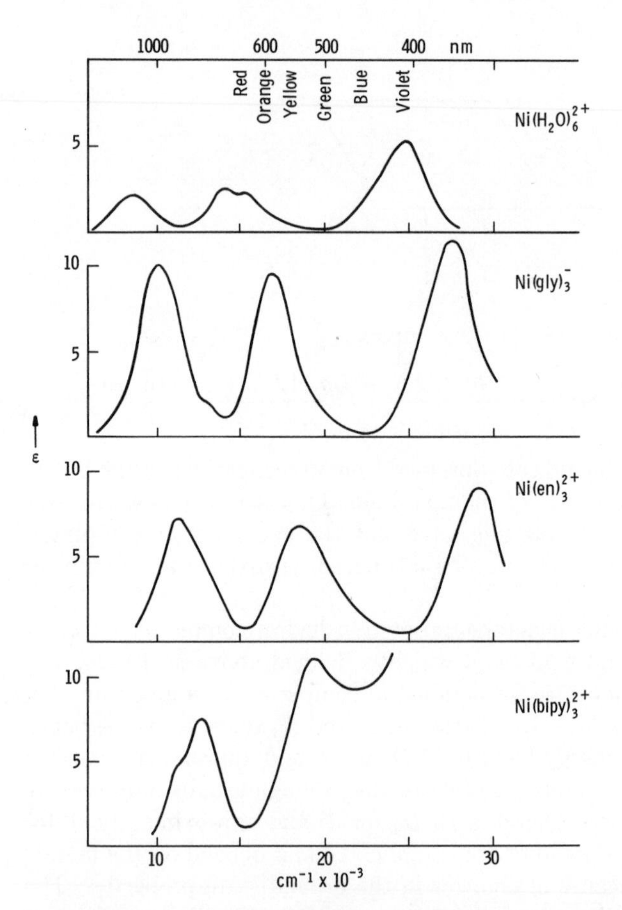

Figure 5.11. The color and spectra of several nickel(II) complexes. The characteristic color is given by the point of minimum absorption in the visible region.

complex transmits green; [Ni(glycinato)$_3$], blue; [Ni(ethylenediamine)$_3$]$^{2+}$, violet; and [Ni(bipyridyl)$_3$]$^{2+}$, red light. The values of Dq taken from the position of the first band ($10Dq$) are [Ni(H_2O)$_6$]$^{2+}$, 850 cm^{-1}; [Ni(glycinato)$_3$]$^{-}$, 1010 cm^{-1}; [Ni(ethylenediamine)$_3$]$^{2+}$, 1120 cm^{-1}; and [Ni(bipyridyl)$_3$]$^{2+}$, 1265 cm^{-1}. The weak shoulders present on some of the bands are due to spin-forbidden transitions.

B. *Molecular Orbital Theory*

The molecular orbital theory of transition metal complexes (26, 27) predicts the presence of the *d-d* (intermetal orbital) ligand field transitions as well as ligand–metal electron transfer (charge transfer) and ligand–ligand bands. The interaction of the metal *d*, *s*, and *p* orbitals with the ligand *s* and *p* orbitals is given in Figure 5.12. The $t_{2g}(\pi^*)$ and $e_g(\sigma^*)$ levels have the same significance as in ligand field theory. These orbitals are filled to levels in accordance with the number of *d* electrons. Transitions are then possible to these orbitals from the ligand-like π orbitals only when these

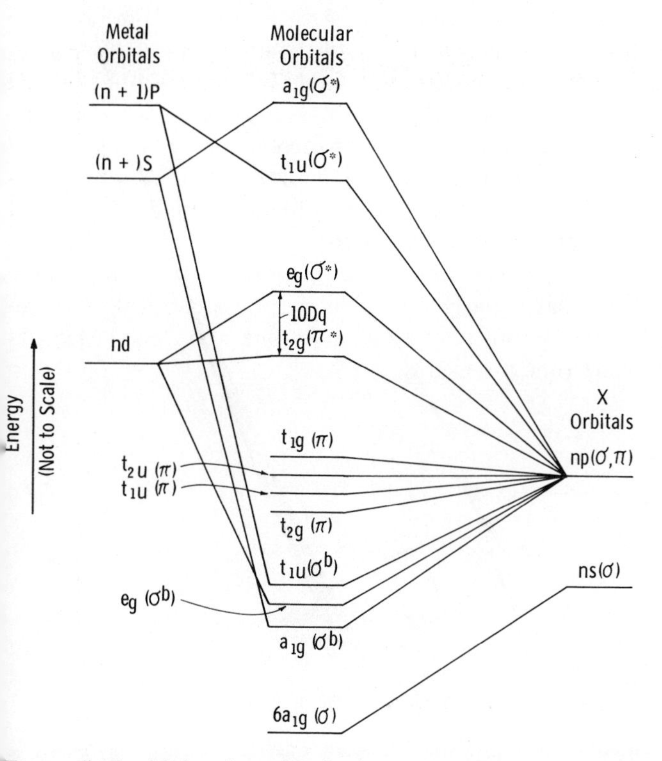

Figure 5.12. Molecular orbital scheme for MX_6^{n-}. complex (X = halide ion).

metal-like orbitals are partly unoccupied; that is, when there are nine or less d electrons for $e_g(\sigma^*)$ and five or less d electrons for $t_{2g}(\pi^*)$. Transitions from the ligand-like π orbitals to the ligand-like σ^* and π^* orbitals usually occur at high energy in the far ultraviolet. Most complexes do have electron transfer bands in the ultraviolet, especially those which involve π bonds. These electron transfer bands are very important in the case of second- and third-row transition metal ions where they may occur in the visible as well as in the UV regions. They also are important in tetrahedral complexes.

III. INSTRUMENTATION

Electronic spectra in the visible and UV regions are usually measured with a spectrophotometer (1), the basic elements (illumination, dispersion, and detection) of which are illustrated in Figure 5.13. This type of spectrophotometer is based on single-beam operation; that is, the sample is examined to determine the amount of radiation absorbed at a given wavelength and the results are then compared with a standard or reference obtained in a separate examination. In double-beam operation the monochromatic beam (after dispersion) is split into two components of equal intensity. One beam passes through the sample, the other through the reference, and the difference between the two components is determined simultaneously. This type of operation is characteristic of the more expensive, high-quality instruments (e.g., Cary Model 17, Figure 5.14), and is desirable for the determination of the entire spectrum of a compound. Single-beam spectrophotometers, considerably less expensive than double-beam instruments, are particularly useful for routine analytical measurements at fixed wavelength, but may be cumbersome when measurements must be made in short time intervals.

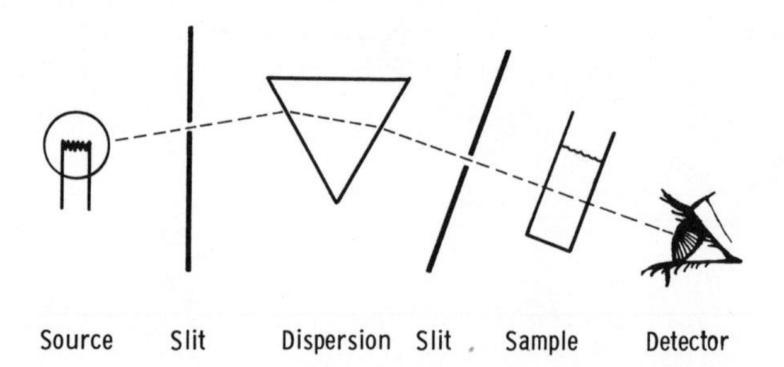

Figure 5.13. A typical spectrophotometer with source, collimators (slits), a means of dispersion into component wavelengths and a detector.

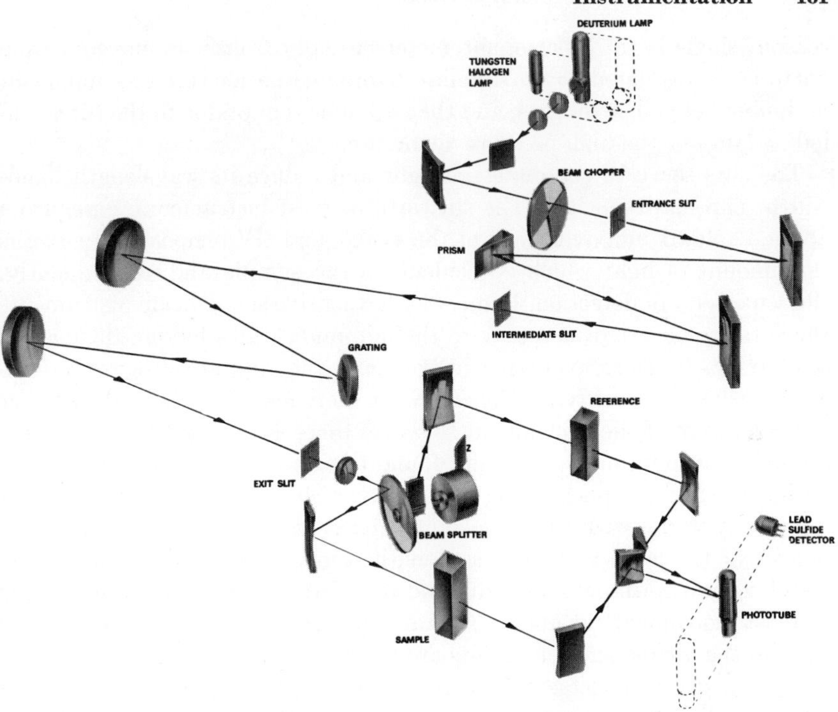

Figure 5.14. The optical diagram of the Cary Model 17 spectrophotometer (reprinted with permission from Cary Instruments.)

The source of radiation depends on the region of the spectrum to be studied. In the visible and near infrared regions, the source is a tungsten bulb which emits continuous radiation between 2500 and 350 nm. The recent availability of tungsten-iodine lamps with quartz envelopes has extended the usable wavelengths of these sources to 300 nm. These new lamps are preferred because they have a longer lifetime and a higher intensity in the visible region. In the UV region high pressure hydrogen or deuterium gas discharge lamps are used as a continuum source in most instruments; they emit continuous radiation in the 380–180 nm region. Deuterium lamps are preferred because of their greater intensity. High pressure mercury lamps or zirconium arcs may also be used in the ultraviolet. A source which may be used in both the visible and UV regions is the high pressure xenon arc, which will probably be available in future instruments. It is presently not used as such because this source is expensive and not very stable.

The stability of the source is very important for single-beam operation since the intensity must not vary in the time of the experiment. For this reason, single-beam spectrophotometers usually utilize dc currents from batteries or regulated power supplies. Lamp fluctuations are less important in double-beam instruments, and they are more adaptable to the tungsten-iodine lamp or the high pressure xenon arc.

The slits serve to collimate the light and reduce its wavelength bandwidth (chromatic purity). The slitwidth of most instruments varies from about 1 to 0.01 mm over most of the visible and UV regions, and governs the amount of light which is incident on the sample and, consequently, the sensitivity of detection. In most less-expensive single-beam instruments the slits are adjusted manually to the minimum width having the desired sensitivity. In more expensive instruments the slits are programmed to give the desired sensitivity. Generally the more narrow the slit, the greater the resolution of the instrument. However there is a slit width below which no further resolution may be gained, and this minimum slit depends on the design of the instrument. The width of the slit used to obtain optimum sensitivity varies with the wavelength and is an indication of the spectral output of the lamp and the wavelength response of the detector. In the visible region optimum sensitivity and therefore minimum slit widths occur in most instruments at about 500 nm. The sensitivity in the UV is lower than in the visible and larger slits are used.

The means of dispersion may be either a prism or a grating. A prism has a nonlinear dispersion and a lower spectral resolving power than a grating, and must be made of a transmitting material such as quartz or fluorite. The latter material is used for the dispersion of UV radiation. Reflection gratings are now produced with high efficiency and quality and are preferred in modern instruments. The resolving power of a spectrophotometer using a grating will depend on the number of grooves on the grating surface, the focal length of the instrument, and the slit width. Most instruments use gratings with 15,000 to 30,000 grooves/in. The wavelength resolution of the best commercially available instruments does not exceed 0.1 nm; that is, bands separated by 0.1 nm or more may be resolved. Optimum resolving conditions are essential when studying narrow band spectra. If the resolution is not sufficient, the band will appear broadened and the measured extinction coefficient will be lower than the actual value. This behavior, therefore, does not obey Beer-Lambert's law.

The most sensitive detectors used in spectrophotometers are quartz-windowed photomultipliers. For each photon absorbed by the detector as many as 10^8 electrons may be emitted by the multiple grid photomultipliers. The spectral response of the photomultiplier varies with the type of surface, but most have their maximum sensitivity at about 500 nm. Red-sensitive photomultipliers are available and may be used to enhance sensitivity or resolution in the 900–600 nm region. Unfortunately these red-sensitive tubes usually have a poor UV response. A variety of photomultipliers with good response from 700 to 200 nm are available. The response of the detector should be linear with the number of photons absorbed. In all instruments the currents from the photomultipliers are amplified before being recorded. Single-beam instruments use dc amplifier circuits.

In most double-beam instruments a low-frequency alternating current is produced by chopping the sample and reference beams. The amplifiers used here are frequency selective and the amplification is carried out with optimum signal to noise ratios.

The spectrum record is achieved by sweeping the wavelengths and recording the currents detected by the photomultiplier. The currents are calibrated to an appropriate absorbance or transmittance scale and recorded.

The samples are normally liquids and are contained in cells which are transparent in the region being studied. Glass or plastic cells may be used for the visible region, and quartz may be used for both the visible and UV regions. Normally the path length of the cell is 1 cm, but since cells of varying thickness may be used, it is important that the sample compartment of the spectrophotometer be large enough to accommodate long cells and other apparatus which may be required to support, heat, or cool the sample.

The Cary Model 17 is a typical example of a high-quality spectrophotometer. The optical diagram of this instrument is given in Figure 5.14. The light source is selected depending on the region to be studied. The beam is first dispersed by a quartz prism which also serves as an order sorter. A grating is used as the second dispersion stage. This double dispersion reduces the stray light (a plague of monochromators) to about one part per million and yields better (purer) monochromatic light. After dispersion the beam is split by a rotating mirror and is alternately allowed to pass through the sample and reference. The difference in the signal detected by the photomultiplier is amplified and recorded. This instrument is also capable of near-infrared measurements by the use of a PbS detector. Because of the low stray light, this instrument is capable of accurate measurements of optical densities greater than 4.0. The wavelength resolution of the Model 17 is 0.1 nm in the visible and UV regions. The instrument is also capable of recording the derivative of the absorption curve, which is a valuable tool in finding hidden bands or shoulders.

Another new type of instrument is the dual-wavelength spectrophotometer made by Aminco and by Perkin-Elmer. The Perkin-Elmer Model 356 has an optical path similar to the Cary Model 17, except that it is a single-pass instrument with a grating monochromator. Two beams are produced

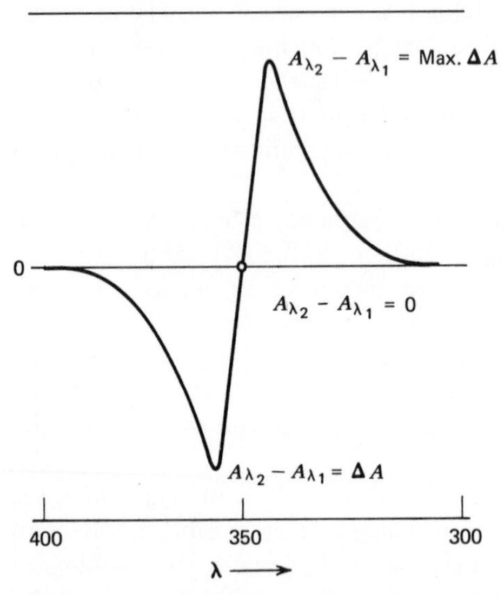

Figure 5.15. Derivative of a Gaussion absorption curve (see Figure 5.5*b*) with maximum at 350 nm. The derivative is equal to zero at maximum value (reprinted with permission from the Perkin-Elmer Corp.).

by a mask at the source and are dispersed by two separate gratings. The two gratings may be operated in tandem and normal spectra are obtained. However the two gratings may also be set to a constant difference (about 1–5 nm) and the instrument behaves like a phase-sensitive detector. In this case the output plot is a derivative spectrum. The derivative of the absorption curve (Figure 5.15) looks like an ESR spectrum. The crossover points at the baseline correspond to a wavelength of maximum absorption; the width of the band is proportional to the wavelength between the inflection points. The peak to trough height, which is proportional to concentration, will obey Beer-Lambert's law.

An example of the utility of derivative spectra in sorting out bands in anthracene is given in Figure 5.16. Each intersection with the baseline corresponds to the position of an absorption maximum; the saddle points of the derivatives are the positions of the shoulders. This technique is excellent for accurate wavelength determination of shoulders and the sorting of vibrational progressions. The dual-wavelength capability of the Model 356 may also be used to monitor two wavelengths and it is therefore possible to follow the behavior of two separate species as a function of time. By this technique, the difference in the rate of reaction and the production of certain species may be followed, from which one can deduce reaction mechanisms and the nature and quantity of the reaction intermediates.

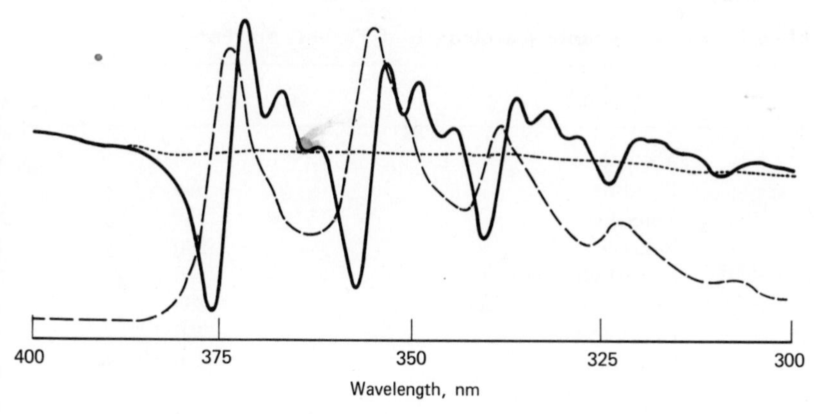

Figure 5.16. A comparison of the derivative (solid line) and absorption (broken line) spectrum of anthracene (reprinted with permission from the Perkin-Elmer Corp.).

IV. TECHNIQUES

Most spectral determinations are carried out in solution. Distilled water is the most frequently used solvent as it does not absorb in the visible or UV regions. Tap water however may contain organic matter or Cl_3^-, which do absorb. Any clear, colorless solvent which does not react with the solute is appropriate for visible studies. Many common organic solvents (e.g., benzene and acetone) absorb in the UV and have limited use. The best solvents for visible light and UV spectroscopy are ethers, alcohols, and saturated hydrocarbons. The UV cutoffs of the more common solvents are given in Table 5.7.

Several mixed solvent systems do not crystallize when cooled but form a highly viscous glassy state. This phenomenon allows "solution" spectra to be obtained at temperatures down to 80°K. For nonpolar solutes, a mixture of ether, ethanol, and isopentane or methylcyclohexane and isopentane is very good. Polar solutes may be dissolved in a mixture of ethanol and methanol or water and dimethylsulfoxide. Ionic solids may be dissolved in 6 *M* magnesium chloride, which also forms a glassy state on cooling.

Turbid solutions tend to scatter the light and give false absorption readings particularly in the UV region. This problem can be somewhat alleviated by placing the sample as close as possible to the photomultiplier.

Single crystals may be studied conveniently in a spectrophotometer by mounting them in the path of the beam. For small crystals it is possible to condense the beam by using mirrors or lenses. The absorption of a crystal approximately follows Beer-Lambert's law; the concentration is defined

Table 5.7. Lowest Usable Wavelength of Various Solvents

Solvent	Cut-off, nm
Acetone	330
Pryidine	305
Toluene	285
Benzene	280
Dimethylformamide	270
Carbon tetrachloride	265
Ethyl acetate	260
Chloroform	250
1,2 Dichloroethane	233
Dichloromethane	230
Glycerol	230
Hexane	220
Isooctane	220
Acetonitrile	212
Methanol	210
Ethanol	210
Ether	210

by the density of the solid. If the chromophore in the crystal has a high extinction coefficient, it is necessary to use very thin crystals which may render the method impractical. Insoluble solids or powders may be pressed into KBr pellets for quantitative studies, or mulled on a quartz plate or filter paper for qualitative studies. The index of refraction of the mulling liquid should be close to that of the solid in order to reduce scattering of the light.

Solids may also be studied by reflectance techniques in which the light reflected from the surface of a crystal or powder is compared with a standard reflector or reference such as MgO (6,11). The difference between the reflected light of the sample and the reference yields a spectrum analogous to the transmission spectrum. The reflecting power of the sample is $R = I/I_0$, in which I_0 is the intensity of the incident and I the intensity of the reflected light. The simplest technique for observing a reflectance spectrum is to mount the sample behind opal glass in such a way as to direct the diffuse reflected light to the photomultiplier (48). This method causes a considerable portion of the light to scatter, which can be acceptable in the visible region but makes observation of reflectance spectra in the UV difficult. A widely used method is attenuated total reflectance, in which the sample is mounted behind a prism such as a Dove prism. In this configuration the light beam undergoes multiple reflections between the surface of the prism and the surface of the sample before it is transmitted to the detector. The beam is attenuated because of losses at the interface of the prism and the sample.

The method may be improved by collecting all the reflected light of the sample and transmitting it to the detector. In this technique, called total reflectance, the sample is mounted inside a nearly spherical mirror which has entrance and exit ports. The light enters and is reflected by the sample, and all the scattered light is collected by the mirror and transmitted to the detector via the exit port.

V. APPLICATIONS

Visible and UV spectroscopy have been applied extensively to the solution of a wide variety of analytical problems. Any one of the roughly 30,000 spectra given in the Sadtler (20) collection could be used as the basis of a qualitative or quantitative analysis. It is impossible to illustrate all the applications, but an attempt will be made to demonstrate a few typical applications.

Water pollution is a typical example of the utility of UV spectroscopy. Pure water has no UV absorption but most natural waters do. The amount of UV absorption is a measure of the organic matter present, and polluted water is always high in organic compounds. However, UV studies alone may not establish the type or types of organic pollution (49).

The geometrical isomers of transition metal complexes are easily distinguished from their visible spectra. The cis isomer of $[Co(en)_2F_2]NO_3$, for example, is violet and the trans isomer is green. The spectra given in Figure 5.17 illustrate these color differences (36). Note that the first band of the complex has two maxima in the trans isomer but only one in the cis. This is attributed to the fact that similar axial substituents split the octahedral levels of the complex twice as much in the trans as in the cis orientation, where the field is more averaged.

Sacconi (47) has shown that the temperature dependence of the spectrum of the Schiff base complex *bis*(*N*-decylsalicylaldimino)nickel(II) indicates planar$\rightleftharpoons$tetrahedral equilibrium. The intensity of the characteristic tetrahedral bands at 7,200 and 11,200 cm^{-1} increases with temperature, while the band at 16,000 cm^{-1}, which is characteristic of a planar complex, decreases in intensity with increasing temperature. The change in color of *cis*-$[Ni(triethylenediamine)(H_2O)_2]^{2+}$ from blue to yellow on adding inert salts to the solution has been shown from spectra by Jorgensen (43) to involve an octahedral$\rightleftharpoons$planar equilibrium. The pressure dependence of

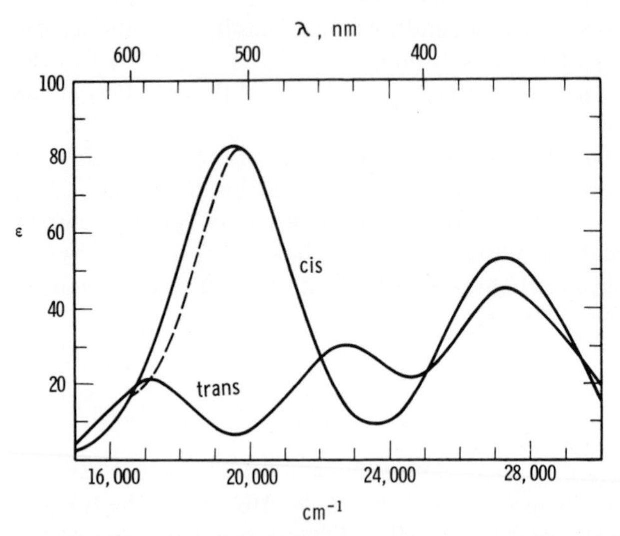

Figure 5.17. The absorption spectrum of *cis*- and *trans*-[Co(ethylenediamine) F_2]NO_3.

the spectrum (38) of salicylaldimine and aminotroponeimineate complexes also displays configurational equilibria. The planar forms have smaller volumes and are favored at higher pressures.

Spectrophotometric titrations can be used to study complex formation reactions, to evaluate equilibrium constants, and to determine the nature of species present. A comparison of the spectrophotometric titration of Cu^{2+} with racemic *d,l*-alanine and *l*-alanine shows a preferential stability of the meso (*d*-alanine plus *l*-alanine) in the bis-chelated species (41). Several nickel dipeptide complexes are blue up to pH 9 and turn yellow as the base content is increased. The formation of the yellow, square-planar species is observed during the deprotonation of the dipeptide (37). Spectrophotometric titrations may be carried out by changing the pH of a large quantity of solution external to the spectrophotometer and then using small amounts in the absorption cell, or by use of an internal cell such as that designed by Auld and French (35).

Metalloporphyrins are important in photosynthesis, nitrogen fixation in plants, and oxygen transport in the blood. The square-planar chelates of porphyrins (29, 30) with divalent metal ions exhibit three bands which are described in Table 5.8. In general, the α and β bands are of moderate intensity. The Soret band, which is characteristic of the porphyrin moiety, is more more intense. The shift of the bands to the red is indicative of lower stability as is a decrease in the ratio of the intensity of the α and β bands. From these data the stability order

$$Ni(II) > Co(II) > Ag(II) > Cu(II) > Zn(II) > Cd(II)$$

is obtained. These results are in agreement with data obtained by thermal analysis.

Table 5.8. Spectra of Metal Porphyrin Complexes in CCl_4 Solution (nm)

Metal Ion	α	β	Soret
Co(II)	561.5	528.5	403
Ni(II)	561	525.5	403
Cu(II)	573	534	409
Ag(II)	570	534	417.5
Zn(II)	579	541	411.5
Cd(II)	587	559	414

The absorption of β-diketones, 8-hydroxyquinoline, pyridine, and bipyridyl on metal thin films has been observed from the spectra of these surfaces (43). The magnitude of the $\pi \rightarrow \pi^*$ transitions of the chemisorbed species differs considerably from the energy of ligand or metal ion complex values, and the bands are more indicative of metal (zero oxidation state) complexes of the substrates. In some cases several modes or sites of attachment have been detected. Oxygen and other reactants remove the substrates so it is certain that the complexes are not those of metal ions. Studies of this nature are important in understanding the nature of metal surfaces which is very valuable in the understanding of catalytic processes.

Visible and UV spectroscopy is a common tool for the study of reaction rates. The analytical applications of these spectroscopic tools in reaction kinetics have recently been reviewed by Mark and Rechnitz (9). Since the rate of a reaction is proportional to the concentration of a reactant, measurements of initial rates may be used to derive concentrations. This technique is used, for instance, in measuring the enzyme activity of a sample. The activity of the enzyme peroxidase is measured by the rate of oxidation of dianisidine at 360 nm (44). Enzyme activity may de defined by the rate of increase of absorbance per unit time above that of a blank.

A different technique has been developed for the determination of proteins in the presence of nucleic acids (40). Nucleic acids absorb in the UV region but there are a number of wavelengths where the absorbance is equal. These are called isoabsorbance wavelengths; for example, 224.0 and

233.3 nm for yeast nucleic acid. In the presence of protein a difference in the absorbance at these wavelengths is observed. Calibration curves of absorbance difference versus concentration for a large number of proteins have been determined.

The sugar ribose degrades in hot 1 *M* HCl to a chromophore with maximum absorbance at 277 nm. This is the basis of a method for its estimation (39). The tranquilizer phenaglycodol can be determined without separation by oxidizing a sample with acid-dichromate and converting to a carbonyl for measurement. The sensitivity of the method may be enhanced by conversion of this carbonyl to the semicarbazone derivative (50).

The pesticide DDT, like other chlorinated hydrocarbons, attacks the central nervous system of insects and mammals and blocks the transport of cations across the nerve membrane (45). The spectrum of DDT has a λ_{max} at 240 nm. In the presence of nerve axons from cockroaches, the spectrum is shifted to 245 nm and a shoulder at 270 nm appears. These results, in conjunction with other data, lead to the suggestion that DDT acts as a charge transfer acceptor with the nerve as a donor. A complex is formed which results in a deactivation of the nerve functions.

The spectra of a series of amino- and nitro-substituted benzene compounds display a very interesting effect [Figure 5.18 (46)]. The mononitroanilines all show one band in the 400-nm region, as does 2,6-dinitroaniline. The compounds 2,4-dinitroaniline and 2,4,6-trinitroaniline both have an additional band at higher energy. These components have been assigned to the 1, 4 (para) intramolecular charge-transfer bands from the amino donor to the nitro acceptor. In the case of 1,3-diamino-2,4,6-trinitrobenzene there are two higher energy components. The band at 411 nm has been assigned to the four nearly equivalent ortho intramolecular charge-transfer transitions, while the higher energy bands at 276 and 326 nm are assigned to "split" components of the para charge-transfer states. The origin of the splitting is attributed to opposite charge displacements involving meta substituents. If the amino groups of 1,3-diamino-2,4,6-trinitrobenzene are alkylated, the bands coalesce to a single lower intensity peak at about 350 nm. This is due to a twisting of the chromophores out of the plane of the ring by the steric bulk of the substituents and a concomitant reduction in intramolecular charge transfer is then observed.

Many other applications have been reported for the use of electronic spectroscopy in studying chemical problems. They range from studies of the purity of polymers to the use of diffuse reflectance spectra of paint pigments and coloring matter in flower petals. The data for these investigations can usually be obtained with a minimum of effort.

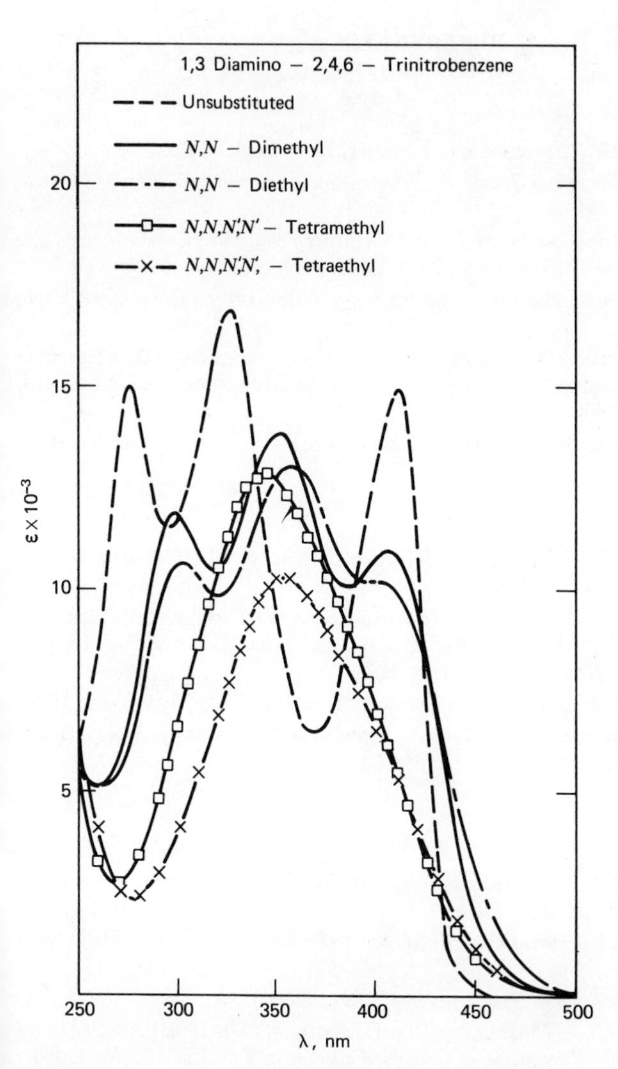

Figure 5.18. Absorption spectra of *N*,*N*-dialkyl and *N*,*N*,*N'*,*N'*-tetraalkyl derivatives of 1,3-diamino-2,4,6-trinitrobenzene (46) (reprinted with permission from the *Journal of the American Chemical Society*).

REFERENCES

1. General

1. R. E. Dodd, *Chemical Spectroscopy*, Elsevier, New York, 1962.
2. J. R. Edisbury, *Practical Hints on Absorption Spectrometry*, Plenum Press, New York, 1967.
3. L. N. Ferguson, *Advances in Analytical Chemistry and Instrumentation*, C. N. Reilley, Ed., Vol. 4, Inteerscience, New York, 1965, p. 411.
4. N. J. Harrick, *Internal Reflection Spectroscopy*, Wiley-Interscience, New York, 1967.
5. G. Herzberg, *Molecular Spectra and Molecular Structure*, Vol. III, Electronic Spectra and Electronic Structure of Polyatomic Molecules, Van Nostrand, Princeton, N. J., 1966.
6. G. W. King, *Spectroscopy and Molecular Structure*, Holt, Rinehart and Winston, New York, 1964.
7. G. Kortüm, *Reflectance Spectroscopy*, Springer-Verlag, New York, Inc., New York, 1969.
8. H. B. Mark and G. A. Rechnitz, *Kinetics in Analytical Chemistry*, Wiley-Interscience, New York, 1968.
9. A. A. Schilt and B. Jaselskis, in *Ultraviolet and Visible Spectrophotometry*, *Treatise on Analytical Chemistry*, Part 1, Vol. 5, I. M. Kolthoff and P. J. Elving, Eds., Interscience, New York, 1964, p. 2943.
10. E. I. Stearns, *Practice of Absorption Spectrophotometry*, Wiley, New York, 1969.
11. W. W. Wendlandt and H. G. Hecht, *Reflectance Spectroscopy*, Wiley-Interscience, New York, 1966.

2. Organic

12. L. J. Andrews and R. M. Keefer, *Molecular Complexes in Organic Chemistry*, Holden-Day, San Francisco, 1964.
13. C. N. Banwell, *Fundamentals of Molecular Spectroscopy*, McGraw-Hill, New York, 1966.
14. R. P. Bauman, *Absorption Spectroscopy*, Wiley, New York, 1962.
15. A. B. F. Duncan, F. A. Matsen, and D. R. Scott, in "Chemical Applications of Spectroscopy," in *Technique of Organic Chemistry*, Vol. IX, W. West, Ed., Interscience, New York, 1968.
16. W. F. Forbes, "Ultraviolet Absorption Spectroscopy," in *Interpretive Spectroscopy*, S. K. Freeman, Ed., Reinhold, New York, 1965.
17. H. H. Jaffe and M. Orchin, *Theory and Applications of Ultraviolet Spectroscopy*, Wiley, New York, 1962.
18. J. N. Murrell, *The Theory of the Electronic Spectra of Organic Molecules*, Methuen, London, 1963.
19. D. J. Pasto and C. R. Johnson, *Organic Structure Determination*, Prentice-Hall, New York, 1969.

20. *Sadtler Standard Spectra, Ultraviolet*, Sadtler Research Laboratories, Philadelphia.
21. A. I. Scott, *Interpretation of the Ultraviolet Spectra of Natural Products*, Macmillan, New York, 1964.
22. R. M. Silverstein and G. C. Bassler, *Spectrometric Identification of Organic Compounds*, 2nd ed., Wiley, New York, 1967.
23. H. Suzuki, *Electronic Absorption Spectra and Geometry of Organic Molecules*, Academic, New York, 1967.
24. S. Walker and H. Straw, *Spectroscopy, Volume II: Ultraviolet, Visible, Infrared and Raman Spectroscopy*, Chapman Hall, London, 1967.

3. Inorganic

25. C. J. Ballhausen, *Progr. Inorg. Chem.* **2,** 251 (1960).
26. C. J. Ballhausen, *Introduction to Ligand Field Theory*, McGraw-Hill, New York, 1962.
27. C. J. Ballhausen and H. B. Gray, *Molecular Orbital Theory*, Benjamin, New York, 1964.
28. T. M. Dunn, R. G. Pearson, and D. S. McClure, *Crystal Field Theory*, Harper & Row, New York, 1965.
29. J. E. Falk and D. D. Perrin, in *Haematin Enzymes*, J. E. Falk, R. Lemberg, and R. K. Morton, Eds., Pergamon, London, 1961.
30. J. E. Falk, *Porphyrins and Metalloporphyrins*, Elsevier, New York, 1964.
31. B. N. Figgis, *Introduction to Ligand Fields*, Interscience, New York, 1966.
32. C. R. Hare, "Visible and Ultraviolet Spectroscopy," in *Spectroscopy and Structure of Metal Chelate Compounds*, K. Nakamoto and P. J. McCarthy, Eds., Wiley, New York, 1968.
33. C. K. Jorgensen, *Absorption Spectra and Chemical Bonding in Complexes*, Addison-Wesley, Reading, Mass., 1962.
34. D. S. McClure, *Solid State Phys.* **9,** 399 (1959).

4. Other

35. D. S. Auld and T. C. French, *Anal. Biochem.* **34,** 262 (1970).
36. F. Basolo, C. J. Ballhausen, and J. Bjerrum, *Acta Chem. Scand.* **9,** 810 (1955).
37. P. Chamberlain, Ph. D. thesis, State University of New York at Buffalo, 1968.
38. A. H. Ewald and E. Sinn, *Inorg. Chem.* **6,** 40 (1967).
39. E. R. Garrett, J. Blanch, and J. K. Seydel, *J. Pharm. Sci.* **56,** 1560 (1967).
40. W. E. Groves, F. C. Davis, and B. H. Sells, *Anal. Biochem.* **22,** 195 (1968).
41. C. R. Hare, B. S. Manhas, T. G. Mecca, W. Mungall, and K. M. Wellman, *Proc. 9th Intern. Conf. on Coord. Chem.*, St. Moritz, Switzerland, 1966, p. 199.
42. C. K. Jorgensen, *Acta Chem. Scand.* **11,** 399 (1957).
43. K. Kishi and S. Ikeda, *J. Phys. Chem.* **71,** 4384 (1967); *ibid.* **73,** 15, 729, and 2559 (1969).

44. S. J. Klebanoff, *Endocrinology* **76,** 301 (1965).
45. F. Matsumura and R. D. O'Brien, *J. Agr. Food Chem.* **14,** 39 (1966).
46. R. R. Minesinger and M. J. Kamlet, *J. Amer. Chem. Soc.* **91,** 4155 (1969).
47. L. Sacconi, *J. Chem. Soc.* **1963,** 4608.
48. K. Shibata, *Methods Biochem. Anal.* **9,** 217 (1962).
49. L. Stanisavlievici, *Waerme* **72,** 98 (1966); *Chem. Abs.* **66,** 5572Z (1967).
50. J. E. Wallace, *J. Pharm. Sci.* **57,** 426 (1968).

NORMAN B. COLTHUP

American Cyanamid Co.
Stamford, Connecticut

VI. Infrared and Raman Spectroscopy

I. INTRODUCTION

The vibrational and rotational energies of molecules can be studied by infrared and Raman spectroscopy. Both techniques have been used in theoretical investigations such as the study of the nature of the molecular force field and the proper characterization of molecular vibrations. Infrared spectroscopy has especially proved to be of tremendous value to the chemist who uses spectroscopy to characterize the structure of molecules in a more or less empirical way. In this type of application infrared can be used alone or in conjunction with available chemical knowledge and, even more powerfully, in conjunction with NMR, UV and mass spectroscopy. With bench chemists IR spectroscopy is currently more popular than Raman spectroscopy. Among the reasons for this are the high cost of Raman instruments and the fact that there have been more restrictions on the type of samples that can be analyzed with the latter technique. Infrared spectra can be run on solids, liquids, or gases whereas Raman spectra have been commonly (though not exclusively) obtained on liquid samples. However with the recent introduction of lasers as sources for Raman spectrometers, the older sampling problems with solids or with samples which fluoresced or were highly colored have been greatly reduced. Water solutions are ideal samples for Raman spectra because water, which is a strong IR absorber, is relatively transparent in the Raman spectrum.

The Raman spectrum is as rich in molecular structure information as is the IR spectrum. Both techniques can be used, in effect, to measure molecular vibrational frequencies. These vibrational spectra can be used directly, simply as molecular "fingerprints" to characterize and identify the molecule, or they can be used as "coded" pieces of information about the molecular structure. While IR and Raman spectra have many similarities, they are not exact duplicates; relative band intensities in each technique may sometimes differ quite markedly (Figure 6.1). The two techniques frequently yield complementary types of information. For a complete vibrational analysis, the use of both methods is generally necessary.

A common application of IR and Raman spectroscopy is the use of these techniques for "group frequency" analyses. In molecules certain functional groups show certain vibrations in which only the atoms in the group move. Since the rest of the molecule is mechanically uninvolved in the vibration, a group vibration will have a characteristic frequency which remains constant no matter what molecule the group is in. This group frequency can be used to reveal the presence or absence of the group in the molecule and this is frequently of tremendous help in characterizing the molecular structure. In addition to the qualitative analyses studies described above, both techniques can be used for quantitative analysis. This is usually carried out by comparison with standard mixtures.

Raman spectroscopy makes use of electromagnetic radiation usually in the visible region between 4000 and 7000 Å. In IR spectroscopy the micron ($\mu = 10^{-4}$ cm) is more often used as the dimension for the wavelength. Using this unit, the visible region is about 0.4–0.7 μ, the near infrared is about 0.7–2.5 μ, the fundamental infrared region is about 2.5–50 μ and the far infrared is about 50 μ to a fraction of a millimeter. The reason for a division at about 2.5 μ is that absorption caused by fundamental vibrational transitions all fall on the long wavelength side of 2.5 μ. The reason for a division at about 50 μ is largely instrumental.

The properties that electromagnetic radiation and molecules have in common are energy and frequency. However, the frequency, ν, in hertz (Hz) in this part of the spectrum is an inconveniently large number so a number which is proportional to frequency is commonly used. This is called the wavenumber, ω (cm^{-1}), which denotes the number of waves in a 1-cm-long wavetrain. This unit is related to the other units by

$$\omega = \frac{\nu}{c} = \frac{1}{\lambda_{\text{cm}}} = \frac{10^4}{\lambda_\mu} \tag{6.1}$$

where λ_{cm} and λ_μ are the wavelengths expressed in cm and μ respectively.

The unit used in the Raman spectrum is the wavenumber expressed in cm^{-1}. In IR spectroscopy both wavenumber (cm^{-1}) and wavelength (μ) have been used. Prism instruments are more commonly seen delivering spectra where the horizontal coordinate is linear with wavelength but grating instruments generally deliver spectra linear with wavenumber. In either case results are more often reported in wavenumbers since these are proportional to molecular properties (frequency and energy), whereas the wavelength is a property of the radiation only.

The vertical coordinate in an IR spectrum is usually presented linear with sample transmittance. Transmittance is defined as the radiant power of the radiation which is incident on the sample, divided into the radiant power transmitted by the sample. Absorption bands are usually presented pointing down. In Raman spectroscopy, the vertical coordinate can be the direct photoelectric intensity of the Raman emission spectrum with the emission bands pointing up.

In Figure 6.1, IR and Raman spectra pairs of certain selected compounds are presented.* The Raman spectra of these liquids were obtained using

* These IR and Raman spectra were kindly provided by H. J. Sloane, Cary Instruments, Monrovia, California.

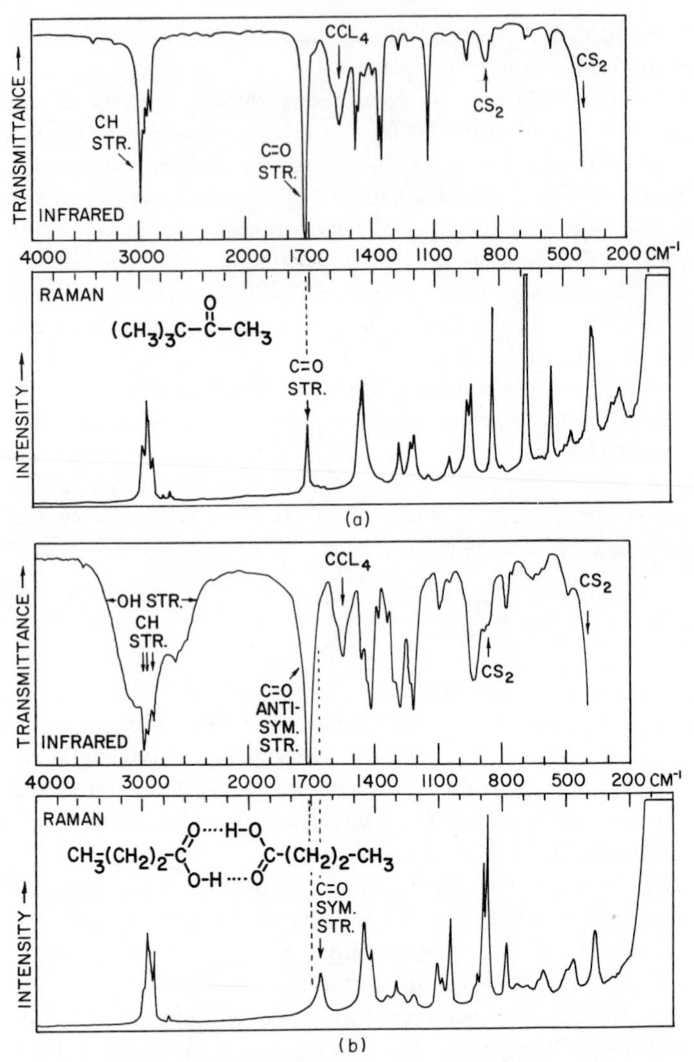

Figure 6.1*a–d*. Infrared and raman spectra pairs.

the He–Ne laser 6328 Å exciting line. The IR spectra were run on 10% solutions in CCl_4 above 1330 cm^{-1} and in CS_2 below 1330 cm^{-1} in a cell about 0.1 mm thick. The uncompensated solvent bands have been marked on the spectra. CCl_4 is seen at 1545 cm^{-1} and CS_2 at 855 and 395 cm^{-1}. These spectra pairs clearly indicate the intensity differences for certain bands in the IR and Raman spectra.

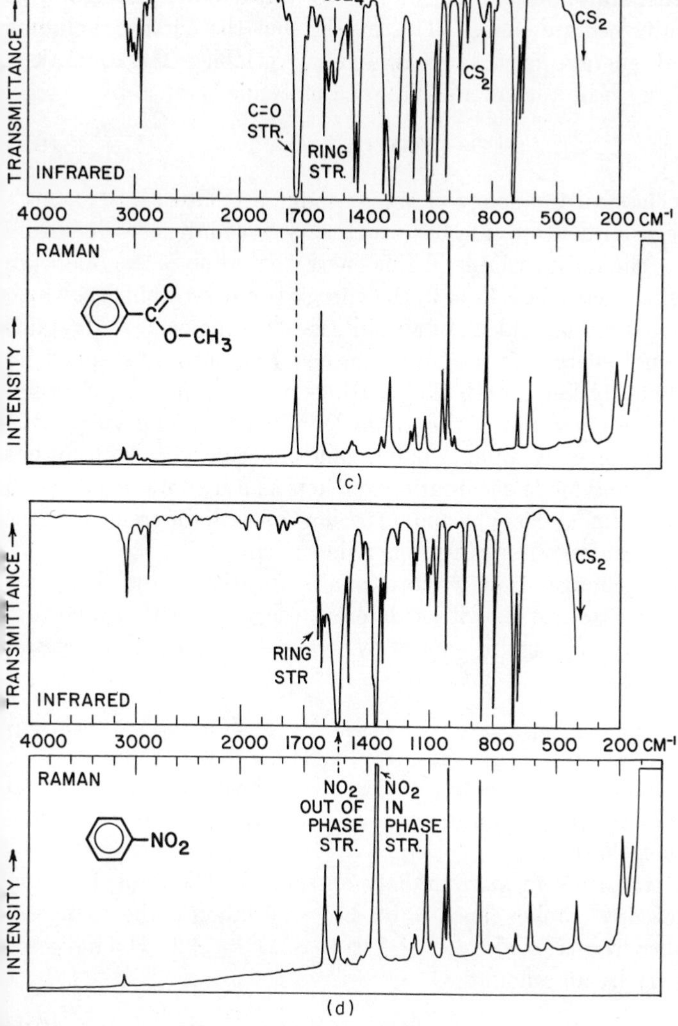

Figure 6.1. (continued)

II. THE HARMONIC OSCILLATOR

1. The Quantum-Mechanical Harmonic Oscillator

Quantum mechanics state that on the molecular level, energy is quantized, in other words a molecule may only exist in certain definite energy states. The molecule may change from one discrete state with its associated energy to another state, but it may not exist for any definite period of time in a state of intermediate energy. This means that the energy, including the vibrational energy, is not continuously variable. The quantized vibrational energy E of a harmonic diatomic molecule is given by

$$E = (V + {}^1/_2)h\nu \tag{6.2}$$

where ν is the classical frequency of the diatomic oscillator, h is Planck's constant and V is a quantum number which may have integer values such as 0, 1, 2, 3, ... The only variable in this energy equation is the quantum number V. Notice that when $V = 0$, the energy is not zero but $\frac{1}{2}h\nu$ which means that even at a temperature of absolute zero the molecule still retains a little vibrational energy. If the vibration could be entirely stopped we would simultaneously know both the locations of the atoms (equilibrium position) and the momentum of the atoms (zero). This would violate the Heisenberg uncertainty principle. A polyatomic molecule with N atoms has more than one normal mode of vibration and acts as if it contained $3N - 6$ (or $3N - 5$ in the linear case) harmonic oscillators with different frequencies, each able to undergo its transitions independently.

If the molecule changes from a vibrational state with energy E with a quantum number V to a state with a different energy E' with a quantum number V', then the change in energy ΔE equals $E' - E$, which is also equal to $(V' + \frac{1}{2})h\nu - (V + \frac{1}{2})h\nu$, so

$$E' - E = (V' - V)h\nu \qquad \text{or} \qquad \Delta E = \Delta V h\nu \tag{6.3}$$

The energy change equals a positive or negative integer (ΔV) times $h\nu$ and may be compensated for by a counter change in the energy of the interacting electromagnetic radiation.

The type of transition most commonly observed in IR absorption and Raman spectroscopy (Stokes lines, to be discussed later) is the change in quantum number from $V = 0$ to $V' = 1$ so that $\Delta V = +1$. The molecule then gains energy by an amount

$$\Delta E = h\nu \tag{6.4}$$

This is the transition which gives rise to the fundamental vibrational frequency observed in IR, Raman, and other branches of spectroscopy (Table 8.1).

Other transitions where $\Delta V = 1$, such as $V = 1 \rightarrow V' = 2$ or $V = 2 \rightarrow V' = 3$, give rise to "hot" bands with frequencies the same as that of the fundamental. However these have much lower intensities than the fundamental because most molecules at room temperature exist in the $V = 0$

level. This makes the hot band transitions which start from the $V = 1$ or possibly $V = 2$ levels much less probable than the fundamental transitions which start from the highly populated $V = 0$ level. However the population of the higher energy levels increases with a rise in temperature. This will increase the intensities of these bands, which is the reason they are called hot bands.

In a transition such as $V = 1 \rightarrow V' = 0$ or $\Delta V = -1$, the molecule loses energy ($\Delta E = -h\nu$) and this process gives rise to emission in the IR spectrum or an anti-Stokes line in the Raman spectrum. A transition such as $\Delta V = +2$ or $+3$ results in what is called an overtone absorption in IR spectroscopy. This phenomenon will be discussed in the next section.

2. The Effect of Anharmonicity

In a harmonic vibration of a molecule a plot of the coordinate of any atom versus time is a sine (or cosine) wave. If this plot departs in shape from the simple sine wave, the vibration is said to be mechanically anharmonic. In the actual anharmonic diatomic molecule the plot of potential energy U versus the change in bond length Δr is no longer exactly the parabola dexcribed by $U = \frac{1}{2}k(\Delta r)^2$, which is applicable to the harmonic oscillator. As a result the energy spacings between the various energy states which were exactly equal in the harmonic oscillator are no longer exactly equal for the anharmonic case. One result is that the hot bands do not have exactly the same frequency as that observed for the fundamental transition. Another result concerns the transition $V = 0$ to $V' = 2$, which gives rise to what are called *overtones*. In an harmonic oscillator the transition $\Delta V = 2$ causes an energy change $\Delta E = 2h\nu$ exactly twice as great as that caused by the fundamental transition. ($\Delta V = 1$, $\Delta E = h\nu$). In an anharmonic oscillator this is no longer true and the overtones, when observed, are not at exactly twice the frequencies of the fundamental.

If the molecular vibrations were entirely harmonic, only fundamentals would be observed in IR and Raman spectroscopy because all transitions except those where $\Delta V = \pm 1$ are strictly forbidden. However if the oscillator is somewhat anharmonic then these transitions are no longer forbidden. Overtones (2ν) and, in polyatomic molecules, combination bands ($\nu_1 + \nu_2$) can then be observed. The overtone intensities depend on the amount of anharmonicity present. Since these intensities are normally quite low but not zero, we conclude that molecular vibrations are only slightly anharmonic and that in the lower vibrational states the potential energy function of displacement departs only slightly from the harmonic type function $U = \frac{1}{2}k(\Delta r)^2$.

III. INFRARED ABSORPTION

1. A Schematic Description of the Infrared Absorption Process

In a typical infrared spectrometer a source simultaneously emits all the infrared frequencies of interest. These radiation frequencies are about the same order of magnitude as the molecular vibrational frequencies. The natural frequency of vibration of the HCl molecule, for example, is about 8.7×10^{13} sec^{-1}(2890 cm^{-1}). When we shine all the infrared frequencies through a sample of HCl and analyze the transmitted radiation with an IR spectrometer, we will observe that some of the radiation which has a frequency about 8.7×10^{13} sec^{-1} has been absorbed by the HCl molecules whereas most of the other frequencies are transmitted. This absorption of radiation has occurred because some HCl molecules have increased their vibrational energy, which in classical terms means that their vibrational amplitudes have been increased. This occurs at the expense of the energy of the IR radiation which has been absorbed. In quantum-mechanical terms, the energy of the absorbed photon E_p is equal to Planck's constant, h, times the frequency of the photon ν_p so that $E_p = h\nu_p$. The change in vibrational energy of the molecule ΔE_m during a fundamental type of absorption transition is $\Delta E_m = \Delta V h\nu_m$ where $\Delta V = +1$ and ν_m is the classical vibrational frequency of the molecule. In order to have conservation of energy in the absorption process $\Delta E_m = E_p$ so that $\nu_p = \nu_m$ for a fundamental vibrational transition. The photon which has the correct energy to increase the quantum number by $+1$ has a frequency equal to the classical vibrational frequency of the molecule.

2. Change in Dipole Moment

When infrared absorption occurs during a fundamental type of transition a radiation frequency matches one of the vibrational frequencies of the absorbing molecules. In addition to the frequency match there is another requirement that must be met which involves the manner in which the molecule is able to absorb the radiation energy. *In order for any infrared absorption to occur, the molecular vibration must cause a change in the molecular dipole moment.* The dipole moment is defined in the case of a simple dipole as the magnitude of either charge of the dipole, multiplied by the charge spacing. Let us consider a molecule such as HCl as a simple dipole with H positively charged and Cl negatively charged. When this molecule is placed in an electric field such as in the case when the molecule is positioned in a beam of electromagnetic radiation, this field will exert forces on the charges in the molecule. Opposite charges will experience forces in opposite directions. This tends to increase or decrease their separation. Since the electric field of the radiation in the environment of a molecule is reversing periodically in polarity, the HCl dipole will experience forces trying to periodically increase and decrease the dipole spacing. Since a vibration of the HCl molecule also increases and decreases the dipole spacing periodically, the radiation forces will try to make the HCl molecule vibrate at the radiation's frequency. This will only be successful if there is a correct frequency match.

In the case of a homonuclear diatomic molecule such as molecular hydrogen, H_2, no IR absorption will occur even if the photon frequency matches the molecular vibrational frequency because this type of molecule has a center of symmetry (which implies a dipole moment of zero) and the molecule retains this symmetry throughout the vibration. In general, for any molecule possessing a center of symmetry when not vibrating, *a vibration which retains the center of symmetry cannot give rise to any infrared absorption* since the dipole moment (which remains zero) does not change. Of course if a molecule has more than two atoms and has a center of symmetry (CO_2, benzene), it has antisymmetrical vibrations which destroy the center of symmetry and these are infrared active. The above is one example of one of the symmetry selection rules for IR absorption which state that for certain types of vibrations the IR absorption intensity is necessarily zero because of the molecular symmetry. In general the IR absorption intensity of a band is proportional to the square of the change in dipole moment.

3. Infrared Vapor-Phase Band Contours

The region of the spectrum where the pure rotational transitions occur is lower than that available to the average chemical spectroscopist (ca. 100 cm^{-1}). The free rotational spectrum can only be seen in the vapor state since in the solid state rotation is restricted and in the liquid state collisions between molecules occur too frequently to allow full free rotation to occur. The molecule changes from a given rotational energy state to a higher rotational energy state by absorbing an infrared photon whose energy is exactly equal to the difference in energy between the two states. In the pure rotational transitions, the molecule remains in the ground vibrational state.

When the vibrational spectrum of a gas is being observed, the molecule is changed from the ground vibrational state to the first excited vibrational state. However when the molecule arrives in the first excited vibrational state, it may be rotating with a greater or smaller angular momentum than that which it had in the initial ground vibrational state. This means that when the vibrational energy is changed, the rotational energy may also be changed simultaneously. As a result, rotational structure is superimposed on the vibrational band which is then called a vibration–rotation band.

The angular momentum is also quantized. If the instrument resolution is sufficiently good and if the molecule has a sufficiently low moment of inertia, the rotational fine structure can be resolved. Usually however a broad contour is seen instead.

To some extent the appearance of the vibration–rotation band contour can reveal information about the direction in which the dipole moment is changing during the vibration (1). This can be of use in making band assignments (Figure 6.2). In a linear molecule such as CO_2, H—C≡C—H, or HCN, a vibration where the dipole moment change is parallel to the molecular axis gives rise to a broad vibration doublet. In the high-frequency wing, the rotational energy has increased during the vibrational transition. In the low-frequency wing the rotational energy has decreased during the vibrational transition. When the dipole moment change is perpendicular to the molecular axis, the vibration–rotation band contour consists of a similar broad doublet but has, in addition, a sharp central peak not present in the parallel band. The central peak results when the rotational energy does not change during the vibrational transition.

In a common type of molecule called an asymmetric top, the molecule has three unequal moments of inertia directed along the three mutually perpendicular axes. When the dipole moment change is parallel to the axis of intermediate inertia, B, a broad doublet band contour with no central peak appears in the gas phase spectrum. If the dipole moment change is parallel to the axis of minimum inertia, A, or the axis of maximum inertia, C, the band contour consists of a broad doublet with a sharp central peak not present in the B-type band. When compared to the doublet, the central peak is usually relatively stronger in the C-type bands than in the A-type bands (Figure 6.2). The exact shape of the contours depends on the relative values of the moments of inertia (1).

Planar molecules such as certain substituted ethylene and benzene

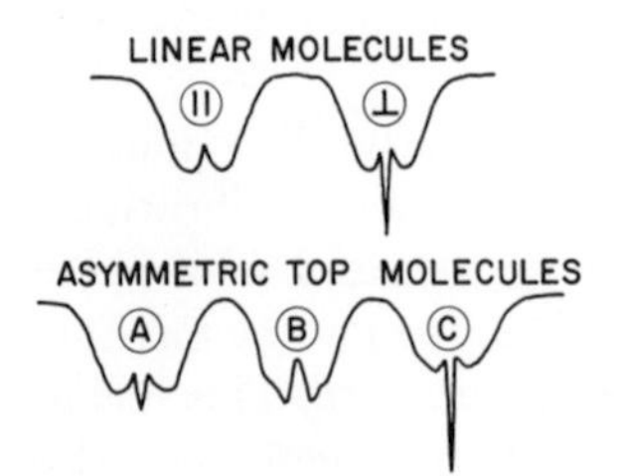

Figure 6.2. Infrared gas phase band contours for linear and asymmetric top molecules. The exact contours for the asymmetric top depend on the relative moments of inertia.

derivatives have out-of-plane vibrations which in the gas-phase spectrum can usually be identified because of their C-type band contours. This is just one application of the use of gas-phase band contours.

IV. THE RAMAN EFFECT

1. A Schematic Description of Raman Scattering

In a typical Raman spectrometer, the sample is generally irradiated by monochromatic radiation in the visible region which is, therefore, considerably higher in frequency than the molecular vibration frequencies, which have frequencies in the IR region. The small amount of radiation scattered by this sample in a direction 90° to that of the source radiation is analyzed by a spectrometer. By far the strongest emission feature in the spectrum of the scattered radiation is that of the unmodified exciting frequency. This effect is called Rayleigh scattering. However on both sides of this exciting frequency certain weak new emission frequencies can be detected. These new features constitute the Raman spectrum. The more important set on the low-frequency side of the exciting line constitutes the Stokes frequencies, and the generally less intense set on the high-frequency side of the exciting line consists of the anti-Stokes frequencies. The Raman effect is named after C. V. Raman who discovered this phenomenon experimentally in 1928.

Rayleigh scattering can be considered as an elastic scattering of a photon which has colllided with a molecule and has "bounced" off without changing the molecular vibrational or rotational energy of the molecule. There is consequently no change in the energy of the photon. Since $E_p = h\nu_p$, there is also no change in the frequency of the photon. The Raman effect can be looked upon as an inelastic collision between a photon and a molecule where the molecule possesses a different vibrational or rotational energy than before the collision. In this case the scattered photon must have a different energy and therefore a different frequency than the incident photon.

In an analogy of the Raman effect, we can imagine a piano hammer (photon) having a certain amount of kinetic energy as it approaches a nonvibrating piano string (molecule in its lowest vibrational state). After rebounding from the string the hammer has less energy than before (the scattered photon has less energy and therefore a lower frequency) and the string (molecule) has increased its vibrational energy by vibrating with increased amplitude at its natural frequencies. This is analogous to the Stokes lines in the Raman spectrum. In our analogy, we can also imagine a slowly moving piano hammer touching a violently vibrating piano string (molecule in an excited vibrational state). After collision, the vibrational energy of the string (molecule) may be reduced; if so the rebounding hammer will have more kinetic energy than before the collision (the scattered photon has more energy and therefore a higher frequency). This is analogous to the anti-Stokes lines in the Raman spectrum. We have of course assumed no frictional losses of energy in our analogy. It can be noted that the frequencies at which the string (molecule) can vibrate do not depend on the energy of the hammer (exciting radiation frequency) which caused the change in vibrational energy.

2. Raman Frequency Shift

The Raman effect can be described in terms of transitions between vibrational energy levels or states. A photon is absorbed by a molecule in the lowest or ground vibrational state. The absorbed photon temporarily raises the molecule's energy to some energy which is higher than the vibrational energy levels. Since this is not a stable energy state for the molecule, two things can occur now. Most probably the molecule returns to its ground vibrational state and emits a photon with the same energy and frequency as the exciting photon. This is Rayleigh scattering. However, some of the excited molecules may return, not to the ground state, but to some excited vibrational state. Such a molecule emits a photon which is lower in energy than the exciting photon, this energy difference being equal to the energy difference between the initial and final vibrational states. This is Raman scattering, Stokes type. If a photon is absorbed by a molecule which is in the first excited vibrational states, then the molecule is again raised to some high, nonstable energy state. Most probably, this molecule then returns to the ground vibrational state, and in doing so emits a photon which is higher in energy than the exciting photon, the energy difference being equal to the energy difference between the two vibrational states. This is Raman scattering, anti-Stokes type. Since the probability of a molecule being in the first excited vibrational state is lower than the probability of the molecule being in the ground vibrational state, the probability of the molecule undergoing an anti-Stokes type transition is lower than that for a Stokes type transition. The intensity of anti-Stokes Raman lines is therefore lower than the intensity of Stokes Raman lines. Furthermore the intensities of the anti-Stokes lines continue to decrease as their frequencies increase since the population of excited states decreases as the frequencies increase.

Imagine a photon whose frequency is 69×10^{13} sec^{-1} (indigo radiation) colliding with an HCl molecule whose natural vibrational frequency is about 8.7×10^{13} sec^{-1} and that this collision results in an increase in the vibrational energy of the HCl molecule. In quantum mechanical terms this vibrational energy change in the molecule is $\Delta E_m = h\nu_m$ for a fundamental type of transition for a Stokes line in the Raman effect. The energy difference ΔE_p between the exciting photon with frequency ν_e and the scattered photon with frequencies ν_s is $\Delta E_p = h\nu_e - h\nu_s$. Since ΔE_m must equal ΔE_p in order to have conservation of energy before and after collision, then

$$\nu_m = \nu_e - \nu_s \tag{6.5}$$

for a Stokes line. A scattered photon is observed with a frequency equal to 60.3×10^{13} sec^{-1} (blue-green radiation). The frequency difference between the exciting and scattered radiation frequencies $(69 - 60.3) \times 10^{13}$ sec^{-1} is about 8.7×10^{13} sec^{-1} (ν_m), and is the same as the IR radiation frequency which was absorbed by the HCl molecule in this case. Of course a different exciting frequency such as 59×10^{13} sec^{-1} could be used in which case the scattered frequency would be 50.3×10^{13} sec^{-1}. The difference would still be about 8.7×10^{13} sec^{-1} as before.

3. Change in Polarizability

Infrared and Raman spectra are not exact duplicates of each other because of the different manner in which the photon energy is transferred to the molecule. A molecule subjected to electromagnetic radiation will be polarized because the electric component of the radiation force field will subject the electrons and protons to forces in opposite direction. The value of the induced dipole moment divided by the strength of the field causing the induced dipole is the *polarizability* of the molecule. This parameter is therefore a measure of the deformability of the electron cloud by the electric field. We have pointed out that the IR absorption intensity depends on the change in the permanent dipole moment of the molecules caused by the incident radiation. In order for this radiation to give rise to any intensity in the Raman spectrum, the radiation must cause a change in the polarizability of the molecule.

Let us consider a molecule positioned in an electromagnetic field with a frequency which is too low to excite any electronic transitions but which is considerably higher than the vibrational frequencies. In this case the nuclei cannot follow the field oscillations, but the electrons, having a much lower mass, are displaced by the alternating force field. An oscillating dipole moment is induced. These oscillating electrons will generate electromagnetic radiation which is scattered in all directions and which has the same frequency as the excitation oscillation.

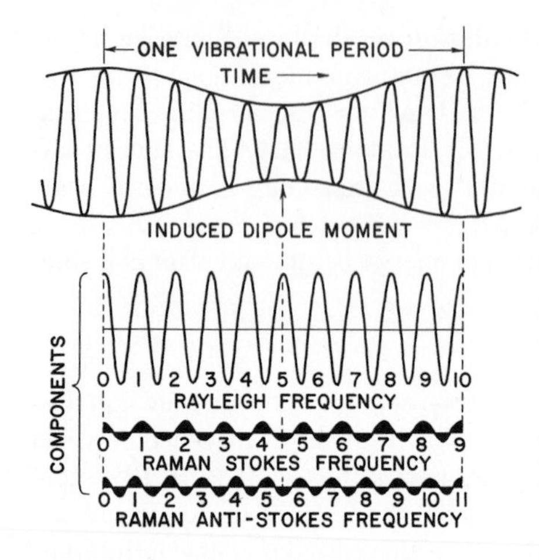

Figure 6.3. Induced dipole moment (vertical coordinate) versus time (horizontal coordinate). The top curve is the amplitude-modulated induced dipole moment oscillation which results when exciting electromagnetic radiation displaces electrons in a vibrating diatomic molecule. The bottom curves are the three derived amplitude components of the emitted electromagnetic radiation.

In a diatomic molecule, the polarizability is different when the molecule is in the stretched or in the contracted condition. In general, if the molecule is vibrating in any manner which changes the polarizability then the rapidly oscillating, externally induced dipole moment in such a molecule has a a slowly varying maximum oscillation amplitude. The maximum oscillation amplitude varies periodically as the vibrational frequency. In other words the induced dipole moment oscillation plotted as a function of time (Figure 6.3) is an amplitude-modulated wave where the modulation frequency is equal to the vibrational frequency. Such a wave can be resolved mathematically or graphically into three constant amplitude waves which have frequencies equal to the exciting frequency, the exciting frequency minus the vibrational frequency, and the exciting frequency plus the vibrational frequency. The first of these gives rise to the Rayleigh scattering frequency and the last two give rise to the Raman Stokes and anti-Stokes frequencies.

If a polyatomic molecule is vibrating in a manner which does not change the polarizability then there can be no amplitude modulation to the induced dipole moment oscillation so that all of the scattered light will vibrate at the exciting radiation frequency and the Raman spectrum intensity will be zero. In general, *the intensity of a band in the Raman spectrum is proportional to the square of the change in polarizability caused by the vibration giving rise to the band.*

4. Raman Depolarization Ratios

In this type of Raman measurement the radiation scattered by the sample is polarized in turn in each of two mutually perpendicular planes and the intensity of each resulting spectrum is measured. The intensity ratio is called the depolarization ratio. If the incident radiation is unpolarized and comes to the sample from one direction only, then the intensity of the scattered radiation polarized in the plane containing both the incident and scattered beam axes is divided by the intensity of the scattered radiation polarized in a plane 90° to this plane to give the depolarization ratio. In this case the depolarization ratio is 6:7 for all Raman lines caused by vibrations which are not totally symmetrical, in other words, the vibrationally distorted molecule is antisymmetrical with respect to at least one symmetry element possessed by the molecule in its equilibrium configuration. If the line is caused by a totally symmetrical vibration then the depolarization ratio can be anywhere between zero and 6:7. If, as in laser excitation, the incident radiation is plane polarized then antisymmetrical vibrations give rise to bands with depolarization ratios of 3:4. Totally symmetrical vibrations give rise to bands with depolarization ratios anywhere between zero and 3:4. This of course is useful when making band assignments because totally symmetrical vibrations can usually be picked out. The laser Raman source has an advantage in that accurate depolarization ratios can be more easily obtained than with other sources.

V. INFRARED INSTRUMENTATION AND SAMPLE-HANDLING TECHNIQUES

1. Infrared Spectrometers

Figure 6.4 shows a schematic diagram of an IR double-beam recording spectrophotometer. The major components of the system are (*a*) the IR source; (*b*) the monochromator; (*c*) the sample; and (*d*) the detector. These separate items will now be discussed.

A. Infrared Spectrometer Source

The source of IR radiation is generally an incandescent solid material which has a radiant energy distribution almost similar to that for the theoretical black-body. At the usual operating temperatures the energy distribution shows a maximum at 2 μ or less and falls off in intensity continuously at longer wavelengths. This causes two problems in instrument design. The decrease in source energy with increasing wavelength must be compensated for. This is achieved by programming a slit to continuously open up to allow more radiant energy to enter the spectrometer as the wavelength under examination increases. The second problem is the possibility of scattered light, namely radiation of the wrong wavelength, reach-

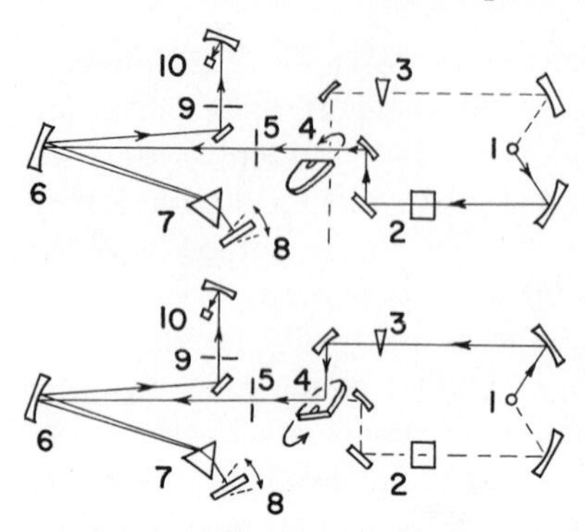

Figure 6.4. **Schematic optical diagram of a type of infrared double beam recording spectrophotometer. In the top figure the rotating mirror (4) puts the instrument into the sample beam cycle. In the bottom figure the spectrophotometer is in the reference beam cycle. Instrument components are: 1, source; 2, sample in sample beam; 3, optical attenuator wedge in reference beam (coupled to recorder pen); 4, rotating semicircular mirror; 5, entrance slit; 6, off-axis paraboloid; 7, prism; 8, rotatable mirror to control the frequency of radiation; 9, exit slit; and 10, detector.**

ing the detector. Since the short wavelength intensity output by the source is so much greater than the longer wavelengths a very small percentage of scattered light of short wavelength can have a strong deleterious effect on readings in the long-wavelength spectral region. Scattered light can be reduced with filters or with a double monochromator.

B. Infrared Monochromator

The second feature that IR spectrometers have in common is a monochromator which provides some means of separating the source radiation wavelengths. This is commonly done with either prisms or gratings. The most common prism material used is NaCl which has a lower limit frequency cut-off at about 650 cm^{-1}. It has good resolution from 1000 to 650 cm^{-1}, and adequate resolution throughout the 4000–1000-cm^{-1} region, the most commonly used IR region. One type of monochromator consists of an adjustable entrance slit, an off-axis paraboloid mirror which renders the radiation from this slit parallel and directs it through the prism after which the radiation hits a rotatable flat mirror. This mirror redirects the radiation back through the prism to the paraboloid which refocuses a spectrally dispersed image of the entrance slit across the exit slit which allows only a small part of the dispersed spectrum through to be focused onto the detector. The wavelength of the spectral region which goes through the exit slit is controlled by the angle of the rotatable mirror. Some instruments use a double monochromator, that is the exit slit of the first monochromator is the entrance slit to the second monochromator. This arrangement is the best way to control scattered light.

In a grating monochromator the optics arrangement can be similar but the prism is left out and the rotatable flat mirror is exchanged for a diffraction grating. Gratings disperse the radiant energy into more than one order but the radiation intensity in the unused orders can be minimized by controlling the shape of the grooves in the grating. However since some energy still remains in the orders other than that being used, more than one wavelength can go through the exit slit at the same spectrometer setting. To separate the undesirable wavelengths which are different by at least a factor of 2 from the desirable wavelengths, a crude additional separation method is sufficient. This separation of orders can be carried out with a low-dispersion prism or with high-quality filters.

C. *Infrared Detector*

Infrared detectors usually change the thermal radiant energy into electrical energy. One type commonly used is a thermocouple which generates an electromotive potential when a junction of dissimilar metals is heated. Another type is a bolometer which changes its electrical resistance with temperature.

D. *Double-Beam Spectrophotometers*

Most IR instruments are of the double-beam type. In these spectrophotometers the source radiation is split up by mirrors into two beams called the sample and reference beam. The sample beam goes through the sample and the reference beam goes through air or through a pure solvent for compensation. The two beams are recombined onto a common axis and are alternately (in time) focused onto the entrance slit of the monochromator. If the intensities of the sample and reference beams are exactly equal then no alternating intensity radiation goes through the slit. If the sample absorbs some radiation, an alternating intensity radiation is observed by the detector which generates an ac component which can be selectively amplified. The alternating current is used to adjust the position of a beam attenuator (usually a comb with tapered teeth) in the reference beam to change the intensity of the reference beam until it matches the sample beam intensity. The recording pen is coupled to this beam attenuator thereby recording sample absorption as a function of spectral frequency. Since only the ac component of the detector signal is used, the instrument is insensitive to changes in the temperature of its environment.

2. Infrared Transparent Optical Materials

Sample cells must of course be constructed from infrared transparent material. Glass and quartz transmit only in the highest frequency IR regions (approximately above 3000 cm^{-1}) which is one reason why mirrors instead of lenses are used in IR spectrometers. The most used single window material is NaCl which transmits down to about 650 cm^{-1}. KBr transmits down to about 400 cm^{-1}, CsBr to about 250 cm^{-1}, and CsI to about 200 cm^{-1}. These materials are all water soluble. Some water-insoluble materials are BaF with a transmission limit of about 850 cm^{-1}, and IR-TRAN-2 (a trademark of Eastman Kodak Company) with a transmission limit of about 750 cm^{-1}. AgCl is also used for window materials but darkens in light eventually and is easy to deform. Linear polyethylene has several strong IR bands but is almost free of absorption below 600 cm^{-1} and can be used as a low-frequency window material between 650 and 33 cm^{-1}.

3. Salt Polishing

The crystal faces of the NaCl or CsBr plates used to hold the sample must be periodically resurfaced. If such a plate has had only light use it may be simply cleaned off with solvent and reused if its surface is good enough. Otherwise it should be repolished. Polishing is usually done using a pair of "laps," one wet and one dry. One type of wet lap may consist of a double layer of nylon cloth held over a 6-in.-diameter flat stainless steel lap indented in the rim to take a large rubber "O" ring which holds the cloth in place. The dry lap may consist of a diaper cloth held over an identical stainless steel plate. The wet lap is moistened with water until damp and coated lightly with polishing abrasive such as aluminum oxide or cerium oxide. The NaCl plate is rubbed on the wet lap 10 to 30 strokes and immediately buffed about five strokes on the dry lap. A lap which is too wet will generate an undesirable convex surface on the plate. For KBr or CsBr an alcohol (methanol, isopropanol) is substituted for water but otherwise the process is the same. If the surface of the plate has been scratched or corroded (with a wet sample, for example), it should be sanded flat on the finest grade abrasive paper before polishing. For CsBr plates, which are both soft and expensive, the sanding should be as delicate as possible. The author uses an abrasive paper wetted with an alcohol for smoothness.

New NaCl plates may be cleaved from NaCl blocks using a single edge razor blade and a small hammer. The razor is held parallel to one of the faces and tapped with the hammer starting a cleavage plane. Continued tapping with the razor in the cleavage plane will eventually result in full cleavage. Blocks can be cleaved in halves or thirds but for smaller slices the block should be cleaved in half and then the halves cleaved in half, and so on, rather than trying to cleave a small slice off a big block. (This usually results in a cracked plate.) The cleaved faces must be sanded flat using a somewhat coarse grade of abrasive paper first and finally polished with a fine grade sandpaper. One should be careful to keep the two faces of the plate parallel to each other. The KBr will cleave even more readily than NaCl. CsBr and AgCl will not cleave at all. These are usually bought preshaped and ready to use.

4. Infrared Sampling of Gases

One of the powerful features of IR spectroscopy is that a wide variety of samples can be examined by this technique. Solids, liquids, or gases can all be handled with ease. Gases are normally run in a gas cell 10-cm long with the ends ground square to which infrared transparent windows are glued on or clamped on with sealing gaskets in place. Two tubes with stopcocks and ground glass joints protrude from the middle of the cell for attaching to a gas handling system. Some commercial gas cells are available which have sides that closely fit the converging IR beam and therefore take less volume of gas than the cylindrical type. For measuring very dilute gases or gaseous components, long path cells are available. Since the sampling area of most spectrometers is restricted in length, mirrors are used to deflect the IR beam and bounce it back and forth within the gas chamber area which can protrude sideways from the spectrometer. The beam is then redirected into the spectrometer after it has gone through a considerable length of gas.

5. Infrared Sampling of Liquids

Liquids can be run in sealed cells of which many commercial types are available. These usually consist of two IR transparent windows separated by a metal foil spacer of the proper thickness which can range from 0.001 to 1 mm. The metal is frequently amalgamated with mercury before assembly to provide a seal for the thinner cells. Thicker cells are sometimes glued together. Two holes through one of the windows provide access into and exit from the sample chamber. Another type of cell simply consists of a cavity (not highly polished) of the proper thickness carved out of the body of a salt crystal from the top. This cell requires no sealing or spacers and ranges in thickness from 0.05 to 4 mm. It is at present not available in 0.03–0.01-mm cell thicknesses.

If the cells have good quality windows, flat and parallel, then the thickness, t, in cm can be calculated from $2t = N/(\omega_1 - \omega_2)$ where N is the number of transmission undulations (interference fringes) between the wavenumbers (in cm^{-1}), ω_1 and ω_2, in the spectrum of the empty cell. Transmission undulations are caused by the fact that the transmitted beam finds itself accompanied by another weaker beam which has been reflected once off each inner wall of the cell and so is retarded by $2t$ relative to the transmitted beam. Interference occurs and interference fringes can be observed.

If the liquid is not too volatile and if reproducibility is not important, then the liquid can be run simply by using two small strips of metal foil as spacers in a sandwich of IR windows with the sample in between.

Typical thicknesses used for pure liquids are 0.01–0.02 mm. If the liquid is water there are several suitable IR transmitting window materials which are water insoluble. These include silver chloride, barium fluoride, and the commercial material IRTRAN-2.

6. Infrared Sampling of Solids

A. Solids Run in Solution

Solids can be run in a variety of ways. If a suitable solution can be made then the solution is run in one of the cells for liquids. While there are definite advantages to running a solid as a solution, suitable solvents are unfortunately limited in number. There is no single solvent which is transparent throughout the IR region. Even in the best case at least two different solvents must be used if the complete spectrum of the solute is desired. A commonly used pair of solvents is CCl_4 for measurements above 1330 cm^{-1} and CS_2 for measurements below 1330 cm^{-1}. A 1% solution in these solvents in a 1-mm cell, for example, will generally yield a good spectrum. A single solvent which is sometimes used is $CHCl_3$. The spectrum is obtained in a 0.1-mm cell with about a 10% solution. In this case the 1216-cm^{-1} region and the 757-cm^{-1} region will for the most part be obscured by the solvent. In double-beam instruments, solutions are commonly run with a cell containing solvent only in the reference beam. If the cell is of the proper matching thickness, the solvent bands in the solution spectrum will be cancelled out. However at frequencies where the solvent transmits less than about 5% of the radiation at the cell thickness used, the recorder, now operating with less than 5% of its usual energy, in effect tends to go "dead." At standard operating speeds the recorder will not respond adequately to solute bands even with perfect solvent compensation. The user of the compensated solvent technique should be aware of the sluggish or dead regions in this type of spectrum.

B. Solid Films

If a solid is amorphous in nature (polymers, waxes, tars, etc.) then a film can be cast from solution on an IR transmitting window. Alternately, materials of this sort may be prepared by melting the sample between salt plates and allowing it to cool into a film. Neither of these techniques are recommended for crystalline compounds because the random crystallite orientations will cause nonreproducible modifications in the spectrum.

C. Mull Technique

Crystalline solids can be handled by the mull or pressed pellet technique, both of which involve grinding the sample down to a particle size ideally smaller than the smallest wavelength used (2μ) and then surrounding the particles with a transparent medium whose index of refraction is similar to that of the sample. In the mull technique the finely ground sample is mixed with Nujol (mineral oil) to make a thick paste which is then spread between IR transmitting windows. Nujol is transparent throughout most of the spectrum, but absorbs strongly in the CH stretching and bending regions. In order to have the complete spectrum a second mull must be prepared using a completely halogenated hydrocarbon such as perfluorokerosene or Halocarbon Oil. These oils have strong carbon–halogen absorption bands, but are clear in the CH regions obscured by Nujol.

D. *Pressed-Pellet Technique*

In the pressed pellet technique (6, 8) a small amount of finely ground sample is intimately mixed with about 100 times as much powdered KBr. The mixture is then placed in a pellet die and subjected to high pressure in a mechanical press. The resulting pellet, if properly made, is transparent and is run as is. Other alkali metal-halides besides KBr have occasionally been used. In the more expensive commercial dies, the powdered material is evacuated before pressing, which results in a somewhat improved disk clarity. The permanence of this clarity, while desirable, is not crucial. Inexpensive dies are commercially available which use no evacuation and have bolts which can be tightened with a hand wrench to supply the necessary pressure within a nut-type die. The KBr pellets are advantageous because they have no KBr bands to obscure the spectrum above about 400 cm^{-1}. However they are plagued with absorbed water bands which cause some absorption in the OH region which may be confused with sample absorption. The pressure involved can also cause polymorphic changes or changes in crystallinity which can complicate the spectrum. However for some applications the KBr pellet technique is very valuable. For example, very small samples are most easily handled by the pellet technique and some polymers yield their best spectra in KBr disks. One technique for handling polymers which are insoluble, unmeltable, and difficult to grind because of their tough nature consists of making a low-temperature KBr disk. The grinding is done at liquid nitrogen temperature which makes a normally tough material brittle. The sample can be placed in a metal capsule with a metal ball inside, cooled in liquid nitrogen, and then placed in a mechanical shaker. This breaks down the sample particles to the desired size.

However because of nonreproducible OH bands and polymorphic difficulties, the author prefers the mull technique as the best general method for running crystalline compounds in the solid state and reserves the KBr pellet method as one of the many special techniques available for special problems.

7. Internal Reflection Technique

The internal reflection technique (2) is another of the special techniques available. In one form of this technique a prism (45°, 45°, 90°) of a material with a suitable high index of refraction such as KRS-5 (thallium bromide-iodide) or silver chloride is used (Figure 6.5). The radiation introduced through one of the small faces emerges from the other small face after it has been totally reflected off the inside face of the prism hypotenuse. Total internal reflection occurs because the radiation approaches the inside face of the hypotenuse of the prism at an angle which exceeds the critical angle. If the sample should be placed in optical contact with the prism at this place, the radiation will penetrate a few microns into the sample before reflection occurs. If the sample absorbs any of the incident wavelengths of radiation the reflected beam will be attenuated in intensity in these regions. The resulting spectrum will be similar to a normal transmission spectrum but they will not be identical. One reason is because relative to a transmission spectrum the band intensity of the reflected spectrum increases with increasing wavelength since the sample penetration increases with increasing wavelength. Some spectral distortion also occurs because the sample penetration is a function of the index of refraction of the sample which changes in the vicinity of an absorption band. This distortion can be minimized by working at some distance from the critical angle. This, however, leads to smaller sample penetration and an undistorted single internal reflection spectrum may be quite weak in intensity. In the most common form of this technique not one but many internal reflections are used which increase the spectral intensity. This is accomplished by using a trapezoidal plate (a prism with the square corner cut off to make a new face parallel to the hypotenuse) (Figure 6.5). The sample can be placed on both faces of this type of plate. A big advantage of the internal reflection technique is that soft rubbery materials, normally difficult to handle can simply be clamped onto the plate faces and run as is. The actual sample thickness is unimportant as long as it is larger than a few microns since the sample penetration

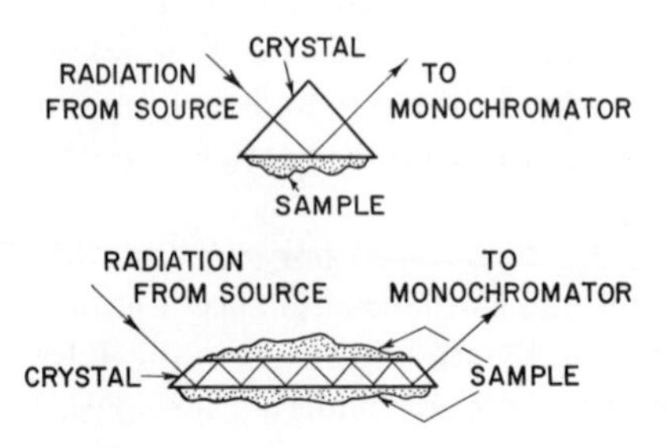

Figure 6.5. Internal reflection crystals, single and multiple internal reflection types.

does not depend on sample thickness beyond this range. Because of the relative softness of the prism materials used in the internal reflection technique, optical contact can even be achieved with flat smooth pieces of hard plastic samples. The internal reflection technique provides a good method of looking at surface films where, for example, an impurity or a coating may be concentrated on the surface of a material. In some bicomponent films the internal reflection technique has been used to selectively identify the two different polymers on the two film faces. Water solutions can be handled by the internal reflection technique since the crystal plates normally used are insoluble in water.

8. Pyrolysis

Another special technique is pyrolysis (5). In this technique the sample is subjected to a controlled amount of heat. Commercial units are available which enable one to analyze both the gases and the nonvolatiles formed. The solids and liquids are collected by condensation on an NaCl or internal reflection plate. The pyrolizate spectrum may differ from a straight transmission spectrum in that nonvolatiles such as inorganic fillers in polymers may be left behind. Polymer pyrolizates, for example, may contain monomers so pyrolizate spectra can be compared to a library of pyrolizates of known materials. The spectra of carbon-filled rubber-type materials can not be conveniently obtained by any other technique.

9. Infrared Quantitative Analyses

A. Beer's Law

The primary relationship used in quantitative analyses in IR and UV spectroscopy is based on the Beer-Lambert law which is most commonly referred to as Beer's law. The absorbance A is given by

$$A = \log \frac{P_0}{P} = abc \tag{6.6}$$

where P_0 is the radiant power of monochromatic source radiation entering the sample, P is the radiant power after this radiation has passed through the sample, b is the sample thickness, c is the concentration, and a is the proportionality constant called the *absorptivity*. The absorptivity is a property of the sample and is different at different frequencies of radiation.

Most instruments today deliver a spectrum where the vertical coordinate is linear with transmittance (P/P_0) but the paper provided may have either a linear transmittance scale or a nonlinear absorbance scale. Some instruments can be adjusted to make the vertical coordinate linear with absorbance.

B. *Cell In–Cell Out Method*

One way to measure absorbance of a sample in solution is called the "cell in–cell out" method. The instrument is set at the desired wavenumber and the cell which is filled with the solution is placed in the sample beam. The reading on linear transmittance paper provides a number proportional to P for the sample. The same cell is then filled with pure solvent and the transmittance reading is noted without having changed any instrument settings. This transmittance reading provides a number proportional to P_0 and Beer's law can be used to calculate the sample absorbance.

If absorbance paper has been used, then the solvent (plus cell) absorbance can be read directly and subtracted from the absorbance of the solution (plus cell) to give directly the absorbance of the sample. If the output of the spectrometer is linear with transmission, the absorbance of the sample has to be calculated from the observed transmittance values.

C. *Baseline Method*

Another popular method of measuring sample absorbance is called the baseline technique (Figure 6.6). This consists of drawing a background line on a spectrum in a manner which approximates the appearance of the spectrum if the band under consideration were not present. In the case of an isolated band this would consist of a line drawn tangent to the background level of the spectrum on either side of the band. If there are overlapping neighboring bands in a multicomponent mixture then some other line tangent to nearby minima in absorption may be chosen. It is desirable that

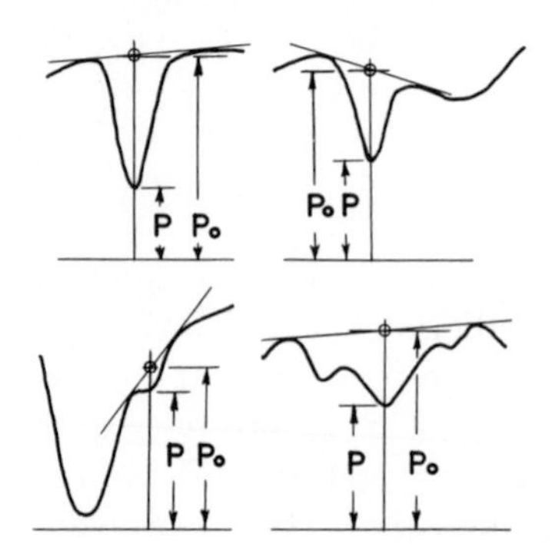

Figure 6.6. Quantitative analysis baseline constructions.

the absorbance of the two wavenumber regions chosen for these inflections points should be relatively insensitive to expected concentration changes. If linear transmittance paper has been used, the transmittance value (P) is measured at the bottom of the band and the transmittance value (P_0) is measured at the same wavenumbers at the chosen background line and the absorbance is calculated. If absorbance paper has been used, absorbances are read directly at these same two positions and subtracted to give the desired working absorbance value. The baseline method automatically corrects for cell absorbance and reflection losses.

D. Beer's Law Deviations

Beer's Law is derived for ideal conditions of monochromatic radiation. However monochromators never deliver truly monochromatic radiation but rather a narrow range of wavenumbers. This range depends on the slit width. Beer's Law is also derived assuming the sample's absorptivity to be independent of the concentration. In very dilute solutions such as that used in UV work, the sample molecules are only surrounded by solvent molecules. In some typical solutions used in IR analyses this may no longer be true because the solutions are more concentrated. Some samples have a tendency to form complexes, one example of which is the formation of a hydrogen bond between solute molecules as the concentration increases. This means that the absorptivity may not really be the same in dilute and concentrated solutions and Beer's law is not valid anymore. In actual practice one usually has to verify whether or not Beer's law holds over the concentration range under consideration. This is best done by preparing a series of standards. If Beer's law does not hold, one solution is to use a working curve, in other words, an actual plot of absorbance versus sample concentration. Departure from Beer's law simply means that this plot will be nonlinear.

Another difficulty that occurs in infrared but not in ultraviolet spectroscopy concerns the cell material. The cells used in infrared, being made of NaCl, usually are of a delicate, less-perfect, and nonpermanent nature. The cell thickness is frequently not known as accurately as would be desired. For these reasons, usually standards of known concentration are first run. The samples to be measured are then run in the same cell. If we assume the cell thickness to be essentially constant it can be incorporated with the absorptivity. In this case Beer's law becomes simply "the absorbance is directly proportional to sample concentration" with a single proportionality constant.

E. *Multicomponent Analyses*

In multicomponent analyses more than one component may absorb at the selected frequency. For three components Beer's law reads:

$$A_1 = a_1bc_1 + a_2bc_2 + a_3bc_3 \tag{6.7a}$$

$$A_2 = a_1'bc_1 + a_2'bc_2 + a_3'bc_3 \tag{6.7b}$$

$$A_3 = a_1''bc_1 + a_2''bc_2 + a_3''bc_3 \tag{6.7c}$$

Since the a values can be determined from standards and the absorbances at three different frequencies A_1, A_2, and A_3 can be measured, one has three equations and the concentration of three unknowns can be computed. Of course it is desirable to select analytical frequencies where the interferences from other components are as small as possible relative to the main band intensities being used.

F. *Solid-State Analyses*

Quantitative analyses are best done in solution or in the liquid state. If necessary they can also be done in the solid state. In a well-prepared solid sample in Nujol or KBr, measurements are frequently carried out using only band absorbance ratios. These values should be independent of sample thickness or total sample concentration. These absorbance ratios are calibrated using absorption coefficient ratios which can be determined from standards.

G. *Gas-State Analyses*

In the analyses of gases, Beer's law reads

$$A = abp \tag{6.8}$$

where p is the pressure or partial pressure of the gaseous component. Notable departures from the linearity predicted by Beer's law can occur because of pressure broadening. Basically this means that a constant partial pressure of a gaseous component can give rise to a variable absorbance at its analytical frequency if the total pressure is changed by introducing another gas which does not absorb at all at that analytical frequency. These modifications in absorbance are produced because of sensitivity to collisions with the molecules of the other gas components. This can cause severe problems in the quantification of the results. For this reason analyses are usually carried out at a constant total pressure. For example, after the selected partial pressures of various components have been added together, the total pressure can be brought up to atmospheric pressure by the addition of a nonabsorbing gas, such as nitrogen.

VI. RAMAN INSTRUMENTATION AND SAMPLE-HANDLING TECHNIQUES

1. Raman Instrumentation—Monochromators and Detectors

As in the IR spectrometer, the radiation path in a Raman spectrometer goes from source through a sample, monochromator, and on to a detector. In the monochromator, both lenses and mirrors have been employed. Prisms have been used but most instruments use a diffraction grating as the dispersing element. A double monochromator is frequently employed when it is desirable to have good control over undersirable scattered light. Scattered light within the monochromator can be a severe problem in a Raman spectrometer. This is due to the great intensity of the Rayleigh line compared to the Raman spectrum, and the even greater intensity of primary light. The latter is light which is not scattered by the sample molecules but which is reflected from the cell wall or from the surface of solid sample particles.

The Raman spectrum can be detected and recorded in two basically different ways. In one arrangement, the spectrum from the prism or grating is simply focused onto a photographic plate. Both the line frequencies and intensities can be measured using external equipment after the plate is developed. A more convenient system uses an exit slit at the focus of the spectrum produced by the monochromator which allows a narrow wavelength region to pass through. This is focused onto a photomultiplier-type detector which, when used with an amplifier and recorder, can provide a usable spectrum directly. In addition to its convenience, the photoelectric method of recording the Raman spectrum has the advantage that the response to the Raman line intensity is a linear. This greatly simplifies intensity measurements and quantitative analysis.

2. Raman Sources

The part of a Raman spectrometer which differs most markedly from an infrared spectrometer is the source and sample area. The source irradiates the sample with an intense monochromatic radiation. The radiation which the sample scatters in a direction perpendicular to the incident radiation path is directed into the monochromator. Since the intensity of the scattered radiation is very weak, an extremely intense source of radiation must be used. One type commonly used is a mercury lamp shaped in the form of a helix surrounding the sample tube. The sample tube in this case can be basically a cylindrical glass tube with a window at one end and some sort of light trap at the other. Between the helical mercury lamp and the sample is a filter jacket which removes unwanted radiation and only allows the 4358 Å mercury line to pass through. This monochromatic radiation irradiates the sample tube from all sides and the light which is scattered out of the end of the tube is introduced into the monochromator. Because the Raman effect is so weak, precautions are taken in instrument design not to waste any usable signal. A device known as an image slicer (used on the Cary Model 81) has been used to change the circularly shaped beam leaving the sample tube into a narrow rectangularly shaped beam which can pass unobscured through the entrance slit with little loss in radiation intensity.

The newer types of Raman sources which have many strong advantages over older type sources are the optical masers or lasers. The energy output from these sources is in the form of a monochromatic narrow parallel beam. In comparison, a mercury lamp emits its radiation in all directions and is therefore less efficient. Two of the types that have been used are the pulsed ruby optical maser and the continuous helium–neon gas laser. The output of the helium–neon gas laser has a wavelength of 6328 Å in the red part of the spectrum. A red source has some disadvantages over an indigo (Hg) source because the Raman intensities are proportional to the fourth power of the radiation frequency and photoelectric detectors are usually less sensitive in the red as compared to the indigo part of the spectrum. However when a laser source is used, these difficulties are not insurmountable. There are as a matter of fact, some very definite advantages in using the long-wavelength red source. Raman spectroscopy using Hg excitation has long been plagued by two serious difficulties which limited its applicability. Usually the Raman spectrum cannot be obtained if the sample absorbs the exciting wavelength. This means that colored samples usually cannot be analyzed. The other difficulty is that if the sample fluoresced when irradiated the Raman spectrum would be obscured by the fluorescence spectrum. The fluorescer is frequently a trace impurity which can sometimes be removed by vacuum distillation. This however involves an extra time-consuming step. Because few materials absorb in the red end of the spectrum, and because red radiation is frequently less effective in generating fluorescence, both of these problems tend to be less troublesome using the red laser source.

A versatile source is an argon–krypton mixed-gas laser with argon lines at 4880 and 5145 Å and krypton lines at 5682 and 6471 Å (blue, green, yellow, red). The spectroscopist can select the line of his choice to avoid fluorescence, to avoid absorption by colored samples, or to make the signal strength as large as possible.

3. Raman Sampling Techniques

A Raman spectrum is most easily obtained using liquid samples. When the helical mercury arc source is used, the radiation enters through the cylindrical wall of the sample tube and the scattered radiation escapes through the end of the tube through an optical window in the end of the tube. In the past one disadvantage of Raman spectroscopy was that a much larger sample was required (a few milliliters) than is required for an infrared spectrum. However modern commercial instruments provide special cells whose design allows good spectra to be obtained on much smaller amounts of sample. When a laser source is used, the laser beam can be simply focused into a capillary tube containing the liquid.

One big advantage that Raman spectroscopy has over IR spectroscopy is that water solutions yield very good spectra. Water, which is such a strong absorber in the IR region, gives a relatively weak Raman spectrum and therefore is an excellent solvent. The Raman spectra of inorganic materials are frequently studied using water solutions (9).

The Raman spectrum of gases can also be obtained with the commercially available special gas cells for use with the mercury source. In this cell the weak spectral intensity is compensated by multiple-pass optics which increase the effective cell length. When a laser source is used the gas sample can be placed inside the laser cavity.

Solid samples previously caused more difficulty in Raman spectroscopy than in IR spectroscopy (3, 4, 7, 10). Solid surfaces can scatter the exciting radiation into the spectrometer and this scattering tends to lower the amount of exciting radiation penetrating into the sample. Both of these effects tend to lower the Raman intensity relative to the intensity of the exciting line. However in modern laser Raman spectrometers with double monochromators, the problems with solid samples have been largely overcome and good spectra can usually be obtained on a few milligrams of a powdered solid sample. In one simple technique the laser beam is focused into a capillary tube containing the powdered solid.

VII. SPECTRA–STRUCTURE CORRELATIONS

When the spectra of a related series of chemical compounds are compared one may recognize that certain bands in the spectrum may be associated with particular functional groups in the molecules. In other words the presence or absence of certain bands in the spectrum may be used to indicate whether a certain chemical functional group is present or absent in the molecule. These frequencies have been called group frequencies, and they have proven to be of immense value to organic chemists. There are of course many bands in a spectrum which are *not* group frequencies. These have been sometimes called fingerprint bands. For example, pentane, hexane, and heptane will have a number of common features in their spectra which are group frequencies, but they will also have some "fingerprint" areas which do not look alike at all. We make use of these fingerprint bands when we want to distinguish these compounds from each other.

If a group frequency vibration is active in both IR and Raman spectroscopy, then the IR and Raman spectral frequency correlations with structure are identical. The spectral intensity correlations, however, may be different, sometimes severely so, so that the correlation may be of value only in one or the other of the two techniques. Indeed there seems to be a tendency that when a group frequency has a weak intensity in the IR it may have a stronger intensity in the Raman spectrum and vice-versa. For example, OH and C═O stretch give rise to strong infrared absorption, but show up weakly in the Raman spectrum. Amino NH, SH, C═C, and C≡C stretch show strongly in the Raman spectrum, but may be weak in the infrared. Of course, if the molecule has a center of symmetry, then symmetrical vibrations including N═N and S—S stretch show up in the Raman spectrum only. Some very useful out of plane bending CH vibration in olefins and aromatics show up strongly in infrared but are quite weak in the Raman spectrum.

There are essentially no fundamental molecular vibrations above 4000 cm^{-1}. The region above 4000 cm^{-1} is called the near infrared region and is populated with overtone and combination bands, usually the result of hydrogen stretching vibrations. In this region glass or quartz cells may be used and the cell thickness is considerably larger than that used in the fundamental infrared region. The near infrared region has some advantages in quantitative analysis and is sometimes used for qualitative analysis. For general qualitative analysis, however, the fundamental infrared spectrum obtained close to the near infrared region is preferable because this spectrum is easier to interpret.

Table 6.1 shows the wavenumbers of the general regions in which certain types of group frequencies are expected to occur. For example, hydrogen stretching vibrations absorb throughout the region of 4000–2000 cm^{-1}. For a compound X—H there is a general tendency that the more electronegative X is, the higher will be the X—H stretching frequency.

1. Categories of Group Frequencies

The simplest and, in many ways, the most closely defined type of group frequency is that derived from a diatomic group X—Y attached at one end

Table 6.1. General Regions for Spectra–Structure Correlations

Spectrum, cm^{-1}	Structure
3800–3100	OH, NH, ≡C—H
3100–3000	Olefinic, aromatic, 3-membered ring CH
3000–2700	CH_2, CH_3, aldehyde CH
3100–2400	Acidic and strongly hydrogen bonded hydrogens
2600–2100	SH, BH, PH, SiH
2300–1900	X≡Y and X=Y=Z
2000–1700	Aryl and olefinic overtones and combination bands
1900–1500	C=O, C=C, C=N
1660–1450	N=O
1660–1500	NH_2, NH_3^+, and =CNH
1620–1420	Aromatic and heteroaromatic rings
1500–1250	CH_3 and CH_2 deformation
1470–1310	B—O, B—N, NO_3^-, CO_3^{2-}, and NH_4^+
1400–1000	SO_2, SO_3^-, SO, and SO_4^{2-}
1300–1140	P=O
1350–1120	CF_3 and CF_2
1350–1150	CH_2, and CH wag
1300–1000	C—O
1000–600	Olefinic and aromatic CH wag
830–500	C—Cl, C—Br, C—I, P=S

to the rest of the molecule, where the frequency of the vibration of the isolated X—Y group is considerably higher than the stretching frequencies of the bonds attaching the group to the rest of the molecule. Examples of this group are the stretching vibrations of (A)—O—H, (A)—C≡N, $(A)_2C$=O, and $(A)_3P$=O, where the X—Y atoms vibrate as they would in an isolated environment. The attached (A) atoms hardly move if they consist of carbon or heavier atoms. This vibration stretches the X—Y bond; to some extent it also simultaneously contracts the X-(A) bond. Because the (A) atoms are almost stationary, the (A)—X—Y vibration is essentially mechanically independent of the nature of the rest of the molecule. This means that once the geometry of the (A)—X—Y group is established, the group frequency can only be appreciably shifted by chemical effects. The latter relates to the shifting of the electrons in the bond and these mesomeric and inductive effects are fairly well understood.

The second common type of group is the XY_2 entity. In this case the two X—Y bonds have identical isolated X—Y frequencies and one cannot be excited without exciting the other. Since they must both vibrate simultaneously there are two alternatives; they can either vibrate in phase or

180° out of phase. This means that we have to consider this group as a whole. In many cases the in-phase and out-of-phase stretching frequencies of the isolated XY_2 group are considerably higher than the stretching frequencies of the bonds attaching the XY_2 group to the rest of the molecule. Again both stretching frequencies of the XY_2 group should be good group frequencies if the rest of the molecule remains stationary as the XY_2 group vibrates. Examples of this type of group are NH_2, CH_2, SO_2, NO_2 and CO_2^-. If the XY_2 scissors deformation vibration has a frequency higher than the frequencies of the bonds attaching the XY_2 group to the rest of the molecule then the XY_2 deformation will be a good group frequency with the rest of the molecule standing still. Examples of this are NH_2 and CH_2 deformations.

The groups CH_3 and SO_3^- are examples of an XY_3 group where three identical X—Y bonds are coupled together. As in the XY_2 case we cannot excite one X—Y bond without exciting the other two as well, so again we must consider the group as a whole. There will be three X—Y stretching frequencies, one in-phase stretch and two out-of-phase stretches. These will yield good group frequencies if the bond connecting the XY_3 group to the rest of the molecule has a definitely lower frequency than the XY_3 vibrations.

In a group such as a benzene ring we have six equal C—C bonds. We cannot excite one bond without exciting all of them. Hence we must consider the ring as a whole. There are some ring stretching vibrations which take place without appreciably moving the substituent atoms and which, therefore, give rise to good group frequencies.

There are certain out-of-plane hydrogen bending vibrations of olefins and aromatic rings which give rise to good group frequencies. These vibrations can take place without moving the substituent atoms because the substituent out-of-plane bending occurs at a much lower frequency than the CH bend frequencies.

In a group such as (A)—N=C=O the isolated CO and CN bonds do not have identical frequencies but they are rather similar. The N=C=O group and other X=Y=Z groups should, therefore, be considered as a whole. The out-of-phase X=Y=Z stretch is an excellent group frequency because it is considerably higher in frequency than the stretching frequency of the bond attaching the group to the rest of the molecule. The in-phase stretch of the X=Y=Z is somewhat less reliable as it tends to come in the single-bond region. It can be rather close in frequency to the isolated single-bond vibration of the bond attaching the group to the rest of the molecule and hence can interact with it.

In strained ring (AB)C=O compounds the C=O frequency goes up as the ring becomes smaller in size. We have already discussed the fact that as long as the A—C and B—C isolated bond frequencies are considerably lower than the isolated C═O frequency, then only the C and O atoms move appreciably, as in a diatomic C═O vibration. In this case the A and B atoms essentially stand still. However, the A—C and B—C bonds are contracted as the C═O bond stretches and this has an effect on the frequency. The contracted A—C and B—C bonds exert forces on the C atom which cooperate in restoring equilibrium with the force on this same atom exerted by the stretched C═O bond. Therefore, the total force exerted on the displaced C atom is greater than if the A—C and B—C bonds were not present. The increase in the force means that the frequency will increase. The amount by which the A—C and B—C bonds cooperate with the C═O bond in restoring equilibrium depends on the A—C═O and B—C═O angle. If this angle is 90° the cooperation is zero since the A—C bond exerts a force at 90° to the C═O bond. As the ring becomes smaller, this angle becomes larger so the cooperation increases and the frequency goes up.

There are some types of group frequencies which result because the X—Y group has a frequency considerably *lower* than the attached (A)—X isolated bond. In this case when the X—Y group stretches, the (A)—X "high-frequency bond" hardly changes in length. The (CH_2—CH_2)—Cl group C—Cl bond stretch is an example of this type of group frequency where the C—C bond has a high frequency relative to the C—Cl bond. One might note that when CO and C—C stretching vibrations take place the "high-frequency" OH or CH bonds hardly change in length.

There are some groups which one might expect would give rise to good group frequencies but which in fact do not. An example of this is the group (AB)C═S where A and B are C, N, or O. Although the C═S and C—N groups look different on paper, their isolated frequencies are rather similar. The group (AB)C═S must, therefore, be considered as a whole with all the atoms moving to some extent. Because both A and B move somewhat, mechanical interaction can occur with the rest of the molecule. This will shift the frequency. Some broad correlations can be made involving the C═S group but a large part of the rest of the molecule must be specified because of mechanical interaction. The P═S group is another such group and again correlations can be made if the balance of the molecule is specified.

There are some group frequencies involving C—O and C—C bonds such as those for ethers and alcohols. We expect the C—O and C—C bonds all to interact since they have similar isolated frequencies. As a result of this the skeletal single-bond vibrations are spread out over a region of the spectrum and generally in all the vibrations all the atoms are moving. Some of these vibrations give rise to some of our fingerprint bands. However it so happens that the high-frequency limit for these interacting C—O and C—C vibrations is rather well defined and also the highest frequency band is usually among the most intense of the C—O, C—C vibrations in the infrared spectrum. The fortunate result is that recognizable group frequencies in C—O compounds do occur which are quite useful in characterizing the molecule.

More complete tables on group frequencies are available in the literature (see Recommended Reading 5, 6, 8, 9, 22, and 25). A detailed discussion of these tables is outside the scope of this chapter, but the scientist who is often involved in making structural assignments in unknown spectra will find these tables quite indispensable in his work.

REFERENCES

1. R. M. Badger and L. R. Zumwalt, *J. Chem. Phys.* **6,** 711 (1938).
2. J. Fahrenfort, *Spectrochim. Acta* **17,** 698 (1961).
3. J. R. Ferraro, J. S. Ziomek, and G. Mack, *Spectrochim. Acta* **17,** 802 (1961).
4. J. R. Ferraro, *Spectrochim. Acta* **20,** 901 (1964).
5. D. L. Harms, *Anal. Chem.* **25,** 1140 (1953).
6. U. Schiedt, *Z. Naturforsch.* **7b,** 270 (1952).
7. B. Schrader, F. Nerdel, and G. Kresge, *Z. Anal. Chem.* **170,** 43 (1959).
8. M. M. Stimson, *J. Amer. Chem. Soc.* **74,** 1805 (1952).
9. R. S. Tobias, *J. Chem. Educ.* **44,** 2 and 70 (1967).
10. M. C. Tobin, *J. Opt. Soc. Amer.* **58,** 1057 (1968).

RECOMMENDED READING

D. M. Adams, *Metal-Ligands and Related Vibrations*, St. Martin's Press, New York, 1968.

G. M. Barrow, *The Structure of Molecules*, W. A. Benjamin, New York, 1963.

G. M. Barrow, *Introduction to Molecular Spectroscopy*, McGraw-Hill, New York, 1962.

R. P. Bauman, *Absorption Spectroscopy*, Wiley, New York, 1962.

L. J. Bellamy, *Infrared Spectra of Complex Molecules*, 2nd ed., Wiley, New York, 1958.

L. J. Bellamy, *Advances in Infrared Group Frequencies*, Methuen, London, 1968; Barnes and Noble, New York, 1968.

W. Brugel, *An Introduction to Infrared Spectroscopy*, Methuen, London, 1962.

N. B. Colthup, L. H. Daly, and S. E. Wilberley, *Introduction to Infrared and Raman Spectroscopy*, Academic Press, New York, 1964.

A. D. Cross, *An Introduction to Practical Infrared Spectroscopy*, Butterworths, London, 1960.

R. O. Crisler, *Anal. Chem.* **42,** 388R (1970) (biannual review).

M. Davies, Ed., *Infrared Spectroscopy and Molecular Structure,* Elsevier, New York, 1963.

M. L. Hair, *Infrared Spectroscopy in Surface Chemistry,* Marcel Dekker, New York, 1967.

J. C. Henniker, *Infrared Spectrometry of Industrial Polymers,* Academic Press, New York, 1967.

G. Herzberg, *Infrared and Raman Spectra of Polyatomic Molecules,* Van Nostrand, New York, 1945.

G. Herzberg, *Spectra of Diatomic Molecules,* Van Nostrand, New York, 1950.

J. H. Hibben, *The Raman Effect and Its Chemical Applications,* Reinhold, New York, 1939.

R. N. Jones and C. Sandorfy, in *Chemical Applications of Spectroscopy,* (W. West, Ed., Interscience, New York, 1956.

D. N. Kendall, Ed., *Applied Infrared Spectroscopy,* Reinhold, New York, 1966.

K. W. F. Kohlrausch, *Ramanspektren,* Akademische Verlagsgesellschaft Becker und Erler, Leipzig, 1943.

L. H. Little, *Infrared Spectra of Absorbed Species,* Academic Press, New York, 1966.

K. Nakamoto, *Infrared Spectra of Inorganic and Coordination Compounds,* Wiley, New York, 1963.

K. Nakanishi, *Infrared Absorption Spectroscopy—Practical,* Holden-Day, San Francisco, 1962.

J. P. Phillips, *Spectra–Structure Correlation,* Academic Press, New York, 1964.

W. J. Potts, *Chemical Applications of Infrared Spectroscopy,* Vol. 1, *Techniques,* Wiley, New York, 1963.

C. N. R. Rao, *Chemical Applications of Infrared Spectroscopy,* Academic Press, New York, 1963.

H. A. Szymanski, Ed., *Raman Spectroscopy, Theory and Practice,* Plenum Press, New York, 1967.

E. B. Wilson, J. C. Decius, and P. C. Cross, *Molecular Vibrations,* McGraw-Hill, New York, 1955.

R. Zbinden, *Infrared Spectroscopy of High Polymers,* Academic Press, New York, 1964.

DALLAS L. RABENSTEIN

Department of Chemistry
University of Alberta, Edmonton, Canada

VII. Nuclear Magnetic Resonance Spectroscopy

I. INTRODUCTION

Nuclear magnetic resonance spectroscopy (NMR) is a form of absorption spectroscopy: radio-frequency radiation is absorbed by the nuclei of certain isotopes when they are in a magnetic field. This phenomenon was first observed in nongaseous materials in 1946 for the hydrogen nuclei of solid paraffin (62) and liquid water (8).

In the relatively short period since its discovery, NMR has become a standard analytical tool in all areas of chemistry. The frequency of radiation required for NMR absorption depends on the isotope and its chemical environment; the number of absorption peaks for magnetic nuclei in a given chemical environment is determined by the spatial positions of neighboring magnetic nuclei, and the intensity of the absorption peaks is proportional to the number of nuclei. These characteristic features are illustrated in the proton NMR spectrum of *p*-ethoxy-acetanilide in Figure 7.1. NMR is a valuable technique in molecular structure elucidation and verification. In some cases the NMR spectrum alone is sufficient for identifying the structure of an unknown compound, while in other applications structural information from NMR complements that of other chemical

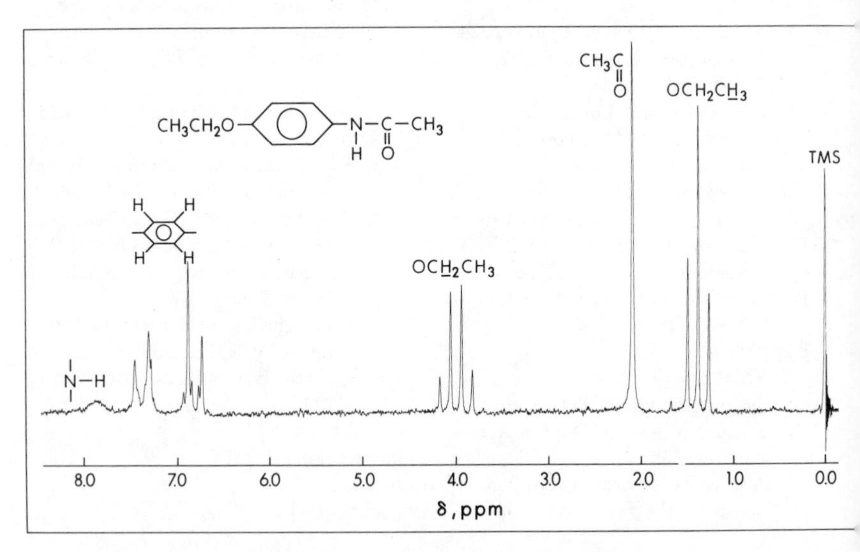

Figure 7.1. Proton NMR Spectrum of *p*-ethoxyacetanilide in $CDCl_3$. Spectrum obtained at a magnetic field of 14,000 G and a radiofrequency of 60 MHz. The chemical shift scale is the delta scale relative to the methyl protons of tetramethylsilane (TMS). Chemical shift scales are discussed in the Experimental section.

and instrumental methods. NMR is also a valuable tool for studying reaction kinetics and chemical equilibria and for quantitatively analyzing mixtures of organic compounds.

It is the objective of this chapter to develop a qualitative treatment of NMR theory pertinent to these applications, to describe the instrumentation and techniques used in the measurement of NMR spectra, and to discuss some representative applications of NMR to chemical problems. For more comprehensive treatments, the reader is referred to the selected list of books and reviews presented at the end of this chapter.

II. BASIC THEORY

In NMR spectroscopy, the sample is placed in a homogeneous magnetic field of typically 10,000–25,000 gauss. The nuclei of some isotopes in the sample have magnetic moments (magnetic dipoles) which interact with the magnetic field in such a way that these nuclei can absorb radio-frequency radiation.

The theory of NMR is in many respects analogous to that of electron paramagnetic resonance (EPR) because the same basic principles are involved. The major differences are in the size of the magnetic moments and the corresponding wavelengths of radiation necessary for resonance to take place. A comparison of the theory given here with the theory on EPR in Chapter 8 is suggested for a better understanding of the underlying principles of these related techniques.

1. Magnetic Properties of Nuclei

Nuclei can be treated as spinning charged particles. A spinning charged particle has angular momentum, p, and generates a magnetic moment, μ. The angular momentum is quantized in units of $\hbar$ where $\hbar$ is Planck's constant divided by 2π. The term $I\hbar$ is the maximum observable component of the angular momentum, where I is the nuclear spin quantum number. Magnetic moment μ is colinear with and proportional to the angular momentum vector p. The relationship between μ and p is

$$\mu = \gamma p \tag{7.1}$$

where γ is the magnetogyric ratio. γ is different for different nuclei. Thus those nuclei that have $I > 0$ have a magnetic moment.

Some of the more commonly occurring isotopes that have a nuclear magnetic moment, along with useful data for discussing the NMR behavior of these isotopes, are listed in Table 7.1. The nuclear magnetic moment is zero for the isotopes ^{12}C, ^{16}O, and ^{32}S, the most abundant isotopes of these elements, and consequently NMR spectra cannot be measured for these chemically important isotopes.

Table 7.1. NMR Properties of Selected Isotopes

Isotope	I	NMR Frequency in a 23,487 Gauss Field, MHz	Natural Abundance, %	Relative Sensitivity at Constant Field: For Equal Numbers of Isotopes	Relative Sensitivity at Constant Field: At Natural Isotopic Abundance
^{1}H	1/2	100.00	99.985	1.00	1.00
^{13}C	1/2	25.14	1.108	0.0159	0.000176
^{15}N	1/2	10.13	0.365	0.00104	0.00000380
^{19}F	1/2	94.08	100.00	0.834	0.834
^{31}P	1/2	40.48	100.00	0.0664	0.0664
^{2}H	1	15.35	0.015	0.00964	0.000145
^{14}N	1	7.22	99.635	0.00101	0.00101
^{23}Na	3/2	26.45	100.00	0.0927	0.0927
^{35}Cl	3/2	9.80	75.4	0.00471	0.00355
^{37}Cl	3/2	8.15	24.6	0.00272	0.000669
^{17}O	5/2	13.56	0.037	0.0291	0.0000108

2. NMR Absorption of Energy

NMR absorption of energy has been treated theoretically by both classical mechanics and quantum mechanics. Both treatments are discussed here because each is useful for describing certain aspects of NMR spectroscopy.

The classical treatment is discussed first. Consider an individual nuclear magnetic moment μ under the influence of a static magnetic field H_0 much larger than μ. If μ lies at an angle relative to H_0, then H_0 exerts a torque on μ causing it to precess around H_0. This motion is analogous to the precession of a spinning gyroscope in the earth's gravitational field. The precession of μ can be resolved into a static component along H_0 and a rotating component in the plane perpendicular to H_0. The frequency of precession, ν_0, is given by Equation 7.2, the Larmor equation.

$$\nu_0 = \frac{\gamma}{2\pi} H_0 \tag{7.2}$$

ν_0 is the Larmor frequency. The Larmor frequencies for typical magnetic fields used in NMR are in the MHz(radio-frequency) range; the frequencies listed in Table 7.1 are the Larmor frequencies for those isotopes in a 23,487 gauss magnetic field. According to quantum theory the individual nuclear magnetic moments in a group of nuclei are distributed among $2I + 1$ allowed orientations relative to H_0. The orientations are not equally populated because their energies are not equal. Thus the ensemble has a resultant macroscopic magnetization. Proton magnetic moments, for example, have two allowed orientations, one orientation has a component in the direction of H_0 while the other has a component that opposes H_0. The orientation of proton magnetic moments in the direction of H_0 is the more populated so that there is a net macroscopic magnetization. If H_0 is along the z axis, the component of the macroscopic magnetization along the z axis, M_z, is nonzero. The component in the xy plane, M_{xy}, is zero because the individual nuclear magnetic moments are randomly precessing around H_0.

A second magnetic field, H_1, rotating in the xy plane can also interact with μ and affect the resultant macroscopic magnetization. Bloch (7) described NMR absorption of energy by a set of equations, the Bloch equations, which relate changes in the resultant macroscopic magnetization to changes in the frequency of H_1. In the Bloch treatment, H_1 flips some nuclear magnetic moments from a lower energy orientation to a higher energy orientation and forces them to precess in phase when the frequency of H_1 is the Larmor frequency. This decreases M_z and creates a nonzero rotating component of magnetization in the xy plane. When the frequency of H_1 does not equal the Larmor freqeuncy, H_1 does not interact with the nuclear magnetic moments and has no effect on the macroscopic magnetization.

A nonzero M_{xy} induces two experimentally observable signals in a coil placed along the x axis when H_1 is applied along the y axis: an absorption signal for the component of M_{xy} out of phase with H_1 and a dispersion signal for the component in phase with H_1. The signals in Figure 7.1 are absorption signals. Dispersion signals have the shapes of but are not the same as the first derivatives of absorption signals. Either the absorption signal or the dispersion signal can be detected with phase sensitive electronics. This procedure is called nuclear induction and is the basis of double-coil NMR spectrometers. H_1 is obtained from electromagnetic radiation that is polarized in the xy plane (such radiation can be resolved into two components rotating around H_0 in opposite directions at the frequency of the radiation).

In the quantum mechanical treatment of NMR, the magnetic nuclei have $2I + 1$ discrete, allowed energy levels in a magnetic field. The energy of these levels relative to the energy in zero field is

$$E_m = -\frac{m\mu H_0}{I} \tag{7.3}$$

where m can have the values $-I, -I + 1, \ldots, I - 1, I$. The energy levels are associated with the $2I + 1$ allowed orientations of μ relative to H_0. There are two orientations for $I = \frac{1}{2}$, for example, of energies equal to $-\mu H_0$ and $+\mu H_0$ relative to the energy in zero field. Electromagnetic radiation of energy equal to the energy difference between adjacent energy levels ($\Delta m = \pm 1$) can cause transitions from a lower to a higher energy level by absorption of radiation and from a higher to a lower energy level by stimulated emission of radiation of the same energy. The energy difference between adjacent levels is

$$\Delta E = \frac{\mu H_0}{I} \tag{7.4}$$

The frequency of radiation of this energy is

$$\nu_0 = \frac{\mu H_0}{hI} \tag{7.5}$$

or in terms of the magnetogyric ratio

$$\nu_0 = \frac{\gamma H_0}{2\pi} \tag{7.6}$$

This result is identical with the result from the classical mechanical treatment.

The probabilities of absorption and stimulated emission of radiation ($E_m \rightarrow E_{m-1}$ and $E_{m-1} \rightarrow E_m$ respectively for $\mu > 0$) are equal and are much greater than the probability of spontaneous emission of radiation at MHz frequencies. The number of absorption transitions is larger than the number of emission transitions because the energy levels are unequally populated so that there is a net absorption of radiation. The absorption of radiation gives an NMR signal whose intensity is proportional to the amount of radiation absorbed; the amount of radiation absorbed is proportional to the total number of nuclei.

The distribution between adjacent energy levels can be predicted by the Boltzman equation. For $I = \frac{1}{2}$, the result is

$$(n_+ - n_-)_{\text{eq}} = \frac{\mu H_0}{kT - \mu H_0} \tag{7.7}$$

where n_+ and n_- are the fractional populations of the lower and higher energy levels. The terms k and T have their usual meanings. The fractional population difference for protons in a magnetic field of 14,000 gauss and at room temperature is approximately 5×10^{-6}. The amount of radiation absorbed is proportional to this small population difference making NMR relatively insensitive compared to other forms of absorption spectroscopy.

3. Spin–Lattice Relaxation

When a sample is placed in the magnetic field of an NMR spectrometer, the allowed energy levels are initially equally populated because they are degenerate when H_0 is zero. The equilibrium distribution is established by transitions between the various energy levels. The transitions occur by transfer of energy between the nuclear spin system and the surrounding lattice; this process is called spin–lattice relaxation and is characterized by the spin–lattice relaxation time T_1. The inverse of T_1 is a first-order rate constant for the rate of attainment of the equilibrium distribution.

The magnitude of T_1 depends on the type of nucleus and the state of the sample. In liquids, T_1 values of nuclei with I equal to one half are usually in the range 10^{-2}–10^2 sec although high viscosities or paramagnetic ions can cause T_1 to be less than 10^{-2} sec. T_1 values of nuclei with I greater than one half are usually less than T_1 values of nuclei with I equal to one half because there is an additional relaxation mechanism for these nuclei that involves interaction of the electric quadrupole moment of the nonspherical nucleus with the surrounding charge gradients. The short spin–lattice relaxation times characteristic of nuclei with I greater than one half dominate their NMR behavior as will be discussed in the section on width of NMR lines.

4. Spin–Spin Relaxation

Neighboring nuclear magnetic moments can exchange spin orientations by an interaction between the magnetic moments with no change in the total energy of the system. Consider the interaction between two neighboring nuclei of $I = \frac{1}{2}$, A and B, that are precessing at the same frequency. Each of the precessing nuclear magnetic moments can be resolved into a static component along H_0 and a rotating component in the plane perpendicular to H_0, in other words, in the plane of H_1. If μ_A is parallel to H_0 and μ_B is antiparallel to H_0, the rotating component of μ_A can force μ_B to flip to the parallel orientation while μ_B simultaneously forces μ_A to flip to the antiparallel orientation. Two transitions have occurred with no net change in the energy of the system. These transitions affect the

NMR signal however because they cause M_{xy} to decay to zero. This process is called spin–spin relaxation and is characterized by the spin–spin relaxation time T_2. The inverse of T_2 is a first-order rate constant for the rate of decay of M_{xy}.

5. Width of NMR Lines

The width of NMR lines is a function of several parameters, including magnetic field inhomogeneity, magnetic fields from neighboring magnetic nuclei, and T_1 and T_2.

If the magnetic field is not homogeneous, the sample will be subjected to a range of magnetic fields, $H_0 + \delta H_0$ where δH_0 is the inhomogeneity, and resonance will occur over a corresponding range of frequencies $\delta\nu$.

$$\delta\nu = \frac{\nu}{2\pi}\,\delta H_0 \tag{7.8}$$

The lower limit to line widths with the best magnets is about 0.1 Hz, while the lower limit for typical magnets is 0.2–0.6 Hz.

The magnetic fields from neighboring magnetic nuclei can also broaden NMR lines. If, for example, nucleus A is located in a position relative to nucleus B such that the total magnetic field at nucleus A is either augmented or decreased by the field from nucleus B, the resonance frequency of nucleus A will not be the resonance frequency predicted for H_0. The field from nucleus B at nucleus A depends on its position relative to nucleus B so that, for a group of nuclei randomly located around nucleus B, resonance actually occurs over a range of frequencies (called magnetic dipole–dipole broadening). This is the situation in most rigid-lattice solids for which proton line widths at half maximum intensity are typically 10^5 Hz. The situation is different in liquids and gases because molecular motion essentially averages out the local magnetic fields from neighboring nuclei. Since all the nuclei are subjected to the same field the resulting spectrum shows much narrower lines and higher resolution. For this reason samples are usually run as solutions.

Line widths are also governed by the relaxation times. According to the Heisenberg Uncertainty Principle, the natural width at half-maximum intensity, $\nu_{1/2}$, is $1/\pi T$ where T is the relaxation time. T_2 is always less than or equal to T_1 so that the relation between the natural line width and the relaxation times is

$$\nu_{1/2} = \frac{1}{\pi T_2} \geq \frac{1}{\pi T_1} \tag{7.9}$$

Experimental line widths for nuclei with short relaxation times, including nuclei in the presence of paramagnetic ions and nuclei that have $I > \frac{1}{2}$, are large according to Equation 7.9. Experimental line widths for nuclei with long relaxation times are usually governed by magnetic field inhomogeneities and not the relaxation times.

The applications of NMR that are the subject of this chapter require high resolution so that we will restrict our discussion to liquid samples or solutions and to nuclei where $I = \frac{1}{2}$.

6. Intensity of NMR Lines

The area (integrated intensity) under an NMR absorption line depends on a number of parameters, including H_0, H_1, T_1, T_2, the temperature, and, as mentioned previously, the total number of nuclei, N, giving the NMR line (46, 67).

The intensity depends on H_0 because the separation of energy levels, hence the relative populations of energy levels, depends on H_0. The area (A) is related to the other parameters by Equation 7.10:

$$A = \frac{KH_1N}{T(1 + (\gamma H_1)^2T_1T_2)^{1/2}} \tag{7.10}$$

According to this equation, the area increases as H_1 increases until $(\gamma H_1)^2 T_1T_2$ becomes appreciable relative to 1, at which point the area decreases as H_1 continues to increase. This can be accounted for qualitatively by the fact that the population difference decreases when the magnitude of H_1 is such that the number of absorption transitions is larger than the number of stimulated emission transitions plus the transitions from the upper to the lower energy level by spin–lattice relaxation. The decrease in signal intensity at large H_1 is referred to as saturation. A consequence of the dependence of the signal intensity on T_1 and T_2 is that different NMR signals in a spectrum will saturate at different H_1 values if T_1 and T_2 are not the same for all the signals. The term K is a constant that depends on the isotope and the experimental conditions. If a single radio-frequency field is used and if selective saturation is not occurring, K is the same for chemically different nuclei of the same isotope. For these experimental conditions, the relative intensities of the different resonances equal the relative concentrations of the different nuclei.

7. Chemical Shifts

The precise frequency at which resonance occurs for a given H_0 is determined not only by the nucleus but also by its chemical environment because the nucleus is shielded from H_0 to a small extent by its electron cloud. In a magnetic field, electrons move in an orbital path around the direction of the magnetic field. The orbital motion of the electrons generates a magnetic field of a magnitude directly proportional to and opposing the externally applied magnetic field which decreases the magnetic field at the nucleus. The magnetic field at the nucleus, H_{local}, is related to H_0 by

$$H_{local} = H_0(1 - \sigma) \tag{7.11}$$

where σ is the shielding constant. The resonance frequency for constant H_0 decreases as the shielding increases according to

$$\nu_0 = \frac{\gamma}{2\pi} H_0(1 - \sigma) \tag{7.12}$$

Such shifts in resonance frequency are called chemical shifts. Qualitatively, the larger the electron density around the nucleus, the more the nucleus is shielded and the larger the chemical shift. Proton chemical shifts cover a range of approximately 1000 Hz for a magnetic field of 14,000 gauss.

It has not been possible to calculate accurate values theoretically for chemical shifts in even the simplest molecules. The theoretical calculations indicate, however, that several factors contribute to chemical shifts, including the electron density around the nucleus, magnetic fields from electron motion in solvent molecules (13, 19), and magnetic fields from electrons localized in chemical bonds in the same molecule. Chemical shifts from the latter source are anisotropic because the magnitude of the induced magnetic field at the nucleus depends on the orientation of the molecule relative to the magnetic field. Thus these magnetic fields can decrease as well as increase the shielding constant. Particularly noteworthy are the induced fields from π electrons, such as the π electrons in C=O, C=C, and C≡C functional groups and in aromatic molecules. Induced magnetic fields in aromatic molecules are attributed to circulation of the π electrons around the ring (ring current) when the plane of the ring is perpendicular to H_0. Induced magnetic fields resulting from the motion of the π electrons in benzene and acetylene are shown in Figure 7.2.

Chemical shifts have been empirically correlated with molecular structure. Such correlations make NMR spectroscopy a powerful tool for elucidating molecular structures, and will be discussed in the section on applications.

8. Spin–Spin Coupling

The previous discussion would predict that individual resonance lines would be observed for the different magnetic nuclei in a molecule. The

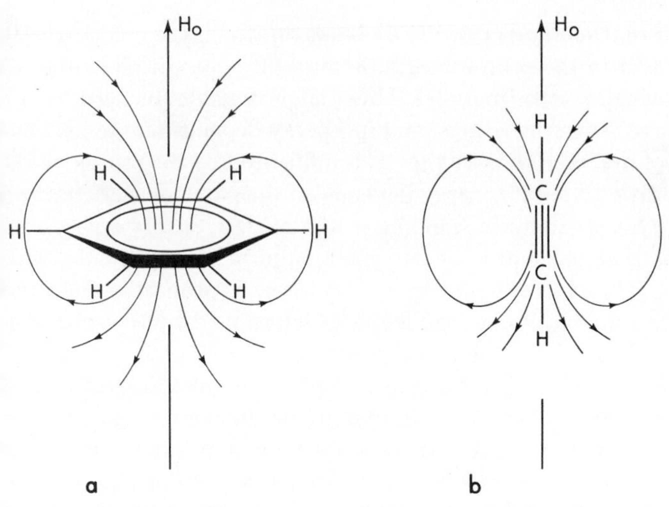

Figure 7.2. Secondary magnetic fields generated by (*a*) π-electron ring current in benzene and (*b*) circulating π electrons in acetylene.

NMR spectrum actually consists of individual lines and groups of lines termed multiplets. There are examples of each of these cases in Figure 7.1. Multiplet patterns arise from through-bond interactions between neighboring nuclear magnetic moments which split the nuclear energy levels; several transitions are possible in place of a single transition expected in the absence of the interaction. This interaction is called spin–spin coupling and provides information on the spatial positions of magnetic nuclei in a molecule.

Splitting of NMR lines due to spin–spin coupling can be qualitatively described in terms of the relative energies of nuclear and electron magnetic moments that are paired either parallel or antiparallel to each other. Consider a system of two nuclei with a spin of one half joined by a two-electron bond, for example, a ^{13}C atom bonded to a hydrogen atom. The Pauli exclusion principle requires that the spins of the two electrons in the bond be opposite so that their magnetic moments are antiparallel. The favored orientations of the nuclear magnetic moments are antiparallel to the adjacent electron magnetic moments. This situation can be pictured as

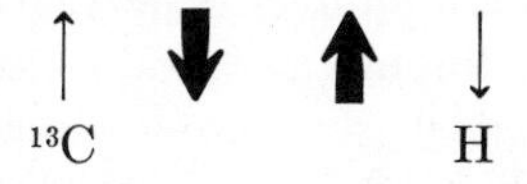

where the darker arrows represent the electron magnetic moments. This situation is only slightly lower in energy than the one in which the ^{13}C magnetic moment is parallel to the proton magnetic moment so that, at room temperature, the number of ^{13}C–H bonds in which the ^{13}C magnetic moment is parallel to the proton magnetic moment is essentially equal to the number in which it is antiparallel. Thus the proton is influenced by the ^{13}C in one of these two orientations, and its energy depends on the particular orientation of the ^{13}C. Likewise the ^{13}C is influenced by the proton in one of its two orientations and its energy depends on the particular orientation of the proton. This results in a splitting of the nuclear energy levels by a constant amount independent of an externally applied magnetic field. This is in contrast to the chemical shift which is directly proportional to the applied field. The shift in the energy levels is determined by the spin–spin coupling constant, J.

Chemical bonding interactions are not as simple as just described, and in some systems parallel orientations of nuclear magnetic moments are actually favored. Such systems are described by a negative spin–spin coupling constant. It is not possible to experimentally determine the absolute signs of coupling constants. Theory indicates, however, that J is positive for coupling of ^{13}C to H through a two-electron bond. The signs of other coupling constants can sometimes be determined relative to the sign of the ^{13}C–H coupling constant. Theory also indicates that spin–spin coupling constants between protons separated by an even number of bonds are negative, and between protons separated by an odd number of bonds, positive.

Nuclear energy levels are split by groups of magnetic nuclei according to the number of nuclei in the group. Consider splitting of the energy levels of the HC═O proton of acetaldehyde by the three protons of the methyl group. The aldehyde proton is influenced by a methyl group having one of four unique combinations of methyl proton magnetic moments

$$\begin{array}{cccc} & \alpha\alpha\beta & \alpha\beta\beta & \\ \alpha\alpha\alpha & \alpha\beta\alpha & \beta\alpha\beta & \beta\beta\beta \\ & \beta\alpha\alpha & \beta\beta\alpha & \end{array}$$

where α and β represent orientations in the direction of and opposed to H_0. The energy of the nuclear spin system for each of these four combinations is different. Thus the methyl protons split the aldehyde proton resonance into four lines (a quartet) and the relative intensities of these lines are 1:3:3:1 corresponding to the number of ways each of these combinations can be formed. Likewise, the methyl protons are influenced by two different orientations of the aldehyde proton so that the methyl resonance is split into two lines (a doublet). Spin–spin coupling also occurs between the methyl protons, but this has no effect on the NMR spectrum.

The actual shape of a multiplet pattern depends not only on the number of coupled nuclei but also on the relative magnitudes of the spin–spin coupling constant and the chemical shift difference between the coupled nuclei (both in Hz). Simple doublets, quartets, and other multiplet patterns whose multiplicity and intensity can be predicted by the above methods will be observed only in those systems where the chemical shift differences are much larger than the spin–spin coupling constants. A further requirement is that the nuclei in each group must be chemically and magnetically equivalent. Nuclei are chemically equivalent if they have the same chemical shift and are magnetically equivalent if they are equally coupled to other nuclei. Such systems are said to be weakly coupled and the spectra are described as first-order spectra. The spectrum of acetaldehyde is first order because the chemical shift between the methyl and aldehyde protons is large and each of the methyl protons is equally coupled to the aldehyde proton.

If $\Delta\nu$ is comparable to J, or if the nuclei are chemically but not magnetically equivalent, the system is said to be strongly coupled and the resonance patterns will be systems of lines often having little or no regularity of spacing or intensity. The aromatic protons of *p*-ethoxyacetanilide (Figure 7.1) are strongly coupled because the chemical shift separation between H_A and H_B is not large relative to J_{AB}, and H_A is coupled differently to H_B and H_B'. Multiplet patterns for the two-spin system are shown in Figure 7.3 as a function of $\Delta\nu/J$. A characteristic feature of such spectra is that, as $\Delta\nu/J$ decreases, the intensity of the outer lines decreases while the intensity of the inner lines increases. Such patterns can be predicted only by considering the effect of the spin coupling interaction on the nuclear energy levels using quantum mechanics.

A conventional nomenclature is used to describe spin coupled systems: (*1*) weakly coupled groups of chemically and magnetically equivalent nuclei are represented by letters from different parts of the alphabet and the number of nuclei in each group is denoted by a subscript, for example, A_3X_2 and AMX; (*2*) strongly coupled chemically and magnetically equivalent groups of nuclei are represented by letters from the same part of the alphabet, for example, A_3B_2; (*3*) systems in which there are both weak and strong coupling are represented by a combination of the above, for example, ABX; (*4*) magnetic nonequivalence in a chemically equivalent group of nuclei is represented by using the same letter for each of the nuclei in the group to indicate chemical equivalence and primes to indicate magnetic nonequivalence, for example, AA′A″B and AA′BB′.

Coupling of nuclear spins is a through-bond interaction, and the energy of the interaction decreases as the number of intervening bonds increases. The spin-spin coupling constant J is small, usually less than 0.5 Hz, for nuclei separated by more than three bonds, although J may be larger if one or more of the intervening bonds is a π bond. Coupling through

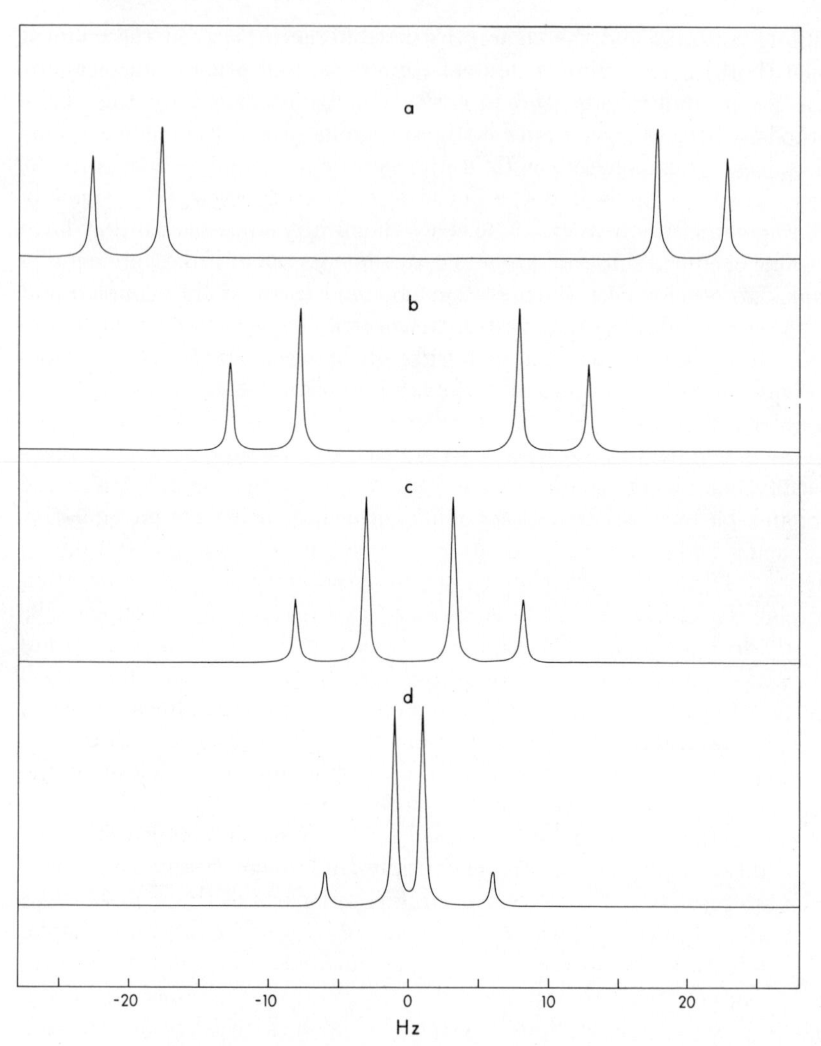

Figure 7.3. Calculated AB spectra for $J = 5.0$ Hz and $J/\Delta\nu$ equal to (*a*) 0.125 (*b*) 0.25 (*c*) 0.5 (*d*) 1.00.

more than three bonds is referred to as long-range coupling and is useful in stereochemical studies (9). A convenient way of representing spin–spin coupling is to indicate the number of intervening bonds by a superscript and the coupled nuclei by a subscript. Proton–proton coupling in the system H—C—C—H would be represented by $^3J_{HH}$.

Spin–spin coupling to nuclei with a spin greater than one half usually is not observed because of the short relaxation times for these nuclei. Chlorine nuclei ($I = \frac{3}{2}$) are not coupled to protons and nitrogen nuclei ($I = 1$) are seldom coupled to protons.

9. Analysis of NMR Spectra

The information contained in an NMR spectrum includes chemical shifts, spin–spin coupling constants, and sometimes relative signs of the coupling constants. Chemical shifts and coupling constants are extremely sensitive to molecular structure and form the basis for the application of NMR to molecular structure elucidation. The signs of coupling constants are of interest in theoretical studies of chemical bonding. Systematic methods have been developed for extracting chemical shifts and coupling constants from NMR spectra; absolute signs of coupling constants cannot be determined but in some cases it is possible to deduce relative signs.

First-order spectra can be analyzed by applying the following rules.

1. A group of n_1 chemically and magnetically equivalent nuclei with spins of one half splits a resonance into $n_1 + 1$ lines. (The general equation is $2nI + 1$, for example, the deuterium in -CHD- would split the hydrogen resonance into three lines.)
2. The relative intensities of the lines in each multiplet pattern due to splitting by nuclei with spins of one half are equal to the coefficients of the binomial expansion: 1:1 for a doublet, 1:2:1 for a triplet, 1:3:3:1 for a quartet, and so on.
3. Coupling to a second group of n_2 chemically and magnetically equivalent nuclei with spins of one half splits each of the $n_1 + 1$ lines in the multiplet pattern into $n_2 + 1$ lines giving a total of $(n_1 + 1)(n_2 + 1)$ lines in the multiplet pattern.
4. The separation between adjacent lines in the multiplet pattern is equal to the coupling constant.
5. The chemical shift at the center of the multiplet pattern is the chemical shift of the group giving the multiplet pattern.

Spectra can be satisfactorily analyzed by these rules when $\Delta\nu/J$ is greater than 7 but still show deviations in relative intensities until $\Delta\nu/J$ is greater than 20. The signs of spin coupling constants cannot be determined from first-order spectra. First-order splitting is schematically illustrated in the spectrum of ethanol in Figure 7.4. The methyl protons split the methylene protons into a quartet while the methylene protons split the methyl protons into a triplet. Each of the resonances in the methylene quartet is further split into a doublet by the hydroxyl proton while the hydroxyl proton is split into a triplet by the methylene protons.

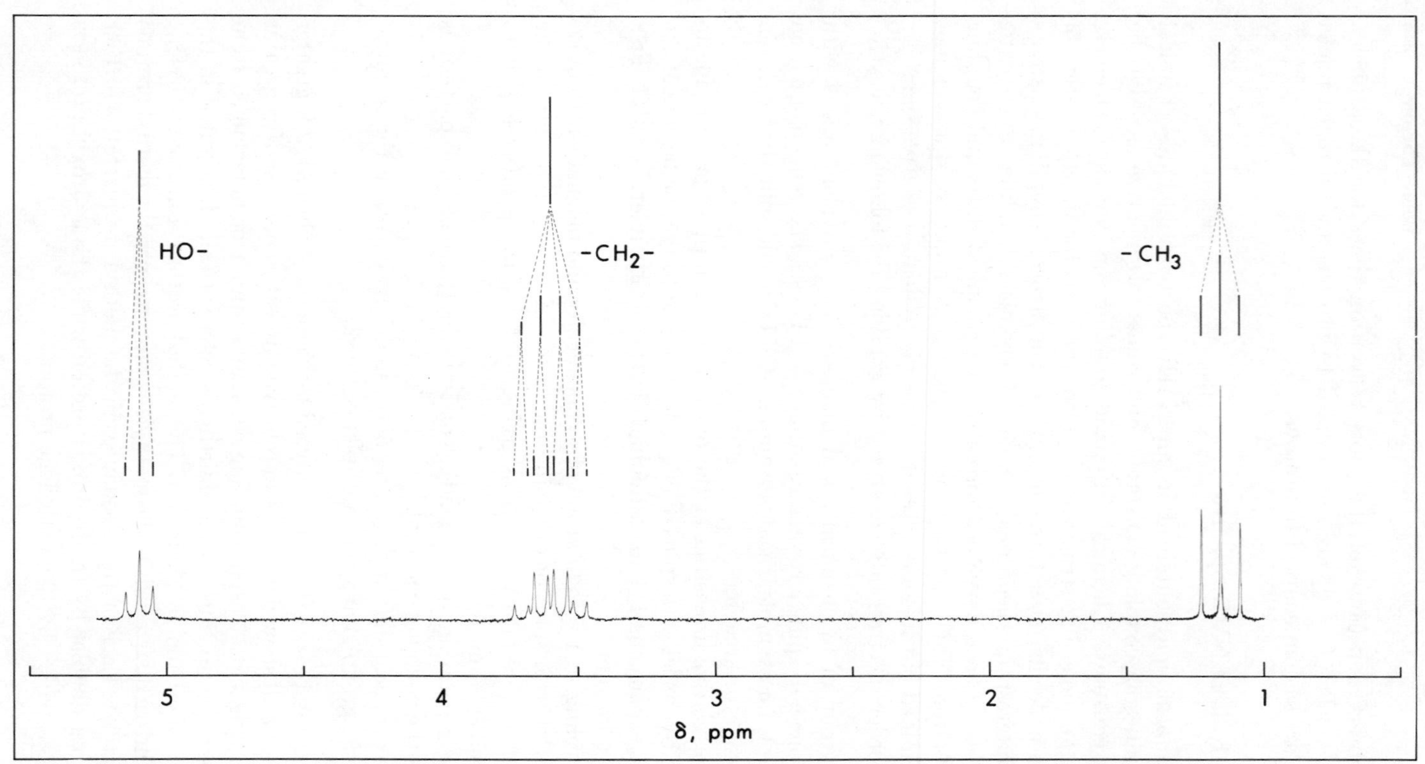

Figure 7.4. Proton NMR spectrum of neat ethanol and a schematic representation of the first order spin–spin coupling in ethanol. Spectrum obtained at a magnetic field of 23,500 G and a radio-frequency of 100 MHz.

Second-order spectra must be analyzed by applying a quantum mechanical treatment of spin coupling. The quantum mechanical treatment will not be discussed in detail here, but a description of some aspects is necessary for a qualitative understanding of the methods used to analyze such spectra (68). To summarize the treatment, nuclear energy levels and stationary state wave functions are derived in terms of chemical shifts and spin coupling constants using Hamiltonian operators which account both for the externally applied magnetic field and the through-bond spin–spin coupling interactions. Selection rules and relative probabilities, which are proportional to resonance intensities, are then determined for transitions between these energy levels when radio frequency energy is applied.

Straightforward algebraic equations for the energy levels, transition frequencies, and intensities have been derived in this way for many spin systems, including the AB, AB_2, AB_3, ABX, AA′XX′ and A_2B_2 spin systems. It is not possible to derive such straightforward equations for some systems, including the ABC system, and such spectra must be analyzed by comparing computer-calculated spectra with experimental spectra. Spectral parameters are varied until the calculated spectrum matches the experimental spectrum. Caution must be exercised in doing spectral analyses in this way, however, because often more than one set of parameters will give a calculated spectrum which matches the experimental spectrum within experimental error. Spectra measured at two different frequencies (e.g., 60 MHz and 100 MHz) are often of use in these cases. Excellent discussions on the analysis of second-order spectra are available (31, 68).

10. Aids for the Analysis of NMR Spectra

Several techniques can aid in the analysis of spectra, including increasing $\Delta\nu/J$ with a larger magnetic field, spin decoupling with a second radiofrequency field, isotopic subsititution, and paramagnetic metal ion induced chemical shifts.

Spin-coupled spectra become more first order as the applied magnetic field increases because the chemical shift is directly proportional to the magnetic field strength while the spin coupling constant is independent of it. This is illustrated by the proton NMR spectra of *N*-sec-butylaniline, measured at 14,000 gauss (60 MHz) and 52,000 gauss (220 MHz), in Figure 7.5. For ease of analysis, spin-coupled spectra should be measured with the largest magnetic field available.

Spin-coupled spectra can be simplified by selectively decoupling some spin–spin interactions by the double resonance technique. In this technique, the sample is simultaneously irradiated with two radio-frequency fields; one frequency is the frequency which gives the resonance pattern of interest

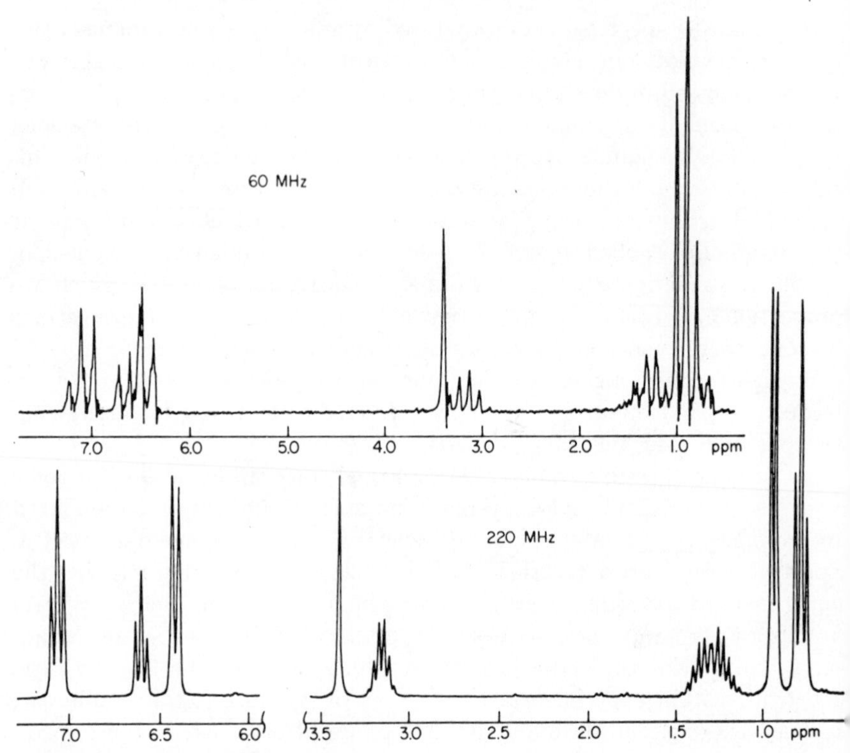

Figure 7.5. Proton NMR spectra of *N*-sec-butylaniline at 14,000 G (60 MHz) and 52,000 G (220 MHz). From Ferguson and Phillips (29) (copyright 1967 by the American Association for the Advancement of Science).

and the other frequency is the resonance frequency of the coupled nuclei. When the second H_1 is of a high enough intensity, the spin–spin coupling collapses giving simpler spectra. Chemical applications of the double resonance technique have been reviewed by Baldeschwieler and Randall (3).

Substitution of an isotope that has a spin of zero or a spin greater than one half for an isotope that has a spin of one half also simplifies spectra. For example, substitution of deuterium for hydrogen often gives spectra having little or no observable coupling to the deuterium because of its small magnetogyric ratio. Substitution may result in spectra that are reduced in complexity so that they can be analyzed by straightforward procedures. Figure 8.14 shows this effect in ESR.

The resonances for N—H, O—H, and other exchangeable protons can be identified by running the spectrum of the sample in a solvent other than

D_2O, then adding several drops of D_2O and rerunning the spectrum after shaking (27). The exchangeable protons are replaced by deuterium and will be observed in the residual H_2O resonance.

Complexation of the organic molecule by certain paramagnetic metal ions can sometimes result in useful changes in the NMR spectrum. Specifically, the resonance frequencies are shifted due to local magnetic fields generated at the magnetic nuclei by the unpaired electrons. If the relaxation time of the electrons is sufficiently short, the resonances are not extensively broadened and the spin–spin coupling is not collapsed. The result is spectra that are more nearly first order. The local fields are the result of through-bond (contact) and through-space (pseudocontact) interactions. It has recently been demonstrated that europium and praesodymium induce large shifts in the proton NMR spectra of organic molecules that coordinate to the metal, for example, molecules that have certain functional groups containing oxygen or nitrogen (14, 39, 69). The magnitude of the induced shift for a given proton depends on the distance between the site of complexation and the proton. The chloroform soluble tris(dipivalomethanato)europium (III) complex, $Eu(DPM)_3$, has been used as a "shift reagent" to simplify the proton NMR spectra of a number of compounds, including alcohols (14, 23, 69), steroids, and terpenoids (24). The shift reagents are an important discovery because they allow otherwise unresolved spectra to be resolved and interpreted without loss of spin–spin coupling information.

11. Chemical Exchange Effects on NMR Spectra

The NMR spectrum of a solution containing several species of different chemical shifts in equilibrium is sensitive to the rate of exchange between the various species. Detailed information about chemical equilibria can often be obtained from spectra of such systems (11).

Consider a nucleus exchanging between two environments, or sites, A and B ($A \rightleftarrows B$). The average lifetimes in the two sites are τ_A and τ_B and the fractional populations are P_A and P_B. ($P_A + P_B = 1$). If the resonances for each site are single lines of chemical shifts ν_A and ν_B, the NMR spectrum for the system depends on the magnitude of $\tau_A(\nu_A - \nu_B)$. When $\tau_A(\nu_A - \nu_B)$ is large, the spectrum is simply a superposition of the individual spectra for A and B. As $\tau_A(\nu_A - \nu_B)$ decreases, the individual resonance lines broaden, approach each other, coalesce, and ultimately give a single line. These cases are referred to as slow exchange, intermediate exchange, coalescence, and fast exchange.

Exchange effects can be quantitatively treated to obtain the kinetic parameters by modifying the Bloch equations to include the contribution of chemical exchange to the total relaxation times of each species and the changes in magnetization at each site from chemical exchange (35, 54). Two-site or multisite exchange situations can be treated with modified Bloch equations. Experimental exchange rates can be determined by matching experimental spectra with spectra calculated as a function of exchange rate.

Multiplet patterns due to spin coupling are also affected by exchange if one of the partners in the coupling interaction is exchanging at a rate comparable to or greater than the coupling constant. The quartet and doublet patterns for the hydroxyl proton and the methyl protons respectively in the spectrum of methanol collapse to single resonances, for example, when the rate of hydroxyl proton exchange between methanol molecules is comparable to the coupling constant. Modified Bloch equations derived for exchange between a number of sites equal to the number of lines in the multiplet pattern are valid for treating the collapse of multiplet patterns in first-order spectra. Second-order spectra must be treated by other methods.

III. EXPERIMENTAL

1. NMR Spectra

NMR spectra are measured by varying the magnetic field while holding the RF frequency constant, or vice versa. Resonance occurs for each magnetically active nucleus when Equation 7.12 is satisfied. At constant RF frequency, the magnetic field required for resonance increases as the shielding increases. At constant magnetic field, the resonance frequency decreases as the shielding increases. Spectra measured by varying the magnetic field while holding the frequency constant (field sweep) are usually displayed with magnetic field increasing from left to right (thus shielding increases from left to right). A resonance occurring at a smaller magnetic field than a second resonance is to the left of and is described as being downfield from the second resonance. Spectra measured by varying the frequency while holding the field constant (frequency sweep) are displayed with frequency decreasing from left to right so that they are directly comparable with those from field-sweep measurements.

2. Basic Instrument

Considering the similarities in the theory of NMR and ESR one would expect the instrumentation for these two techniques to be much the same. This is only true in a general sense. The much lower irradiating frequencies used in NMR can be transmitted through pliable, conducting cables. In

ESR, on the other hand, bulkier and less versatile hollow guides have to be used to transport the microwaves used in the irradiation of the sample. Much more elaborate control functions and multiple resonance techniques are, therefore, possible with NMR.

The basic components of an NMR spectrometer are a magnet, a radio-frequency transmitting and detecting system, a means of varying either the magnetic field or the RF frequency, and a recorder or oscilloscope for displaying the spectrum. Permanent magnets, electromagnets, and magnetic fields from superconducting solenoids have all been used in NMR instruments; the maximum fields that have been obtained in commercial instruments using these magnets are approximately 14,100 gauss for permanent magnets, 24,000 gauss for electromagnets, and 70,500 gauss for superconducting solenoids. Still larger fields are feasible with superconducting solenoids. Large magnetic fields are desirable because spin coupling is more nearly first order and signal intensity is proportional to the square of the magnetic field strength.

The probe is located in the magnet gap and contains a sample holder, a sample spinner, and one or two RF coils depending on the detection system. In the single-coil probe, the crystal-controlled transmitter coil is in a bridge network which balances out the large transmitter signal thus making it possible to detect the absorption of energy from the RF field as an out-of-balance electromotive force across the bridge. In the two-coil probe, the crystal-controlled transmitter coil is placed perpendicularly to the receiver coil; NMR is detected as a signal induced in the receiver coil as predicted by the Bloch equations. The absorption or dispersion signal can be observed using phase-sensitive electronics.

The basic requirements of the magnet are that it be homogeneous and stable. The magnetic field has to be homogeneous and stable to better than 5 parts in 10^9 over the sample volume, for example, to resolve two proton resonances separated by 0.5 Hz with an RF frequency of 100 MHz. Homogeneity is achieved by using a small region of the sample (often 0.05 cm^3), compensating for inhomogeneity with small magnetic fields produced by passing dc currents through small electrical coils (shim coils) located on the pole faces of the magnet, and spinning the sample to effectively average out field gradients in the direction perpendicular to the spin axis. Spinning the sample however modulates the magnetic field giving "spinning side-bands" symmetrically placed about and separated from the main signal by the spinning rate in revolutions per second. They can be identified by changing the spinning rate. Stability is achieved by continuously compensating for small rapid fluctuations in magnetic flux with coils wound around the pole faces and by controlling the temperature of the magnet to minimize thermal fluctuations.

Actually, it is not necessary for the magnetic field and the RF frequency to be this stable; it is sufficient if their ratio remains constant. The required stability of the field–frequency ratio is achieved by continuously adjusting the magnetic field or the frequency to maintain the resonance condition for a control sample. In some probes, for example the Varian A60 probes, the control sample is separated from the analytical sample and has its own RF transmitter–receiver coil. Field-sweep spectra are measured by sweeping only the field affecting the analytical sample. The control sample is usually water. This field–frequency control system is called the external lock system. In the other system, called the internal lock system, the control sample is *in* the analytical sample. In this system, the field–frequency ratio is maintained constant by locking to a resonance of the control sample with one sideband frequency generated by audiomodulating the RF frequency while observing with a second sideband frequency.

Field–frequency control provides magnetic fields that are stable to approximately 1 part in 10^8 per hour for external lock systems and 1 part in 10^9 per hour for internal lock systems. This precise control of the stability makes it possible to use precalibrated chart paper from which chemical shifts can be measured relative to a reference sample peak which serves as a calibration marker. Instrument calibration can be checked by measuring the chemical shifts of standard samples (45), or by measuring the separation of a sideband from the parent peak produced by audiomodulation with a known frequency.

3. Nuclear Magnetic Double Resonance

NMDR is carried out by irradiating the sample with an additional RF frequency. NMDR is most conveniently done in the frequency sweep mode because the resonance for the protons being decoupled can be continuously irradiated while the entire spectrum is scanned. For homonuclear decoupling, a single decoupling frequency is usually used. In heteronuclear decoupling experiments, a single decoupling frequency or a continuous band of decoupling frequencies can be used. The latter technique is useful for eliminating all couplings to a certain isotopic species. A band of decoupling radiation can be obtained by modulating the decoupling RF frequency with random noise of a bandwidth sufficient to cover all the resonances being decoupled. This technique is sometimes called "white noise" decoupling. All proton coupling in ^{13}C spectra of organic molecules, for example, can be eliminated by decoupling the protons with a noise-modulated RF field (42, 77).

4. Fourier Transform NMR

The previously described instruments measure NMR spectra by continuously varying the field–frequency ratio seen by the analytical sample, and is referred to as the continuous-wave or slow-passage method. Spectra can also be measured by applying an RF pulse of typically 0.1-msec duration to the sample and then observing for approximately 1 sec the decay of the nuclear induction signal produced in the receiver coil of a double-coil spectrometer. The range of RF frequencies contained in each pulse is sufficiently wide that all nuclei of a given isotope are irradiated and each produces a nuclear induction signal. The total nuclear induction decay signal is simply the Fourier transform of the normal slow-passage spectrum, which can be obtained from the decay pattern by a Fourier transformation. This method is known as Fourier Transform NMR spectroscopy (FTNMR) (26, 28).

One distinct advantage of FTNMR over continuous-wave NMR is the increased sensitivity that is obtained for an equal amount of instrument time because the entire spectrum is observed with each pulse. This is discussed in more detail in the section on sensitivity enhancement.

5. Intensity Measurements

Most NMR spectrometers have built-in electronic integrators for measuring the relative areas (integrated intensities) under the peaks. The integrated spectrum is a step function, with the height of each step being directly proportional to the area under the peak corresponding to that step. The reproducibility of integrated intensity measurements depends on the signal/noise (S/N) ratio; under optimum conditions a reproducibility of $\pm 2\%$ can usually be achieved.

6. Chemical Shift Measurements

NMR absorption for a nucleus in a particular environment can theoretically be characterized by measuring the field–frequency ratio necessary to satisfy the resonance condition (Equation 7.12). Experimental chemical shift differences for protons range from the minimum resolvable difference (ca. 0.1 Hz) to approximately 1000 Hz when the magnetic field is on the order of 14,100 gauss. Whereas frequencies can be measured to this accuracy (ca. 1 ppb for 0.1-Hz shifts), magnetic fields, on the other hand, cannot. Consequently it is not possible to make an absolute measurement of the field–frequency ratio that satisfies the resonance condition. For this reason, chemical shifts are measured relative to the chemical shift of a reference compound with the same constant and approximately known magnetic field (approximate compared to the accuracy with which the frequencies are known). If chemical shift differences are reported in hertz, the magnetic field or the resonance frequency of the reference compound must be reported for the measurements to be meaningful because relative chemical shifts in these units depend on the strength of the magnetic field.

An alternative and better way of reporting chemical shift data relative to a reference compound is to use a scale that is independent of the magnitudes of the RF frequency and the magnetic field. One such scale, the delta scale, is defined by Equation 7.13

$$\delta = \frac{\nu_s - \nu_r}{\nu_r} \times 10^6 \tag{7.13}$$

where ν_s and ν_r are the resonance frequencies of the sample and reference in hertz. The units of δ are parts per million (ppm). Tetramethylsilane (TMS) is the common reference compound in proton NMR spectroscopy because its protons are more shielded than the protons in most other compounds. The δ's relative to TMS are therefore usually positive and increase as the shielding decreases. Field sweep spectra are usually displayed on a Hz scale, and can be expressed on the δ scale by substituting the constant frequency into the denominator of Equation 7.13.

Another popular scale for reporting proton chemical shifts is the tau scale. This scale is related to the delta scale by

$$\tau(\text{ppm}) = 10.000 - \delta(\text{ppm}) \tag{7.14}$$

The reference compound can be dissolved in the sample (internal reference) if it does not react with the sample constituents. This is the most convenient and accurate method because the sample and reference are then affected equally by induced magnetic fields from the solvent and other compounds in the sample. If, however, the reference compound reacts with the sample or interacts with or is not soluble in the solvent, it must be physically separated from the sample (external reference). The external reference can be contained in a coaxial capillary that is placed in the sample solution. Chemical shifts measured relative to an external reference must be corrected for differences in the magnetic susceptibility of the sample and reference solutions.

Reference compounds that have a single sharp peak are desirable. Tetramethylsilane (TMS) is the most widely used reference in organic solvents. Hexamethyldisiloxane (HMDS) is often used when spectra are being recorded at high temperatures because it is less volatile. The chemical shift of HMDS is 0.06 ppm relative to TMS. Sodium 2,2-dimethyl-2-

silapentane-5-sulfonate (DSS), dioxane, *t*-butyl alcohol, and acetonitrile have all been used as reference compounds in aqueous solutions. The protons of the three methyl groups of DSS give a singlet whose chemical shift is close to that of TMS. One should be careful when comparing chemical shifts of spectra measured relative to different reference compounds and in different solvents because of possible specific molecular interactions.

7. The Sample

To obtain a high-resolution NMR spectrum, the sample must be in the liquid state or in solution; dipole–dipole interaction in the solid state results in line broadening and loss of resolution. Neat liquid samples often show similar line broadening due to high viscosities and consequently are also usually run as solutions or at high temperatures. Ideally, samples should be run as dilute solutions in inert solvents because chemical shifts will depend on concentration, solvent, and temperature if there are solute–solute and solute–solvent interactions. The solvent must not have resonances in the region of interest. Many solvents are now available in the deuterated form for proton NMR spectroscopy, including acetic acid, acetone, benzene, chloroform, deuterium oxide, and dimethylsulfoxide. Acetone-d_6 and DMSO-d_6 give characteristic 5-line multiplets at 2.05 and 2.50 ppm due to splitting of residual hydrogen by two deuterium nuclei. A mixed solvent of $CDCl_3$ and DMSO-d_6 is often useful because it has solvating properties similar to but is less viscous than DMSO-d_6.

Particulate matter in the sample solution can cause loss of resolution, and random and nonreproducible changes in the appearance of the spectrum. This is particularly true if the particle is ferromagnetic, for example, a tiny piece of iron. Particulate matter can be removed from the solution with microfilters.

Line broadening and loss of resolution also occurs if the solution contains paramagnetic impurities such as Cu^{2+}, Fe^{3+}, and Mn^{2+} because the paramagnetic species decrease the relaxation times. Paramagnetic species present at concentrations as low as 10^{-6} *M* can have an appreciable line-broadening effect.

NMR sample containers are usually cylindrical tubes of 5-mm outside diameter for which approximately 0.5 ml of sample solution is required. A signal distinguishable from the noise can be observed for a single proton at a concentration of about 0.01 *M* with most 60-MHz NMR spectrometers if it is not split by spin–spin coupling. Usually sample concentrations of at least 0.2 *M* are required with the minimum concentration for a particular sample depending upon the amount of spin–spin splitting in the spectrum.

8. Sensitivity Enhancement

Often the sample solution cannot be made as concentrated as 0.2 *M* because either the sample is not soluble or there is insufficient sample. In the first case, the sensitivity can be increased by using thin-walled sample tubes to increase the filling factor (the fraction of the volume within the receiver coil occupied by sample solution), by increasing the temperature to increase the solubility, and by using larger sample tubes with some instruments. Use of larger tubes may be accompanied by a decrease in resolution because the sample is subjected to a larger and possibly more inhomogeneous magnetic field. A decrease in resolution will reduce the potential increase in sensitivity with larger tubes.

Special sample tubes (microcell tubes) are available for small amounts of sample. The microcell tubes are designed so that all the sample is contained within the approximately 5-mm length of sample tube surrounded by the receiver coil. More concentrated solutions can be run with a small amount of sample because less solution is required; NMR spectra can be obtained on 25 μl of solution with microcell tubes. Microtechniques for small samples have recently been reviewed (50).

The S/N ratio can be increased at the expense of additional scanning time by scanning the spectrum a number of times and summing up the individual spectra. The S/N ratio theoretically increases as the square root of the number of scans (36). This technique requires a computer to sum and store the spectra; a small "computer of average transients" or "CAT" is often used. Fourier transform NMR has a distinct advantage over slow passage NMR when multiple scanning is done because the instrument time required in FTNMR is less. For example, if a series of 0.1 msec pulses is used and the observation time between pulses is 1 sec, 500 spectra can be accumulated in 500 sec by FTNMR whereas only one spectrum can be obtained by the continuous-wave method using a 500-sec sweep time. The enhancement of the S/N ratio achieved by FTNMR in this example is predicted to be $\sim$22. Experimentally a factor of $\sim$10 has been realized. The FTNMR spectrum of a 0.011 *M* solution of progesterone obtained by summing the 1-sec decay patterns from 500 0.1 msec pulses is shown in Figure 7.6. Also shown is the slow passage NMR spectrum of the same solution using a 500-sec sweep time.

9. Measurement of NMR Spectra of Nuclei Other Than Hydrogen

The methods discussed above for the measurement of NMR spectra of hydrogen are applicable to other magnetically active nuclei. Additional considerations associated with the measurement of NMR spectra of several important isotopes will be mentioned.

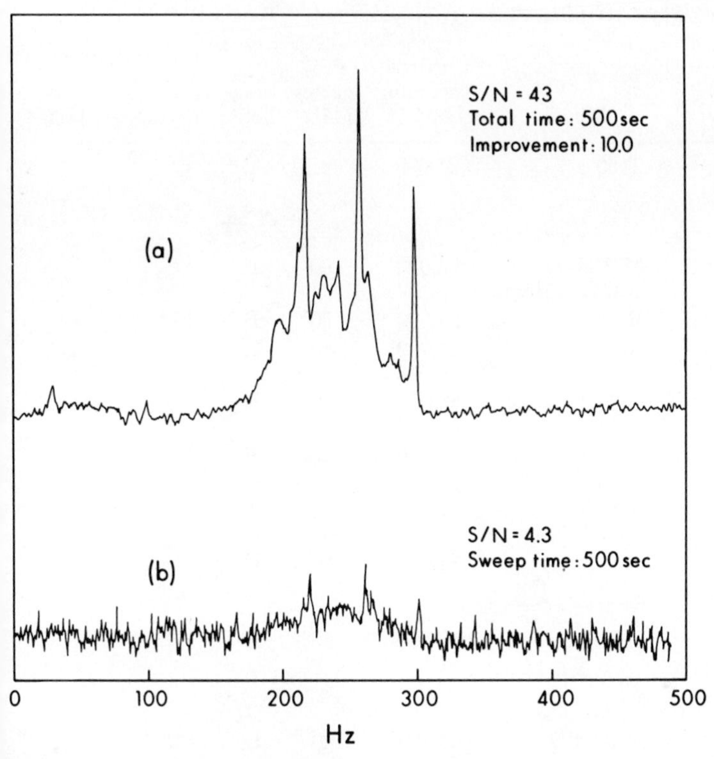

Figure 7.6. Proton NMR spectra of a 0.011 *M* solution of progesterone measured (*a*) by the Fourier transorm NMR method and (*b*) by the slow passage method. The Fourier transform spectrum is from the free induction decay signal obtained by summing the free induction decay signals from 500 successive pulses of 0.1 msec duration. The decay signal from each pulse was recorded for one second. From Ernst and Anderson (26).

Carbon-13 NMR spectra are more difficult to obtain than proton NMR spectra because of several factors. ^{13}C, the only isotope of carbon having a spin of one half, is only 1.59% as sensitive as hydrogen for an equal number of ^{13}C nuclei and hydrogen nuclei at the same magnetic field. Combining the low natural abundance with the inherently low sensitivity, ^{13}C NMR is only 0.018% as sensitive as proton NMR for an equal number of natural abundance carbon and hydrogen nuclei at the same magnetic field. The effective sensitivity is reduced even further and the spectral interpretation is complex because of spin–spin coupling to hydrogen nuclei up to three bonds removed from the ^{13}C nucleus. ^{13}C-enriched organic compounds can be prepared to improve the sensitivity. However for ^{13}C

Table 7.2. **Principal Commercially Available NMR Spectrometers (5)**

Company	Model	Approximate Price Including Duty, $	Frequency MHz	Magnet [a]	Sweep [b]	Lock [c]
Varian	T-60	21,000	60	P	FI [h]	[i]
JEOLCO	MH-60-II (MiNiMar)	22,000	60	E	FQ, FI	I, E
Perkin-Elmer	R-12	23,000	60	P	FI [h]	[i]
Varian	A-60-D	31,000	60	E	FI	E
JEOLCO	C-60HI	35,000	60	E	FQ, FI	I, E
Hitachi (Perkin-Elmer)	R-20A	36,000	60	P	FQ	E
JEOLCO	MH-100	40,000	100	E	FI [h]	I, E
Bruker	HX-60	44,000	60	E	FQ, FI	
Bruker	HX-90	59,000	90	E	FQ, FI	I
Varian	HA-100D	59,000	100	E	FQ, FI	I
JEOLCO	PS-100	70,000	100	E	FQ, FI	I, E
Varian	XL-100	80,000	100	E	FQ	I
Bruker	HFX-4	88,000	90	E	FQ, FI	I
Varian	HR-220	160,000	220	S	FI	[i]

[a] Magnet types: *E* = electro-; *P* = permanent; *S* = superconducting.
[b] *FQ* = frequency sweep; *FI* = field sweep.
[c] *I* = internal; *E* = external.
[d] As determined by using the tallest peak in the CH_3 quartet of 1% ethylbenzene.
[e] Full width at half-maximum amplitude, using a specified peak in the spectrum of o-dichlorobenzene.

Sensitivity [d]	Resolution [e]	Other Nuclei	Largest (o.d.) Spinning Tube [f]	Standard Accessories [g]	Other Accessories Available
18:1	0.5	yes	5	dc	I lock; VT (−100° to +100°); stand-by battery
18:1	0.4	no	5	dc; VT (−50° to +150°)	
18:1	0.5	yes	5		dc; I lock
18:1	0.3	no	5		dc; VT (−150° to 200°)
25:1	0.3	yes	8	dc; VT (−150° to 200°)	
25:1	0.3	yes	10	dc; VT (−120° to 200°)	
30:1	0.5	no	5	VT (−70° to 170°); dc	Auto. Y shim
30:1	0.5	yes	15	wide line; VT (−150° to 200°)	
40:1	0.3	yes	15	wide line; VT (−150° to 200°); auto. Y shim	18 in. magnet
40:1	0.3	yes	8	VT (−150° to 200°) auto. Y shim; F^{19}	15 in. magnet
50:1	0.3	yes	12	dc; VT (−150° to 200°); auto. Y shim; F^{19}; C^{13}	
50:1	0.3	yes	8	VT (−150° to 200°); dc (Homo- and Hetero-); auto. Y shim; F^{19}; C^{13}	15 in. magnet; E lock
50:1	0.25	yes	15	VT (−150° to 200°); auto. Y shim; wide line; P^{31}; C^{13}	18 in. magnet
55:1	1.1	yes	5	VT (−50° to 150°); dc	FQ Sweep; Pulse; auto. homo.

[f] Using standard size magnet. The larger magnets all permit the use of larger spinning tubes.

[g] *VT* = variable temperature probe; *dc* = spin decoupler.

[h] For decoupling a frequency sweep can be simulated.

[j] Because of the magnet stability, no lock is needed on routine scans.

NMR to realize its full potential, NMR instruments have to measure ^{13}C NMR spectra on the ^{13}C present at natural abundance.

The sensitivity enhancement techniques discussed previously, especially large sample tubes, large magnetic fields, and time averaging of spectra, are all useful for increasing the sensitivity in ^{13}C NMR spectroscopy. The sensitivity can be further enhanced and the spectrum simplified by collapsing all the proton couplings by the white-noise spin decoupling technique (42,77). The decoupled ^{13}C spectrum then consists of a single resonance for each chemically nonequivalent carbon atom in molecules having no other nuclei of spin equal to one half (the probability of two ^{13}C nuclei being bonded together in the same molecule is about 1 in 10^4). It is feasible to do ^{13}C NMR routinely at the natural abundance level by combining white-noise proton decoupling with time averaging of spectra. Both continuous-wave and FTNMR methods can be used, however, the FTNMR method requires significantly less time (28, 41).

Another feature of spectra obtained with the double resonance technique is the increased intensity of some resonances due to the nuclear Overhauser effect. Dipole–dipole interaction between the saturated, spin-decoupled nuclei and nearby nuclei provides a relaxation mechanism for the nearby nuclei that causes their population distribution to be non-Boltzman. Specifically, for proton–proton and proton–^{13}C couplings, the lower energy levels are more populated giving an enhancement of the signal (43). Nuclear Overhauser enhancement is a valuable aid in stereochemical studies; this application has been reviewed recently (58).

There is no standard reference as yet for ^{13}C chemical shifts; $^{13}CS_2$, $CH_3{}^{13}COOH$, $^{13}C_6H_6$, and aqueous solutions of $K_2{}^{13}CO_3$ have all been used. Tetramethylsilane has been proposed as a standard with a shift convention identical to that for protons.

Fluorine-19 is suited for NMR study because it has a spin of one half, a natural abundance of 100%, and is 0.834 times as sensitive as protons for an equal number of nuclei at constant field. The most common ^{19}F reference compounds are CCl_3F (internal reference) and CF_3COOH (external reference).

Phosphorus-31 is also suited for NMR study because it has a spin of one half and a natural abundance of 100%, although its inherent sensitivity is low. At constant field, ^{31}P NMR is only 0.0664 times as sensitive as proton NMR for an equal number of nuclei. The most common reference compounds for ^{31}P chemical shifts are P_4O_6 and H_3PO_4 (85% in water); both are used as external references.

10. Commercial Instrumentation

The principal commercially available NMR spectrometers as of December 1969 are listed in Table 7.2 (5). Each vendor was requested to verify the data presented for his instruments. Variations on these instruments are available. In some cases, accessories are indicated by additional numbers or letters in the model designation.

IV. APPLICATIONS

1. Structural and Conformational Studies by Proton NMR

Proton NMR has been widely used in problems of molecular structure elucidation and verification. The proton NMR spectrum of a compound is a fingerprint of that compound, and agreement with the spectrum of a known compound often serves to identify a compound. There are several compilations of NMR spectra which are useful for structure determinations (10, 15, 76). More frequently, however, structural information is derived from NMR spectra by using empirical correlations between NMR spectral parameters and molecular structures. These correlations allow one to predict empirically proton chemical shifts in a given molecular environment.

Approximate chemical shift ranges for protons in a variety of functional groups are given in Figure 7.7. These data are intended only to illustrate the general dependence of proton chemical shifts on functional groups; more extensive and detailed correlations are available (25, 55, 71, 72, 73).

Structural information can be systematically obtained from proton NMR spectra. The first step is to determine whether close-lying peaks are caused by protons of different chemical shifts or are part of a multiplet pattern. First-order splitting patterns can easily be identified by the spacing and intensities of the lines. More complex splitting patterns can be identified by measuring the spectrum at two different magnetic fields and by comparison with calculated spectra. The volume of calculated spectra by Wiberg and Nist is particularly useful for this purpose (79). After the peaks have been identified as either chemically shifted peaks or peaks in multiplet patterns, the approximate chemical shift of each chemically shifted peak and of each spin multiplet is determined. The resonances are then assigned to specific proton groupings and functional groups on the basis of their chemical shifts; this can be done without a detailed analysis of the spectrum to determine the exact chemical shifts. The absence of resonances in char-

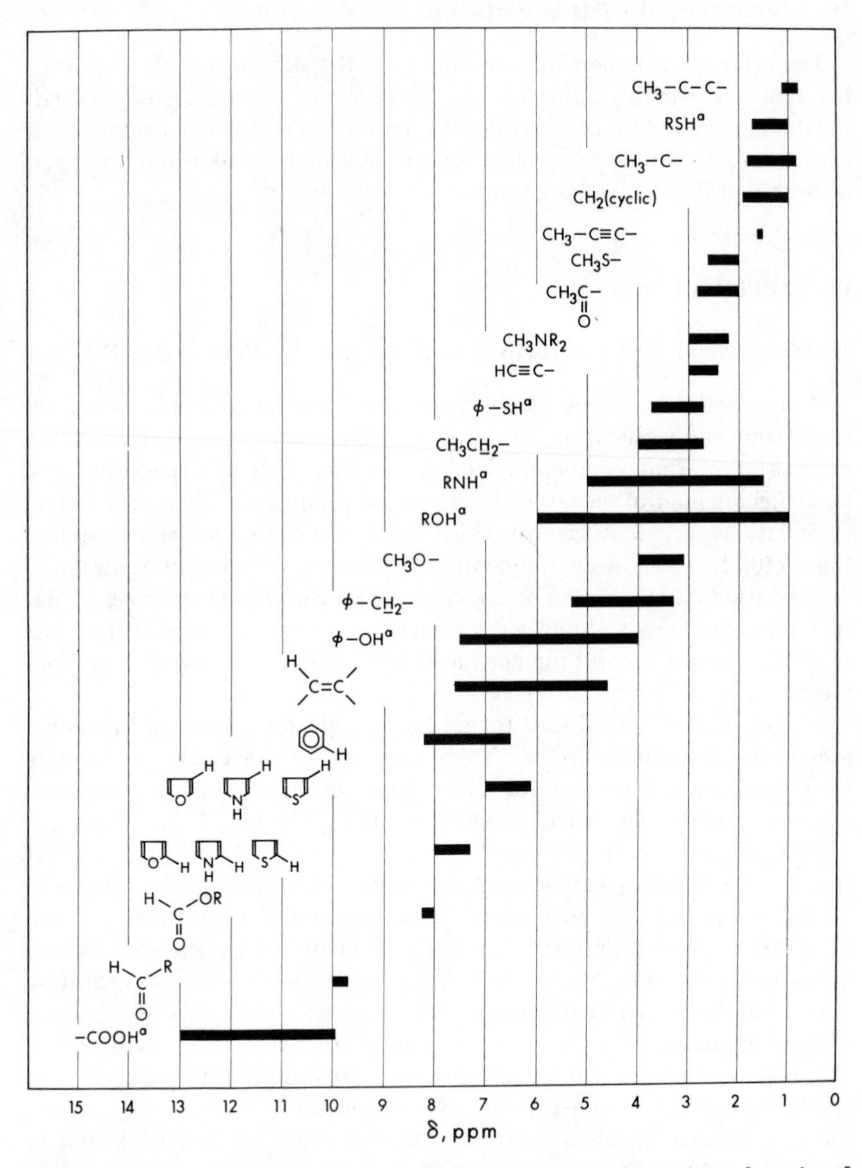

Figure 7.7. Approximate chemical shifts for protons in various functional groups relative to TMS.

[a] **Chemical shift highly dependent on hydrogen bonding and exchange effects.**

acteristic regions of the spectrum is useful in eliminating possible structures. The intensities of the resonances are then measured to determine the relative numbers of protons in each group. It is frequently possible to determine the total number of protons in the molecule (proton counting) if the number of protons responsible for any resonance is known. For example, if a certain resonance has been assigned to a methyl group, the number of protons in the molecule can be determined by comparing the intensities of the other resonances to the intensity of the methyl resonance. Finally, the spatial relation between some of the proton groupings can be established from the multiplicity and intensity of the multiplet patterns. For example, the quartet and triplet patterns in Figure 7.1 can be assigned to the methylene and methyl protons of an ethyl group because the spacing of the lines in each first order pattern is the same and the multiplicity is in agreement with the relative intensities of two to three. More detailed procedures for extracting structural information from NMR spectra have been described (4).

In some cases, a complete structure determination is possible from such an analysis of a proton NMR spectrum. However, this information is more often used in conjunction with results from other chemical and instrumental measurements.

Some examples of molecular structure problems that have been solved by NMR will be mentioned. The structure of the hydrocarbon produced in the reaction of 2,5-diacetoxy-2,5-dimethyl-3-hexyne with benzene in the presence of aluminum chloride was narrowed down to structures (**1**) and (**2**) by its chemical and physical properties (37).

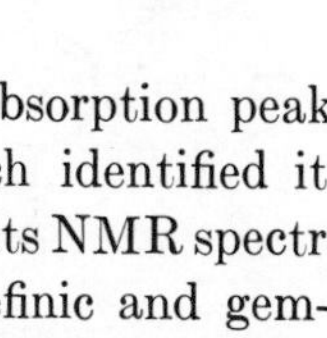

H_3C CH_3 H_3C CH_3

(1)

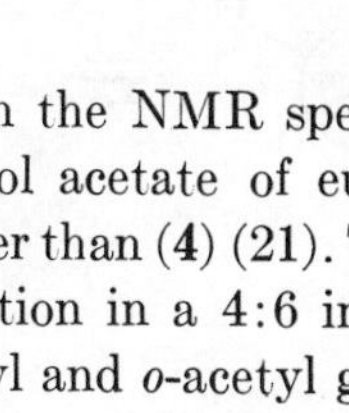

H_3C CH_3 H_3C CH_3

(2)

No olefinic absorption peak was observed in the NMR spectrum of the product which identified it as (**1**). The enol acetate of eucarvone was identified by its NMR spectrum to be (**3**) rather than (**4**) (21). The spectrum contained olefinic and gem-dimethyl absorption in a 4:6 intensity ratio as well as two singlet peaks for allylic methyl and *o*-acetyl groups. There were no peaks that could be assigned to the two tertiary hydrogens in (**4**).

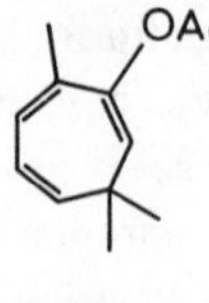

(3) 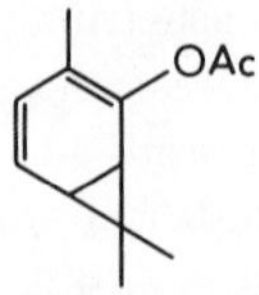 (4)

Proton NMR is useful for studying the conformations of molecules. The axial and equatorial substituents of cyclic molecules are shielded differently because of the magnetic anisotropy of the ring and the other substituents. Correlations of axial and equatorial chemical shifts with molecular structure have been useful in studies on the conformations of cyclic molecules and the structures of steroids (25). Also the magnitude of proton–proton coupling constants depends on the stereochemical relation between the coupled protons, thus providing structural and conformational information (9, 75). To illustrate, the dependence of vicinal coupling constants on stereochemistry will be discussed.

Vicinal coupling refers to coupling between magnetic nuclei bound to adjoining atoms, for example the protons in H—C—C—H. The magnitude of this coupling is a function of the angle between the two C—H bonds when viewed down the C—C bond, the dihedral angle. A valence-bond treatment of vicinal coupling between protons on adjacent sp^3 hybridized carbon atoms by Karplus predicts that J_{vic} should be at a maximum at dihedral angles of 0 and 180° and at a minimum at 90°. Experimental data on the geometrical dependence of vicinal coupling constants has amply verified the trends predicted by the Karplus treatment (9). These studies have shown, however, that substituents, especially electronegative substituents, can also cause substantial changes in the coupling constants.

In acyclic systems, rotation around carbon–carbon bonds is usually rapid so that observed vicinal coupling constants are the average of the coupling constants in the three staggered conformations. The staggered conformations for the molecule CHMNCHXY are

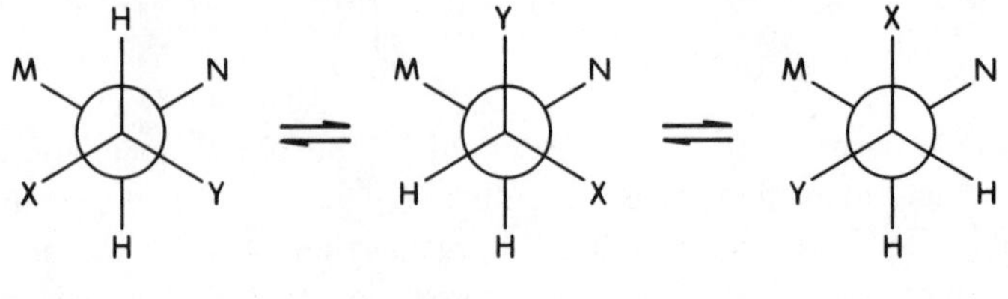

A B C

The observed coupling constant is weighted according to the relative populations.

$$J_{\text{obs}} = P_A J_t + P_B J_g^B + P_C J_g^C \tag{7.15}$$

where P_A, P_B, and P_C are the relative populations of the three conformers ($P_A + P_B + P_C = 1$) and J_t, $J_g{}^B$, and $J_g{}^C$ are the indicated trans and gauche coupling constants.

In an elegant application, Anet (1) showed that *dl*-malic acid from the enzymatic hydration of the deuterated *dl*-fumaric acid (**5**) is erythro (**6**)

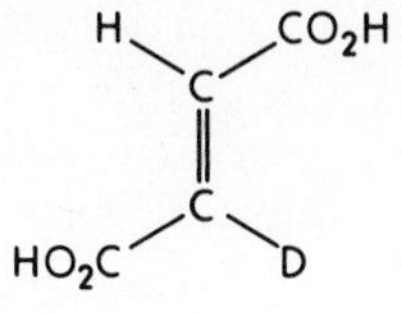

(**5**) 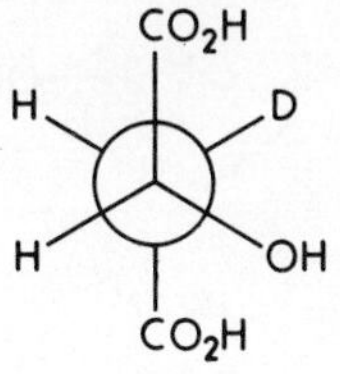 (**6**) 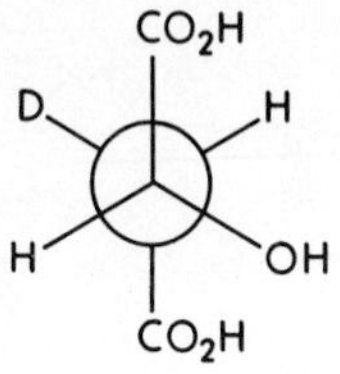 (**7**)

by comparison of the vicinal coupling constant of the product of the enzymatic hydration with the vicinal coupling constant of threo-*dl*-malic acid (**7**) synthesized by an unequivocal route. Averaged vicinal coupling constants also indicate conformational preferences. For example, acetyl choline was found to exist predominantly in conformation (**8**) (22).

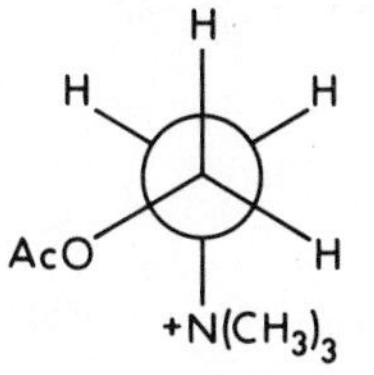

(**8**)

The conformations of cyclic systems can sometimes be established from the magnitudes of vicinal coupling constants if the dihedral angles between ring protons are different in the possible conformations; specifically the ring hydrogens are identified as being in axial or equatorial positions.

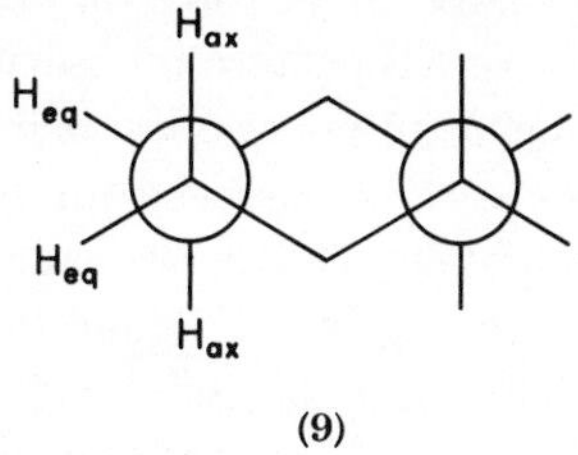

(**9**)

Typically, $J_{\text{ax-ax}}$ is about 10 Hz while $J_{\text{ax-eq}}$ and $J_{\text{eq-eq}}$ are both about 3 Hz. The chair conformation (**10**) was assigned to 2,3-benzocyclohepten-1,4-

diol because the axial 1- and 4- protons appear as doublets coupled only to the trans axial protons on the adjacent carbon atoms; the small cis coupling defines the dihedral angle and indicates the chair conformation (16).

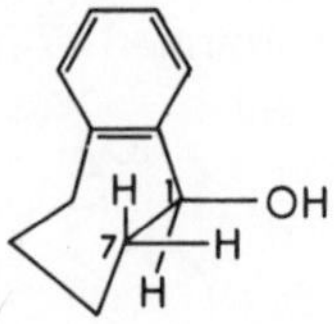

(10)

The conformations of steroids and other rigid ring systems can often be determined from vicinal coupling constants. This area has recently been reviewed (33).

Vicinal coupling constants are also useful in molecular structure elucidation. The product resulting from the methoxide methanolysis of 3-methoxycyclohexane-α-oxide followed by acetylation was shown to be (**11**) rather than (**12**) (48). The resonance for H_x is a 1:2:1 triplet which indicated equal coupling to two identical vicinal protons in a trans, diaxial relationship with H_x.

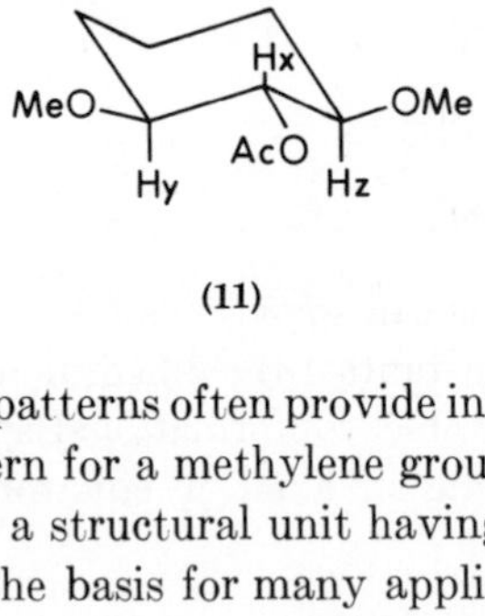

(11) 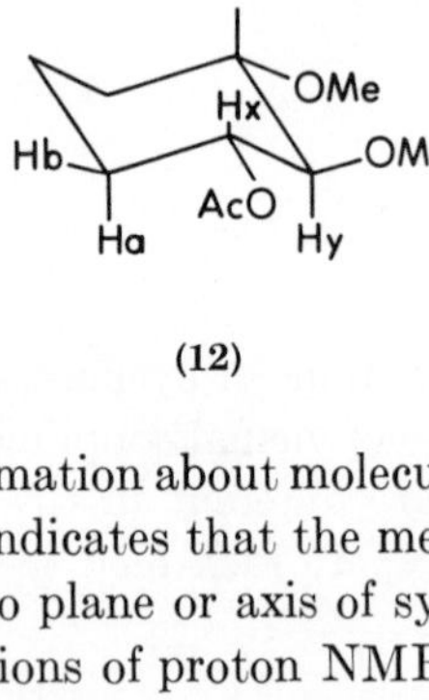 **(12)**

Multiplet patterns often provide information about molecular symmetry. An AB pattern for a methylene group indicates that the methylene group is bonded to a structural unit having no plane or axis of symmetry. This has formed the basis for many applications of proton NMR to structural studies of vinyl polymers. An AB pattern is observed for the -CH_2- protons in the isotactic triad sequence of polymethyl methacrylate (structure *a* in Figure 7.8), a single resonance is observed for the -CH_2- protons in the syndiotactic triad (*b*), and both single resonances and AB patterns are observed for the -CH_2- protons in the heterotactic triad (*c*) (11). The α-methyl protons in the isotactic and syndiotactic triads also have different chemical shifts due to their different environments. The resonance lines for α-methyl groups in heterotactic triads have chemical shifts intermediate between those for the methyl groups in isotactic and syndiotactic triads. A

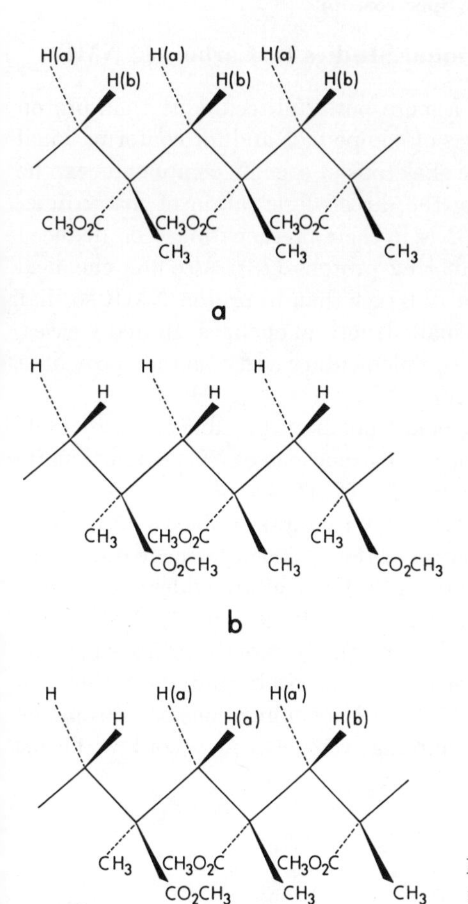

Figure 7.8. Triad sequences of polymethyl methacrylate polymers: (*a*) isotatic (*b*) syndiotactic (*c*) heterotactic.

number of NMR studies of polymer configuration have been reported, and the subject has been reviewed recently (11, 70).

Spin coupling patterns for protons on substituted benzene rings are useful for establishing the positions of substituents. The four aromatic protons of a disubstituted aromatic will give a symmetrical AA′BB′ pattern for ortho substitution, an unsymmetrical A_2BC pattern for meta substitution and a single resonance for para substitution when the two substituents are the same. If the two substituents are different, the para isomer can be identified by the symmetrical AA′BB′ pattern characteristic of para substitution (the aromatic pattern in Figure 7.1 is an example). Ortho and meta substitution give ABCD spectra and the position of substitution cannot be determined from the symmetry of the pattern alone.

2. Structural and Conformational Studies by Carbon-13 NMR

Carbon-13 NMR is potentially a more powerful technique than proton NMR for elucidating the structures of compounds and for conformational studies on known compounds. The skeletons of organic compounds can be observed by ^{13}C NMR permitting the direct observation of many functional groups, whereas in proton NMR their presence must be deduced from the chemical shifts of neighboring protons (78). Also the chemical shift range in ^{13}C NMR is inherently larger than in proton NMR so that ^{13}C spectra are more sensitive to small structural changes. In many cases, ^{13}C NMR and proton NMR are complementary and together provide a powerful probe of molecular structure.

As with proton chemical shifts, it has not been possible to predict ^{13}C chemical shifts theoretically. Empirical correlations of ^{13}C chemical shifts with molecular structure are available (25, 57, 66, 75).

Several applications of ^{13}C NMR to structural problems have been reported. As an example, the structure of the cyclooctatetraene dimer was established from its ^{13}C NMR spectrum (32). Chemical evidence had been presented to support each of the following structures. The ^{13}C NMR spectrum of the compound consisted of four equally intense peaks when the protons were noise-decoupled. The number of peaks and their chemical shifts confirmed structure (**14**). The ^{13}C—H coupling constants from the ^{13}C spectrum without proton decoupling were also in accord with this structure.

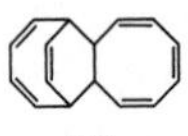

(13) 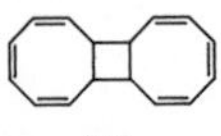 (14) 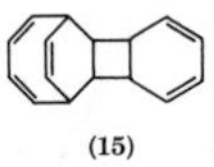 (15)

Carbon-13 chemical shifts are also sensitive probes of the conformation of known compounds (18). The chemical shift of the carbinol carbon in cyclohexanol, for example, depends on the orientation of the hydroxyl group (17). A single ^{13}C resonance was observed for the carbinol carbon because of rapid conformational interconversion; the chemical shift of the averaged resonance indicates that in 74% of the cyclohexanol the hydroxyl is equatorial corresponding to an equilibrium constant of $\sim$2.8 and a $\Delta G°$ value of 0.6 kcal/mole.

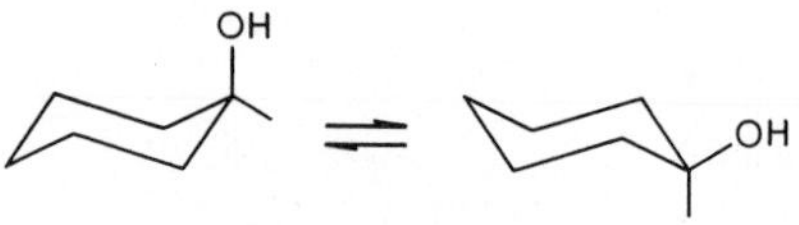

OH OH

(16)

Carbon-13 NMR should also find many applications as an isotopic tracer technique for the study of reaction mechanisms using ^{13}C-enriched compounds.

3. Structural Studies by NMR of other Nuclei

The NMR spectral parameters for magnetic nuclei other than protons and ^{13}C also provide structural information. Empirical chemical shift–molecular structure correlations are available for other nuclei, including ^{19}F (25, 26) and ^{31}P (25, 44).

4. Investigation of Chemical Equilibria by NMR

NMR spectra of molecules involved in chemical equilibria are dependent on the relative populations of and rates of exchange between the various species. Equilibrium constants and kinetic parameters have been determined by NMR for a number of systems at chemical equilibrium.

The determination of equilibrium constants involves measuring the concentrations of the different species at equilibrium. When exchange between the different species is slow and separate resonances are observed for each species, the relative intensities of the various peaks provides a measure of the concentrations. Separate resonances are observed for the protons of acetylacetone in the keto and enol forms from which it has been determined that acetylacetone is a 4:1 mixture of keto and enol forms at room temperature (64). Measurement of the equilibrium concentrations over a range of temperatures gave a value of 2.7 ± 0.1 kcal/mole for the enthalpy of the keto–enol conversion. When the exchange between the various species is rapid and one averaged resonance is observed, the chemical shift of the averaged resonance provides a measure of the relative concentrations of the various species. The stoichiometry and formation constants of the nitrilotriacetic acid (NTA) chelates of Cd^{2+} and Zn^{2+} were determined from the dependence of the -CH_2- chemical shift on the metal ion concentration and the solution pH (63). Formation constants for the $M(NTA)^-$ and $M(NTA)_2^{4-}$ complexes formed by the reactions

$$M^{2+} + NTA^{3-} \rightleftarrows M(NTA)^-$$

$$M(NTA)^- + NTA^{3-} \rightleftarrows M(NTA)_2^{4-} \tag{7.15}$$

were determined. Hydrogen bonding equilibria have been studied using the averaged chemical shift of the hydrogen atom in the hydrogen bond. The equilibria between the hydrogen bonded clusters and their components are rapid and reversible, and detailed information has been obtained on many hydrogen bonded systems. Conformational equilibrium constants can also be determined from averaged chemical shifts. The conformational equilibrium constant of cyclohexanol (**16**) was determined from the averaged chemical shift of the carbinol carbon (17).

The kinetics of a number of intermolecular and intramolecular equilibrium reactions have been studied by using the dependence of the spectral line shapes on the rates of exchange between the various forms (the NMR line-broadening method). Some of the early studies include the protonation kinetics of methylamine (34)

$$CH_3NH_3^+ \rightleftarrows CH_3NH_2 + H^+ \tag{7.16}$$

and the proton exchange kinetics of alcohols (51).

$$ROH + \overset{*}{R}O\overset{*}{H} \rightleftarrows RO\overset{*}{H} + \overset{*}{R}OH \tag{7.17}$$

The kinetics of the interconversion of cyclohexane between the two chair forms have been studied by observing the exchange of ring protons between

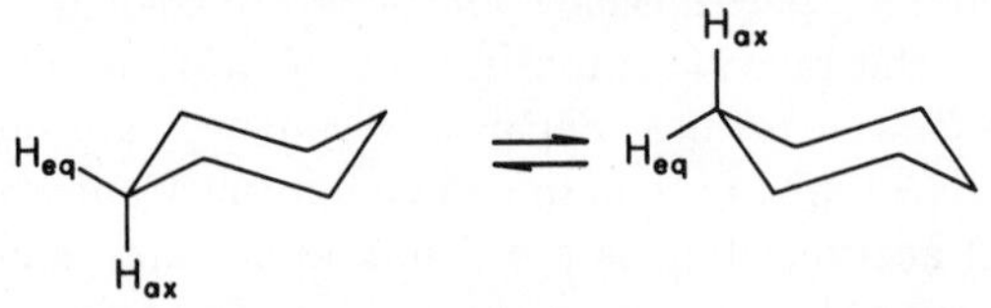

axial and equatorial environments (12). The rates of rotation about single bonds with partial double bond character have been studied by NMR. The classic example is the study by Phillips of the rotation about the C–N bond in dimethylformamide (61). The resonance for the methyl group trans to the carbonyl group is downfield from the resonance for the cis methyl group; at room temperature rotation about the C–N bond is slow and separate resonances are observed for the two methyl groups. The rate of rotation is increased at higher temperatures and the resonances collapse to a single averaged resonance. Measurement of the kinetics of conformational interchange by NMR has recently been reviewed (6).

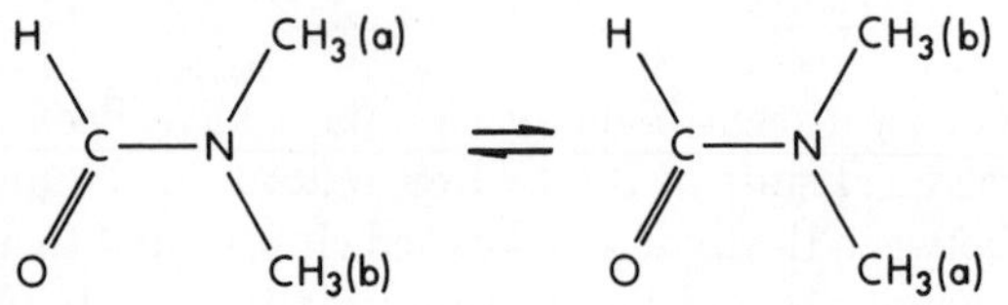

5. Quantitative Analysis by NMR

The area (integrated intensity) under each resonance in an NMR spectrum is directly proportional to the number of nuclei giving the resonance (Equation 7.10). The use of relative intensities in the identification of a pure compound by NMR was discussed previously. The subject of this section is the use of NMR intensity measurements for quantitative analysis (46, 59).

If a single radio-frequency field is used and if the intensity of the radio-frequency field is small enough that selective saturation is not occurring, the proportionality constant K in Equation 7.10 is the same for chemically different nuclei of the same isotope. For these conditions, quantitative measurements can be made without determining K because the relative areas under the resonances for chemically different nuclei of the same isotope are equal to the relative numbers of nuclei giving the resonances. This may not be the case however, if the spectrum is spin decoupled by the double resonance technique because some resonances may be enhanced by the nuclear Overhauser effect. Nuclear Overhauser enhancement is particularly important in ^{13}C NMR when proton decoupling is used. In such a case, the relative numbers of nuclei can still be determined if the nuclear Overhauser enhancement factors are known. Alternatively, nuclear Overhauser enhancement can be eliminated in noise-decoupled ^{13}C spectra by adding appropriate free-radical species to the solution (80). Although no quantitative determinations by ^{13}C NMR have been reported, many analytical problems will undoubtedly be solved by this technique as more laboratories become equipped with ^{13}C NMR instrumentation. The remainder of this discussion will be concerned with quantitative determinations from spectra measured with a single radio-frequency field under experimental conditions where no saturation takes place.

There are several factors which should be kept in mind when doing quantitative analysis by NMR. Resonances used for quantitative analysis (analytical resonances) should preferably be derived from carbon-bonded protons and other nonexchangeable protons. Nitrogen-bonded, oxygen-bonded, and other exchangeable protons will often be observed as an averaged resonance that also usually includes the resonance from any water in the sample or solvent. Exchangeable proton resonances are sometimes broad and may overlap the analytical resonances. This interference can usually be eliminated by proper choice of solvent and temperature, or by adding a small amount of acid. The area of the standard analytical resonance should be approximately equal to the area of the analytical resonance for the compound being determined. The errors involved in determining the concentrations of minor components when the standard resonance is from a major component of the sample are usually larger than the ±2% integral reproducibility quoted previously. Satellite peaks from spin–spin coupling to ^{13}C nuclei and spinning side bands should be added to the analytical resonance intensities for quantitative determinations. ^{13}C satellite peaks can easily be accounted for because the natural abundance of ^{13}C is accurately known; spinning side bands can be minimized by proper adjustment of the magnetic field homogeneity and careful selection of spinning rate and sample tube.

The relative concentrations of individual components in a mixture can be determined from the relative intensities of analytical resonances for each of the compounds. The relative concentrations of the following isomers in dinitrotoluene isomer mixtures have been determined from the relative intensities of the methyl resonances (53). The relative concentrations of

CH_3 NO_2 NO_2 — CH_3 NO_2 NO_2 — CH_3 NO_2 NO_2 — CH_3 NO_2 NO_2 — CH_3 NO_2 O_2N

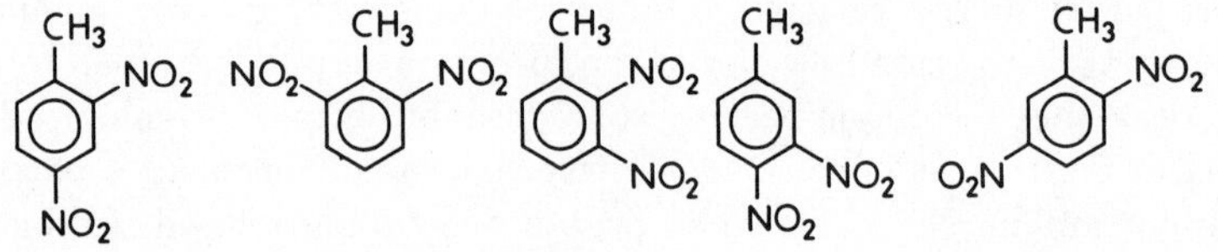

primary, secondary, and tertiary alcohols have been determined by measuring the relative intensities of the dichloroacetyl proton resonances of the dichloroacetate esters (2) and the relative intensities of the trifluoroacetyl fluorine resonances of the trifluoroacetate esters (52). These latter two determinations demonstrate the use of derivatives in quantitative analysis by NMR. In both cases the chemical shift of the derivative depends on whether the alcohol is primary, secondary, or tertiary. The relative concentrations of isomers in ortho–para alkylphenol mixtures have been determined by measuring the relative intensities of the acetyl proton resonances of the acetate esters (49). Waters of crystallization in diamagnetic metal complexes have been determined by comparing the intensity of the HDO resonance of a D_2O solution of the metal complex to the intensity of the resonances for the nonexchangeable carbon-bonded ligand protons (47).

Absolute concentrations can be determined with an intensity standard. The standard compound can be added to the sample as an internal standard or can be in a separate tube as an external standard. A NMR method which uses an internal standard for quantitatively determining exchangeable hydrogen has been reported (60). The sample and standard are dissolved in D_2O, and the exchangeable hydrogen is determined from the relative intensities of the HDO and standard resonances. Exchangeable hydrogens of the type NH_2, COH, COOH, C≡C—H, $CH_3C{=}O$ have been determined by this method. In another application, organic compounds in the 0.1–5.0 mole % concentration range have been determined by using ^{13}C satellite peaks from the solvent as internal standards (20).

If an external standard is used, the external standard solution can be placed in a separate NMR tube or in a precision coaxial tube of the type discussed previously for external reference compounds. If a separate tube is used for the external standard, both the sample tube and the standard tube must be of precision-bore tubing so that the effective volume within the receiver coils will be the same for both solutions. A method for the analysis of commercial analgesic preparations containing aspirin, phenacetin, and caffeine (APC), which uses an external standard in a separate tube, has been reported (40). If the external standard is contained in a precision coaxial tube, the coaxial tube is placed in the sample tube and the external standard and sample are run simultaneously. This method requires a calibration curve of sample intensity/standard intensity ratio versus sample concentration for a given coaxial tube–NMR tube pair. Water in deuterium oxide, *t*-butyl vinyl ether in CCl_4, and weight percent hydrogen in organic compounds have been determined in this way (38). The results for hydrogen analyses on several organic compounds are listed in Table 7.3. Also listed for comparison are the hydrogen contents as determined by combustion analysis.

As an analytical tool NMR has the advantage that the compounds to be analyzed do not necessarily have to be pure. This is important if the compound is unstable or difficult to obtain in the pure form. NMR is rapid and nondestructive. It is, in principle, applicable to the analysis of any

Table 7.3. Hydrogen Analysis by NMR and by Combustion [a]

Compound	Calculated [b]	NMR Method [b]		Combustion Method [b]	
		Found	Error	Found	Error
Polystyrene	7.72	7.60	−0.12	7.80	0.08
PBVE [c]	7.50	7.67	0.17	7.30	−0.20
P(p-Me-BVE) [d]	8.16	8.03	−0.13	8.12	−0.04
Methanol	12.58	12.51	−0.07		
DMF [e]	9.65	9.32	−0.33		

[a] From Hatada, Terawaki, Okuda, Nagata, and Yuki (38).
[b] Wt. % hydrogen.
[c] Poly(benzyl-vinyl ether).
[d] Poly(*p*-methyl benzyl vinyl ether).
[e] Dimethylformamide.

compound that contains protons or other magnetic nuclei. Quantitative analysis by proton, ^{19}F, and ^{31}P NMR has been reported.

The major disadvantage of the technique is that it is relatively insensitive. It is not suited for trace analysis. Quantitative measurements are usually based on area determinations obtained with the electronic integrator built into most commercial instruments. The precision of such determinations is usually limited by the reproducibility of the integral measurements.

V. SUMMARY

The rapid development of NMR spectroscopy from an area of research in theoretical physics to a standard analytical tool in all areas of chemistry has been characterized by a number of significant advances in instrumentation and theory. Chemists initially became interested in NMR when the chemical shift and spin–spin coupling effects were reported; the observation of these effects was made possible by improvements in the homogeneity of magnetic fields. Advances in magnet technology have made available magnets with homogeneity such that it is now possible to measure high-resolution NMR spectra routinely. Advances in magnet technology have also increased the available field strengths from 7.04 to 70.5 kilogauss (with a corresponding increase in the resonance frequency of protons from 30 to 300 MHz). This has led to many new applications of NMR because of the increased sensitivity and the larger chemical shifts at the higher magnetic fields. The theoretical methods which have been developed for extracting chemical shifts and coupling constants from NMR spectra together with the extensive correlations of these parameters with molecular structure make possible detailed studies on the identity, structure, and solution chemistry of molecules. While initially applicable to only relatively simple molecules, high-resolution NMR can now be used effectively to study such large and structurally complex molecules as natural products, synthetic polymers, proteins, and nucleic acids.

Continued developments in instrumentation are expected to further increase the scope and power of NMR spectroscopy. The most recent advances in NMR instrumentation include instruments employing magnetic fields from superconducting solenoids, Fourier transform NMR spectrometers, white-noise decoupling, and the use of on-line computers. The white-noise decoupling technique and Fourier transform NMR are especially significant because they make possible the routine measurement of ^{13}C spectra at the natural abundance level. Computers are finding many applications in NMR spectroscopy, ranging from instrument operation and control to data acquisition and treatment to spectral interpretations and compound identifications. These advances in instrumentation greatly increase the power of NMR in the applications discussed in this chapter and will undoubtedly make possible many new applications of NMR.

ACKNOWLEDGMENTS

It is a pleasure to acknowledge the courtesy of the American Chemical Society for permission to publish Table 7.3; the Society for Applied Spectroscopy for permission to publish Table 7.2; the American Association for the Advancement of Science for permission to publish Figure 7.5; and Dr. W. A. Anderson, Varian Associates, for permission to reproduce Figure 7.6.

REFERENCES

1. F. A. L. Anet, *J. Amer. Chem. Soc.* **82,** 994 (1960).
2. J. S. Babiec, Jr., J. R. Barrante, and G. D. Vickers, *Anal. Chem.* **40,** 610 (1968).
3. J. D. Baldeschwieler and E. W. Randall, *Chem. Rev.* **63,** 81 (1963).
4. R. H. Bible, Jr., *Interpretation of NMR Spectra, An Empirical Approach,* Plenum Press, New York, 1965.
5. R. H. Bible, Jr., *Appl. Spectroscopy* **24,** 326 (1970).
6. G. Binsch, in *Topics in Stereochemistry,* Vol. 3, N. L. Allinger and E. L. Eliel, Eds., Interscience, New York, 1968, p. 97.
7. F. Bloch, *Phys. Rev.* **70,** 460 (1946).
8. F. Bloch, W. W. Hansen, and M. Packard, *Phys. Rev.* **69,** 127 (1946).
9. A. A. Bothner-By, in *Advances in Magnetic Resonance,* Vol. 1, J. S. Waugh, Ed., Academic Press, New York, 1965, p. 195.
10. F. A. Bovey, *NMR Data Tables for Organic Compounds,* Interscience, New York, 1967.
11. F. A. Bovey, *Nuclear Magnetic Resonance Spectroscopy,* Academic Press, New York, 1969.
12. F. A. Bovey, E. W. Anderson, F. P. Hood, and R. L. Kornegay, *J. Chem. Phys.* **40,** 3099 (1964).
13. J. H. Bowie, D. W. Cameron, P. E. Schutz, D. H. Williams, and N. S. Bhacca, *Tetrahedron* **22,** 1771 (1966).
14. J. Briggs, G. H. Frost, F. A. Hart, G. P. Moss, and M. L. Staniforth, *Chem. Comm.* **1970,** 749.
15. W. Brugel, *NMR Spectra and Chemical Structure,* Academic Press, New York, 1967.
16. G. L. Buchanan and J. M. McCrae, *Tetrahedron* **23,** 279 (1967).

17. G. W. Buchanan, D. A. Ross, and J. B. Stothers, *J. Amer. Chem. Soc.* **88,** 4301 (1966).
18. G. W. Buchanan, J. B. Stothers, and S. Wu, *Can. J. Chem.* **47,** 3113 (1969).
19. A. D. Buckingham, T. Schaefer, and W. G. Schneider, *J. Chem. Phys.* **32,** 1227 (1960).
20. F. F. Caserio, *Anal. Chem.* **38,** 1802 (1966).
21. E. J. Corey, H. J. Burke, and W. A. Remers, *J. Amer. Chem. Soc.* **77,** 4941, (1955).
22. C. C. J. Culvenor and N. S. Ham, *Chem. Comm.* **1966,** 537.
23. P. V. Demarco, T. K. Elzey, R. B. Lewis, and E. Wenkert, *J. Amer. Chem. Soc.* **92,** 5734 (1970).
24. P. V. Demarco, T. K. Elzey, R. B. Lewis, and E. Wenkert, *J. Amer. Chem. Soc.* **92,** 5737 (1970).
25. J. W. Emsley, J. Feeney, and L. H. Sutcliffe, *High Resolution Nuclear Magnetic Resonance Spectroscopy*, Vol. 2, Macmillan (Pergamon), New York, 1965.
26. R. R. Ernst and W. A. Anderson, *Rev. Sci. Instr.* **37,** 93 (1966).
27. H. M. Fales and A. V. Robertson, *Tetrahedron Lett.* **1962,** 111.
28. T. C. Farrar, *Anal. Chem.* **42,** 109A April (1970).
29. R. C. Ferguson and W. D. Phillips, *Science* **157,** 257 (1967).
30. R. Freeman and R. C. Jones, *J. Chem. Phys.* **52,** 465 (1970).
31. E. W. Garbisch, Jr., *J. Chem. Educ.* **45,** 311, 402, 480 (1968).
32. K. Grohmann, J. B. Grutzner, and J. D. Roberts, *Tetrahedron Lett.* **12,** 917 (1969).
33. A. Grouiller, *Bull. Soc. Chim. France* **1966,** 2405.
34. E. Grunwald, A. Loewenstein, and S. Meiboom, *J. Chem. Phys.* **25,** 1228 (1956).
35. H. S. Gutowsky, D. W. McCall, and C. P. Slichter, *J. Chem. Phys.* **21,** 279 (1953).
36. G. E. Hall, in *Annual Review of NMR Spectroscopy*, Vol. 1, E. F. Mooney, Ed., Academic Press, New York, 1968, p. 227.
37. J. E. H. Hancock and D. R. Scheuchenpflug, *J. Amer. Chem. Soc.* **80,** 3621 (1958).
38. K. Hatada, Y. Terawaki, H. Okuda, K. Nagata, and H. Yuki, *Anal. Chem.* **41,** 1518 (1969).
39. C. C. Hinckley, *J. Amer. Chem. Soc.* **91,** 5160 (1969).
40. D. P. Hollis, *Anal. Chem.* **35,** 1682 (1963).
41. W. Horsley, H. Sternlicht, and J. S. Cohen, *J. Amer. Chem. Soc.* **92,** 680 (1970).
42. L. F. Johnson and M. E. Tate, *Can. J. Chem.* **47,** 63 (1969).
43. A. J. Jones, D. M. Grant, and K. F. Kuhlmann, *J. Amer. Chem. Soc.* **91,** 5013 (1969).
44. R. A. Y. Jones and A. R. Katritzky, *Angew. Chem. Intern. Ed. Engl.* **1,** 32 (1962).
45. J. L. Jungnickel, *Anal. Chem.* **35,** 1985 (1963).
46. J. L. Jungnickel and J. W. Forbes, *Anal. Chem.* **35,** 938 (1963).
47. R. J. Kula, D. L. Rabenstein, and G. H. Reed, *Anal. Chem.* **37,** 1783 (1965).

48. R. U. Lemieux, R. K. Kullnig, and R. Y. Moir, *J. Amer. Chem. Soc.* **80,** 2237 (1958).

49. L. P. Lindeman and S. W. Nicksic, *Anal. Chem.* **36,** 2414 (1964).

50. R. E. Lundin, R. H. Elsken, R. A. Flath, and R. Teranishi, *Appl. Spectroscopy Rev.* **1,** 131 (1967).

51. A. Luz, D. Gill, and S. Meiboom, *J. Chem. Phys.* **30,** 1540 (1959).

52. S. L. Manatt, *J. Amer. Chem. Soc.* **88,** 1323 (1966).

53. A. Mathias and D. Taylor, *Anal. Chim. Acta* **35,** 376 (1966).

54. H. M. McConnell, *J. Chem. Phys.* **28,** 430 (1958).

55. E. Mohacsi, *Analyst* **91,** 57 (1966).

56. E. F. Mooney and P. H. Winson, in *Annual Review of NMR Spectroscopy,* Vol. 1, E. F. Money, Ed., Academic Press, New York, 1968, p. 243.

57. E. F. Mooney and P. H. Winson, in *Annual Review of NMR Spectroscopy,* Vol. 2, E. F. Mooney, Ed., Academic Press, New York, 1969, p. 153.

58. G. Moreau, *Bull. Soc. Chim. France* **1969,** 1770.

59. P. J. Paulsen and W. D. Cooke, *Anal. Chem.* **36,** 1713 (1964).

60. P. J. Paulsen and W. D. Cooke, *Anal. Chem.* **36,** 1721 (1964).

61. W. D. Phillips, *J. Chem. Phys.* **23,** 1363 (1965).

62. E. M. Purcell, H. C. Torrey, and R. V. Pound, *Phys. Rev.* **69,** 37 (1946).

63. D. L. Rabenstein and R. J. Kula, *J. Amer. Chem. Soc.* **91,** 2492 (1969).

64. L. W. Reeves, *Can. J. Chem.* **35,** 1351 (1957).

65. L. W. Reeves and W. G. Schneider, *Trans. Faraday Soc.* **54,** 314 (1958).

66. H. J. Reich, M. Jautelat, M. T. Messe, F. J. Weigert, and J. D. Roberts, *J. Amer. Chem. Soc.* **91,** 7445 (1969).

67. C. A. Reilly, *Anal. Chem.* **30,** 839 (1958).

68. J. D. Roberts, *An Introduction to the Analysis of Spin–Spin Splitting in High-Resolution Nuclear Magnetic Resonance Spectra,* W. A. Benjamin, New York, 1961.

69. J. K. M. Sanders and D. H. Williams, *Chem. Comm.* **1970,** 422.

70. P. R. Sewell, in *Annual Review of NMR Spectroscopy,* Vol. 1, E. F. Mooney, Ed., Academic Press, New York, 1968, p. 165.

71. R. M. Silverstein and G. C. Bassler, *Spectrometric Identification of Organic Compounds,* 2nd ed., Wiley, New York, 1967.

72. G. Slomp and J. G. Lindberg, *Anal. Chem.* **39,** 60 (1967).

73. F. C. Stehling and K. W. Bartz, *Anal. Chem.* **38,** 1467 (1966).

74. J. B. Stothers, *Quart. Rev.* **19,** 144 (1965).

75. W. A. Thomas, in *Annual Review of NMR Spectroscopy,* Vol. 1, E. F. Mooney, Ed., Academic Press, New York, 1968, p. 43.

76. *Varian High-Resolution NMR Spectra Catalog,* Vols. 1 and 2, Varian Associates, Palo Alto, Calif., 1963.

77. F. J. Weigert, M. Jautelat, and J. D. Roberts, *Proc. Natl. Acad. Sci. U.S.* **60,** 1152 (1968).

78. E. Wenkert, A. O. Clouse, D. W. Cochran, and D. Doddrell, *J. Amer. Chem. Soc.* **91,** 6879 (1969).

79. K. B. Wiberg and B. J. Nist, *The Interpretation of NMR Spectra*, W. A. Benjamin, New York, 1962.

80. G. N. LaMar, *J. Amer. Chem. Soc.* **93,** 1040 (1971).

RECOMMENDED READING

E. D. Becker, *High Resolution NMR*, Academic Press, New York, 1969.

F. A. Bovey, *Nuclear Magnetic Resonance Spectroscopy*, Academic Press, New York, 1969.

J. A. Pople, W. G. Schneider, and H. J. Bernstein, *High-Resolution Nuclear Magnetic Resonance*, McGraw-Hill, New York, 1959.

J. W. Emsley, J. Feeney, and L. H. Sutcliffe, *High-Resolution Nuclear Magnetic Resonance Spectroscopy*, Vols. 1 and 2, Pergamon Press, New York, 1965.

CHARLES P. POOLE, JR.
Department of Physics
University of South Carolina, Columbia, S. C.

VIII. Electron Spin Resonance

I. INTRODUCTION

The field of electron spin resonance (ESR) was founded 25 years ago and since then has grown to be one of the standard instrumental techniques of the physical and biological sciences. An extensive literature has accumulated on the various aspects of this type of instrumentation. In comparison to nuclear magnetic resonance (NMR), its sister technique, the scope of application of ESR to actual problems in analytical chemistry may not be as broad and diverse but in those cases where unpaired electrons are involved, such as in organic free radicals, odd electron molecules, biradicals, triplet states, transition metal ions, color centers, and paramagnetic molecules such as O_2 and NO, ESR may often offer the unique solution to the problem. Modern ESR techniques can detect the presence of unpaired electrons down to very low levels while characterizing them precisely with regard to their location and energy states.

The phenomenon of ESR is based on the fact that an electron is a charged particle which constantly spins around its axis with a certain angular momentum. Associated with the intrinsic spin is a magnetic moment, the value of which is called the Bohr magneton.

If an external magnetic field is impressed on the system, the electron will align itself with the direction of this field and precess around this axis. This behavior is analogous to that of a spinning top in the earth's gravitational field. Increasing the applied magnetic field will induce the electron to precess faster.

In practice, the magnetic field will split the electrons into two groups. In one group the magnetic moments of the electrons are aligned with the magnetic field, while in the other group the magnetic moments are aligned opposite or antiparallel to this external field.

Quantum-mechanically we can state that the spin quantum number of an electron is equal to $\frac{1}{2}$. Quantum conditions rule that resolved components of quantum numbers along an axis of quantization must differ by 1. The two possible orientations of these electrons in the applied field correspond to the projections $M_S = \pm\frac{1}{2}$ along the magnetic field direction. Each orientation is associated with a different energy, the one with the spins antiparallel to the external field ($M_S = -\frac{1}{2}$) being in the lower energy state. These two levels, where the quantum number M_S is either $+\frac{1}{2}$ or $-\frac{1}{2}$, are often referred to as the $+\frac{1}{2}$ or the $-\frac{1}{2}$ states.

If a second weaker alternating magnetic field is now applied at right angles to the main field, for example, by the use of a high-frequency microwave resonant cavity, an electron can be "tipped" over when the precession frequency is equal to the incident microwave frequency. Basically the phenomenon of ESR can also be described quantum-mechanically by noting that the quanta of the incident microwaves may induce transitions between the two states of the unpaired electron. When the energy $h\nu$ of these quanta coincides with the energy level separation $E_{1/2} - E_{-1/2}$ between the two states, resonance absorption takes place.

In practice, the applied radio-frequency is maintained at a certain value, and the magnetic field strength is varied to locate those values where resonance occurs. The incoming radiation $h\nu$ absorbed by the electrons in the lower energy level will induce these electrons to jump into the higher energy state. The incoming radiation is, however, also absorbed by the electrons in the higher energy level. The absorbed energy results in stimulated emissions by these electrons. Since the coefficients of absorption and stimulated emission are equal, no net value would be observed if the spin population would be equally distributed between these two levels. In general, however, n_1, the population of the ground state, exceeds n_2, the population of the excited state, and a net absorption of microwave radiation takes place. This signal is proportional to the population difference $n_1 - n_2$.

The population ratio of these two states can in most cases be described by the Boltzmann distribution

$$\frac{n_1}{n_2} = e^{-h\nu/kT} \tag{8.1}$$

The sensitivity of this technique is, therefore, improved by using a high applied frequency. In practice, the most common frequencies in ESR are those in the well-developed radar wavebands. In addition, low operating temperatures are also beneficial for signal enhancement.

In comparison to NMR, the electron has a much smaller mass and larger magnetic moment that a proton. For a given magnetic field the precession frequency is, therefore, much higher. In a field of say 3600 Oe, the frequency of the precession is around 10^{10} Hz. This corresponds to microwaves with a wavelength of about 3 cm.

In most substances chemical bonding results in the pairing of electrons because electron(s) are either transferred from one atom to another atom to form an ionic bond; or electrons are shared between different atoms to form covalent bonds. The spins and magnetic moments of paired electrons point in opposing directions and there is no external spin paramagnetism. But in a paramagnetic substance where an unpaired electron is present resonance occurs at definite values of the applied magnetic field and incident microwave radiation.

At first glance one would expect the resonance spectrum of an unpaired electron to be always the same. But, as with the proton resonance spectrum in NMR, the magnetic behavior of the electron is modified by the magnetic fields in its surroundings. It is in this deviation from the standard behavior that we can learn about the structure of the substances under study.

To carry out an ESR measurement, one places the sample under study in a resonant cavity located between the pole pieces of an electromagnet. The microwave frequency is adjusted to a matched condition with the

Table 8.1. Summary of the Various Branches of Spectroscopy

Branch	Frequency, Hz	Wavelength	Typical Energy Unit: Name[a]	Typical Energy Unit: Value, J
Static	0–60		Joule	1
			Calorie	4.186
Low or audio frequency	10^3–10^5	3–300 km	Kc	6.626×10^{-31}
Radio frequency	10^6–10^8	300–3 m	Joule	1
			Cm^{-1}	1.986×10^{-23}
Microwaves	10^9–10^{11}	30 cm to 3mm	Mc	6.626×10^{-28}
Infrared	10^{12} to 3×10^{14}	300–1μ	Cm^{-1}	1.986×10^{-23}
			kk	1.986×10^{-20}
			Kcal/m	4.186×10^3
			Joule	1
Visible, ultraviolet	4×10^{14} to 3×10^{15}	0.8–0.1μ	Erg	1×10^{-7}
			eV	1.602×30^{-19}
			Mc	6.626×10^{-28}
X-rays	10^{16}–10^{19}	30–0.03 mμ	eV	1.602×10^{-19}
			KeV	1.602×10^{-16}
γ-rays	10^{19}–10^{22}	3×10^{-9} to 3×10^{-12} cm	MeV	1.602×10^{-13}
Low energy, nuclear	10^{19}–10^{23}	3×10^{-9} to 3×10^{-13} cm	MeV	1.602×10^{-13}
High energy, nuclear	10^{23}–10^{26}	3×10^{-13} to 3×10^{-17} cm	BeV	1.602×10^{-10}
			GeV	1.602×10^{-7}
High-energy cosmic rays	$>10^{25}$		BeV	1.602×10^{-9}
			GeV	1.602×10^{-6}

[a] A kilokayser (kk) is 1000 cm^{-1}.
[b] NQR denotes nuclear quadrupole resonance, NMR signifies nuclear magnetic resonance, and ESR means electron spin resonance.
[c] TWO means travelling wave oscillator.

Phenomenon[b]	Typical Radiation Generator	Typical Detector
	Battery	Ammeter, voltmeter
Dielectric absorption	Mechanical	Ammeter, voltmeter
NQR, NMR, Dielectric absorption	Tuner circuit, crystal	Antenna
Molecular rotations, ESR	Klystron, Magneton, TWO[c]	Antenna, crystal, bolometer
Molecular vibrations	Heat source	Bolometer, PbS cell
Electronic transitions	Incandescent lamp	Photocell, photographic film
Electronic transitions	Discharge tube	Photocell
Inner shell electronic transitions	Heavy element bombardment	Geiger counter, photomultiplier
Nuclear energy level transitions	Naturally radioactive nuclei	Scintillation detector
Strange particle creation	Accelerator (e.g., synchrotron)	Bubble chamber, spark chamber
Extraterrestrial	Star, magnetic field in galaxy	Extensive shower detector

aid of a tuning device. The magnetic field is then varied through the resonant conditions, and the amount of microwave energy absorbed by the sample is plotted on a recorder. The resulting spectrum is analyzed to determine the mechanisms associated with the unpaired electron's interaction with the external magnetic field and its local environment.

The subject of electron spin resonance has several more or less equivalent names. The three most prevalent ones with their associated abbreviations are the following.

1. Electron Spin Resonance (ESR)
2. Electron Paramagnetic Resonance (EPR)
3. Paramagnetic Resonance

The first two are currently the most widely used, while the third was more prevalent in the earlier literature. Our choice of the first name is more or less a matter of personal taste—*de gustibus non est disputandum*. The sister field of NMR does not suffer from this affliction of polyappellation.

1. Spectroscopy

The following remarks place the position of ESR among the other spectroscopic techniques. The general field of spectroscopy is subdivided into several regions depending on the energy involved in a typical quantum jump. A summary of the various branches of spectroscopy is presented in Table 8.1. Historically, they developed as separate fields of research; each employed particular experimental techniques, and these instrumentation differences just happened to coincide with different physical phenomena such as the progressively increasing energies associated with rotational, vibrational, and electronic spectra.

ESR is frequently considered to be in the microwave branch of spectroscopy, and NMR is usually classified in radio-frequency spectroscopy, but these are merely instrumental characterizations based, for example, on the last two columns of Table 8.1. In terms of the observed phenomena, ESR studies the interaction between electronic magnetic moments and magnetic fields. Occasionally electron spin resonance studies are carried out with NMR instrumentation using magnetic fields of several gauss rather than several thousand gauss. The splitting of energy levels by a magnetic field is customarily referred to as the Zeeman effect, and so we may say that ESR is the study of direct transitions between electronic Zeeman levels while NMR is the study of direct transitions between nuclear Zeeman levels. In concrete terms, it may be said that ESR and NMR study the energy required to reorient electronic and nuclear magnetic moments, respectively, in a magnetic field.

Straight microwave spectroscopy uses apparatus similar to that employed in ESR, but in contrast to ESR, it measures molecular rotational transitions directly; and when it employs a magnetic field, it is usually for the purpose of producing only a small additional splitting of the rotational energy levels. In fact, in this branch of spectroscopy, it is much more customary to produce Stark effect splittings by means of an applied electric field. In ESR, on the other hand, a strong magnetic field is an integral part of the experimental arrangement.

There is also a certain analogy between NMR, ESR, infrared (IR) and ultraviolet spectroscopy (UV). The basic differences are given in Table 8.2.

Table 8.2. Comparison of NMR, ESR, IR, and UV Spectroscopy Showing Typical Values of Several Parameters

Parameter	NMR	ESR	IR	UV
ΔE between upper and lower levels, cal/mole	10^{-3}	1	300	10^{5}
Wavelength of observed and emitted quanta, cm	3×10^{3}	3	10^{-2}	3×10^{-5}
Population in excited state, %	49.9999	49.9	20	Negligible
Lifetime of excited state, sec	10	10^{-3}		10^{-8}
Relative shift of energy levels by magnetic field	Large	Large	Small	Negligible

2. The Larmor Condition

Before proceeding further it is well to say a few words about the phenomenon of magnetic resonance. This consists of the behavior of electronic (or nuclear) moments in a constant magnetic field. In a typical experiment the sample is placed in a high Q resonant cavity (see Section II.5) at the microwave frequency ω_0, and the magnetic field is varied until resonance occurs at the value H_0 given by

$$\omega_0 = \gamma H_0 = \frac{g\beta}{\hbar} H_0 \tag{8.2}$$

In this equation γ is the gyromagnetic ratio, a distinct and fixed ratio of the electron's magnetic moment $(L \pm 2.0023S) = gJ$ to its angular momentum $(L \pm S) = J$, and g is the "g-factor" which is a dimensionless constant and a physical property of the electron. The total angular momentum, J, is the vector sum of the orbital L and spin S parts. Planck's constant h divided by 2π is denoted by $\hbar$, and β is the unit magnetic moment of the spinning electron called the Bohr magneton.

$$\beta = \frac{e\hbar}{2mc} = 0.92731 \times 10^{-20} \text{ erg/gauss} \tag{8.3}$$

where e is the charge of the electron, m is the mass of the electron, and c is the velocity of light. Equation 8.2 constitutes the Larmor condition. The first part of this equation is ordinarily used in NMR where gyromagnetic ratios are tabulated. In ESR it is customary to measure the g-factor. In liquids and solids the orbital motion, L, is usually "quenched" and only the spin motion is observed, so most values of g are fairly close to g = 2.0023 which characterizes a free electron. The conversion $\gamma = g\beta/\hbar$ is easily made.

3. ESR Spectra

Most ESR spectra are more complex than is implied by Equation 8.2. The resonant absorption is not infinitely narrow, which would be the case if absorption occurs at a precise value of the applied magnetic field, but rather it has a finite width $\Delta\omega$ which varies from $\sim$.01 gauss for radical ions to $\sim$3 gauss for exchange-narrowed free radicals to over 1000 gauss for some transition metal compounds.

The reason for the finite line width is that electrons not only interact with the externally applied magnetic field but also interact in a more or less random manner with the magnetic fields in their environment. By observing this line width and line intensity, one can obtain information about the spin environment. Electron spin exchange between identical and nonidentical molecules, chemical exchange between the paramagnetic molecule and its environment, and the interactions of neighboring molecules having spin are some examples of environmental effects which can influence the line width and intensity in the ESR spectrum.

An observed spectrum frequently contains several lines because it exhibits hyperfine structure due to interactions with nuclear spins. The electronic spin of a transition metal ion usually interacts with its own nuclear spin, but in some systems, such as in aromatic molecules, the unpaired electron circulates among several atoms; and the resultant hyperfine structure is the result of the interaction of this electronic spin with several nuclear spins. The possible energy levels of an unpaired electron interacting with two nuclear spins are shown in Figure 8.1.

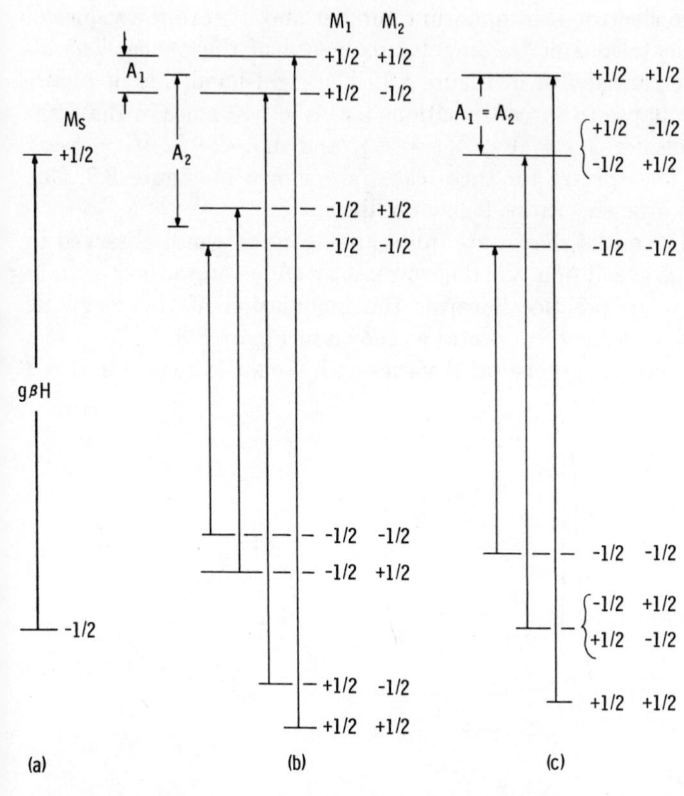

Figure 8.1. Energy level diagram for an unpaired electron in the absence of hyperfine structure (*a*), with two unequal hyperfine coupling constants (*b*), and with two equal coupling constants (*c*). The corresponding ESR spectra are shown on Figures 8.2. and 8.3.

The simplest case is one in which there is no hyperfine interaction. By placing the unpaired electron in a magnetic field, the number of energy levels is increased from one ($E = E_0$) to two ($E = E_0 \pm \frac{1}{2}\, g\beta H$). This is shown in the left portion of Figure 8.1. The center of the figure shows the case where two nuclear spins interact with the electron. A measure for the amount of interaction is given by the hyperfine coupling constant, A_i.

Each nuclear spin splits each original level into two levels. Hence, for $A_1 \neq A_2$, the original two levels are split into 8 levels as shown. Although there are 8 levels, only 4 transitions are possible because of the following selection rules.

$$\Delta M_S = \pm 1 \tag{8.4}$$

$$\Delta M_I = 0 \tag{8.5}$$

where M_S is the electron spin quantum number and M_I are the values of the possible orientations of the angular momenta of the nuclei. The allowed transitions are shown in Figure 8.1. The right-hand side of Figure 8.1 shows the collapse to three transitions for $A_1 = A_2$, since in that case the energy levels for $M_1 = +\frac{1}{2}$, $M_2 = -\frac{1}{2}$, and $M_1 = -\frac{1}{2}$, $M_2 = \frac{1}{2}$ are superimposed. The spectra for these cases are shown in Figure 8.2. One should note the intensity ratios below the lines.

The absorption peaks shown are quite analogous to those observed in UV, IR, and high-resolution NMR spectroscopy and even to those seen in chromatography. In practice however the modulation of the magnetic field produces first-derivative spectra as shown in Figure 8.3.

The hyperfine coupling constant A varies with the nuclear species. It is a

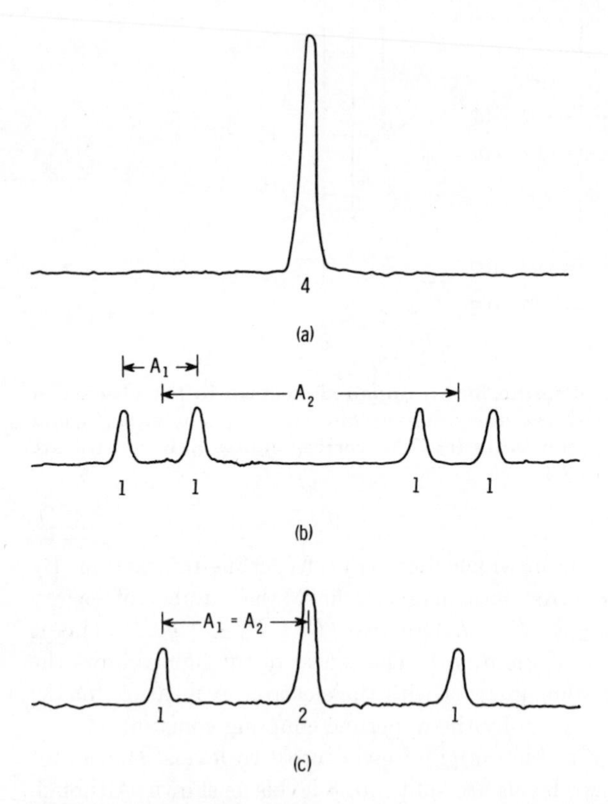

Figure 8.2. ESR absorption spectra for (*a*) a singlet, (*b*) a quartet due to hyperfine structure with two protons where $A_2 > A_1$ and (*c*) the same spectrum when $A_1 = A_2$. The intensity ratios are given below the lines, the power absorbed is plotted as the ordinate, and the magnetic field is represented on the abscissa.

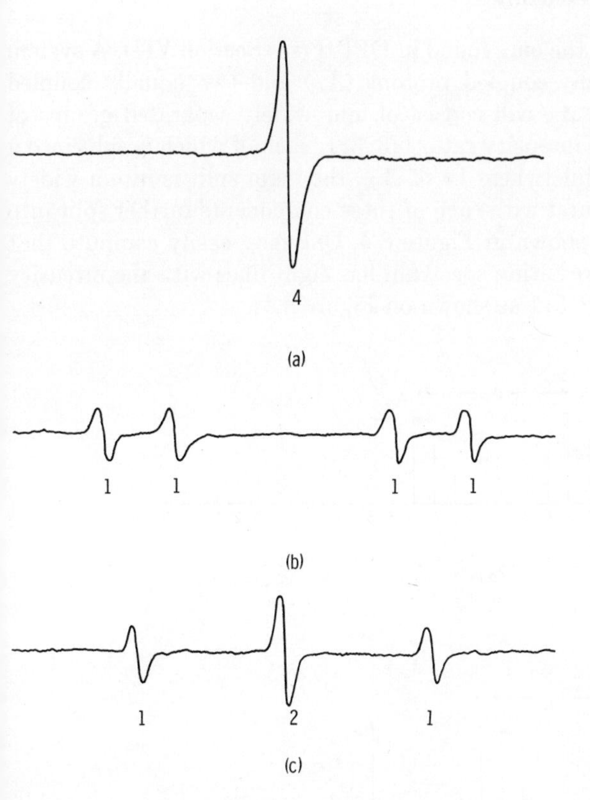

Figure 8.3. First derivative spectra arising from the energy levels of Figure 8.1. (*a*) $A_1 = A_2 = 0$; (*b*) $A_2 \gg A_1$; (*c*) $A_2 = A_1$.

measure of the strength of the interaction between the nuclear and electronic spins. When studying transition group metals where the electron interacts almost exclusively with a single atom, hyperfine splitting of the order of 100 gauss or more can be observed. In organic compounds, where the electron may interact with several nuclei, the hyperfine splitting between the individual lines may be as low as a few milligauss with a total splitting between the terminal lines of about 25 gauss. To resolve these small differences, large demands are placed on the applied magnetic field since it must have a homogeneity which exceeds this line width.

When several $I = \frac{1}{2}$ nuclei are equally coupled (i.e., have the same A_i), the resultant absorption peaks show an intensity ratio that follows the binomial coefficient distribution. As shown in Tables 8.3 and 8.4 the intensity ratio of 1:3:3:1 is obtained with the methyl radical CH_3($I = \frac{1}{2}$) while the ratio of 1:2:3:2:1 is obtained with two equally coupled nitrogen nuclei ($I = 1$) such as the ones found in DPPH (see Section VII). A system containing three equally coupled protons (A_P) and two equally coupled nitrogens with $A_P \gg A_N$ will consist of four widely separated groups of lines with the relative intensity ratio 1:3:3:1, each of which is split into a 1:2:3:2:1 quintet, while when $A_P \ll A_N$, the main split is into a widely spaced 1:2:3:2:1 quintet with each of these components further split into a 1:3:3:1 quartet as shown in Figure 8.4. One may easily compute that when $A_P = A_N$, the resulting spectrum has eight lines with the intensity ratio 1:5:12:18:18:12:5:1 as shown on Figure 8.4*c*.

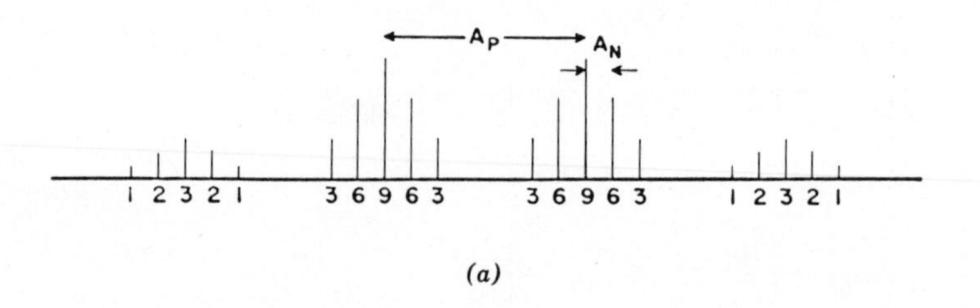

(*a*)

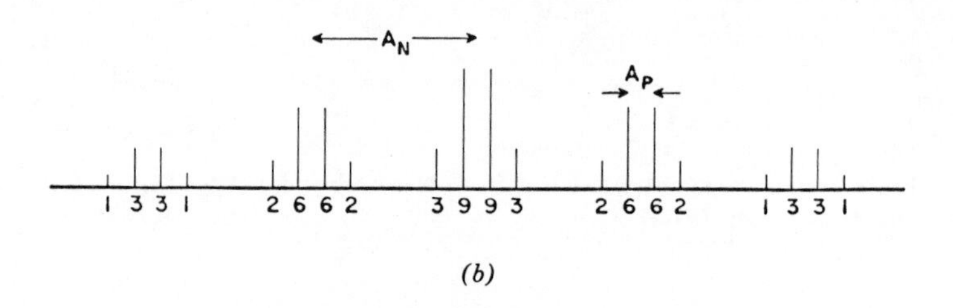

(*b*)

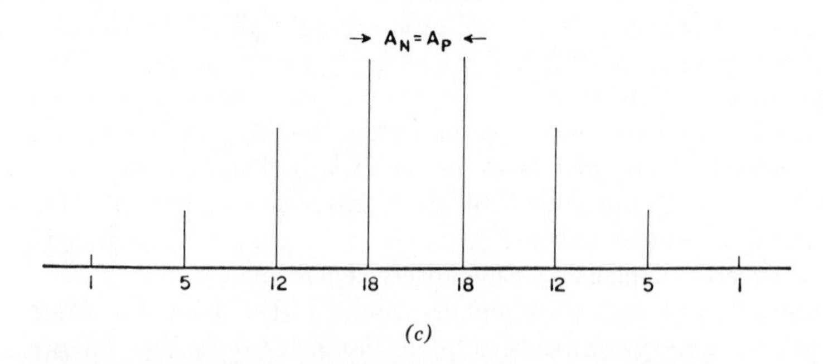

(*c*)

Figure 8.4. Hyperfine structure patterns for three equally coupled $I = \frac{1}{2}$ nuclei with coupling constant A_P and two equally coupled $I = 1$ nuclei with coupling constant A_N.

Table 8.3. Determination of Hyperfine Structure Intensity Ratios for Three Equally Coupled $I = \frac{1}{2}$ Nuclei (e.g., Protons)

Spin Configuration	m_1	m_2	m_3	$M = m_1 + m_2 + m_3$	Intensity Ratio
↑ ↑ ↑	$\frac{1}{2}$	$\frac{1}{2}$	$\frac{1}{2}$	$\frac{3}{2}$	1
↑ ↑ ↓	$\frac{1}{2}$	$\frac{1}{2}$	$-\frac{1}{2}$		
↑ ↓ ↑	$\frac{1}{2}$	$-\frac{1}{2}$	$\frac{1}{2}$	$\frac{1}{2}$	3
↓ ↑ ↑	$-\frac{1}{2}$	$\frac{1}{2}$	$\frac{1}{2}$		
↑ ↓ ↓	$\frac{1}{2}$	$-\frac{1}{2}$	$-\frac{1}{2}$		
↓ ↑ ↓	$-\frac{1}{2}$	$\frac{1}{2}$	$-\frac{1}{2}$	$-\frac{1}{2}$	3
↓ ↓ ↑	$-\frac{1}{2}$	$-\frac{1}{2}$	$\frac{1}{2}$		
↓ ↓ ↓	$-\frac{1}{2}$	$-\frac{1}{2}$	$-\frac{1}{2}$	$-\frac{3}{2}$	1

Table 8.4. Determination of Hyperfine Structure Intensity Ratios for Two Equally Coupled $I = 1$ Nuclei (e.g., Nitrogen) Such as the Ones Found in DPPH

Spin Configuration	m_1	m_2	$M = m_1 + m_2$	Intensity Ratio
↑ ↑	1	1	2	1
↑ →	1	0	1	2
→ ↑	0	1		
↑ ↓	1	−1		
→ →	0	0	0	3
↓ ↑	−1	1		
→ ↓	0	−1	−1	2
↓ →	−1	0		
↓ ↓	−1	−1	−2	1

If there are n nuclei with $I = \frac{1}{2}$ contributing to the hyperfine structure then there will be 2^n different hyperfine components if all of the coupling constants differ and no degeneracy occurs. If there are n nuclei with the nuclear spin I, then there will be $(2I + 1)^n$ hyperfine components. For several nuclei with the individual values, I_i and n_i, the total number of hyperfine components, N_{hfs}, will be

$$N_{\text{hfs}} = \Pi_i(2I_i + 1)^{n_i} \tag{8.6}$$

where the symbol Π_i denotes the formation of a product. For example, the system depicted in Figure 8.4 has

$$I_p = \tfrac{1}{2} \qquad n_p = 3 \tag{8.7}$$

$$I_N = 1 \qquad n_N = 2 \tag{8.8}$$

with the result that

$$N_{\text{hfs}} = [2(\tfrac{1}{2}) + 1]^3(2 + 1)^2 = 72 \tag{8.9}$$

This may be checked by adding the intensities shown on Figure 8.4*c*:

$$1 + 5 + 12 + 18 + 18 + 12 + 5 + 1 = 72 \tag{8.10}$$

When some nuclei are equivalent to others, the resulting degeneracy has the effect of decreasing the number and increasing the amplitude of the components in the hyperfine pattern without affecting the overall integrated intensity. Usually all of the hyperfine components have the same line width and shape, but sometimes relaxation mechanisms cause deviations from this rule (38, 61).

Many ESR spectra are more complicated than those shown in Figures 8.2 and 8.3 due to the presence of zero field splittings, anisotropic hyperfine coupling constants, additional hyperfine splittings, saturation effects, and so on. Some of these situations will be treated in later sections.

II. ESR SPECTROMETER OPERATION

For ESR measurements the spectrometer is generally operated at a constant frequency while the magnetic field is varied through the region of interest. The magnetic field sweep should be linear and calibrated to obtain a satisfactory spectrum. A sensitive detection system with its preamplifier and lock-in detector is used to remove noise and to obtain high-quality spectra.

Figure 8.5 presents a block diagram of a typical electron spin resonance spectrometer. It is composed of:

1. A large electromagnet and its associated power supply.
2. A microwave bridge which houses a klystron that generates microwaves. Attached to this bridge is a waveguide arm at the end of which is found a waveguide cavity suspended between the magnet polepieces.
3. Electronic circuitry which detects and amplifies the microwave signal, removes noise, and displays the results on a chart recorder.

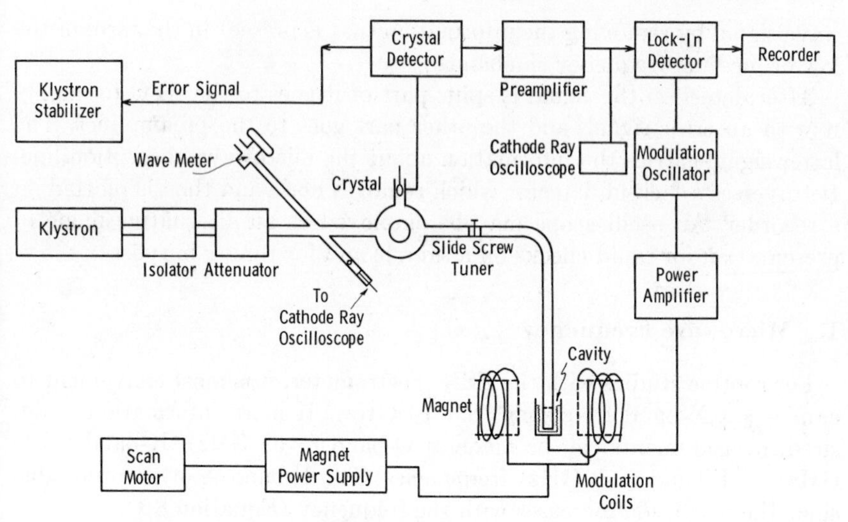

Figure 8.5. Block diagram of ESR spectrometer.

The magnet contains a field-dial or Hall-effect device which provides a sweep that is linear in magnetic field strength. This device removes the nonlinearities associated with magnet saturation. The sweep permits the magnetic field to be varied over a desired range (e.g., 5000 gauss, 1000 gauss, 100 gauss, 5 gauss) with a preselected center field such as 3400 gauss. Hence both broad and narrow lines can be conveniently recorded.

The magnetic field may be modulated at a low frequency (e.g., 20 or 400 Hz) or at a medium frequency (e.g., 100 kHz). The use of field modulation permits detection and amplification at the modulation frequency, which provides greater sensitivity than a dc detection scheme. The modulation coils are ordinarily located on the sides of the resonant cavity, although some workers mount them directly on the polepieces. The latter method will work only at low frequencies (100 Hz, 1 kHz) since in practice one cannot provide enough power at 100 kHz for such a modulation scheme.

The microwave power is generated in a klystron which is suspended in an oil bath for greater stability. The usual automatic frequency control (AFC) is provided by a 10 kHz sinewave voltage impressed on the reflector of the klystron. The isolator impedance matches the klystron to the waveguide system. The frequencymeter (i.e., wavemeter) measures the microwave frequency. A directional coupler side-arm is provided for monitoring the klystron output. The circulator is a device for impressing the incident microwave power on the resonant cavity and for directing the reflected signal to the crystal detector. The detector demodulates the incident microwave signal by removing the information that is present in the form of the low or medium frequency sidebands.

After detection the signal is split: part of it goes to the AFC to supply it with an error signal, and the other part goes to the preamplifier. The latter signal carries the information about the microwave absorption line. It traverses a lock-in detector which removes noise and then is plotted on a recorder. An oscilloscope may be employed as an alternative mode of presentation for rapid checks on a sample and for tuning purposes.

1. Microwave Frequency

For routine studies with the ESR spectrometer, it is most convenient to employ an X-band frequency (8.5–10 GHz). If more information is desired, measurements can be made at Q-band (~35 GHz), K-band (~24 GHz), or S-band (~3 GHz) frequencies. The R-band is often preferable since the resolution increases with the frequency (Equation 8.1).

The principal object of varying the frequency is to sort out the frequency-dependent and the frequency-independent terms in the Hamiltonian. Hyperfine structure intervals will be independent of frequency, and anisotropic g-factors will produce separations which vary linearly with the frequency.

Another reason for varying the microwave frequency is to ascertain the principal line-broadening mechanisms. If the line-width is due to dipole–dipole broadening, exchange narrowing, or unresolved hyperfine structure, this parameter will not change with frequency, while if it is due to unresolved g-factor anisotropy it will increase with increasing frequency. Therefore it is much easier to interpret an ESR spectrum which has been measured at two or more microwave frequencies instead of merely at one.

2. Modulation Frequency and Amplitude

It is important to employ a modulation frequency, f_m, which is much less than the line-width ΔH_{pp} expressed in frequency units Δf_{pp}

$$f_m \ll \Delta f_{pp} = \frac{\gamma}{2\pi} \Delta H_{pp} = \frac{g\beta}{h} \Delta H_{pp} \tag{8.11}$$

A modulation frequency of 100 kHz may be employed for line-widths as narrow as $\frac{1}{10}$ gauss, but this modulation frequency will appreciably distort lines less than this width.

As the modulation amplitude H_m is gradually increased, the observed line-width will be unchanged as long as H_m is much less than the line-width

ΔH_{pp}. When H_m becomes close to ΔH_{pp}, then the observed line will begin to broaden and distort. When $H_m \ll \Delta H_{pp}$, the amplitude of the ESR signal will increase linearly with the modulation amplitude, while after H_m exceeds ΔH_{pp}, the amplitude will begin to decrease with increasing H_m. The ESR signal will reach a maximum near the point $H_m \sim 3\ \Delta H_{pp}$. In order to determine ESR spin concentrations and line-shapes, it is best to have $H_m \ll \Delta H_{pp}$, and a practical criterion is to keep $H_m \leq \Delta H_{pp}/5$. When one attempts to detect very weak signals, then one may set $H_m \sim 3\ \Delta H_{pp}$ for maximum sensitivity.

A practical technique to use in determining the desired value of H_m is to record the ESR signal with successively doubling modulation amplitudes. One may obtain a reasonably undistorted spectrum by using $\frac{1}{4}$ of the lowest value of H_m that does not produce a doubling of the ESR signal amplitude. Maximum sensitivity, of course, corresponds to the H_m, which furnishes the greatest ESR signal amplitude: this point may also be determined by successively doubling the modulation amplitude.

3. Magnet Scan

An ESR spectrum can either be spread out over several thousand gauss or confined to a small fraction of a gauss, and in searching for unknown resonances one must "guess" at the linewidth, and set the scan accordingly. Sometimes it will be advantageous to make several scans covering different ranges of gauss, and perhaps centered at different field values. If a narrow scan is employed to record a very broad line, then the resonance will manifest itself as a sloping baseline, and if a broad scan is used to record a very narrow resonance, the absorption will be unusually weak and distorted. The use of too broad a scan with a weak, sharp resonance may even render it undetectable.

Once the overall features of a complex resonance are known, it may be necessary to use especially selected narrow scans to resolve particular features of the spectrum. Even when a narrow resonance is properly recorded, it is often worthwhile to make a broad scan (e.g., over a range of 1000 gauss) to ensure that one has not overlooked an additional broad background resonance. The scanning rate dH/dt is the rate at which one varies the magnetic field. It is conveniently expressed as the number of gauss per minute (gauss/min). The response time or time constant τ is a measure of the inability of the narrow band detector to pass without distortion ESR signals which are scanned through in a time shorter than τ. For an undistorted signal, it is necessary that τ be less than one-tenth the time that it takes to scan between the two first-derivative peaks of the resonant line.

If the time constant τ is too long, then the observed line-shape will be distorted.

The amplitude of the noise on the recorder decreases as the response time increases, so from the viewpoint of noise it is always desirable to increase τ. As a rule of thumb, one may say that the strongest undistorted line is obtained when the peak-to-peak scanning time is ten times τ, while the best signal-to-noise ratio occurs when the peak-to-peak scanning time equals τ. The former should be employed routinely for most applications, while the latter may be resorted to for signals that are close to the limit of detectability where one does not mind obtaining a distorted line-shape. The use of a computer for average transients for signal-to-noise enhancement may be employed if sensitivity is a problem.

An observed spectrum may be tested for the proper response time by recording it with successively increasing settings until one reaches the value of τ where the amplitude decreases. The next lower time constant may be used for recording undistorted line-shapes with high sensitivity. When one studies a complex spectrum containing both broad and narrow absorption lines, it may be desirable to use several scans with different time constants in order to resolve properly all of the component resonances.

4. Temperature

Since many chemical and physical phenomena of interest occur only in a definite temperature range, temperature control of the sample is advisable in any spectrometer system. In addition, the population distribution of the different spin states is, as already indicated in Equation 8.1, a function of the temperature. Low-temperature operation may be necessary in many systems to increase the sensitivity to detectable levels. In general, the g-factor, hyperfine interaction constants, and other terms in the Hamiltonian are independent of the temperature. The line-shape, line width, and relaxation times T_1 and T_2 are the principal quantities which are sensitive to the temperature.

The standard method for obtaining variable sample temperatures in the 100–500°K range is to pass a controlled stream of cooled or heated nitrogen gas past the sample (64). A thermocouple placed in the nitrogen stream before it impinges on the sample container is used to monitor the temperature. For subambient operation, an overpressure of dry nitrogen is maintained within the resonant cavity to prevent the condensation of moisture. The output of the thermocouple can, of course, also be connected to an automatic temperature controller which regulates the temperature of the gas blowing past the sample. Another method for obtaining high temperatures consists in placing the resonant cavity within an oven flushed with nitrogen gas (57).

Cryogenic temperatures in the liquid-helium range are generally obtained by immersing the resonant cavity into a liquid-helium bath. This requires the use of a double Dewar with liquid nitrogen in the outer chamber and liquid helium in the inner chamber. If stainless steel Dewars are used, a radiation shield maintained in contact with the liquid-nitrogen bath can be substituted for the outer chamber.

Variable-temperature studies between 4–100°K may be carried out by inserting a small heater between the cavity and the liquid helium. A more recent method employs the Joule-Thompson effect to cool helium gas in the vicinity of the resonant cavity and sample.

Much interesting relaxation-time work has been carried out in this low-temperature region. Low-temperature operation is also important because the sensitivity of the ESR signal increases with decreasing temperature. Some systems such as Co^{2+}, for example, are often difficult to observe at higher temperatures.

Most work at cryogenic temperatures requires the use of a low-power bridge since the relaxation times are so long that saturation occurs in the milliwatt range. Microwatt power levels are produced by stabilizing the klystron off a reference cavity instead of the sample cavity. Superheterodyne detection is often employed in this temperature range.

Most of the terms in the Hamiltonian are independent of temperature, while the line-width and relaxation times are often strongly dependent on the temperatures. For a rigid lattice, the dipole–dipole broadening, exchange interaction, and g-factor anisotropy mechanisms do not change appreciably with the temperature. If any of the electron spins or nuclei which are responsible for the line-broadening mechanism undergo changes in their translational, rotational, or vibrational motion with temperature, then such changes will produce strongly temperature-dependent line-widths and relaxation times. In general, rapid motion can average out the other line-broadening mechanisms, and narrow an ESR resonance absorption line. As a result, solids ordinarily have much broader lines than fluids, and highly viscous liquids and glasses are intermediate between these two. Sometimes variable temperature studies can sort out the range where certain groups such as methyl groups stop rotating. Variable temperature studies supply detailed information about crystallographic phase transitions, and about such magnetic phenomena as the onset of paramagnetism, ferromagnetism, and antiferromagnetism.

Low-temperature relaxation-time measurements (54) often reveal a strong dependence of the spin lattice relaxation time T_1 on the temperature (e.g., $T_1 \propto 1/T$ or $T_1 \propto T^{-7}$). Such information allows one to deduce the principal relaxation process (e.g., the direct process or the Raman scattering process). Some spin systems are only detectable at very low temperatures.

The intensity of ordinary ESR spectra depends inversely on the temperature when $g\beta H \ll kT$, and otherwise exponentially on $1/T$ in accordance with the Boltzmann distribution (Equation 8.1).

5. Sample Cavity

There are two main types of cavities in current use, the cylindrical TE_{011} and the rectangular TE_{012} cavities. Both are transverse electric (TE) modes, which means that the electric field lines are confined to the plane perpendicular to the longitudinal axis. No such restriction exists for the magnetic field lines. For a general cylindrical TE_{mnp} mode, the subscripts m, n, and p give the number of half-cycle variations along the angular, radial, and longitudinal directions. Since in the present case $m = 0$, it follows that the cylindrical TE_{011} mode has axial symmetry. The TE_{012} and the TE_{112} modes shown in Figure 8.6 illustrate the orientation of the electric and magnetic fields. The TE_{011} cavity is half as long as the TE_{012} cavity.

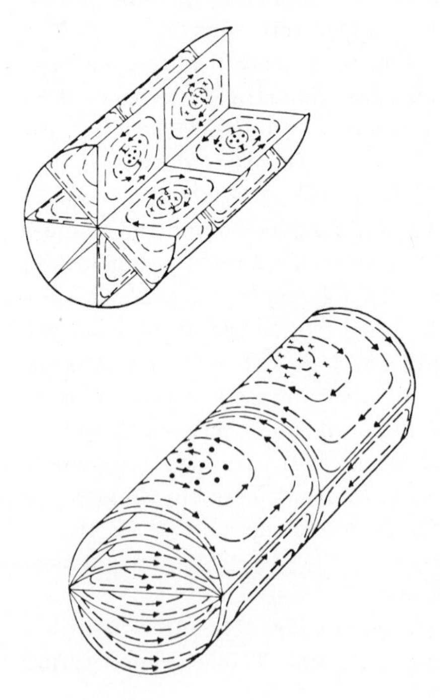

Figure 8.6. Diagrammatic sketch of the TE_{012} (above) and TE_{112} (below) cylindrical resonant cavity modes. The rf magnetic field lines (———) and electric field orientations (dots and x's) are shown (52).

It is always desirable to locate the sample at a point of maximum RF magnetic field strength. For sample tubes the best position in the cavity is along the axis with the sample centered in the middle. This sample position is also at a minimum of the RF electric field, and so dielectric losses are minimized. These losses become appreciable if the sample tube has too large a diameter. Excessive losses manifest themselves by a lowering of the cavity quality factor, Q, and a decrease in the sensitivity.

Most ESR studies make use of a standard 3-mm ID, 4-mm OD pure quartz sample tube. High-purity quartz is necessary because glass usually contains traces of Fe^{3+}. For maximum sensitivity, this tube may be filled to a height of 2.5 cm for a rectangular cavity and 5 cm for a cylindrical cavity. This is convenient to do if one is studying a powder or low-dielectric loss solution.

Liquids with high dielectric losses such as water would load down a resonant cavity by reducing the quality factor Q if they were placed in a standard 3-mm ID quartz tube. This is because the sample extends too far into the region of appreciable microwave electric field strength. To obviate this difficulty, a thin quartz cell is used. This considerably reduces the amount of sample in the region of appreciable electric field strength. Instead, it spreads out the sample in the transverse plane of a rectangular TE_{011} cavity where the microwave magnetic field strength is still a maximum. As a result, a high sensitivity can still be obtained with lossy samples.

A liquid mixing cell in which two liquids flow together and then enter the resonant cavity is commercially available. This permits one to study chemical reactions just as they occur in a cavity. To study photolytic and photochemical systems, a cavity provided with slots in the endplate can be used. Such a cavity is also commercially available.

Single crystals of dimensions $1 \times 1 \times 2$ mm are convenient for ESR studies, and larger ones may be used if they are not too lossy. Anisotropic effects may be investigated either by mounting the crystal on a goniometer and rotating it within the cavity, or by rotating the magnet around the cavity. It is usually desired to carry out a rotation about three mutually perpendicular crystal planes in order to be able to calculate the Hamiltonian parameters.

To compare measurements carried out with different cavities, use is made of the quality factor Q, which is defined by

$$Q = \frac{2\pi(\text{stored energy})}{\text{energy dissipated per cycle}} \tag{8.12}$$

$$= \frac{\omega_0}{\Delta\omega} \tag{8.13}$$

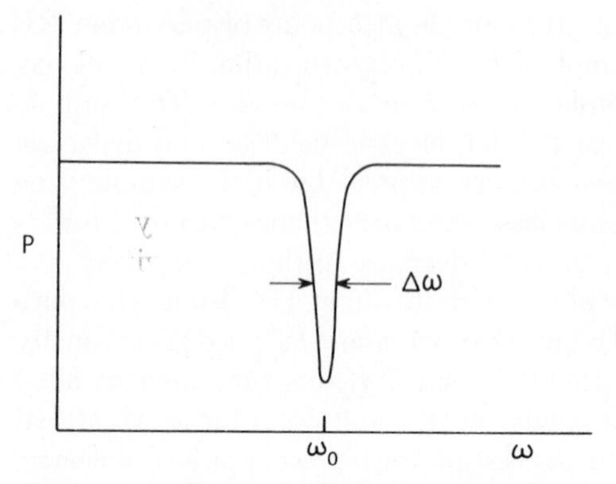

Figure 8.7. Power P reflected from a microwave resonant cavity as a function of the frequency ω. The resonant frequency is ω_0, the full width at half amplitude is $\Delta\omega$ and the loaded $Q = \omega_0/\Delta\omega$.

The width $\Delta\omega$ is defined in Figure 8.7. Equation 8.13 allows Q to be estimated from the width of the cavity dip. Typical values are 4000 for the rectangular and 12,000 for the cylindrical cavities discussed above. Some samples such as carbon, strong acids, and aqueous solutions are quite lossy and produce a marked lowering of Q. To reduce the sample size, these may be measured in capillary tubes or in a flat quartz cell. In addition to the cavity, one also has to consider the filling factor, η, which is a measure of the efficiency with which the RF magnetic field is concentrated in the sample (52). The filling factor is strongly dependent upon the sample position because H_{RF}^2 varies throughout the cavity. The positions described earlier correspond to maximum filling factors.

6. Detector

The last electronic system to be discussed is the crystal detector–preamplifier–lock-in detector combination. The phenomenon of ESR entails the absorption of microwave energy in a magnetic field adjusted to satisfy the Larmor condition (Equation 8.2). Any absorption which takes place in the sample will be detected as a dip in the output of the crystal detector, and can be displayed on an oscilloscope. In practice, a more sophisticated approach is used to measure this signal quantitatively.

Since the magnetic field is modulated at the frequency ω_{mod}, the absorption of microwave energy also occurs at this frequency. The crystal detector demodulates the microwaves by extracting the ESR signal oscillating at the frequency ω_{mod} and passes it on to the preamplifier. The object of the preamplifier is to amplify the signal from the crystal without introducing additional noise.

In the next step, the ESR signal enters the lock-in detector which is sometimes referred to as a coherent detector or a phase-sensitive detector. In this instrument, the ESR signal is compared with a reference signal at the frequency ω_{mod}, derived from the same oscillator which drives the modulation coils. Only that small part of the noise which is at the same frequency and in-phase with the reference signal is allowed to traverse the lock-in detector, most other noise being eliminated. This considerably enhances the sensitivity. Additional noise is eliminated by employing a long time constant filter which averages out noise at frequencies greater than the reciprocal of its time constant, τ_0. The result of this process is that the ESR signal which entered the lock-in detector at the modulation frequency has been converted to a dc signal for display on the recorder.

A more sophisticated method of noise elimination entails the use of a computer for average transients (CAT) for the repeated traversal of the ESR signal and the electronic addition of the various superimposed spectra. This signal enhancement technique through continuous averaging by a computer makes use of a multichannel pulse-height analyzer. This efficiently removes both low- and high-frequency noise. In comparison, the long time constant filter mentioned above does not remove low-frequency noise. Signal enhancement is very helpful with weak signals, but this approach is also quite expensive.

III. ANCILLARY INSTRUMENTAL TECHNIQUES

1. Relaxation Times

The population distribution between the spin states is a strong function of the energy density of the incident radiation. By increasing the radiation intensity, the probability of transitions will increase. The signal from a typical paramagnetic spin system will, at first, increase in amplitude, reach a maximum, and then show a decrease at greater power levels. The latter phenomenon is referred to as saturation. The onset of saturation is accompanied by a gradual broadening and distortion of the resonance line. In general, it is best to study spin systems below saturation.

By switching off the incident microwave power, the system will return to equilibrium because of thermal effects. This is called relaxation; it is an exponential process whereby the spin populations return to their thermal equilibrium value defined by the Boltzmann distribution of Equation 8.1, and it is strongly influenced by the environment around the electron.

Accordingly, information about the environment can be obtained from relaxation data. This decay process occurs with a time constant, T_1, referred to as the spin-lattice or longitudinal relaxation time. These names are derived from the fact that the spin system has to equilibrate with the "lattice" or the environment of the system, and because the energy is defined along the direction of the external magnetic field.

Another relaxation process which systems of electron spins undergo is spin–spin relaxation or transverse relaxation, commonly denoted by the characteristic time constant T_2. In this case the individual spins lose phase coherence with each other in the process of decaying into the random orientations. The increase in the variations of the local magnetic field results in the broadening of the ESR line.

There are two general methods for measuring relaxation times: the saturation method and the pulse method. The former is convenient for use with a commercial spectrometer. It is suitable for some spin systems, but in general the information acquired thereby is not as satisfactory as that obtained by the more sophisticated pulse methods.

A. Saturation Method

In the saturation method, the magnitude of saturation is measured by the saturation factor s:

$$s = \frac{1}{1 + \gamma^2 H_1^2 T_1 T_2/4} \tag{8.14}$$

where the radio-frequency field $H_{RF} = H_1 \cos \omega_0 t$. Above saturation, s becomes very small, while at very low power levels s is quite close to unity $(H_1^2\gamma^2 T_1 T_2/4 \ll 1)$. The terms T_1 and T_2 are the longitudinal and transverse relaxation times. The relaxation times may be evaluated by plotting $1/s$ versus H_1^2 using measurements of either the line-width or the amplitude as a function of power, where the power, P, is proportional to H_1^2. The slope of the straight line in each case is $\gamma^2 T_1 T_2/4$ and $1/s = 1$ when $H_1 = 0$. Thus this plot determines the ratio $T_1\!:\!T_2$. To determine the separate contributions, the spin–spin relaxation time, T_2, is assumed to be proportional to the line width below saturation, in other words,

$$T_2 = \frac{2}{\gamma \Delta H_{pp}^0 \sqrt{3}} \tag{8.15}$$

$$= \frac{1.3131 \times 10^{-7}}{g \Delta H_{pp}^0} \tag{8.16}$$

where ΔH_{pp}^0 is the limiting line-width below saturation. The term ΔH_{pp}^0 is the distance in gauss measured between the two peaks of the absorption derivative line-shape. Both T_1 and T_2 are denoted in seconds.

B. Pulse Methods

Relaxation times are routinely determined by pulse methods in the field of nuclear magnetic resonance (3, 10, 43, 58, 63). The pulse method (18, 37, 41, 42, 50, 51, 62, 67) consists of: (1) exposing the sample to a very high-power, short-duration pulse of microwave energy; and (2) measuring the strength and decay rate of the induced magnetization. In this method T_1 is measured directly without the use of Equation 8.15.

To perform a typical pulse measurement, a high-power pulse of known intensity and duration must be generated and applied to the sample while the detector is shielded. After the completion of the pulse, a low-power klystron may be used to perform a regular ESR experiment which monitors the saturation factor of the strongly saturated resonance. The magnitude of the ESR signal at this low power will depend upon the population difference of the energy levels, and the relaxation of these population differences back to the Boltzmann thermal equilibrium values will be reflected in the magnitude of the detected signals as a function of time.

Another pulse-type experiment makes use of the spin echo technique (28, 29, 46, 47). Two or three pulses of microwave energy are applied to the spin system and an echo is detected at a fixed interval later. The echo is studied as a function of the time interval between the pulses. One may "burn holes" of simple form and controllable width in the line and observe the rate at which these holes are filled in.

2. Double Resonance

In double resonance experiments the sample, which is already subjected to a magnetic field and an incident microwave frequency, is subjected to an additional resonance frequency from a supplementary circuit. Most commercial ESR spectrometers do not permit the routine use of double resonance techniques, although in NMR this technique is routinely applied through the use of spin decouplers. Special expensive accessories are, nevertheless, commercially available for ESR double resonance work.

The principal double resonance experiments that have been carried out with ESR spectrometers are those which simultaneously irradiate the unpaired spins with a microwave frequency ($\sim 10^{10}$ Hz) while subjecting the nuclear spins to a radio frequency ($\sim 10^7$ Hz). Many of these are nuclear polarization schemes, which means that nuclear spin levels become populated with a Boltzmann factor characteristic of electron spins. This can produce a thousand-fold increase in the NMR resonance amplitude. A number of such polarization schemes have been used, and several will be briefly described.

A. The Overhauser Effect

This was originally studied in metals where the dominant nuclear spin–lattice relaxation mechanism is via the conduction electrons through the isotropic Fermi contact interaction (2, 17, 49). The populations of the electronic spin system may be equalized by saturating the ESR transition, and as a result the nuclear spins distribute themselves among the Zeeman levels in accordance with the electronic Boltzmann factor. Consequently the NMR signal is enhanced by the factor $g\beta/g_n\beta_n$ where the numerator contains the electronic g-factor and magneton, and the denominator contains their nuclear counterparts. The Overhauser effect can also be observed in semiconductors and other nonmetals. A complementary polarization scheme called the Underhauser effects produces half the polarization as the Overhauser effect, but in the opposite sense. It can occur in dipolar coupled solids.

B. Electron Nuclear Double Resonance (ENDOR)

This is also called the fast passage effect and was first reported by Feher (22–24, 30, 60). This technique places no requirements on the relaxation processes. It was first described in a system consisting of an unpaired electron with $S = \frac{1}{2}$ interacting with a nucleus with spin $I = \frac{1}{2}$ to produce a hyperfine ESR doublet. To carry out the experiment, an adiabatic fast passage is carried out on one hyperfine component to invert the populations. A net nuclear polarization is established by carrying out an NMR adiabatic fast passage on either the upper or lower NMR transition. The term adiabatic fast passage means a sweep through resonance slow enough so that the alignment of the spins along the radio-frequency field is not disturbed, and fast enough so that the spins do not have time to relax.

C. The Methods of Parallel Fields

The dynamic polarization of forbidden transitions or method of parallel fields makes use of radio-frequency fields parallel to the main field to induce the forbidden transition $\Delta M = \pm 1$, $\Delta M_S = \pm 1$. Such forbidden transitions disturb the Boltzmann population distribution in such a manner that a net polarization results (34).

D. The Abragam and Proctor Method

The solid effect, double effect, or Abragam and Proctor method is used with two spin systems which couple via an interaction such as the dipolar one (1, 15, 21, 40, 48). The frequency $\omega = \omega_e \pm \omega_n$ is employed to simultaneously flip an electronic and nuclear spin at a microwave power level adjusted so that the transitions $\omega_e \pm \omega_n$ are induced at a rate faster than

$1/T_{1n}$ and slower than $1/T_{1e}$, where T_{1n} and T_{1e} are the nuclear and electronic spin–lattice relaxation times, respectively. As a result, the nuclear spins acquire an electronic Boltzmann distribution.

E. *Electron–Electron Double Resonance (ELDOR)*

This occurs when the two frequencies are in the microwave region (31). It employs a double frequency resonant cavity. High microwave power at one frequency saturates one transition while low power at another frequency monitors its effect on another hyperfine component or another part of the same resonant line. This is a recently developed technique which should prove very useful in sorting out cross relaxation and other phenomena.

F. *Acoustic Electron Spin Resonance*

In an ordinary electron spin resonance experiment, the spins absorb microwave energy and then relax by passing on this energy to the lattice vibrations which are "on speaking terms" with the spins. The inverse process is also possible whereby the sample is irradiated with ultrasonic energy at the resonant frequency. One may observe the effect on the ESR absorption due to the simultaneous irradiation with ultrasonics. Both pulsed and continuous wave (cw) spectrometers have been employed for this work (5–8, 11, 13, 20, 35, 65).

IV. SENSITIVITY

The amplitude, Y', is related to the number of spins, N, in a sample through the expression

$$Y_m' = \frac{N}{N_{\min}} \tag{8.17}$$

where $N_{\min}$ is the minimum detectable number of spins. By $N_{\min}$ we mean the number of spins which produces a signal-to-noise ratio of one. This number will depend upon such things as the line-width, line-shape, number of hyperfine components, sample volume, spectrometer characteristics, cavity Q, temperature, microwave power, noise, and so on.

The reduction in sensitivity because of hyperfine splitting, for example, is given by a multiplicity factor, D, which takes into account the fact that in a hyperfine multiplet the intensity is spread over several lines, so that even the strongest line is lower in amplitude. For example, the standard DPPH (α,α'-diphenyl-β-picryl hydrazyl) in benzene solution has a 5-line hyperfine pattern with intensity ratios 1:2:3:2:1. Since the three center lines of a multiplet overlap and thus give misleading relative intensities,

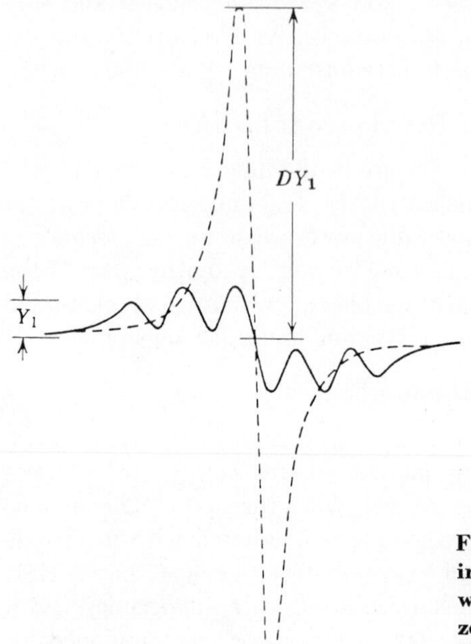

Figure 8.8. Spectrum of DPPH showing the hypothetical singlet which would be obtained if I were equal to zero for ^{14}N (courtesy O. F. Griffith III; Ref. 52).

calibrated spin measurements should be made on the outer lines of a multiplet. In this case, each outer line contains $\frac{1}{9}$ of the total integrated area, so the multiplicity factor is 9 for each of these lines. Figure 8.8 gives the spectrum of DPPH together with the hypothetical singlet that would result if all of the lines collapsed to a single one of the same width.

The minimum detectable number of spins is proportional to the inverse square root of the microwave power, $P^{-1/2}$. This dependence assumes that the power level is sufficiently low so that the sample does not saturate.

If the various spectrometer parameters are omitted, the number of spins can be said to be proportional to the area under the absorption curve, which is given by

$$\text{area} = 3.63Y_{m'}(\Delta H_{pp})^2 \qquad \text{Lorentzian shape} \tag{8.18}$$

$$\text{area} = 1.03Y_{m'}(\Delta H_{pp})^2 \qquad \text{Gaussian shape} \tag{8.19}$$

for a singlet. Other line shapes have constants that can be determined by direct integration.

A general expression for the minimum detectable number of spins $N_{\min}$ in an unsaturated, undermodulated line is

$$N_{\min} = K \frac{V_s T_s D \Lambda' (\Delta H_{pp})^2}{Q \eta g^2 S(S+1) \omega_0{}^2 H_{\text{mod}}} \left\{ \frac{T_d \Delta f}{P} [F_k - 1 + (t + F_{\text{amp}} - 1)L] \right\}^{1/2} \tag{8.20}$$

where

V_s = volume of sample, cm^3
K = constant characteristic of spectrometer
Q = quality factor of cavity
D = multiplicity factor
η = filling factor
g = g-factor
S = spin
ω_0 = microwave angular frequency
ΔH_{pp} = peak to peak full linewidth in gauss
H_{mod} = peak to peak modulation amplitude in gauss
Λ' = line shape factor (Λ' = 3.63 for a Lorentzian line, 1.03 for a Gaussian line)
T_d = detector temperature in °K (usually 300°K)
T_s = sample temperature
P = microwave power incident on the resonant cavity in watts
Δf = effective receiver–detector bandwidth in Hz
F_k = klystron noise figure
t = detector noise temperature
F_{amp} = preamplifier noise figure
L = detector insertion loss (reciprocal of detector conversion gain).

The derivation of Equation 8.20 and the definitions of the various parameters are given elsewhere (52). The formula assumes that the resonant cavity is well matched to the waveguide, that the power level is below saturation, that the line is not overmodulated, and that the temperature is sufficiently high so that the population difference in equilibrium is proportional to $h\nu/kT$.

V. INFORMATION GAINED BY ESR

A typical electron spin resonance spectrum is characterized by the position, intensity, and shape of each component line. The position of the main line or the center of gravity of a hyperfine pattern provide the g-factor of Equation 8.2.

$$E = \hbar\omega = g\beta H M_S + A M_S M_I \tag{8.21}$$

and the spacing A between the lines of a hyperfine multiplet provides the hyperfine coupling constant A. This is illustrated in Figures 8.2 and 8.3 for the more complicated case with two separate hyperfine coupling constants A_1 and A_2. A proton magnetometer may be employed to calibrate the magnet scan to provide accurate line positions and spacings for determining these quantities.

If the curve is symmetric, it will usually be either Lorentzian or Gaussian in shape. An unsymmetric shape is characteristic of an anisotropic g-factor. The Lorentzian line shape is usually observed in low-viscosity liquids where the Bloch equations are satisfied, while dipole–dipole broadening in typical solids produces a Gaussian shape. In solids with very high spin concentrations such as solid DPPH, exchange narrowing occurs, and the shape appears Lorentzian in the center with a more Gaussian contour in the wings. Figure 8.9 compares Lorentzian and Gaussian shapes.

The number of spins in the sample is proportional to the intensity of the resonant line. For a singlet the intensity is the integrated area under the absorption curve, and for first derivative presentation this area can be obtained by a double integration.

$$\text{Intensity} = \int_{-\infty}^{+\infty} dH \int_{-\infty}^{H} Y'_{(H'-H_0)} dH' \tag{8.22}$$

$$= \int_{-\infty}^{+\infty} (H_0 - H) Y'_{(H-H_0)} dH$$

In Equations 8.18 and 8.19, Y_m' is the peak-to-peak amplitude, ΔH_{pp} is the peak-to-peak full line-width and the area factor is 3.63 for a Lorentzian and 1.03 for a Gaussian shape. The presence of structure may be taken into account through the multiplicity factor D (see Section IV).

The number of spins may be determined by measuring a known and unknown sample and evaluating the ratio: $\text{area}_{\text{unknown}}/\text{area}_{\text{known}}$. This procedure is only valid if the unknown and known samples are similar types of spin systems (e.g., both free radicals); otherwise corrections from Equation 8.20 must be used.

The object of an ESR measurement is often the determination of the identity of a paramagnetic species. This can be best accomplished by comparing one's spectra with those obtained from various sources. The g-factor, the hyperfine structure, the line-width, intensity, and shape anisotropy are frequently characteristic of particular species. We shall describe a few important paramagnetic spin systems to illustrate the method of identifying unknown spectra.

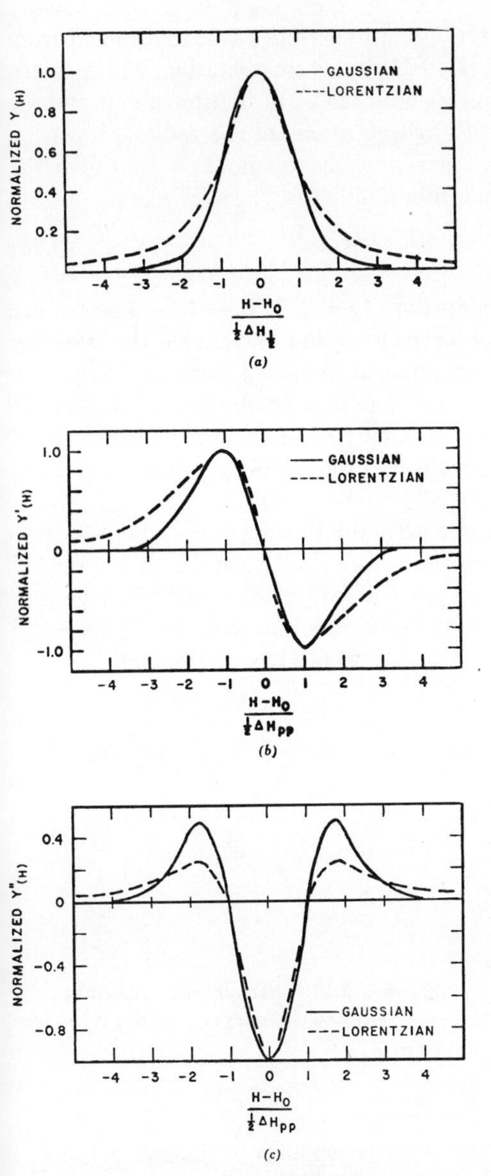

Figure 8.9. **(*a*) Lorentzian and Gaussian absorption curves with the same half amplitude line-width, (*b*) Lorentzian and Gaussian absorption first derivative curves with the same peak-to-peak line-width, (*c*) Lorentzian and Gaussian absorption second derivative curves with the same peak-to-peak line-width (52).**

VI. TYPICAL SPECTRA

In Figure 8.8 we presented the 5-line hyperfine pattern obtained from α,α'-diphenyl-β-picryl hydrazyl (DPPH) in benzene solution. This pattern arises from the unpaired electron spending most of its time about equally divided between the two central nitrogen atoms of the radical shown on Figure 8.10. The two nitrogen nuclei each have spin $I = 1$ to give the orientations of the total spin quantum number M

$$M = M_1 + M_2 \tag{8.23}$$

shown earlier in Table 8.4.

The number of ways of forming the $M = 2, 1, 0, -1, -2$ states are 1, 2, 3, 2, 1, respectively, so the observed hyperfine pattern has the intensity ratios 1:2:3:2:1. If oxygen is scrupulously removed from the sample by the freeze-pump-thaw technique, then one may resolve the much smaller splitting constants from the ring protons (19), as shown in Figure 8.11. This occurs because the relaxation effect of the paramagnetic oxygen molecules is removed.

The spectra illustrated on Figures 8.8 and 8.11 were obtained in solution. Solid DPPH exhibits an exchange narrowed singlet similar to the dotted line curve on Figure 8.8, but somewhat narrower (8) ($\Delta H_{pp} \sim 2.7$ gauss). The strong exchange interaction between adjacent radicals in the solid state averages out the hyperfine structure to produce the singlet.

```
H     H
 \    |
  C = C                 O
  |     \               ||
  C      C - H        O=N         H
 /  \\   ||             \        /
H     C - C              C = C
     /      \           /      \      O
    H        \         /        \   //
              N - Ṅ - C          C-N
             /         \\       //  \\
    H       /            C - C        O
     \     /             |   |
      C - C            O=N   H
    //    ||             ||
H - C     C - H          O
    |    /
H - C = C
        |
        H
```

Figure 8.10. Structural formula for α,α'.-Diphenyl-β-picryl hydrazyl free radical (DPPH).

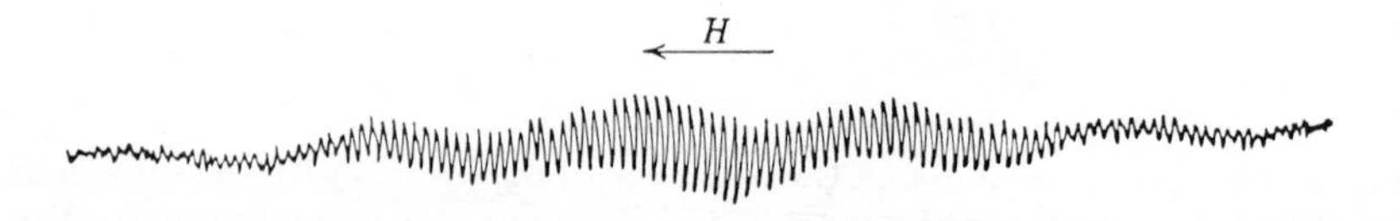

Figure 8.11. High-resolution electron spin resonance spectrum of DPPH in tetrahydrofuran after the removal of dissolved oxygen (from Ref. 19).

The spectrum of DPPH stretches over only a relatively narrow frequency range. Figure 8.12 shows a spectrum of a chromia alumina catalyst that extends from zero field to beyond 6000 gauss (55, 56). It results from the superposition of spectra from three phases, a broad line centered at $g = 2$ arising from clumped or clustered Cr^{3+} ions, a low field line peaking at $\sim$1600 gauss due to isolated Cr^{3+} ions, and a sharp singlet near $g = 2$ attributed to Cr^{5+} ions. This spectrum illustrates the possibilities of studying catalysts (56) and distinguishing valence states by ESR.

Irradiated organic single crystals exhibit strongly orientation-dependent hyperfine patterns in their ESR spectra. The irradiated crystal *l*-α-alanine produced the free radical

```
     H     O
     |  .  ||
H—C—C—C—OH
     |   |
     H  H
```

which exhibited different hyperfine patterns at liquid nitrogen and room temperatures (32) as shown on Figure 8.13. Second-derivative line-shapes are shown since the lock-in detector had a first harmonic reference signal. Both spectra shown on the figure were strongly orientation-dependent. Along the c axis the spectrum was particularly simple since the α-proton bonded to the carbon with the unpaired electron and the β protons on the

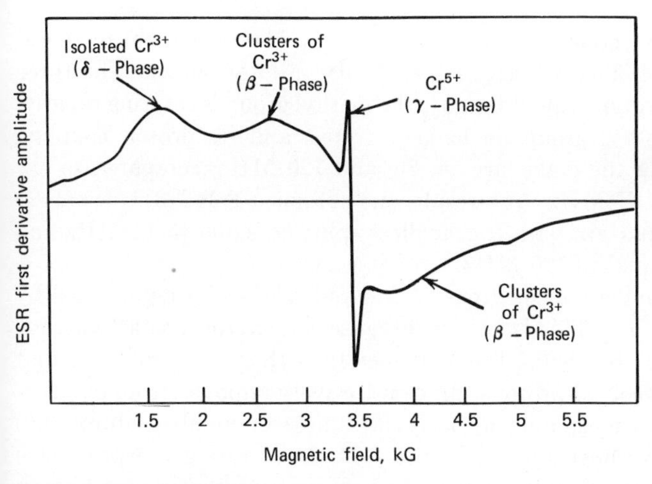

Figure 8.12. Observed ESR spectrum of 5.3 mole % Cr_2O_3 chromia alumina catalyst (from Ref. 55).

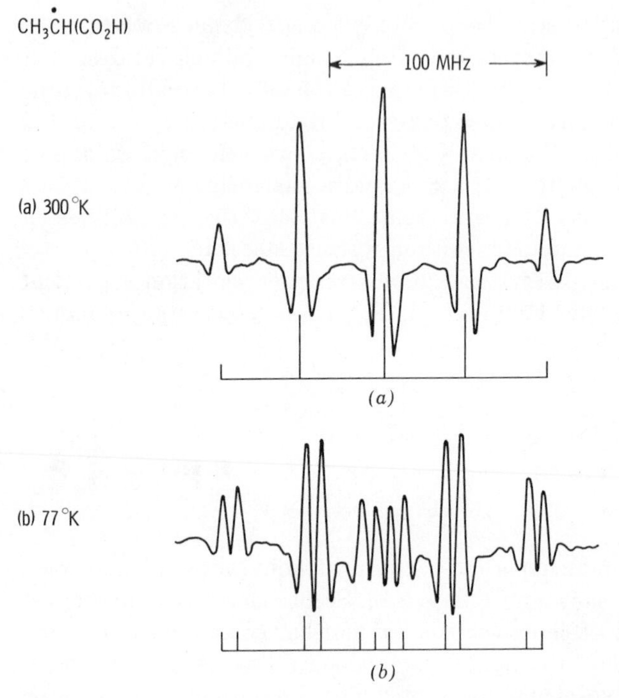

Figure 8.13. Second derivative spectra of γ-irradiated single crystal of *l*-α-alanine; *H* parallel to *c* axis: (*a*) at 300° K, (*b*) at 77° K (32).

methyl group have equal coupling constants (77 MHz) at room temperature. Thus one obtains a 1:4:6:4:1 intensity ratio as shown. The three methyl protons are equivalent because the methyl group is rotating rapidly. At 77°K the methyl group no longer rotates, and its proton coupling constants β_i along the *c* axis are 14, 76, and 120 MHz, compared to the α-proton value of 77 MHz. As a result, one obtains a 1:1:2:2:1:1:1:1:2:2:1:1 hyperfine pattern with the smallest splitting equal to 14 MHz and the largest one equal to 120 MHz.

The origin of hyperfine splittings on free radicals is best determined by selective deuteration. This is helpful because a deuteron has a hyperfine splitting constant that is 6.5 times smaller than that of a proton so that resolved and unresolved protons can be selectively removed from influencing the spectrum. An example of this is the radical anion dibenzofuran (66) whose spectra are illustrated in Figure 8.14. Deuterating the proton on position 9 causes the lines to collapse in pairs as a comparison of Figures 8.14*a* and *c* indicates. Deuterating positions 2 and 4 and the hydroxy

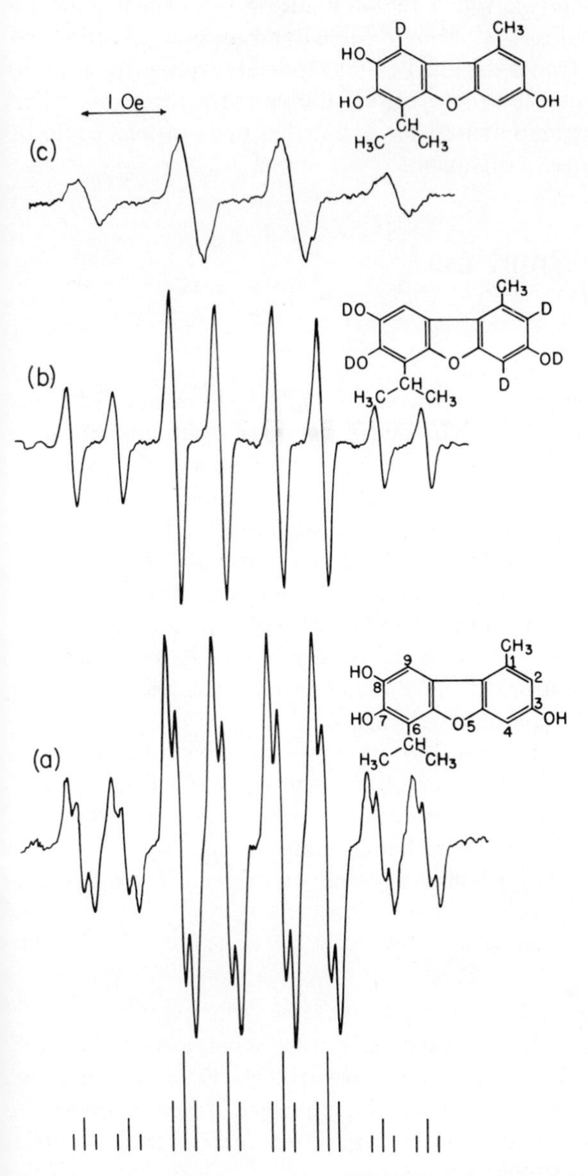

Figure 8.14. **ESR spectra from radical anions of (*a*) 1-methyl-3,7,8-trihydroxy-6-isopropyldibenzofuran and its reconstructed spectrum. (*b*) same as (*a*) except for the substitution of deuterium for hydrogen on positions 2, 3, 4, 7 and 8. (*c*) same as (*a*) except for a deuteron on position 9 instead of *H* (66). (Courtesy of the authors and of Pergamon Press Ltd.)**

group produces the well-resolved octet of Figure 8.14*b*, which indicates that these protons produce the barely resolved and unresolved structure. The former is a 1:2:1 triplet assigned to magnetically equivalent protons 2 and 4. The 1:3:3:1 quartet structure is attributed to the three equivalent protons of the methyl group at position 1. Further deuterations could be employed to confirm these assignments.

VII. SYSTEMS STUDIED BY ESR

1. Biological Systems

ESR has been applied quite extensively to biological systems (12). One can follow the variations that occur under changing environmental conditions by monitoring the intensity of a free radical signal. For example, the presence of free radicals has been studied in healthy and diseased tissue. If a transition metal ion is present, as in hemoglobin (Fe), then its valence state changes may be studied by ESR. Early concrete evidence that free radical activity is linked to photosynthesis was demonstrated by ESR. By irradiating chloroplast containing cells with light in the same wavelength range that produces photosynthesis, a sharp ESR resonance line was observed. When the incident light was turned off, the resonance soon weakened or disappeared completely.

An inconvenience when biological samples are analyzed is the presence of water in the sample. This will result in high dielectric losses, which make it necessary to use a flat quartz sample cell. Some typical systems which have been studied by ESR are hemoglobin, nucleic acids, enzymes, chloroplasts when irradiated, riboflavin (before and after UV irradiation), and carcinogens.

2. Chemical Systems

A number of chemical substances such as synthetic polymers and rubber contain free radicals. The nature of the free radical depends upon the method of synthesis and on the history of the substance. Recent refinements in instrumentation have permitted the detection of free radical intermediates in chemical reactions.

Typical chemical systems that have been studied by ESR are polymers, catalysts, rubber, long-lived free radicals, free radical intermediates, charred carbon, and chemical complexes, especially with transition metals.

3. Conduction Electrons

Conduction electrons have been detected by both conventional electron spin resonance methods, and by the related cyclotron resonance technique. The latter employs an ESR spectrometer and the sample is located in a region of strong microwave electric field strength. In the usual ESR arrangement, on the other hand, the sample is placed at a position of strong microwave magnetic field strength. Conduction electrons have been detected in solutions of alkali metals in liquid ammonia, alkaline earth metals (fine powders), alloys (e.g., small amounts of paramagnetic metal alloyed with another metal), nonresonant absorption of microwaves by superconductors, and graphite.

4. Free Radicals

A free radical is a compound which contains an unpaired spin, such as the methyl radical $\cdot CH_3$ produced through the breakup of methane

$$CH_4 \rightarrow \cdot CH_3 + \cdot H \tag{8.24}$$

where both the hydrogen atom and the methyl radical are electrically neutral. Free radicals have been observed in gaseous, liquid, and solid systems. They are sometimes stable, but usually they are short-lived intermediates in chemical reactions.

Free radicals and radical ions ordinarily have g-factors close to the free electron value of 2.0023 (e.g., for DPPH $g = 2.0036$). In low-viscosity solutions, they exhibit hyperfine patterns with a typical overall spread of about 25 gauss. The scrupulous removal of oxygen often reveals hitherto unresolved structure. In high-concentration solids, a single exchange narrowed resonance appears ($\Delta H_{pp} \sim 2.7$ gauss for DPPH prepared from benzene solution). In irradiated single crystals the free radicals may have strongly anisotropic hyperfine interactions and slightly anisotropic g-factors.

Radical ions of many organic compounds can be produced in an electrolytic cell (26, 27), which is usually a flat quartz cell with a mercury pool cathode and a platinum anode. This electrolytic cell may be directly connected to a flat measuring cell located in the resonance cavity. When the applied voltage in the electrolytic system is increased, the current will first increase but soon it levels off to a plateau. Radical ions are formed in this plateau region. Radical formation can sometimes be observed visually because of color changes in the solution. To carry out the experiment the magnetic field is scanned for resonance over a 50-gauss region near the free electron value of $g = 2.0023$. Radical ions can also be formed in flow-through cells (4).

Since oxygen is also paramagnetic, dissolved air must be scrupulously removed prior to the experiment. The best method is to use the freeze-pump-thaw technique where the sample is first frozen and then connected to a high vacuum source. After closing off the vacuum pump, the sample is melted and refrozen. The cycle is repeated until no air is released during the solid–liquid transitions. The difference in spectra when dissolved oxygen is present and absent is dramatically shown in Figures 8.8 and 8.11 for a DPPH solution in benzene. For this type of experiment, very low power levels and low modulation amplitudes are necesary since line-widths are typically in the 50–100-milligauss range.

Experimentally, the following have been detected by ESR: (1) stable solid free radicals (a single exchange-narrowed resonance); (2) stable free radicals in solution (hfs obtained); (3) free radicals produced by irradiation (often at low temperature; sometimes single crystals); (4) condensed discharges (free radicals produced in a gas and condensed on a solid at low temperature); (5) biological systems; (6) biradicals; (7) electrochemical generation of ion-radicals (polarography); (8) triplet states; (9) paramagnetic molecules (e.g., NO, NO_2, and ClO_2); and (10) intermediates in chemical reactions. A tabulation of experimental data has been given by Bowers (14) and Fischer (25).

5. Irradiated Substances

A considerable amount of work has been done on free radicals and color centers which are produced by irradiation. Most irradiations are carried out with x-rays, γ-rays, or electrons whose energies far exceed chemical bond energies. Paramagnetic spins can also be produced by less energetic ultraviolet light (photolysis) or neutrons.

Most ESR spectra are obtained *after* the sample is irradiated. Many paramagnetic centers are sufficiently long-lived to warrant such a procedure. More sophisticated experimental techniques entail simultaneous irradiation and ESR detection. This is particularly popular when the irradiation source is an ultraviolet lamp. Low-temperature irradiation and detection can reveal the presence of new centers which can be studied at gradually increasing temperatures to elucidate the kinetics of their recombination. Routine spectrometers are satisfactory for most radiation-damage investigations.

Details on irradiation procedures have been given by Motchane (48). Some typical systems that have been studied are (1) ionic crystals (e.g., alkali halides, *F* centers, and other centers); (2) solid organic compounds;

(3) liquid organic compounds; (4) organic single crystals; (5) polymers; (6) semiconductors (e.g., Ge and Si); and (7) photoconductors (e.g., dyes).

6. Naturally Occurring Substances

Most of the systems studied by ESR are synthetic or man-made. Nevertheless, from the beginning of the field, various naturally occurring substances have been studied, such as: (1) minerals with transition elements [e.g., ruby (Cr/Al_2O_3), dolomite $Mn/(Ca, Mg\ (CO_3)]$; (2) minerals with defects (e.g., quartz); (3) hemoglobin (Fe); (4) petroleum; (5) coal; (6) rubber; and (7) various biological systems.

7. Semiconductors

Commercially useful semiconductors tend to be naturally or intentionally doped with impurities, so they lend themselves to ESR studies. Examples are (1) Ge, Si, and InSb (by cyclotron resonance); (2) doped semiconductors (e.g., Si with As, Sb, or P); (3) irradiated semiconductors; and (4) graphite.

8. Transition Elements

A large percentage of ESR studies have been carried out with transition element compounds, particularly with the first transition series. The second and third transition series, and also the rare earths and transuranic elements have been investigated. Some of the most favorable and widely studied valence states are $V^{4+}(3d^1)$, $Cr^{3+}(3d^3)$, $Mn^{2+}(3d^5)$, $Fe^{3+}(3d^5)$, $Co^{2+}(3d^7)$ at low temperature, $Cu^{2+}(3d^9)$, $Eu^{3+}(4f^7)$ and $Gd^{2+}(4f^7)$. The spectra in doped single crystals exhibit such effects as hyperfine structure (V^{4+}, Mn^{2+}), zero field splitting, D, (Cr^{3+}, Mn^{2+}, Fe^{3+}, Co^{2+}) and an anisotropic g-factor (Co^{2+}, Cu^{2+}). Typical systems that have been studied include single crystals (1% in diamagnetic crystal; anisotropic g-factors, and hfs constants evaluated); relaxation time studies (mostly liquid He temperature; low power); chelates and sandwich compounds; and alloys.

Table 8.5 is a summary of the principal valence states of the first transition series. Most of the common valence states may be observed at room temperature or below. Some like Co^{2+} ordinarily require very low temperatures for detection. Ions with an even number of electrons are only observable under special conditions. The zero field splittings and hyperfine coupling constants are given in the units of reciprocal centimeters. The conversion factor

$$1\ \text{cm}^{-1} = 29.98\ \text{GHz} = (21.42/g)\ \text{kilogauss} \qquad (8.25)$$

is useful for transforming to the units of gauss which are directly measured in an experiment.

Table 8.5. Summary of Typical ESR Data on the First Transition Series[a]

Number of Electrons	Ions	Spectroscopic State	g	$\lvert D\rvert$, cm^{-1}	$\lvert A\rvert$, cm^{-1}
1	Ti^{3+}, Mn^{6+}, V^{4+}, Cr^{5+}	^2D	1.1–2.0		0.015 (^{55}Mn, ^{51}V)
2	V^{3+}, Ti^{2+}, Cr^{4+}	^3F	1.9	5–10	0.02 (^{51}V)
3	V^{2+}, Cr^{3+}, Mn^{4+}	^4F	2.0	10^{-3}–1.0	0.001 (^{53}Cr)
					0.008 (^{51}V, ^{55}Mn)
4	Cr^{2+}, Mn^{3+}	^5D	2.0	2	
5	Mn^{2+}, Fe^{3+}	^6S	2.0	10^{-3}–0.2	0.006–0.01 (^{55}Mn)
6	Fe^{2+}, Co^{3+}	^5D	0–9	0.2	
7	Co^{2+}	^4F	1.4–7	4.5	0.01–0.03
8	Ni^{2+}	^3F	2.2–2.3	0.1–4	
9	Cu^{2+}	^2D	2–2.5		0.002–0.02

[a] Tabulation of data may be found in

1. B. Bleaney and K. W. H. Stevens, *Rept. Prog. Phys.* **16**, 108 (1953).
2. K. D. Bowers and J. Owen, *Rept. Prog. Phys.* **18**, 304 (1955).
3. G. E. Pake, *Paramagnetic Resonance*, Benjamin, N. Y., 1962.
4. S. A. Al'tshuler and B. M. Kozyrev, in *Electron Paramagnetic Resonance, Transl. Scripta Technica*, C. P. Poole, Jr., Ed., Academic Press, N. Y. 1964.
5. W. Low and E. L. Offenbacher, *Solid State Phys.* **17**, 135 (1965).
6. G. E. König, in *Electron Paramagnetic Resonance*, Group 2, Vol. 2, Landolt-Bornstein, Springer Verlag, Berlin, 1965.

The table shows the wide variation in nuclear spins and g-factors that occurs for various ions. These are useful for identification purposes. Several tabulations of data for specific matrices and compounds have been compiled and the reader may refer to them for assistance in making particular assignments (8, 39, 44, 45).

REFERENCES

1. A. Abragam and W. G. Proctor, *C. R. Acad. Sci.* **246,** 2253 (1958); *Phys. Rev.* **109,** 1441 (1958).
2. A. Abragam, *Phys. Rev.* **98,** 1729 (1955).
3. A. Abragam, *The Principles of Nuclear Magnetism*, Clarenden, Oxford, 1961.
4. J. Q. Adams and J. R. Thomas, *J. Chem. Phys.* **39,** 1904 (1963).
5. S. A. Al'tshuler, *Dokl. Akad. Nauk. SSSR* **85,** 1235 (1952); *Zh. Eksper. Teor. Fiz.* **28,** 49 (1955).

6. S. A. Al'tshuler, M. M. Zaripov, and L. Ya. Shekun, *Izv. Akad. Nauk. Ser. Fiz.* **21,** 844 (1957).
7. S. A. Al'tshuler and J. D. Bashkirov, *Conference on Paramagnetic Resonance*, Kozan, 1960, p. 78.
8. S. A. Al'tshuler and B. M. Kozyrev, in *Electron Paramagnetic Resonance*, trans. by Scripta Technica, C. P. Poole, Ed., Academic Press, New York, 1964.
9. P. W. Anderson and P. R. Weiss, *Rev. Mod. Phys.* **25,** 269 (1953).
10. E. R. Andrew, *Nuclear Magnetic Resonance*, Cambridge Univ. Press, Cambridge, 1956.
11. P. Averbuch and W. G. Proctor, *Phys. Lett.* **4,** 221 (1963).
12. M. S. Blois et al., Eds., *Free Radicals in Biological Systems*, Academic Press, New York, 1961.
13. D. I. Bolef, J. DeKlerk, and R. B. Gosser, *Rev. Sci. Instr.* **33,** 631 (1962).
14. K. W. Bowers, *Advances in Magnetic Resonance*, Vol. 1, Academic Press, New York, 1965.
15. J. Burget, M. Odelhnal, V. Petricek, and J. Sacha, *Arch. Sci. Spec.* **14,** 487 (1961).
16. A. Carrington and A. D. McLachlan, *Introduction to Magnetic Resonance*, Harper & Row, New York, 1967, Chap. 11.
17. T. R. Carver and C. P. Slichter, *Phys. Rev.* **102,** 975 (1956).
18. C. F. Davis, M. W. P. Strandberg, and R. Kyhl, *Phys. Rev.* **111,** 1268 (1958).
19. Y. Deguchi, *J. Chem. Phys.* **32,** 1584 (1960).
20. J. DeKlerk, *Ultrasonics* **2,** 137 (1964).
21. E. Erb, J. L. Motchane, and J. Uebersfeld, *C. R. Acad. Sci.* **246,** 1833, 2121, 3050 (1958).
22. G. Feher, C. S. Fuller, and E. A. Gere, *Phys. Rev.* **107,** 1462 (1957).
23. G. Feher and E. A. Gere, *Phys. Rev.* **103,** 501 (1956); **114,** 1245 (1959).
24. G. Feher, *Phys. Rev.* **103,** 500, 834 (1956); **114,** 1219 (1959); *Phys. Rev. Lett.* **3,** 135 (1959).
25. H. Fischer, *Magnetic Properties of Free Radicals*, Group 2, Vol. 1, Landolt-Bornstein, Springer Verlag, Berlin, 1965.
26. D. H. Geske and A. H. Maki, *J. Chem. Phys.* **30,** 1356 (1959).
27. D. H. Geske and A. H. Make, *J. Amer. Chem. Soc.* **82,** 2671 (1960).
28. J. P. Gordon and K. D. Bowers, *Phys. Rev. Lett.* **1,** 368 (1958).
29. B. Herzog and E. L. Hahn, *Phys. Rev.* **103,** 148 (1956); E. L. Hahn, *ibid.* **80,** 580 (1950).
30. J. S. Hyde, *J. Chem. Phys.* **43,** 1806 (1965).
31. J. S. Hyde, R. C. Sneed, and G. H. Rist, *J. Chem. Phys.* **51,** 1404 (1969).
32. A. Horsfield, J. R. Morton, and D. H. Wiffen, *Mol. Phys.* **4,** 425 (1961); **5,** 115 (1962).
33. D. J. E. Ingram, *Spectroscopy at Radio and Microwave Frequencies*, Butterworths, London, 1955.
34. C. D. Jeffries, *Progr. Cryogenics* **3,** 129 (1961); C. D. Jeffries, *Dynamic Nuclear Orientation*, Interscience, New York, 1962.

35. E. H. Jacobsen, N. S. Shiren, and E. B. Tucker, *Phys. Rev. Lett.* **3,** 81 (1959).

36. C. S. Johnson, Jr., *Adv. Magnetic Res.* **1,** 33 (1965).

37. D. E. Kaplan, *J. Phys. Radium Suppl.* **23** (3), 21A (1962).

38. D. Kivelson, *J. Chem. Phys.* **33,** 1094 (1960); **45,** 1324 (1966).

39. E. König, *Electron Paramagnetic Resonance*, Group 2, Vol. 2, Landolt-Bornstein, Springer Verlag, Berlin, 1966.

40. K. D. Kramer and W. Müller-Warmuth, *Z. Naturforsch* **18A,** 1129 (1963).

41. O. S. Leifson and C. D. Jeffries, *Phys. Rev.* **122,** 1781 (1961).

42. P. M. Llewellyn, P. R. Whittlestone, and J. M. Williams, *J. Sci. Instr.* **39,** 586 (1962).

43. A. Lösche, *Kerninduktion*, Veb. Deutscher Verlag Der Wissenschaften Berlin, East Germany, 1957.

44. W. Low, *Paramagnetic Resonance in Solids*, Academic Press, New York, 1960.

45. W. Low and E. L. Offenbacher, *Solid State Phys.* **17,** 135 (1965).

46. W. B. Mims, K. Nassau, and J. D. McGee, *Phys. Rev.* **123,** 2059 (1961).

47. W. B. Mims, *Proc. Royal Soc.* **283,** 452 (1965); *Rev. Sci. Instr.* **36,** 1472 (1965).

48. J. L. Motchane, *Ann. Phys.* (*France*) **7,** 139 (1962).

49. A. W. Overhauser, *Phys. Rev.* **89,** 689 (1953); **92,** 411 (1953); **94,** 1388 (1954).

50. P. P. Pashinin and A. M. Prokhorov, *Zh. Eksper. Teor. Fiz.* **40** (33), 49 (1961).

51. D. H. Paxman, *Proc. Phys. Soc.* **78,** 180 (1961).

52. C. P. Poole, Jr., *Electron Spin Resonance*, Interscience, New York, 1967.

53. C. P. Poole, Jr. and R. S. Anderson, *J. Chem. Phys.* **31,** 346 (1959). See also Lebedev et al., *Atlas of Electron Spin Resonance Spectra*, Consultants Bureau, New York, Vol. 1, 1963; Vol. 2, 1963.

54. C. P. Poole, Jr. and H. A. Farach, *Relaxation in Magnetic Resonance*, Academic Press, New York, 1971.

55. C. P. Poole, Jr., W. L. Kehl, and D. S. MacIver, *J. Catal.* **1,** 407 (1962).

56. C. P. Poole, Jr., and D. MacIver, *Adv. Catal.* **17,** 223 (1967).

57. C. P. Poole, Jr., and D. E. O'Reilly, *Rev. Sci. Instr.*, **32,** 460 (1961).

58. J. A. Pople, W. G. Schneider, and H. J. Bernstein, *High-Resolution Nuclear Magnetic Resonance*, McGraw-Hill, New York, 1959.

59. M. L. Randolph, *Rev. Sci. Instr.* **31,** 949 (1960).

60. G. H. Rist and J. S. Hyde, *J. Chem. Phys.* **49,** 2449 (1968).

61. R. N. Rogers and G. E. Pake, *J. Chem. Phys.* **33,** 1107 (1960).

62. R. H. Ruby, H. Benoit, and C. D. Jeffries, *Phys. Rev.* **127,** 51 (1962).

63. C. P. Slichter, *Principles of Magnetic Resonance*, Harper & Row, New York, 1963.

64. L. S. Singer, W. H. Smith, and G. Wagoner, *Rev. Sci. Instr.* **32,** 213 (1961).

65. E. B. Tucker, *Phys. Rev. Lett.* **6,** 183 (1961).

66. A. C. Waiss, Jr., J. A. Kuhnle, J. J. Windle, and A. K. Wiersema, *Tetrahedron Lett.* **50,** 6251 (1966).

67. G. M. Zverev, *Pribory Tekh. Eksper.* **5** (930), 105 (1961).

RECOMMENDED READING

R. S. Alger, *Electron Paramagnetic Resonance, Techniques and Applications*, Interscience, New York, 1968.

S. A. Al'tshuler and B. M. Kozyrev, in *Electron Paramagnetic Resonance*, Transl. Scripta Technica, C. P. Poole, Jr., Ed., Academic Press, New York, 1964.

A. M. Bass and H. P. Broida, Eds., *Formation and Trapping of Free Radicals*, Academic Press, New York, 1960.

G. B. Benedek, *Magnetic Resonance at High Pressure*, Interscience, New York, 1963.

M. S. Blois, H. W. Brown, R. M. Lemmon, R. O. Lin, and M. Weissbluth, Eds., *Free Radicals in Biological Systems*, Academic Press, New York, 1961.

A. Carrington and A. D. McLachlan, *Introduction to Magnetic Resonance*, Harper & Row, New York, 1967.

H. Fischer, *Properties of Free Radicals*, Group 2, Vol. 1, Landolt-Bornstein, Springer Verlag, Berlin, 1965; an excellent tabulation of data.

H. M. Hershenson, *NMR and ESR Spectra* (Index 1958–1963), Academic Press, New York, 1965.

D. J. E. Ingram, *Spectroscopy at Radio and Microwave Frequencies*, Butterworths, London, 1955.

D. J. E. Ingram, *Free Radicals as Studied by Electron Spin Resonance*, Butterworths, London, 1958.

C. D. Jeffries, *Dynamic Nuclear Orientation*, Interscience, New York, 1963.

E. Konig, *Electron Paramagnetic Resonance*, Group 2, Vol. 2, Landolt-Bernstein, Springer Verlag, Berlin, 1966; an excellent tabulation of data on transition metal ions.

Ya. S. Lebedev et al., *Atlas of ESR Spectra*, Consultants Bureau, New York, 1963.

William Low, "Paramagnetic Resonance in Solids," *Solid State Phys. Suppl.* **2**, Seitz and Turnbull, Eds., Academic Press, New York, 1960.

G. J. Minkoff, *Frozen Free Radicals*, Interscience, New York, 1960.

G. Pake, *Paramagnetic Resonance*, Benjamin, New York, 1962.

C. P. Poole, Jr., *Electron Spin Resonance, A Comprehensive Treatise on Experimental Techniques*, Interscience, New York, 1967.

C. P. Poole, Jr., Ed. *Magnetic Resonance Review*, Gordon and Breach, New York, an annual review of the literature.

C. P. Poole, Jr. and H. A. Farach, *Relaxation in Magnetic Resonance*, Academic Press, New York, 1971.

C. P. Slichter, *Principles of Magnetic Resonance*, Harper & Row, New York, 1963.

T. L. Squires, *An Introduction to Electron Spin Resonance*, Academic Press, New York ,1964.

S. V. Vonsovskii, Ed., *Ferromagnetic Resonance*, U.S. Dept. of Commerce, Washington, D. C., 1964.

J. S. Waugh, Ed., *Advances in Magnetic Resonance*, Academic Press, New York, an annual review series.

ROBERT A. FLATH

Western Utilization Research and Development Division
Agricultural Research Service
U.S. Department of Agriculture, Albany, California

IX. Gas Chromatography–Mass Spectrometry*

* Reference to a company or product name does not imply approval or recommendation of the product by the U.S. Department of Agriculture to the exclusion of others that may be suitable.

I. INTRODUCTION

Gas chromatography has had a tremendous impact upon the examination of complex mixtures since its development in the early 1950s. The technique requires only a very small quantity of sample mixture, much smaller than could conceivably be separated into its constituents by any other nonchromatographic technique. Gas chromatography can provide excellent separation of a mixture's components; even with extremely complex mixtures containing components with a wide range of polarities and boiling points, an investigator can get nearly complete separation by properly choosing operating and column parameters. It is potentially useful with nearly any compound having some volatility and sufficient thermal stability to survive the operating conditions of the gas chromatograph.

The major shortcoming of GC becomes apparent when it is applied to examination of a mixture of unknown components because the technique does not provide direct identifications of the separated compounds. The only information presented for each component is its retention behavior in comparison with that of the other constituents under a given set of GC operating conditions. If the mixture is quite simple, if the source and perhaps the history of the mixture is known, and especially if some other information about the compounds, such as spectral data or chemical behavior, is available, educated guesses about the identities of these components can often be made. Such "identifications" remain just that, however—educated guesses.

The most reliable method of identification is to isolate the compound in some way (often via gas chromatographic separation and trapping). This compound is then examined, using spectral and/or chemical approaches.

1. Chemical Examination

The types of chemical treatment available as aids to structure determination usually involve the preparation of derivatives, hydrogenation, oxidation (of many kinds), and acid and/or base treatment. The results of such chemical treatment are followed closely by suitable spectral techniques. A major complication is the small sample size usually available when individual components have been isolated from a complex mixture, for example, by preparative gas chromatography.

2. Spectral Examination

Most workers are at least aware of the kind of information obtainable with the common spectral methods of IR, UV, NMR, Raman, and mass spectrometry; but application of such techniques to the examination of a minuscule quantity of isolated compound requires considerable care and refined technique. Infrared spectroscopy is a widely used spectral method and may be applied to quantities of sample as small as tens of micrograms. "On-the-fly" gas chromatography-infrared is currently only useful in very limited situations, largely due to the slow scan rates and sensitivity limitations of even "fast-scan" instruments. The Fourier Transform interferometric-type instrument holds considerable promise for the future of direct combination with a gas chromatograph but is at present very expensive. The comments concerning IR largely hold for Raman spectrometry as well. The sensitivity and applicability of UV spectrometry varies considerably with the presence and nature of the sample molecule's chromophores.

Nuclear magnetic resonance spectrometry is extremely useful in structure elucidation; however, it has roughly one-tenth the sensitivity of IR, even when used with a time-averaging computer. As with all the other methods mentioned above, NMR is also a batch technique.

Mass spectrometry is the most sensitive of the spectral methods mentioned; mass spectra of medium resolution ($M/\Delta M$ = 500–1000, 10% valley; see Chapter 10, Section III, for resolution definition) may be obtained from most instruments with as little as 10^{-9}–10^{-10} g of sample. The instrument can accommodate any material sufficiently volatile and thermally stable to be gas chromatographed. The mass spectrum contains considerable *potential* information about the sample molecule. At the very least, a mass spectrum can be used by the neophyte in much the same way he might use an infrared spectrum, as a "fingerprint."

In the discussions throughout this chapter, low-to-medium resolution is inferred wherever reference to mass spectrometry or a mass spectrometer is made. When high resolution ($M/\Delta M > 10{,}000$) is being considered, it will be so stated.

II. BATCH SAMPLING

1. Sample Collection and Transfer

Even though the mass spectrometer requires only very small amounts of sample, efficient trapping is desirable, especially when attempting to isolate those components present in very low concentrations. Once trapped, a purified compound must be protected from contamination and degradation of any kind. The most obvious cause of decomposition of labile materials is exposure to air and light, particularly when the sample is likely to be spread in a thin film or fog of droplets on the inner walls of a trap. Ideally, sample manipulations should be minimized. The chances for decomposition, evaporation, polymerization, or absorption into rubber- or plastic-lined vial closures increase considerably when trapped samples are stored for any significant period of time. Unfortunately, batch handling of a complex mixture's components makes some time lag between trapping and mass spectral examination unavoidable.

The volatility of a given sample determines the choice of trap to be employed. If the sample is very volatile, or if it "fogs" and then is swept out of the trap, total eluant collection may be necessary. An evacuated collection flask is connected through a needle valve to the gas chromatograph column exit when the desired peak starts to appear. The entire column effluent is then collected in this flask by opening the needle valve.

There are a multitude of trap designs, both homemade and commercial, for use with compounds of moderate to low volatility. However because introduction of a sample into a mass spectrometer involves vacuum vaporization of the sample at some point, several traps which have been found useful will be reviewed here. Amy et al. (1) have described the use of small glass tubes containing a short column (1–5 mg) of solid support coated with gas chromatographic stationary phase, usually the same as that found in the gas chromatography column. Activated coconut charcoal has been recommended as an adsorbent by Damico et al. (13). The sample band eluted from the column dissolves in the cooler trap stationary phase and is retained. The individual traps are then sealed for storage. For mass spectral examination the tube is cut and the solid support and sample are inserted into the mass spectrometer via the solid inlet probe. Moderate heating of the probe in the ion source vacuum serves to vaporize the sample from the comparatively nonvolatile stationary phase.

McFadden (35) has described a simple "lock"-type inlet with provisions for cooling and heating which may be useful with any small trap. Using such an inlet, collected sample need not be manually removed from the trap at all. Other more conventional traps and transfer techniques have also been described (8, 11, 42, 62).

A somewhat more direct arrangement, which avoids sample handling and long term storage, involves the use of a trapping manifold between the gas chromatograph and mass spectrometer (7, 16, 41). Basically it involves trapping a compound from the effluent by passing the carrier gas stream through a cooled trap mounted on the manifold. The trap is then isolated by diverting the gas stream through another trap or vent. The refrigerant is removed from the first trap and the sample is bled into the mass spectrometer inlet line by appropriate valving. Slow scan rates can then be employed, with the mass spectrometer operated in a conventional batch mode. This approach is probably most useful with fairly simple mixtures, as is the simple batch method itself.

2. Advantages of Batch Technique

There are several decided advantages to employing a batch technique, particularly for fairly simple mixtures, and when one is not particularly concerned with minor components. One of the major advantages is that little or no modification of the mass spectrometer inlet system is required—this may be a deciding factor if the only instrument available belongs to some other group or individual in the organization.

Batch sample introduction permits the operator to optimize a number of mass spectrometer parameters to make best use of each sample. This is particularly important when trying to determine the concentration of an impurity in a sample by looking at the relative intensities of ions characteristic of each component.

3. Disadvantages of Batch Technique

This section would be unnecessary if there were no significant drawbacks to the batch technique. Each component must first be trapped (if possible), a process which may be quite inefficient. Once trapped, it must be stored for variable periods of time, then must be removed from the trap and inserted into the mass spectrometer inlet system. During much of this procedure, the sample is subject to decomposition, evaporation, and contamination. The biggest problem, however, is the handling of the large number of samples which might be of interest. The length of time during which the samples must be stored without suffering change or loss is directly proportional to the number of components in the sample mixture. For routine use, this discontinuous method becomes quite unwieldy with anything more complex than a 10–20 component mixture.

III. DIRECT COMBINATION OF THE GAS CHROMATOGRAPH AND MASS SPECTROMETER

The direct introduction of a gas chromatograph's effluent stream into the mass spectrometer is a logical step, for the two instruments have many operating parameters in common. Both require samples in the vapor state, and both perform quite well with extremely small samples. With the current state of the art, it is well to recognize GC–MS as an entity in itself and not as a loose combination of two different instruments.

The elution interval for a component band in a gas chromatograph is on the order of 5–60 sec. This is a rather short period of time relative to the typical scan times of most spectrometric instruments and is one of the major problems encountered in attempts to obtain "on-the-fly" IR spectra with a combination of a gas chromatograph and IR spectrometer. Fortunately, most modern low and medium resolution mass spectrometers can scan a useful m/e interval (e.g., 10–300) in 1–4 sec, if necessary, although with some loss in precision and signal-to-noise ratio at the faster scan rates.

A major advantage of such a "marriage" (51) is the elimination of sample manipulation and storage. As an added dividend (perhaps a mixed blessing), mass spectra of more mixture components may be recorded per unit time, in other words, the data output rate is much higher from a combination gas chromatograph–mass spectrometer than from isolated operation of the two instruments.

The most obvious problem in combining these two instruments is the considerable difference in operating pressures. A gas chromatograph is usually operated with the column exit at atmospheric pressure and with carrier-gas flow rates of 0.5–50 ml/min, depending upon the type of column used. Most mass spectrometers, in contrast, tolerate ion source pressures no higher than 10^{-5}–10^{-4} torr (1 torr = 1 mm Hg). It might be mentioned in passing that the type of pressure gage most commonly used for monitoring pressure within a mass spectrometer is calibrated for nitrogen (or air). The carrier gas usually employed in gas chromatography, at least in the United States, is helium, to which such a pressure gage is less sensitive. Therefore the indicated pressure is lower by approximately a factor of five times than the actual helium pressure within a mass spectrometer coupled directly with a gas chromatograph (15). Many authors do not indicate whether the pressures reported are corrected.

1. Total Effluent Introduction

A simple calculation, ignoring any gas temperature changes, reveals that 0.5–50 ml of carrier gas/min at 1 atm corresponds to approximately 127–12,700 liter/sec at 5×10^{-5} torr. The net pumping speed of the ion source vacuum system of most instruments is rarely more than 100–200 liters/sec, so direct introduction of the total column effluent stream into a mass spectrometer is limited to situations where the smaller bore open tubular columns are used. A 0.01-in. ID capillary is operated at flow rates of 0.5–1.0 ml/min, so most of the gas flow can be accommodated by a mass spectrometer. The required pressure drop is maintained by using a suitable length of additional capillary tubing as restrictor between the column exit and mass spectrometer ion source (14). If the ion source pumping system cannot handle the entire flow, a portion (usually ½ or less) of the gas flow is pumped away by an auxiliary pumping system before it reaches the ion source region. In a variation of this approach, the 0.01-in. ID column itself is used as the capillary restrictor, with the inlet pressure decreased by 1 atm, and with the exit at approximately 10^{-3}–10^{-5} torr (36–38, 56, 58). There is generally no loss in column performance at reduced pressure. Because a smaller quantity of carrier gas per unit time is required to maintain the same average linear velocity through the column when it is operated at reduced pressure, correspondingly smaller demands are made upon the mass spectrometer's pumping system. Unfortunately, the sample capacity of a 0.01-in. ID column is limited to a maximum of about 2–5 μg/component (55). Assuming a range of component concentrations in a typical mixture of 1 to 10^3–10^5, the smaller components are likely to be below or at the sensitivity limits of the mass spectrometer. By replacing the 0.01-in. ID capillary column with a longer 0.02-in. ID column, the sample load can be increased approximately five-fold without loss of gas chromatographic resolution. Unfortunately, a four-fold flow rate increase must also be dealt with. Most mass spectrometers cannot accommodate 4 ml/min of column effluent without discarding part of the flow (and sample) through an auxiliary pumping system. However the 0.02-in. ID column can also be operated with its exit under vacuum; the resultant decrease in carrier gas load may be sufficient to permit operation in this manner with a given mass spectrometer.

2. Effluent Introduction with Splitting

Packed columns of various kinds are probably used for the majority of all gas chromatographic separations. However because the exit flows of most analytical packed columns vary between approximately 15–50 ml/min (1 atm), their total flow cannot be introduced into any conventional mass spectrometer. Before 1964 most workers used some sort of valving or splitting arrangement between the two instruments, which permitted them to feed only a portion of the effluent into the mass spectrometer, usually through a restrictor (18, 30). Split ratios were often in the order of 50:1 or even 100:1, with most of the sample being discarded. As a result, the overall sensitivity of the gas chromatograph–mass spectrometer combination was quite poor. Such systems are still used occasionally, but have largely been replaced by more efficient instrument linkages.

3. Enrichment Devices—Requirements and Parameters

A. Performance Requirements

The carrier gas portion of a gas chromatographic column's effluent has served its purpose once the sample mixed with the carrier is eluted from the column. At that point, it becomes a complicating factor in a gas chromatograph–mass spectrometer linkage. Ideally the effluent stream could be passed through a device of some sort which would strip the sample from the carrier gas, discard the carrier, and introduce the sample into the mass spectrometer. Preferably this device should not decompose any of the sample; it should remove nearly all of the carrier gas, but none of the sample; it should perform satisfactorily with materials having wide ranges of polarity, volatility, and stability (chemical and thermal); and it should be rugged and easily maintained.

B. *Performance Parameters*

Two parameters are commonly used in discussing enrichment device (molecular separator) performance (19).

The first of these, yield, Y, is simply the ratio of the amount of sample passed from a separator into the mass spectrometer to the amount of sample introduced into the separator:

$$Y = \frac{\text{amount of sample into MS}}{\text{amount of sample into separator}}$$

called "efficiency" by Kreuger et al. (28). In discussions, this value is usually converted to a percentage. The higher the yield, the better; in practice, any separator with a consistent yield of 40–50% is quite satisfactory, for, as pointed out by Ryhage (47), no matter what is done to improve such a separator, the performance can only be increased by a factor of 2. Yield only describes part of a separator's performance however, for a simple connecting line between the column exit and mass spectrometer has a yield of 100% but is not a very useful interface device. The second performance term, enrichment, N, provides a measure of the sample concentration accomplished by passage of the effluent stream through the separator.

$$N = \frac{\text{amount of sample to MS}}{\text{amount of carrier to MS}} \Bigg/ \frac{\text{amount of sample to separator}}{\text{amount of carrier to separator}}$$

Using the definition of Y above, Grayson and Wolf expressed the relationship as:

$$N = Y \frac{\text{carrier flow to separator}}{\text{carrier flow to MS}}$$

Neither yield (Y) nor enrichment (N) alone provides a really valid description of an enrichment device's performance capabilities. A high yield does not necessarily mean a useful separator. Similarly, a high enrichment alone does not guarantee good results; the ratio of carrier gas flow entering a separator to carrier flow entering the mass spectrometer could be very high, but if much of the sample is lost with the bulk of the carrier gas the yield might be only a few percent. In such a case, the enrichment value might still be 10–20, a respectable value with most analytical columns. It should be noted that N may be artificially "inflated" by using relatively high flow rate columns. As long as a given separator can tolerate the increased flow without suffering an appreciable drop in Y, N will be larger. The basic question about any separation device is: How much of an injected sample reaches the mass spectrometer's ion source while a sufficiently low pressure is maintained in the ion source region?

4. Interface Designs

A. Jet

Ryhage was the first investigator to report using an enrichment device (46, 48, 49). The separation device, based upon a concept proposed by Becker (4) for isotope separation, is depicted in its two-stage form in Figure 9.1. Two stages are usually used, although a modified single-stage device has been employed with small-bore capillary columns (50). Basically each stage consists of an evacuated chamber connected to a pumping system. The column effluent enters the separator through a very fine jet

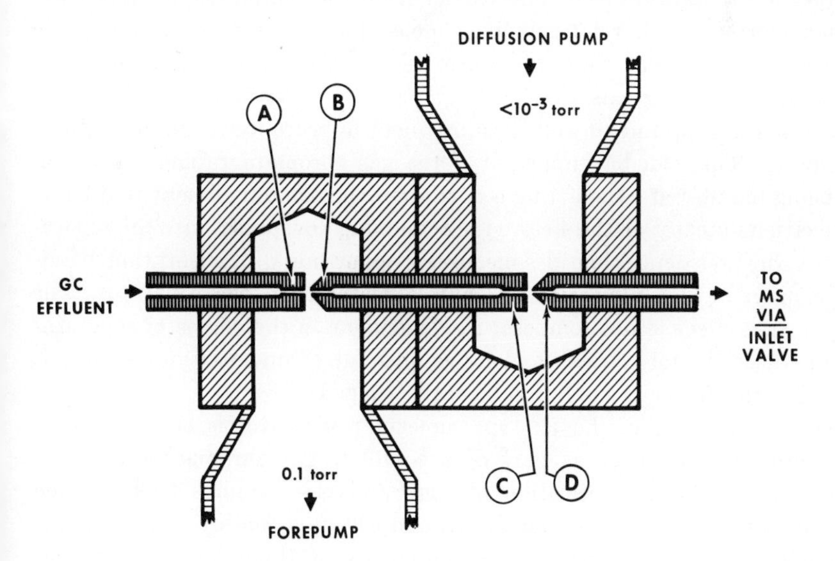

Figure 9.1. Two-stage jet separator.

(approximately 0.004–0.009-in. ID) which is directly aligned with a small exit orifice (0.009–0.012-in. ID) positioned a short distance (0.005–0.020 in.) from the entrance jet (47). In operation, the column effluent enters the separator as a very high speed stream of gas, the entrance jet acting as a restrictor between the exit of the gas chromatographic column and the evacuated separator. The first-stage chamber pressure is maintained at approximately 0.1 torr (160 liters/min forepump). The low-molecular-weight carrier gas (commonly helium) has a higher rate of diffusion than does the higher-molecular-weight sample. Therefore the helium stream tends to diffuse away from the line of sight between the jet, A, and exit orifice, B, much more than does the sample portion of the column effluent stream. The carrier gas molecules which have diffused away from this line of sight are pumped away by the first-stage pumping system, while the bulk of the sample molecules, along with the remaining carrier gas, pass into the exit orifice, then on to the second stage, where the process is repeated at jet C and orifice D. The pressure in this stage is maintained by a diffusion pump (150 liters/sec) at less than 10^{-3} torr.

This type of separator is quite effective in removing helium carrier gas preferentially from the effluent stream. Using methyl palmitate as the model compound, yields of 65–75% and enrichments of 78–90 have been reported.

The jet-type separator has a number of desirable features. It provides very good separation of helium carrier from the sample vapor, especially when working with relatively high-molecular-weight materials. Organic samples have a very short residence time in the separator; therefore no noticeable time lag exists between the time of appearance of the sample band at the column exit and its subsequent appearance in the mass spectrometer. This can be important if the gas chromatographic separation is being monitored by splitting a portion of the column effluent to a flame ionization detector (FID) before the remaining flow passes into the separator. A short residence time also means less opportunity for adsorption, band-spreading, or decomposition of sample material in the separator. The separation effect is not temperature dependent so the choice of separator temperature is not critical, as long as no sample condensation can occur. Finally, the device is small and basically rugged.

On the debit side, the jet-type molecular separator is currently only available commercially as part of a complete gas chromatograph–mass spectrometer unit. Fabrication of such a device requires sophisticated design detailing, precision machining, and care in assembly because of the critical spacing and alignment requirements of these jets and orifices. The jets are reportedly quite easily plugged with either solid particles or liquid stationary phase. Although samples do have short separator residence times, sample decomposition has been reported at elevated interface temperatures (43).

B. Porous Tube

This interface device was originally developed as a split and enrichment system for use with a high-resolution mass spectrometer (59, 60, 63). This separator, depicted in Figure 9.2, is basically a length of porous, fritted-glass tubing (approximately 1-μ pore size) with restrictors at each end. A continuously-pumped glass envelope surrounds the fritted tube. The chromatographic column is connected directly to the separator entrance.

The preferential removal of carrier gas from the fritted tube depends on the faster rate of effusion of the low-molecular-weight carrier gas through the small 1-μ pores of the frit. To establish suitable effusion conditions within the porous tube, the envelope pressure should be held as low as 0.1 to several torr (59). The fixed-glass restrictors of the original separator design are usually replaced by a length of metal capillary tubing on the inlet side and a metering needle valve downstream (19). The latter is especially useful; solvent bands may be kept out of the mass spectrometer by throttling the flow, and the mass spectrometer inlet flow may be adjusted for optimum separator performance when different types of columns are attached.

Some variation in dimensions is reported by different authors, but usually an 8-in. length of porous glass tube is used, 7–8-mm OD by 4–5-mm ID. The separator is commercially available in these dimensions from various suppliers. On examination of a batch of fritted tubes from the major supplier, it was found that the inner bore of many samples was displaced considerably toward one side of the tube so that the wall thickness varied

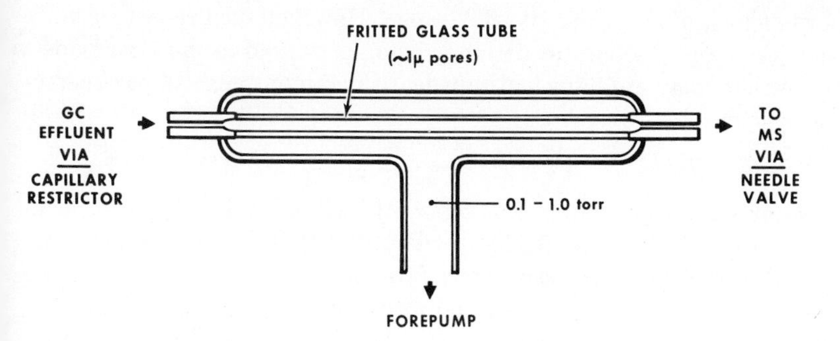

Figure 9.2. Porous tube separator.

by as much as a factor of two around the circumference. It is not known whether this has any deleterious effect upon the separator performance. Optimum dimensions are often arrived at by the "healthy guess" method; they depend in part on the range of column flows likely to be used with a particular separator. Common flow rates are 5–20 ml/min. Ten Noever de Brauw et al. (54) successfully coupled two separators in series when working with higher flow rate columns.

Kreuger and McCloskey (28) have described an attractive modification of the porous-tube design, using porous stainless steel tubing which eliminates the breakage problems of an all-glass design. One version is designed to be inserted through a mass spectrometer solid inlet lock; this is a very desirable arrangement if a researcher has only occasional access to the mass spectrometer.

The glass enrichment device is by far the most commonly employed because it is easily fabricated, easily modified, relatively inert and, most important, is the type supplied by most mass spectrometer manufacturers. The device works well with a wide range of samples; and the enrichment process does not appear to be appreciably temperature dependent, as long as no sample condensation occurs. With isononane, the yield at a given flow rate only drops from about 65 to 42% over a temperature range of 75–300°C (19). There is no appreciable time lag in the separator (less than 1 sec); and band spreading is not usually a problem, particularly with relatively low resolution columns, where minor smearing would not be detectable.

Several disadvantages are that the glass system is fragile, so stresses must be avoided when attaching metal plumbing to both ends of the tubes; and very polar compounds do tend to be adsorbed on the interior surfaces of the separator, but this problem can be largely surmounted by thorough silanization of all surfaces, followed by periodic retreatment. The major complaint of most authors about the porous-tube design is that its yield is somewhat low, often in the 10–30% range. However such reports usually appear when a new separator design is being presented in the same paper, so the authors may not have had sufficient time to optimize all parameters of the system.

C. *Teflon Tube*

An interesting third design was reported in 1966 by Lipsky et al., who mounted a fine Teflon FEP 0.020-in. OD × 0.010-in. ID × 7-ft long tube in the interface shown in schematic form in Figure 9.3. (25, 31). The geometry of this separator is essentially the same as that of the fritted or porous tube interface. The permeable tube is enclosed by an evacuated jacket (stainless steel tee fitting), and restrictors are used at each end of

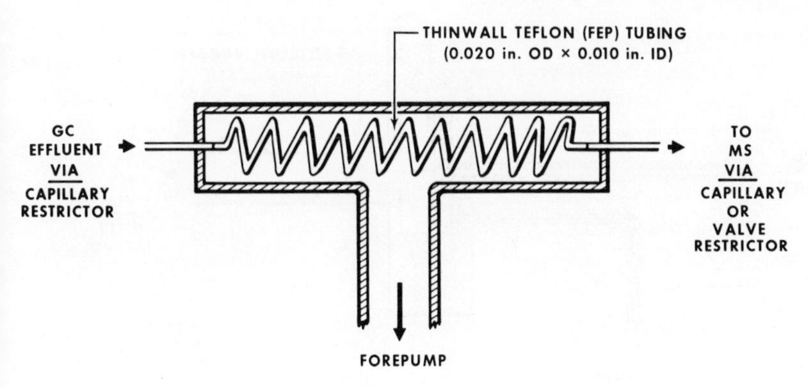

Figure 9.3. Teflon tube separator.

the device to regulate the flow. The column effluent enters the separator, which is maintained at approximately 270–280°C, through the upstream restrictor. The helium carrier gas preferentially passes through the hot Teflon walls, leaving the bulk of the sample and remaining carrier to enter the mass spectrometer through a downstream restrictor. The operating temperature is quite critical. At approximately 320°C the total helium and sample flow is lost through the tubing walls; at approximately 220°C, no sample is lost, but too much helium is also retained. The "melting point" of FEP Teflon is reported to be approximately 290°C; actually, the polymer does not liquefy in this region but changes from a translucent solid to a clear, more flexible material.

The Teflon tube-type separator apparently produces fairly good yields. The device is compact, rugged, and operates at high temperatures. The narrow operating temperature range does not result in any particular problem, unless the sample is unstable at such temperatures. The major drawback of this interface is the long residence time and smearing of sample bands within the Teflon tube. This severely limits the usefulness of this device. Grayson et al. (19) and Blumer (10) report 20–30-sec time lags between entrance of a sample band and its subsequent appearance at the mass spectrometer. Considerable smearing and peak distortion, which becomes worse at higher flow rates, occurs in this time interval.

D. Membrane

Llewellyn and Littlejohn examined the possible use of a selective membrane for separating sample vapor from carrier gas (32). Figure 9.4 is a diagram of a single-stage version of their separator based upon this concept. It basically consists of a small sheet of thin (0.001 in.) polydimethyl-

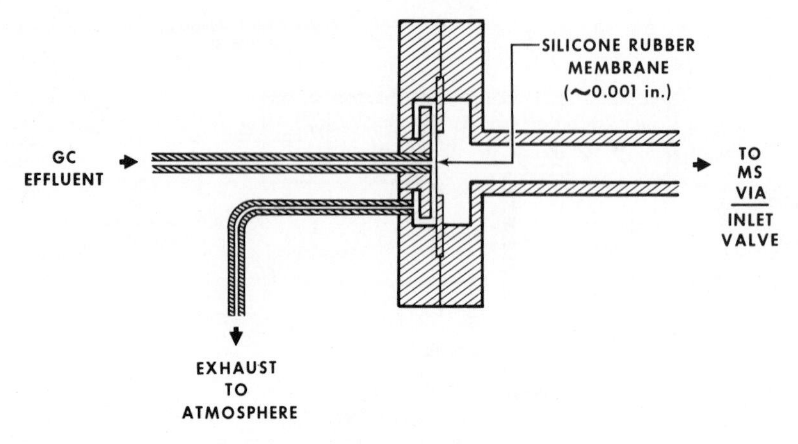

Figure 9.4. Single-stage membrane separator.

siloxane silicone rubber separating the gas chromatographic effluent at 1 atm from a high vacuum line leading to the mass spectrometer ion source. The separator body is designed to spread the effluent stream in a thin sheet across the membrane face for maximum sample exposure to the polymer. At the optimum operating temperature organic vapors preferentially dissolve in the membrane surface, diffuse through the film to the low pressure side and reevaporate into the high vacuum. The sample-depleted carrier stream is then vented to the atmosphere. Permanent gases such as helium are only slightly soluble in such membranes. The separator was originally conceived as a two-stage device, but this design is rather unfavorable for general use. There is a small but finite time required for organic molecules to permeate each membrane, so the two-stage device tends to cause appreciable band tailing. In addition, the overall yield is the product of the yield at each stage, so some sample is wasted unnecessarily. Most mass spectrometers can tolerate the helium flow through a single-stage device without difficulty.

A similar device of somewhat different geometry, but based upon the concept of Llewellyn and Littlejohn, has been constructed in this author's laboratory for use with a quadrupole-type mass spectrometer fitted with a 285 liter/sec diffusion pump. This interface also uses a polydimethylsiloxane polymer membrane (General Electric Company, Schenectady, N.Y. 12305), with an area of 2.5 cm^2. In order to evaluate the performance of this device, 0.040 μl of a test mixture composed of five components (*n*-hexanol, *n*-pentyl acetate, octanal, *n*-decane, and limonene) with boiling points ranging from 149 to 178°C was injected onto a 1000-ft $\times$

0.030-in. ID open tubular column (methyl silicone OV-101) maintained at 125°C. With a column exit flow of 11 ml/min and a separator temperature of 135°C, yields varied from 80 to 95%, enrichment from 19 to 22. The mass spectrometer pressure was maintained at 6×10^{-6} torr (3×10^{-5} torr corrected for helium).

A somewhat different membrane was examined by Black et al. (9) who painted a dilute solution of a silicone rubber polymer on a thin, porous, sintered silver sheet. Yields of 30–50% were obtained with a 5-cm^2 membrane when the 5-component test mixture described above was used.

Recently a report of an all-glass device appeared with a geometry quite similar to that described by Black (22). This separator does not require any accurate machining work in its construction. The polydimethylsiloxane membrane, 0.005-in. thick, is merely clamped between the two halves of the device, and the edge is sealed with silicone rubber adhesive.

The membrane separator is quite small in size, probably potentially the smallest of all interfaces, when all related plumbing and pumping systems of the other devices are considered. The commercially available version is only 1.3-in. OD. A unique feature is that the column exit is always at 1 atm, no matter how much the column flow rate is varied (at least below 50–100 ml/min). This means that gas chromatographic operating conditions need not to be changed at all when switching from a FID or TC detector to a GC–MS arrangement. This is particularly advantageous for the researcher whose access to a mass spectrometer is limited.

One of the main disadvantages of the membrane concept is that its performance is temperature dependent. At low temperatures, the yield is nearly quantitative; but considerable tailing of the bands occurs. As the temperature is increased, the yield observed with a given compound gradually drops off; but the diffusion rate through the membrane increases, reducing the tailing. With a wide boiling range mixture, the separator can be temperature programmed (22); or the temperature can be set at an optimum value for the higher boiling material; usually sufficient low-boiling material still permeates the membrane. The device should not be operated above 300°C, but for maximum membrane life (6–12 months) the temperature should be kept below 225–250°C. These membranes are reportedly also susceptible to damage by large concentrations of sorbed material (45).

Relatively little published information is available on the effectiveness of membrane-type separators with a wide range of sample volatilities and functionalities. Most reports have dealt with alkanes, esters, alcohols, terpenoids, and sesquiterpene hydrocarbons (9, 17, 22); but Hawes did mention success with chlorinated biphenyls and organophosphates (22).

Green (21) has indicated that he has successfully tested this type of interface with a wide range of compounds.

There appears to be an inconsistency between the manner of operation of the silicone rubber membrane separator and the Teflon-tube design. Both are apparently polymer membrane devices, but in the Teflon separator the permanent gas (helium) passes through the membrane more or less preferentially (depending on the temperature). In contrast, the silicone rubber film passes organic molecules preferentially.

Passage of a condensable vapor through a permeable membrane requires five steps: condensation on the membrane surface, solution in the membrane, diffusion through the membrane, dissolution at the opposite face, and evaporation from the surface (57). A permanent gas (a gas above its critical temperature) bypasses the first and last step. The diffusion step is quite rapid for both permanent gases and condensable organic vapors in silicone rubber; the big difference between the two is apparently the much lower solubility of permanent gases in the polymer (2, 3). This is also true with most other polymer films. The decrease in permeability of dimethyl silicone rubber polymer to organic compounds with increasing temperature is discussed by Barrer et al. (2). Teflon (FEP or TEF) is apparently a poor solvent for condensable vapors at moderate temperatures so no detectable quantity of sample passes through the tubing walls. The helium through-put is also nil. At high temperatures (Lipsky, 270°C; Grayson and Wolf, 280°C) the helium flow through the wall becomes appreciable, more or less suddenly. If the tubing was actually behaving as a continuous membrane, organic vapors might be expected to pass through the walls even better than the helium, unless the solubility of organic material is still so low that little, if any, sample dissolves in the tube wall. An alternative explanation would be the one implied by Grayson et al. and by Blumer (19, 10); that the tube actually becomes porous as the "melting point" is approached or passed, and the device then performs in an analogous manner to the fritted-tube separator.

A very different kind of membrane separator has been described by Lucero and Haley (33). They propose using a heated palladium-silver alloy membrane with hydrogen as the carrier gas. Hydrogen permeates such an alloy quite easily, but the membrane is impervious to other materials. The device would then selectively remove hydrogen from the carrier stream quite effectively. Unfortunately such an arrangement would also be excellent for hydrogenating and hydrogenolyzing unsaturated or susceptible functional groups in the samples, such as conjugated dienes, α,β-unsaturated aldehydes, ketones, and nitriles. However this interface does not catalyze the hydrogenation of many kinds of potentially susceptible functional groups; when hydrogenation does occur, the reaction is nearly quantitative (53). This separator design otherwise shows excellent efficiencies and a quantitative delivery of most sample substances. If hydrogenation is desired, this can be more easily accomplished by using a microhydrogenator (5, 6) and a more conventional unreactive separator between the column exit and the inlet to the mass spectrometer.

E. Miscellaneous Designs

Several interface designs have appeared in the recent literature or have become known by word of mouth. They are mostly devices incorporating one or more of the design concepts outlined above (20). Cree, for example, described a device which is apparently designed to operate on the effusion principle of the porous tube separator (12) and also incorporated elements of the jet separator as well. The unit consists of a disk-shaped cavity enclosed by a thick copper ring with thin porous silver walls brazed to each face. The region outside each of the silver walls is kept under vacuum. Entrance and exit lines are short lengths of capillary tubing, directed toward one another through opposite edges of the copper ring. The porous wall area (approximately 28–30 cm^2) is within the same range as that of a conventional fritted-tube separator.

Cree reports a range of enrichments from about 3 for acetone to 150 for 1,2-dibromopropane. This separator design is presently offered by the General Electric Company as an accessory for their monopole mass spectrometer.

Blumer (10) recently reported his results with a simply constructed interface which he assembled from readily available parts. His separator is basically a stainless steel tee fitting, with the gas chromatograph effluent entering at one arm, and with a line to the mass spectrometer via a metering needle valve connected to the other arm. A mechanical forepump is attached to the leg of the tee, which has a small porous sintered silver disk fitted across its bore. Blumer considers the device to be a variation of the effusion-type fritted-tube separator, with the porous wall area drastically reduced to an area of only 0.096 cm^2 (pore size = 3 μ average). In operation, the needle valve is adjusted to establish a flow of 0.1–0.3 ml/min (1 atm) into the mass spectrometer. The remaining column effluent flow is pumped away through the tee leg. This device has been used in the temperature range of 100–250°C, but it should be operable considerably beyond this range. The Teflon gaskets used with the porous silver disk are the temperature limiting features, but these could be replaced with gold gaskets for high-temperature use. It is somewhat surprising that the pumping rate through the small-area porous silver disk is great enough to maintain a pressure in the tee sufficiently low for molecular flow to exist, assuming that enrichment occurs by an effusion process (63). On the basis of somewhat limited data, the separator appears to work quite well. No time lag or band spreading was observed, using 0.055-in. ID packed columns, except when the needle valve to the ion source was nearly closed. Although this particular device performs best at flows of approximately 10 ml/min, larger tees and disks of different porosities can easily be substituted. An interface assembly incorporating this design is now available from Bendix Corporation (34).

Most mass spectrometer manufacturers offer a fritted-glass separator as an accessory; it works fairly well, most potential customers are familiar with the design, and there are apparently no patent complications. The LKB gas chromatograph–mass spectrometer is supplied with a jet separator (LKB Instruments, Inc., Rocksville, Md. 20852); General Electric offers the Cree-type, and Varian markets the membrane-type device. Several other concerns offer enriching devices which bear some superficial resemblance to a jet device. One, of metal, has capillaries inserted in the two arms of a stainless steel tee fitting, with a vacuum line attached at the tee leg. The two capillary ends are closely spaced and roughly aligned. No performance data are available; if any enrichment occurs, it is probably best at low column flows, when the vacuum pump can reduce the pressure in the tee sufficiently for molecular flow to occur. Any combination of capillary ID, spacing, and bore alignments which might provide jet-type enrichment would likely be fortuitous (and momentary), because of the distorting effects of uneven heating and system vibration. The same problems might be expected with a system fabricated of glass.

5. Separator Selection

When working with a simple mixture of hydrocarbons or relatively nonpolar compounds with boiling points (at 1 atm) below 200–250°C, most of the interfaces will perform adequately. If, on the other hand, the mixture tends to be complex, having wide ranges of component concentrations, polarities, and volatilities, then a separator exhibiting uniformly high yield, good enrichment, minimal band-spreading, and no significant adsorption is desirable. Unfortunately, the data available for the various separator designs are not sufficiently complete to permit an unambiguous selection.

The yields provided by the jet, porous tube, Teflon, and membrane separators appear to be adequate, with all of them performing within a factor of approximately 2–8 of one another. In general, the membrane-type probably exhibits the best overall yield, followed by the Teflon tube, and with the porous tube and jet-type following close behind. Yield is more important in some fields such as natural products work than in others, so its "weight" in a selection depends upon the particular use of the GC–MS combination.

All of the major designs can accommodate effluent flows of 20–25 ml/min without significant yield losses. Since the optimum flows of the most useful gas chromatographic columns for GC–MS work are no greater than 25 ml/min, all of the interfaces discussed should be adequate in this respect. The porous tube type performs satisfactorily at 4–5 ml/min (10, 19). The silicone rubber membrane device is quite flow insensitive; the commercial 0.45-cm^2 version performs best from 1 to 20 ml/min but reportedly tolerates flows to 100 ml/min. Performance of the Teflon tube separator is reportedly best at low flow rates (5 ml/min), but is it still quite good at flows up to approximately 20 ml/min.

Any interface will tend to degrade gas chromatographic band tightness if "dead" pockets are present in the connecting lines, including the high vacuum line to the mass spectrometer ion source (29). The higher the system pressure at some point in the interface system, the more likely that a pocket at that point will cause band spreading.

Adsorption is the second major cause of band spreading; it is potentially a problem with any interface design, especially with polar compounds such as alcohols, phenols, and free acids. This is usually counteracted by thoroughly cleaning all surfaces, then treating the interior with a silanizing agent. Usually, the deactivated surface must be periodically retreated to maintain its passivity.

6. System Sensitivity

Components are rarely present in equal amounts in such mixtures; usually a considerable range of concentrations is encountered. Those constituents present in higher concentration present no particular problem, but very often the researcher is at least as interested in the minor components. Here one rapidly becomes aware of the sensitivity limits of gas chromatography–mass spectrometry. The basic question is then: How small an amount of a compound present in a complex mixture will yield a potentially interpretable mass spectrum?

Much of the discussion of instrument sensitivity in the literature expresses a mass spectrometer's sample requirements in terms of fractions of a gram. This is more valid when considering the batch introduction of a sample via a liquid inlet or solid probe, particularly when spectrum recording is on a photoplate, in which case no "scanning" is needed; the ions are dispersed and continuously strike appropriate positions on the photoplate. In contrast, mass spectra must be taken "on the fly" in a gas chromatograph–mass spectrometer system. Therefore, the *rate* of sample introduction in g/sec during the mass spectrometer's scan interval is really the important parameter. Ideally a gas chromatograph would deliver all of a given component to the mass spectrometer during its scan interval at a constant rate (g/sec). Unfortunately gas chromatographic peaks do not have vertical sides and flat tops, but instead roughly approximate a Gaussian distribution. Therefore several parameters must be balanced off to get maximum sensitivity with a given quantity of sample in a gas chromatographic band. If scanning started with the appearance of the peak, the mass spectrum would show very low fragment ion intensities at one end and quite intense ones at the other end, relative to a mass spectrum of the same material recorded at a constant introduction rate. On the other hand, if the scan interval is very short, one cannot introduce a sufficient time constant into the mass spectrometer's output circuitry for adequate noise filtering so the smaller fragment peaks of weak mass spectra become difficult to distinguish from baseline noise. In general, with 0.02–0.03-in. ID open tubular column separations, scan times (m/e = 10–300) of 1–2 sec are sufficiently short; with SCOT and high-efficiency, small-bore packed columns, 2–4 sec is adequate with a low–medium resolution instrument.

An additional problem should be mentioned; very fast scan mass spectral acquisition is notably lacking in precision. Even assuming that the total frequency response of a mass spectrometer's electronics is sufficiently fast to permit rapid recording of a spectrum (e.g., m/e 10–500 in 1 sec) without smearing or distorting the data, statistical variations in the ion stream intensity striking the mass spectrometer's detector may cause a significant error in intensity measurement at a given m/e (35). Therefore the scan rate employed should be no faster than necessary to avoid significant skewing of the spectrum because of changing sample concentration in the column effluent stream.

Sample consumption rates of approximately 10^{-8}–10^{-10} g/sec during a scan interval usually will provide a useful mass spectrum. The actual minimum rate required depends upon the actual mass spectrometer used and upon the nature of the sample, in particular its molecular weight and fragmentation pattern. McFadden (35), for example, was able to identify methyl chloride from the mass spectrum of a peak containing an estimated 10^{-11} g of material because of its very characteristic and simple spectrum. In contrast, mass spectrometric identification of highly branched hydrocarbons or terpenoid materials might require considerably more sample. A number of alternatives have already been described to increase the amount of a component entering the mass spectrometer. In this respect one should also consider the preliminary removal of the major constitutents by preparative gas chromatography, especially in the case of trace analysis.

In qualitative terms a fast-scanning GC–MS system is roughly as sensitive as a "dead" chromatographic FID detector and perhaps an order of magnitude more sensitive than a good thermistor (TC) detector operated at its limit of sensitivity at 150–175°C.

7. Operating Modes and Data-Handling Methods

The column effluent is monitored by using the gas chromatograph's detector. An FID requires only a small proportion of the effluent flow, but the splitting device used must be linear. A TC detector is simplest to use, but it is not very sensitive or linear in response. Sample vapors passed through it are also exposed to hot filaments. A second possible source of a gas chromatographic signal can be obtained from either a second ion source operated below the ionization energy threshold of helium or by the use of a total ion current monitor. The latter intercepts a portion of the total ion beam and converts this current to a voltage output, which may be displayed on a potentiometric recorder. A chromatogram obtained via the mass spectrometer is of considerable value, in combination with the gas chromatographic detector output, for evaluating the degree of band spreading and sample adsorption in an interface system. Because a finite time interval can be established between the arrival of a sample band at the gas chromatographic detector and its subsequent arrival at the mass spectrometer, the FID or TC output can provide a preview of the mixture separation. This is particularly useful when a mass scan is desired near the top of the peak. If a total ion current monitor or second ion source output is not available, the mass spectrometer can be set to scan repetitively; and the mass spectra generated are observed on an oscilloscope. An oscillographic record is then made when the observed mass spectrum reaches a maximum. Some record of the gas chromatographic separation is very desirable so that the recorded mass spectra may be correlated with the chromatographic run.

At this point in the typical GC–MS run relatively little effort has been expended by the investigator. However in the one to two hours required for this run, a prodigious amount of raw data can be generated. As many as several hundred mass spectra might conceivably be recorded during the examination of a complex natural product mixture.

Component identification hopefully follows by comparison of each spectrum with files of reference spectra and by examination of the mass spectral fragmentation pattern. The lack of precision in measuring and recording fast scan mass spectra discussed earlier is one of several factors making comparison of low-resolution mass spectra obtained under different conditions and on different mass spectrometers somewhat uncertain. Not only may one (or both) of the samples examined have been impure, but the different mass spectrometers may give somewhat different mass spectra even with identical pure samples. For example, in comparison with the mass spectrum of a given sample measured with a conventional magnetic deflection mass spectrometer, a quadrupole instrument tends to yield spectra with somewhat reduced ion intensities toward the higher end of a given mass scan range. In contrast, a time-of-flight mass spectrometer often seems to yield a molecular ion of somewhat increased intensity. In addition, magnetic deflection-type instruments from various manufacturers, or even from the same manufacturer, may provide mass spectra of somewhat different appearance. Such differences are usually of a quantitative nature, with the relative intensities of some ions differing. Occasionally an ion fragment may be undetected with one instrument, although another mass spectrometer may show the same fragment as an appreciable peak.

It is obvious that the volume of raw data handling can severely limit the practical usefulness of GC–MS. The steps in this data manipulation, background subtraction, normalization, tabulation, and spectrum comparison are all simple, but tedious—a logical application for a computer. Raw data can be taken from the mass spectrometer and stored on analog (FM) tape (26). Such a system can accept data rapidly, then produce it again at a slower rate for insertion into an off-line computer by running the recorder at a slower tape speed. In the simplest application of such an approach in GC–MS, the operator still monitors the gas chromatographic separation and triggers a mass scan when desired. The only changes from the original method described above is in the data recording and manipulation, subsequent to triggering the mass spectrum scan by hand.

The system may be made considerably more sophisticated by placing the mass spectrometer under greater computer control. In order to do so, a sufficiently high capacity computer must be dedicated to GC–MS use. For example, the mass spectrometer might be triggered to take repetitive scans once every 5 or 10 sec. The data would be fed to the computer for data reduction. Besides simple background subtraction, mass marking, and normalization, the computer might be programmed to extract the individual mass spectra of partially overlapping samples from a combination gas chromatographic band as well. An elegant system has been described by Hites and Biemann (24) for use with a low resolution mass spectrometer. In the near future, unless the cost of suitable computers decreases considerably, it is not likely that most small research groups planning to assemble a gas chromatograph–mass spectrometric combination will rely on computers for much more than off-line handling of raw data in some sort of recorded form, which is of course a considerable improvement in itself. GC–MS is rapidly becoming a common tool in many quite diverse areas of research as more scientists acquire low–medium resolution instruments, which are compatible with a gas chromatograph. In many instances, the necessary computer hardware and software (if it is even available) for full computer control would probably cost as much as the GC–MS system itself. In addition, the task of interfacing and programming the entire system may be too formidable for the neophyte, unless experienced outside help is available. On the other hand, in those situations where the GC–MS system is to be in constant use, computer linkage could increase the usable data output rate considerably.

8. Commercially Available GC–MS Combinations

Most general-purpose mass spectrometers are compatible, or potentially compatible, with a gas chromatograph. An interface of some kind, usually incorporating a fritted tube-type separator, is often offered as an accessory for such instruments. In addition some mass spectrometer manufacturers will supply a gas chromatograph of any make with their instrument, after checking the performance of the package.

At least four instrument manufacturers now offer complete GC–MS systems. These packages vary considerably in both price and versatility, ranging from a relatively simple system at approximately $15,000 to units offering medium resolution, extensive inlet systems, and options which permit their use as general purpose mass spectrometers, at prices of $40,000–$60,000. In most cases the instrument designers have made their GC–MS systems compatible with modern data acquisition methods so that the mass of raw data generated by such systems can be manipulated and refined with a minimum of effort.

Two of the least expensive systems are offered by Finnegan (Finnegan Instruments Corporation, Palo Alto, Calif.). Each incorporates a quadrupole-type mass spectrometer, a Varian Aerograph (Varian Aerograph, Walnut Creek, Calif.) gas chromatograph, and either a simple split or a glass "orifice-type" separator. The simpler of the two units, the Model 3000, is limited to GC–MS use. However it has a m/e range of 10–500; and the company claims a resolution of better than 500 (using the ½ peak height definition). The instrument costs approximately $15,000. The Model 1015 combination is approximately $8,000–$10,000 more expensive but includes a more generally useful mass spectrometer, with a batch liquid inlet and a solid probe. In its standard form, the instrument has a m/e range from 1 to 750, with a resolution of 1000 at m/e 500 (again using the ½ peak height definition for resolution). The operating parameters of a quadrupole-type instrument make it easily adaptable to computer control so Finnegan offers fairly comprehensive systems for both instrument control and data acquisition and reduction.

Bendix has combined one of their time-of-flight (TOF) mass spectrometers with a gas chromatograph of their manufacture, using a separator interface, to provide a third GC–MS system costing $18,000–$20,000. TOF mass spectrometers are quite amenable to use in GC–MS systems; unfortunately the particular spectrometer selected for use in the present combination offers unit resolution of approximately 150 (however, this value is expressed in terms of the more stringent 10% valley definition for resolution). In some laboratory applications, this rather low specification would tend to limit the usefulness of the Bendix system. The instrument does scan to m/e 500. Various batch inlets are available, making the system useful with lower-molecular-weight batch samples.

Perkin-Elmer's (Perkin-Elmer Corporation, Norwalk, Connecticut) Model 270 ($40,000–$50,000) is based upon a Nier-Johnson double-focusing mass spectrometer and incorporates a built-in modified version of their Model 900 gas chromatograph. A fritted-tube separator is used at the interface. A sample probe is included for the introduction of solid or liquid batch samples into the mass spectrometer, and a heated sample inlet chamber is also available. In standard form the instrument has an m/e range of 2–1000 with unit resolution at 850 (10% valley definition). The Model 270 should be useful, both as an integrated GC–MS unit and as a general purpose laboratory mass spectrometer.

LKB (U.S. subsidiary—LKB Instruments, Inc., Rockville, Maryland) offers what is probably the best known of the GC–MS systems, their Model 9000. At $50,000–$65,000, depending upon choice of options, this unit is considerably more expensive than the others described above; but it provides both wide range (2–2000 maximum) and moderate resolution (1200 standard, 5000 maximum, 10% valley definition) in a very versatile combination. This instrument includes a Ryhage jet-type molecular separator, a direct probe, and a heated inlet for volatile batch sample introduction. Numerous options, including a peak matcher unit and a high speed data acquisition system, increase the usefulness of the instrument.

A new integrated GC–MS system, the MAT 111, has recently been announced by Varian/MAT (Palo Alto, Calif.). It incorporates a single-column gas chromatograph with linear temperature programmer and a small sector-type mass spectrometer having a resolution of 600 (standard; 10% valley). Because the instrument package has been designed specifically for GC–MS use, its operation has been made as simple as possible. A direct inlet is also included, making it a more generally useful mass spectrometer. One of the interesting features of the system is the interface, which operates on the effusion principle of the fritted tube-type separator. Instead of a frit it substitutes a knife edge, carefully positioned a short distance from a glass flat. The resulting fine slit constitutes the region through which the carrier gas preferentially effuses. A membrane-type separator is reportedly in preparation. The mass spectrometer has a double ion source operating at 20 and 80 eV to provide both a continuous gas chromatogram (without ionizing helium) and mass spectra. Visual monitoring of the mass spectra is provided on a panel-mounted oscilloscope. Mass scan outputs are linear, with scan rates up to 300 mass units/sec. The combination is presently being offered at approximately $20,000.

9. Specialized GC–MS Combinations

Figure 9.5 shows an interesting but rather specialized noncommercial GC–MS combination being developed at the Jet Propulsion Laboratory, California Institute of Technology (52). This system, part of the projected payload for the 1975 unmanned Viking probe to Mars, is designed to provide both qualitative and quantitative information about the planet's atmosphere. A short, small-bore packed column provides a partial separation of atmospheric gases; the separation need only be sufficient to resolve

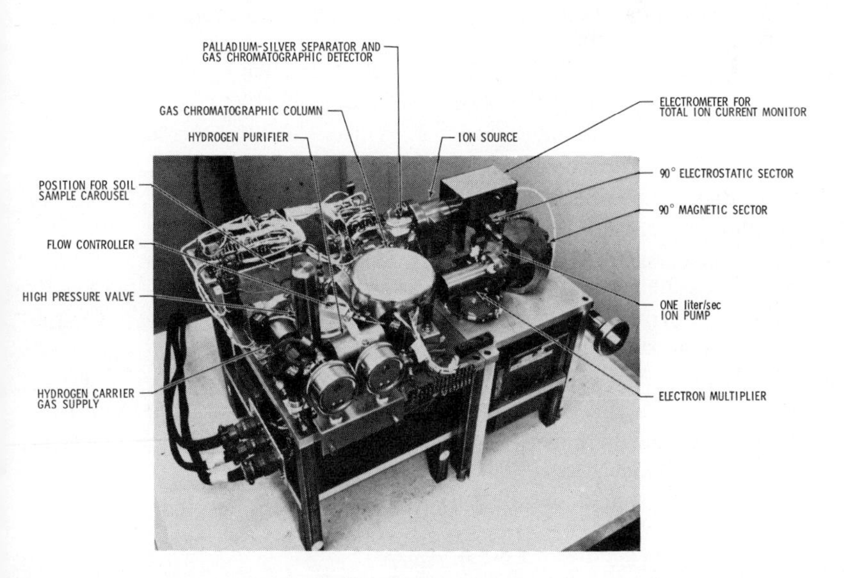

Figure 9.5. Prototype GC–MS system for planetary atmosphere analysis. Under development for the Viking mission to Mars by NASA's JPL (courtesy Dr. Peter G. Simmonds, Jet Propulsion Laboratory).

those gases whose low resolution mass spectra would interfere with one another (for example, CO and N_2). The column effluent then passes into a heated palladium–silver alloy-type separator (33, 53) which effectively removes the hydrogen carrier gas from the effluent. The concentrated effluent then passes into a small double-focusing mass spectrometer for analysis. Probably the most noteworthy features of the system, other than its very compact design, is the choice of interface. In the space probe application, no organic compounds are expected so hydrogenation effects possible with a Pd–Ag surface in a hydrogen atmosphere should be no problem. Of the gases likely to be present in the planet's atmosphere, only oxygen is affected by the separator, being converted quantitatively to water. The estimated weight of the assembly is 23 lb. The size of the unit is $14\frac{1}{4} \times 10 \times 9\frac{1}{2}$ in.

REFERENCES

1. J. W. Amy, E. M. Chait, W. E. Baitinger, and F. W. McLafferty, *Anal. Chem.* **37,** 1265 (1965).
2. R. M. Barrer, J. A. Barrie, and N. K. Raman, *Polymer* **3,** 595 (1962).
3. R. M. Barrer and H. T. Chio, *J. Polymer Sci., Part C* **10,** 111 (1965).
4. E. W. Becker, *Separation of Isotopes,* H. London, Ed., George Newnes, Ltd., London, 1961, Chapter 9.
5. M. Beroza and B. A. Bierl, *Anal. Chem.,* **39,** 1131 (1967).
6. M. Beroza and R. Sarmiento, *Anal. Chem.* **38,** 1042 (1966).
7. J. H. Beynon, R. A. Saunders, and A. E. Williams, *J. Sci. Instr.* **36,** 375 (1959).
8. B. A. Bierl, M. Beroza, and J. M. Ruth, *J. Gas Chromatogr.* **6,** 286 (1968).
9. D. R. Black, R. A. Flath, and R. Teranishi, *J. Chromatogr. Sci.* **7,** 284 (1969).
10. M. Blumer, *Anal. Chem.* **40,** 1590 (1968).
11. K. R. Burson and C. T. Kenner, *J. Chromatogr. Sci.* **7,** 63 (1969).
12. R. F. Cree, "An Efficient Device for the Removal of Helium from Chromatographic Carrier Streams," presented at the Pittsburg Conference on Analytical Chemistry and Applied Spectroscopy, March 1967.
13. J. N. Damico, N. P. Wong, and J. A. Sphon, *Anal. Chem.* **39,** 1045 (1967).
14. J. A. Dorsey, R. H. Hunt, and M. J. O'Neal, *Anal. Chem.* **35,** 511 (1963).
15. S. Dushman and J. M. Lafferty, *Scientific Foundations of Vacuum Technique,* 2nd ed., Wiley, New York, 1962, p. 324.
16. A. A. Ebert, Jr., *Anal. Chem.* **33,** 1865 (1961).
17. R. A. Flath and R. R. Forrey, *J. Agr. Food Chem.* **18,** 306 (1970).
18. R. S. Gohlke, *Anal. Chem.* **31,** 535 (1959).
19. M. A. Grayson and C. J. Wolf, *Anal. Chem.* **39,** 1438 (1967).
20. M. A. Grayson and C. J. Wolf, *Anal. Chem.* **42,** 426 (1970).

21. D. Green, Varian Analytical Instrument Division, Palo Alto, Calif., private communication, 1969.
22. J. E. Hawes, R. Mallaby, and V. P. Williams, *J. Chromatogr. Sci.* **7,** 690 (1969).
23. D. Henneberg and G. Schomburg, in *Gas Chromatography* **1962**, M. Van Swaay, Ed., Butterworths, London, 1962, p. 202.
24. R. A. Hites and K. Biemann, *Anal. Chem.* **40,** 1217 (1968).
25. C. Horvath and W. J. McMurray, *NASA Report N66-35180*, NASA Scientific and Technical Information Facility, Washington, D.C., June 30, 1966.
26. P. Issenberg, Massachusetts Institute of Technology, Cambridge, Mass., private communication, 1967.
27. P. Issenberg, A. Kobayashi, and T. J. Mysliwy, *Agric. Food Chem.* **17,** 1377 (1969).
28. P. M. Kreuger and J. A. McCloskey, *Anal. Chem.* **41,** 1930 (1969).
29. R. L. Levy, H. D. Gesser, and J. B. Westmore, *J. Chromatogr.* **32,** 740 (1968).
30. L. P. Lindeman and J. L. Annis, *Anal. Chem.* **32,** 1742 (1960).
31. S. R. Lipsky, C. G. Horvath, and W. J. McMurray, *Anal. Chem.* **38,** 1585 (1966).
32. P. M. Llewellyn, and D. P. Littlejohn, "The Separation of Organic Vapors from Carrier Gases," presented at the Pittsburg Conference on Analytical Chemistry and Applied Spectroscopy, February 1966.
33. D. P. Lucero and F. C. Haley, *J. Gas Chromatogr.* **6,** 477 (1968).
34. A. J. Luchti and D. C. Damoth, *American Lab.*, September 1970, p. 33.
35. W. H. McFadden, *Separation Sci.* **1,** 723 (1966).
36. W. H. McFadden and R. Teranishi, *Nature* **200,** 329 (1963a).
37. W. H. McFadden, R. Teranishi, D. R. Black, and J. C. Day, *J. Food Sci.* **28,** 316 (1963b).
38. W. H. McFadden, R. Teranishi, J. Corse, D. R. Black, and T. R. Mon, *J. Chromatogr.* **18,** 10 (1965).
39. W. J. McMurray, B. N. Greene, and S. R. Lipsky, *Anal. Chem.* **38,** 1194 (1966).
40. C. Merritt, M. L. Bazinet, and W. G. Yeomans, *J. Chromatogr. Sci.* **7,** 122 (1969).
41. D. O. Miller, *Anal. Chem.* **35,** 2033 (1963).
42. R. K. Odland, E. Glock, and N. L. Bodenhamer, *J. Chromatogr. Sci.* **7,** 187 (1969).
43. R. L. S. Patterson, *Chem. Ind. (London)* **1969,** 48.
44. J. Roeraade and C. R. Enzell, *Acta Chem. Scand.* **22,** 2380 (1968).
45. C. E. Rogers, in *Engineering Design for Plastics*, E. Baer, Ed., Reinhold, New York, 1964, p. 640.
46. R. Ryhage, *Anal. Chem.* **36,** 759 (1964b).
47. R. Ryhage, *Ark. Kemi* **26,** 305 (1966).
48. R. Ryhage and J. Sjövall, *Biochem. J.* **92,** 2P (1964a).
49. R. Ryhage and E. von Sydow, *Acta Chem. Scand.* **17,** 2025 (1963).

50. R. Ryhage, S. Wikstrom, and G. R. Waller, *Anal. Chem.* **37,** 435 (1965).
51. F. E. Saalfeld, *Ind. Res.* **11,** 58 (1969).
52. P. G. Simmonds, *Amer. Lab.*, October 1970, p. 9.
53. P. G. Simmonds, G. R. Shoemake, and J. E. Lovelock, *Anal. Chem.* **42,** 881 (1970).
54. M. C. Ten Noever de Brauw, and C. Brunnée, *Z. Anal. Chem.* **229,** 321 (1967).
55. R. Teranishi, *J. Dairy Sci.* **52,** 816 (1969).
56. R. Teranishi, R. G. Buttery, W. H. McFadden, T. R. Mon, and J. Wasserman, *Anal. Chem.* **36,** 1509 (1964).
57. S. B. Tuwiner, *Diffusion and Membrane Technology*, Reinhold, New York, 1962, p. 215.
58. P. F. Várade and K. Ettre, *Anal. Chem.* **35,** 410 (1963).
59. J. A. Völlmin, W. Simon, and R. Kaiser, *Z. Anal. Chem.* **229,** 1 (1967).
60. J. T. Watson and K. Biemann, *Anal. Chem.* **36,** 1135 (1964).
61. J. T. Watson and K. Biemann, *Anal. Chem.* **37,** 844 (1965).
62. W. D. Woolley, *Analyst* **94,** 121 (1969).
63. P. D. Zemany, *J. Appl. Phys.* **23,** 924 (1952).

RECOMMENDED READING

F. W. Karasek *Res. Develop.* **20** (8), 34 (1969).

F. A. J. M. Leemans and J. A. McCloskey, *J. Amer. Oil Chemists' Soc.* **44,** 99 (1967).

F. E. Saalfeld, *Ind. Res.* **11,** 58 (1969).

W. S. Updegrove and P. Haug, *American Laboratory*, February 1970, p. 8.

J. T. Watson, in *Ancillary Techniques in Gas Chromatography*, L. S. Ettre and W. H. McFadden, Eds., Interscience, New York, 1969.

DENNIS H. SMITH

University of Bristol
School of Chemistry*
Bristol, England

X. Mass Spectrometry

*Present address: Department of Chemistry, Stanford University, Palo Alto, California 94305

I. INTRODUCTION

Previous chapters have dealt with methods of separation and/or characterization of materials based on their bulk properties. Mass spectrometry provides a means for studying these materials at the molecular level in more detail. Although the technique is basically a structure-identifying tool, it is not as specific for the detection and determination of functional groups as is, for example, IR or NMR spectroscopy. Increasing sophistication in methods of data handling and interpretation will make mass spectrometry more versatile; but, for some time to come, interpretation of results will still depend to a great extent upon chemical knowledge derived from other sources or techniques.

As the term mass spectrometry implies, the technique involves the determination of mass; specifically, it ascertains the mass-to-charge (m/e) ratios of gaseous ions. In mass spectrometry, the sample to be analyzed is first introduced into the instrument source, a high-vacuum chamber where the sample is ionized, usually by bombardment with a beam of electrons. In this process, charged particles are formed which may be elemental, molecular, and/or fragmental in nature. The mass spectrometer then separates these ions according to their m/e ratios and determines the relative abundance of these ions.

Quantitative information can be derived from the relative abundance of each species; qualitative conclusions can be deduced from the observed m/e ratios. From the observed fragments of the sample, it is sometimes possible to reconstruct the structure of the parent molecule(s).

Mass spectrometry, like UV, IR, and NMR spectroscopy, works best with single compounds. When large amounts of other compounds are present, it is in most cases very difficult to do structural elucidation of the unknown based on the mass spectrum alone. Prefractionating techniques are therefore important adjuncts to mass spectrometry. The most successful approach involves the coupling of a gas chromatograph with a mass spectrometer (see Chapter 9).

Besides being an important structure-elucidating tool, mass spectrometers can be used to detect very low levels of specific compounds and elements. Another application of considerable significance is in the very accurate determination of masses. This is important in modern nuclear physics because of the information it can supply on nuclear binding energies and shell structures. Accurate mass measurements also form the basis of sophisticated dating techniques of geological specimens billions of years old.

An important advantage of this technique is its high sensitivity and accuracy. Mass spectrometry is capable of providing more specific information per given amount of material than any other analytical technique. In addition, even with minute amounts of material (10^{-6}–10^{-9} g), this information can be provided in a reproducible and accurate manner. Because of this high sensitivity and accuracy, mass spectrometry is frequently the only method of analysis that can be used in many applications where sample size is limited.

Possible exceptions in the analysis of trace elements are the use of neutron activation and emission or absorption spectroscopy. Compared to mass spectrometry, the two latter techniques are less expensive to acquire and to operate. On the other hand, the mass spectrometer is also able to measure isotopic ratios of the elements, which is not possible with an emission or absorption spectrometer.

Because of the explosive development of this field and the summary nature of these discussions, the author attempts to touch briefly only on the more common types of instruments and applications. The field has diversified to the point where a discussion of all specialized developments is clearly impossible in this presentation. The references cited at the end provide sources for more detailed information on the various topics covered.

II. INSTRUMENTATION

The field of mass spectrometry is so diverse that it is not possible to discuss instrumentation by describing only one representative type of mass spectrometer. For this reason it is preferable to view the mass spectrometer as composed of a set of building blocks. The manner in which these blocks are put together depends on the information sought by the investigator.

The mass spectrometer may be pictured in a highly schematic fashion as in Figure 10.1. The sample introduction system provides a means for admitting the sample to the ion source. In some cases, the introduction system is a portion of the ion source itself; in others it is some external system. The sample is ionized, and the beam of ions which is generated in this process is directed into the mass analyzer, The mass analyzer separates and focuses the ions according to their m/e ratios. The detection system senses the mass-separated ion beams. The output device translates the signal provided by the detection system into an output which can be further interpreted by the investigator.

One contribution to the complexity of the instrumentation is the requirement for moderate-to-high vacuum in the ion source, mass analyzer, and detector. The pressure must be low enough to minimize the possibility of collision of the ions with other ions or molecules in the spectrometer. In

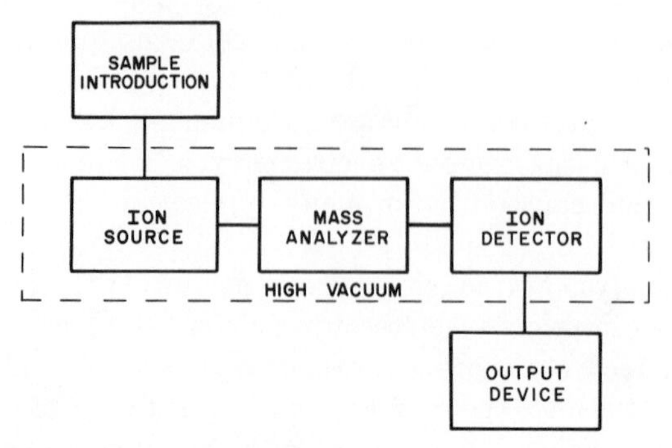

Figure 10.1. Block diagram of a mass spectrometer.

most instances, this means that pressures of the order of 10^{-5} torr or less have to be maintained. Under these conditions, the mean free path is long compared to the distance the ion travels from formation to detection. No attempt will be made to discuss good vacuum techniques, but it is important to point out that cleanliness of the system and choice of suitable pumps are important in order to minimize background interference from foreign material and pumping fluids.

1. Sample Introduction Systems

A. External Sample Introduction Systems

Certain types of ion sources require that the material to be analyzed be in the gaseous state. Systems designed to vaporize samples for these sources are referred to as external sample introduction systems.

a. Glass or Metal Inlet. This type of introduction system, used primarily for organic chemical work, is diagrammed in Figure 10.2. The heated inlet system, operated at moderate vacuum ($\sim 10^{-4}$ torr), is useful for materials with relatively high vapor pressures. It can be fitted to accommodate either gas or liquid samples via a septum or gallium cup/frit inlet. Solid samples are introduced via a probe assembly. The system is constructed of glass if catalytic decomposition of the sample on metal surfaces is anticipated. Stainless steel is otherwise used as the construction material. The system is heated to a temperature which is sufficiently high to vaporize the sample and to ensure rapid pump-out when the analysis is complete. The gaseous sample is admitted to the ion source through a molecular leak to provide a constant flow of sample to the ionization chamber. Sample requirements are generally in the range of $0.1–1.0 \times 10^{-3}$ g, although smaller or greater amounts may be accommodated, depending on the

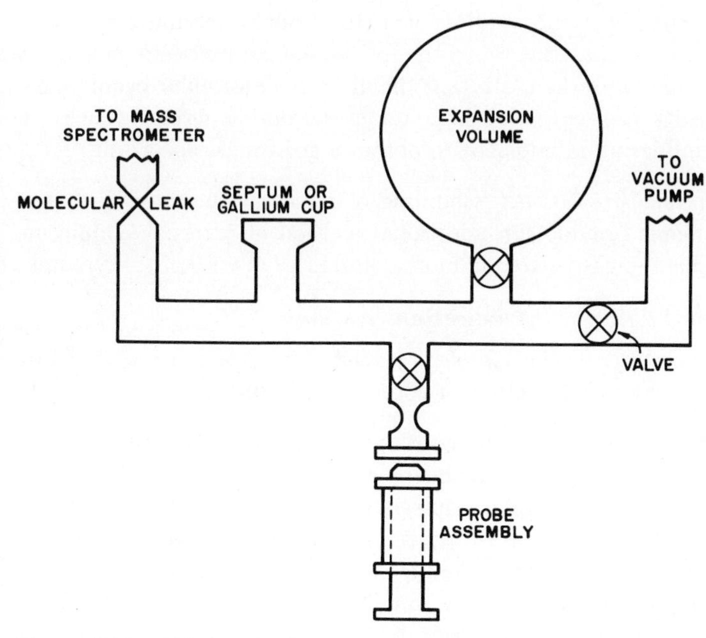

Figure 10.2. Schematic diagram of a heated glass or metal inlet system.

internal volume of the introduction assembly and the size of the molecular leak.

b. Direct Introduction. This system provides for insertion of the sample directly into the ion source. A sliding-seal, high-vacuum probe assembly may be used, with the tip fitting into the ion source. The sample is thus placed very near to the ionization chamber. Depending on the source design however, alternate mechanical arrangements are also possible. The primary goal is to get less volatile materials, which may be subject to thermal decomposition, into the high vacuum in close proximity to the ionization chamber. This approach will minimize decomposition while maximizing sensitivity. Many probe designs possess heating and/or cooling capabilities to attain steady sample evaporation rates. The direct introduction technique is used for many organic and inorganic materials of sufficient volatility at temperatures in the range of 100–600°C. Sample requirements are in the range of 10^{-4}–10^{-9} g, although smaller amounts may be analyzed under favorable circumstances.

c. Knudsen Cell. This method employs a furnace constructed of a suitable refractory material so that the cell can be heated to the high temperatures necessary in studies of many inorganic materials. The furnace is usually mounted close to the ion source, and gaseous molecules are allowed to pass into the ionization chamber as a molecular beam. Sample requirements are variable; but, under favorable circumstances, a fraction of a milligram is sufficient to obtain a good mass spectrum.

d. Gas Chromatograph Inlet. Coupling of a gas chromatograph to the mass spectrometer provides an additional method of external sample introduction. This topic is treated in more detail in Chapter 9 of this volume.

B. Internal Sample Introduction Systems

In these arrangements, the sample is placed within the ionization chamber either as part of the ion source or coated on a filament.

a. Spark Electrode. Many inorganic samples may be analyzed by fabricating spark gap electrodes from the sample itself if it is conducting. For nonconductors, the sample is mixed with some conducting material such as graphite. This method of sample introduction is used mainly for impurity analysis. The required sample size is of the order of 100 mg or more, depending on the construction details of the electrodes.

b. Filament Coating. A separate assembly to vaporize the sample is not required when a thermal ionization source is used. The sample is coated on a filament which, when heated, yields positive ions directly from the solid. This method is particularly valuable in studies of isotopic abundances and can be used with sample sizes of the order of a microgram. With modern, high-sensitivity instruments, it is even possible to use considerably less if careful handling procedures are observed to avoid contamination by foreign material.

2. Ion Sources

Mass analysis generally is based on different deflections of ion beams in various arrangements of magnetic and electric fields. In principle, either positive or negative ions may be analyzed. In practice however, most mass spectrometers are designed to operate with positive ions. This is primarily because positive ion formation is considerably more efficient than comparable negative ion formation processes. Excellent discussions on negative ion formation studies may be found in the work by Field and Franklin (25) and in a chapter by McDowell (39). This chapter only treats those ion sources which are designed for positive ion formation. A major design goal of these sources is the formation of a maximum number of ions for the amount of sample consumed.

A generalized ionization process is shown below. The process requires the absorption of energy by the atom or molecule A.

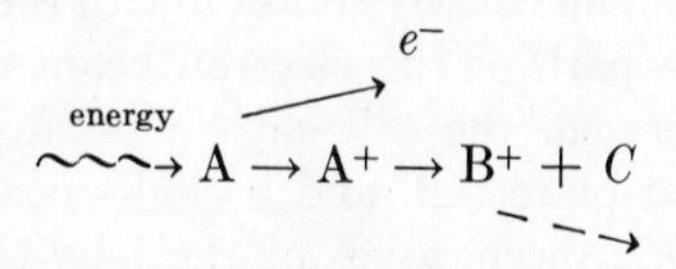

The energy input differs for the various types of ion sources available. Energy transfer may be in the form of electrons, photons, electric fields, electrical discharge, or heat.

When there is sufficient energy to eject an electron from a molecular orbital of A, a positive ion A^+ results. In the case of molecules, sufficient energy may remain in A^+ to cause a bond rupture. This leads to the formation of a new ion B^+ and a fragment C. Ion B^+ may fragment even further. These fragmentation processes are discussed in more detail later in this chapter.

A. Electron Impact Source

This source was developed originally by Bleakney (8) and by Nier (45, 46), but many variations to achieve ion formation by this principle have since been reported. Most current ion sources are electron impact sources.

The general configuration of the electron impact source is shown in Figure 10.3. Electrons are emitted by the heated cathode filament, F, and accelerated by the potential between the filament and the plate, P. This potential difference is generally variable from 0–150 V. The electrons then pass through the ionization chamber to anode, A. A magnetic field parallel to the electron beam is used to collimate the beam. Samples to be analyzed

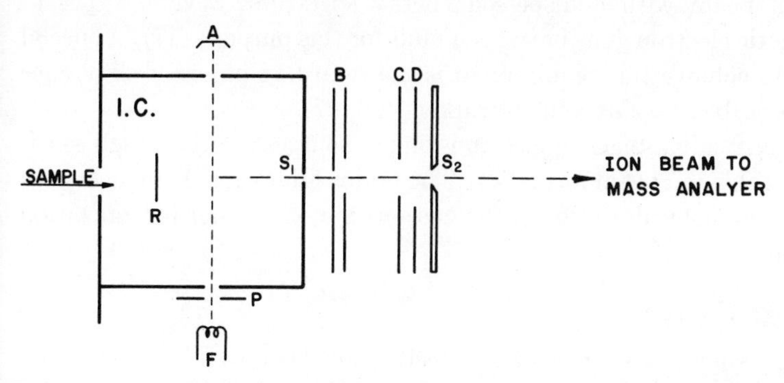

Figure 10.3. Electron impact ion source. See text for description.

are first introduced into the ionization chamber, *IC*, via any of the external methods of sample introduction mentioned previously. Sample molecules in the gas phase are then ionized by collision with these energetic electrons. Ions formed along the path of the electron beam are removed from the ionization chamber through the exit slit S_1, because of field penetration from one or more field plates, *B*, and a slight positive potential (a few volts) on a repeller, *R*. Acceleration of the ions takes place due to the potential between plates *B* and *C* (normally a few thousand volts). There may be one or more plates, *D*, to align and focus the ion beam. The ions finally exit from the source to the mass analyzer through the (generally adjustable) exit slit, S_2.

The electron current is generally controlled by the temperature of the filament, while the electron energy is controlled by the filament potential. The electron current is stabilized by regulation of the anode electron current. Tungsten, rhenium, or thoriated iridium are generally used for filament materials.

Ion sources can be either open or closed in construction. The latter results in a sensitivity about 10 times that obtained with an open source. However the open source can be cleaned out much faster. The necessary pumping speed for this type of source is about 10–20 times higher than that for a closed system. In any case, the openings for the sample inlet, the electron beam, and the ion slit are usually designed to be as small as possible. The potentials involved in the source are adjusted to minimize the energy spread of the resulting ion beam, a consideration that will be discussed when describing mass analyzers.

Many variations on the electron impact source have been reported depending on the particular application. The most important configurations are those sources which are used to obtain information on the energetics of ionization and fragmentation. In this application, it is desirable to obtain an electron beam with a dispersed energy spectrum. Several types of monoenergetic electron guns have been built for this purpose (17). A special technique to achieve this requirement is the retarding potential difference method, described by Fox and co-workers (26, 27).

The primary advantages of electron impact sources are their high sensitivity and stable ion beam formation. The major limitation is the necessity for the sample molecules to be in the gaseous phase in order for ionization to take place.

B. *Spark Source*

The spark source was originally conceived by Dempster (20) as a tool for isotopic studies of the elements. Subsequent work by Dempster (21) and Hannay (29) has led to improvements of the technique for general application to the analysis of solid material (63).

The configuration of a spark source is relatively simple. Samples to be analyzed are formed into either rod- or cylinder-type electrodes. A simple configuration is to form a pair of closely spaced 0.001-in. rods. If the sample is conducting, the rods may be formed from the material itself. If the sample is nonconducting, the sample is mixed with a conductor such as graphite and then formed into a rod. A high-voltage, high-frequency spark is generated between the electrodes. Heating of the electrode tips vaporizes the sample, and ions are formed in the discharge. These ions then drift into an acceleration region that is less complex than that of the electron impact source described earlier. Because of the high energy spread in the beam formed, there is little to be gained by complex focusing systems. In general, a collimating plate is used at the ion entrance, with a second collimating plate and an exit slit at the entrance to the mass analyzer.

A major advantage of this technique is that the ion output of the source for different elements does not differ more than one order of magnitude. This is in contrast to other methods of solid material analysis, such as observed with thermal methods of ionization, in a Knudsen cell or in a gas discharge. This makes the technique particularly attractive for studies on a wide range of solids. In trace element investigations for example, analyses may be carried to the parts per billion level. The major disadvantages are the problems in insulation of electrical leads and feedthroughs at the high voltages necessary, the difficulty in maintaining a relatively steady discharge as the rods burn away, the need for high-speed vacuum pumping in the source region, and the wide ion energy spread, which necessitates a double-focusing mass analyzer.

C. Thermal Ionization

The phenomenon of positive ion formation on heating of solids on a metal filament is called thermal ionization. This method of ionization, which was reported as early as 1918 (19), finds its largest application in isotopic abundance studies of the heavier elements.

This source consists of a filament assembly with one or more filaments, generally of the ribbon or wire type, upon which the sample is placed. The overall design may be patterned after the electron impact source, with the thermal ionization filament assembly occupying the position of the electron beam (33). Different configurations of focusing plates and fields have also been used to meet the special requirements of the method (24).

The number of ions generated in such a source is strongly dependent on the ionization potential of the element being studied. The rate of ion formation from the heated surface of the filament is inversely proportional to the ionization potential of the element. For this reason, the technique has little merit in quantitative studies of mixtures. On the other hand, since the ionization potentials for different isotopes of the same element may be assumed to be the same, thermal ionization is quite applicable to measurements of isotopic ratios. Very accurate isotopic abundance measurements have been reported with this method of ion formation.

The most important considerations in the efficient use of this type of source are in the design of the filament assemblies and in the sample preparation technique. The filament assemblies are generally detachable for convenient cleaning. Multiple-filament sources have been used to increase ion current yields. Ion current yields have also been increased by different methods of sample preparation. The element, in the form of its oxide, is generally placed as a paste on the filament and allowed to dry. Another method is to mix the sample with cement binders or to fabricate a borax bead with the sample, which is then placed around or on the filament.

A prime advantage of this method is the high sensitivity that can be attained. However, several elements, such as carbon and silicon, cannot be processed by this method because of their high ionization potentials. Other disadvantages are the required high pumping speeds at the source and the long-lasting memory effects from previous samples in the system.

D. Photoionization

Although the process of ion formation by photon impact has been known for some time, the impetus to apply this principle to modern research is to a large extent due to the work of Lossing and Tanaka (38) and of Inghram et al. (32, 33). Design of the photoionization source is generally similar to that of the electron impact source, with the photon beam substituting for the electron beam. One arrangement (24) that allows the use of either ionization technique is shown in Figure 10.4. Light produced in a discharge tube filled with an appropriate gas, such as hydrogen, is directed into the source chamber through a window, which is generally made of lithium fluoride. The absorptive properties of lithium fluoride require the sample to be ionized to have an ionization potential below 11.8 eV. The window may be eliminated, but this creates severe vacuum problems in the system. Because much of the research carried out on photoionization involves studies at low energies, the lithium fluoride window technique is generally quite adequate. Ion currents obtainable from such a source are considerably lower than those obtained by electron impact. An advantage of this source is that at the threshold of ionization only a small number of ions are formed. This is in contrast to what is observed with an electron impact source. Operation of electron impact sources at low electron voltages is also much

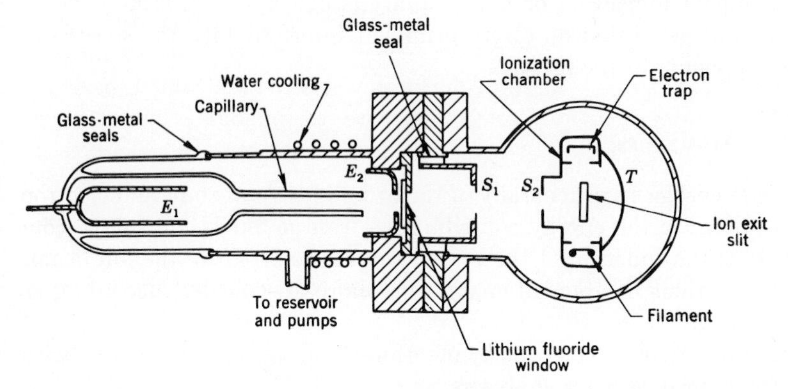

Figure 10.4. Photoionization source: diagram of the arrangement of light source and ionization chamber. The E_1 and E_2 are the discharge electrodes; S_1 and S_2 are collimating slits; a magnetic field prevents the escape of photoelectrons from the "photon trap" T. The auxiliary filament and electron trap may be used for electron impact (27) (courtesy of McGraw-Hill Book Company).

more difficult. This advantage makes photoionization particularly attractive for studies on ionization and appearance potentials.

E. *Field Ionization*

This method of ionization is based on the fact that gaseous molecules are ionized in intense electric fields. Much of the pioneering work in the development of this technique for routine chemical analysis (general organic chemical applications) has been carried out by Block (9) and by Beckey and co-workers (3).

In practice, a sharp edge such as a razor blade, a sharp point, or fine wires are used to create the high fields necessary for ionization to take place. The mechanism of ion formation at these sharp areas is not well understood, and empirical methods of source construction and chemical pretreatment of the field-forming edge or point are used to optimize ion yields. Under the best circumstances, ion currents are considerably lower than observed with electron impact sources. The advantage is that little or no fragmentation of the ions takes place. This results in the formation of an ion beam which is composed almost entirely of intact molecules (molecular ions).

Combination electron impact and field ionization sources can be constructed with relative ease. One such source has been described by Schulze, Simoneit, and Burlingame (51). This allows for the possibility of analyzing the same sample with both methods of ionization. This approach is particularly useful in the analysis of compounds displaying no molecular ions with electron impact ionization or in the analysis of complex mixtures where fragment ions generated on electron impact interfere with the spectra of other components in the mixture.

3. Mass Analyzers

Ion beam energies are generally of the order of several thousand electron volts to minimize the energy contributions from nonideal ion source conditions. It is the function of the mass analyzer to separate the ion beams according to their component masses or, more specifically, according to their m/e ratios.

In general, any of the previously mentioned ion sources can be used with any of the described mass analyzers.

A. Magnetic Analyzers—Single (Direction) Focusing

This method of mass analysis is based on the deflection of the ion beam in a uniform magnetic field. Two geometries are shown in Figure 10.5. The left-hand figure depicts the 180° deflection geometry used by Dempster in one of his early instruments (19). Ions from the source, accelerated through a potential, V, pass through a slit, S_1, into the magnetic field, which is oriented perpendicular to the direction of ion travel. These charged particles pass through the field in a circular orbit with radius r described by the equations below and pass through slit S_2 to the ion detector.

The geometry shown in Figure 10.5*b* was introduced by Nier (45). It is termed a magnetic sector analyzer. The 180° deflection geometry is simply a special case of this sector analyzer geometry. Again, the ions emerge from the source slit, S_1, describe a circular path through the analyzer, and impinge upon the detector through slit S_2.

The equations governing the behavior of the ions in the magnetic field

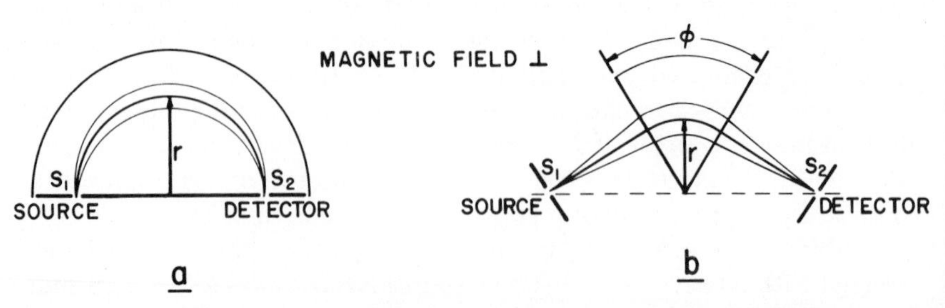

Figure 10.5. (*a*) Dempster's 180° magnetic analyzer; (*b*) Nier's sector analyzer geometry; S_1 and S_2 are slits, *r* is the radius of curvature of the ions in the magnetic field and Φ is the sector angle.

are quite simple. Exiting from the source, the ions have a kinetic energy equal to eV, where e is the electron charge. The energy is equal to $mv^2/2$ assuming that other contributions to ion energies are small; m is the mass of an ion, and v is its resulting velocity. Hence

$$eV = \frac{mv^2}{2} \tag{10.1}$$

All ions of the same charge have the same energy. Ions of different mass have proportionately different velocities. The magnetic induction is given as B. An ion in this field experiences a deflecting force of evB perpendicular to both the magnetic field and the direction of motion of the particle. The result is a circular trajectory of radius r, where the centrifugal force equals the deflecting force.

$$evB = \frac{mv^2}{r} \tag{10.2}$$

or

$$r = \frac{mv}{eB} \tag{10.3}$$

The magnetic sector separates, therefore, according to the *momentum* of the particles. If v is eliminated between Equations 10.1 and 10.3, one obtains

$$r = \left(\frac{2mV}{eB^2}\right)^{1/2} \tag{10.4}$$

This equation determines the radius of the trajectory for ions of various m/e ratios. Assuming V and B to be constant, ions are deflected according to the square root of their mass to charge ratios. Another convenient way of expressing Equation 10.4 is as follows:

$$m/e = \frac{r^2B^2}{2V} \tag{10.5}$$

It may be noted that the two geometries described in Figure 10.5 have detectors at a fixed value of r. This means that under ideal conditions, the m/e ratio of any ion can be determined by measuring the V and B necessary to bring the ion to the detector through slit S_2. By programming the magnitude of either B or V, ions of successively larger (or successively smaller) m/e ratios will impinge on the detector as a function of time.

The most important feature of the sector geometry is its focusing properties. As indicated in Figure 10.5, the ion beam diverges as it leaves the ion exit slit. The magnetic sector refocuses this divergent beam and causes it to converge again at the collector slit. This property is known as direction focusing, and such analyzers are referred to as single- or direction-focusing analyzers. In this configuration, the source slit, the apex of the sector, and the collector slit have to lie on the same line as shown in Figure 10.5*b*. Ion optical theory (30, 36) shows that the angle Φ may have any value as long as the ion beam enters the field at right angles.

It is important at this point to describe the measure of performance of any mass analyzer. This is usually given by an instrumental parameter, called the *resolution*. The resolution is a measure of the ability of the analyzer to separate ion beams of different masses from one another. This is expressed in numerical form through the equation $m/\Delta m$, where m is the mass and Δm is the difference in mass. This equation has no significance unless the degree of resolution is specified. There are many criteria used for resolution, but one of the most common is the 10% valley, or the 5% contribution definition. This states that two peaks of equal height are considered resolved if the height of the valley between them is less than 10% of the peak height. For example, an instrument would have a resolution of 100 if two peaks with a mass difference of 1 part in 100 (e.g., m/e 100 and 101) were resolved to the 10% level. Low-resolution mass spectrometers typically show maximum resolution values between 300–1000. High-resolution mass spectrometers are capable of attaining resolutions well in excess of 10^4. The obvious advantage of a high-resolution mass spectrometer lies in the capability to resolve ions with very small differences in mass. Very little resolution is needed to separate C_4H_7 from C_4H_8, but a very good mass spectrometer is needed to differentiate between C_2H_4 (mass 28.0313) from $C^{13}CH_3$ (mass 28.0268). An example of the same mass spectra obtained at two different levels of resolution is shown in Figure 10.9.

The primary limitation of the single-focusing magnetic analyzer is one of resolution. This can be due to many factors, such as the non-normal beam entry into the field, magnetic field inhomogeneities, and the presence of fringe fields at the entry and exit boundaries. These factors can be overcome to some extent by careful design and construction. One can, for example, attempt to correct for fringe fields by the use of magnetic shunts. Another limitation to resolution that cannot be corrected for in the geometry is the fact that the ion beam does not emerge from the source with uniform energy. Every ion source contributes to some degree to an energy spread of the ion beam. For carefully designed electron impact sources, this spread may be 2–3 eV. Other sources may show an energy distribution of 10–30 eV. This energy spread manifests itself as peak broadening at the collector, thus reducing the attainable resolution. Most single-focusing instruments can attain resolutions of the order of 1000. Some of the latest models can attain resolutions of up to 10,000 but at an extreme decrease in

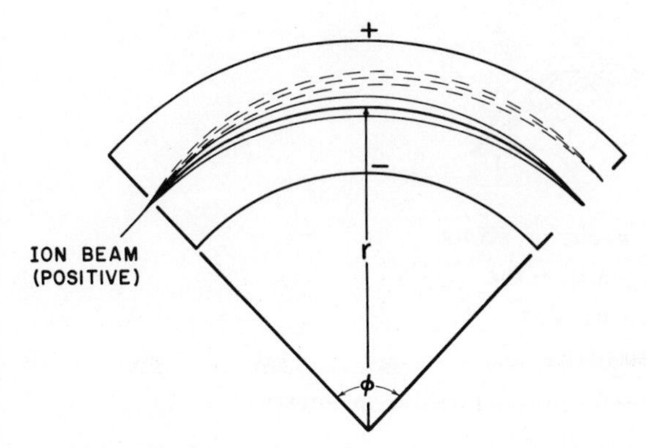

Figure 10.6. Deflection of ions in an electrostatic sector. Paths for ions of two different energies are shown, one with dashed lines; the other with solid lines.

sensitivity due to the narrow slit widths that must be used. These lower resolutions are often unacceptable for many studies. It is for this reason that the double-focusing geometry has been developed.

B. Magnetic Analyzer—Double Focusing

The purpose of a double-focusing geometry is to obtain higher resolutions by correcting for the energy spread of the ion beam. This is accomplished by passing this beam through a cylindrical condenser or electrostatic analyzer, as shown in Figure 10.6. This arrangement acts as an energy filter as may be seen by examining the equations governing the ion beam. As in the magnetic field, the ion beam traverses the condenser in a circular orbit of radius r. The force exerted by the electric field, E, is eE. This is balanced by the centrifugal force, mv^2/r. Hence

$$eE = \frac{mv^2}{r} \quad \text{or} \quad r = \frac{mv^2}{eE} \tag{10.6}$$

Ions are thus deflected according to their *kinetic energy*. Direction focusing is also achieved by this arrangement in the sense that ions of the same energy diverging at the entrance to the sector are refocused. The choice of angle Φ determines whether the focii lie at the field boundary or outside the field. Ions of the same mass but differing slightly in velocity (energy) are thus focused at different points.

Certain specific arrangements of electrostatic and magnetic analyzers which are called *double-focusing* analyzers are capable of refocusing this velocity-dispersed beam at the detector, thus correcting for the energy

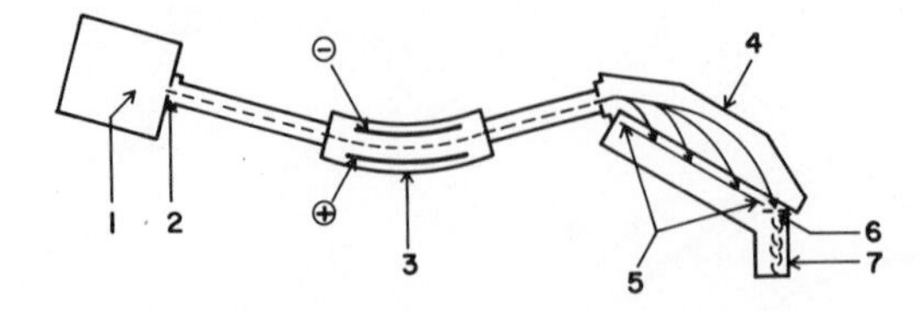

Figure 10.7. Mattauch-Herzog double focusing geometry.

spread in the original ion beam. One such geometry, developed by Mattauch and Herzog (42, 43) is shown in Figure 10.7.

Ions leaving the 31°51′ ($\pi/4\sqrt{2}$ radians) angle electric sector are focused at infinity. A slit or stop arrangement allows ions in a selected energy range to enter the magnetic analyzer, a 90° sector, where m/e separation and direction focusing take place. The condition of double focusing is achieved when the energy dispersion of the ions from the electric sector is exactly compensated for by the direction-focusing conditions in the magnetic sector. In the Mattauch-Herzog geometry, the double-focusing condition is satisfied simultaneously for all masses at the plane indicated in Figure 10.7. A photographic plate detector can be placed in this plane to record all masses under static magnetic field conditions. A rather interesting description of the development of this mass spectrograph has been given recently by Herzog himself (28).

Another double-focusing geometry that is widely used is termed the Nier-Johnson geometry, which is based on the ion-optical theory developed by Johnson and Nier (34). This geometry is pictured schematically in Figure 10.8. This figure shows a 90° electric sector and a 90° magnetic sector, although magnetic sectors of smaller angles may be used. This geometry satisfies the double-focusing conditions only at one point, where an electrical ion detector of some type is used. This configuration has the advantage of being able to achieve second-order direction focusing.

This capability is related to the point source origin of the ion beam. This beam actually enters the analyzer through a slit of finite width, causing a divergence which is dependent on the slit width. The focusing properties of the Nier-Johnson geometry allow for a correction of this second-order effect; the theoretically attainable resolution is thus increased.

Double-focusing mass spectrometers range from $45,000 to $120,000 in price. Routinely attainable resolutions range from 1000 to 40,000, with a mass range capability of more than 2000 mass units.

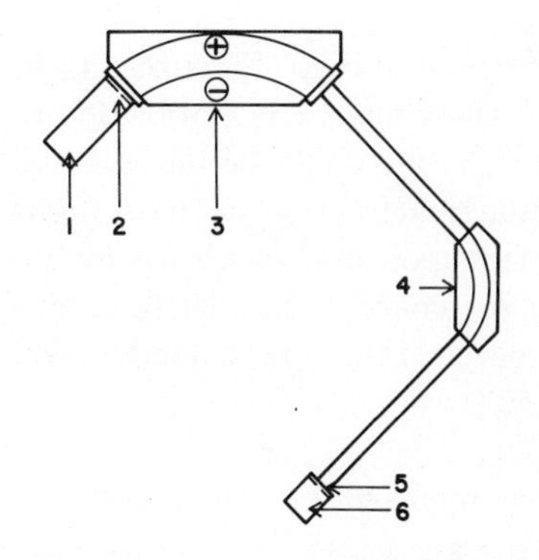

Figure 10.8. Nier-Johnson double focusing geometry.

C. Ion-Cyclotron Resonance Analyzers

This specially constructed analyzer, which was initially developed by Sommer, Thomas, and Hipple (55), is particularly useful for studying ion–molecule reactions since the ions can be maintained in resonance for as long as 100 msec (1). This allows the study of a number of molecular reactions in the resonance region of the analyzer since the quantities of both reactant and product ions can be monitored by this technique.

D. Quadrupole Analyzers

The basic quadrupole analyzer was developed by Paul and co-workers (15, 48). The method is based on mass separation in a quadrupole RF electric field. This field is established on a set of four precision rods, ideally hyperbolic in cross-section. Both a dc voltage and an RF alternating voltage are applied to these electrodes. Conventional instruments use rods of circular cross-section because the rigid requirements in uniformity and alignment present formidable machining problems. These rods are mounted parallel to each other such that the cross-section of the analyzer tube shows the centers of the four circular cross-sections of the rods to be on the four corners of a square. By varying the RF component of the field, ions of various masses inserted into the field perform stable oscillations and pass through the length of the analyzer tube to a detector. At a specific value of the dc voltage and the RF alternating voltage, ions with only a certain m/e ratio can go through the analyzer tube without striking an electrode. The spectrum is obtained by sweeping the applied RF alternating frequency and measuring the detector current as a function of time.

The maximum resolution attainable with this analyzer is around 1000.

At lower resolutions, the transmission is extremely high. This results in high sensitivities. Another advantage is that these analyzers can be miniaturized easily because they require no magnet. They can be designed to dissipate little power, making them particularly attractive in space flight applications (11, 61). A special case of the quadrupole analyzer is the monopole analyzer. Consisting of only one rod, it is somewhat simpler to manufacture; but its performance to date has not matched that of the quadrupole.

E. *Time-of-Flight Analyzers*

Time-of-flight (TOF) analyzers have been reported by many workers. One of the better discussions on the detailed design of these instruments has been presented by Wiley and McLaren (66). Using specially developed electronic switching circuitry, the ion beam from the source is allowed to pass into a field-free drift tube in pulses. This can be accomplished by switching on the electron gun in an electron impact source for a fraction of a microsecond. The gun is switched off, and the ions formed are accelerated out of the ion source by the application of an electric field. This pulse of ions passes into the drift tube and is mass analyzed by the fact that ions of different m/e ratios have different velocities and thus different flight times through the analyzer. The time of flight, t, is proportional to the length of the flight path, l, and to the square root of $m/2eV$,

$$t = l\sqrt{m/2eV} \tag{10.7}$$

From this equation one can compute that for a singly charged ion with a mass of 100, a flight path of 50 cm, and a potential drop of 1000 V, the time of flight is 11.39 μsec. For a singly charged ion with a mass of 101, the flight time is 11.44 μsec. Elaborate electronics are obviously necessary to detect these small differences in flight times. The output of a suitable detector versus time is used to record the mass spectrum.

Because of its capability to analyze ion pulses very rapidly (up to 50,000 spectra/sec), the time-of-flight analyzer is particularly useful in monitoring moderate-to-fast reaction kinetics. Many specialized ion sources have been constructed for this purpose, including those adapted for studies of shock-wave reactions and explosions. One of the major limitations of this analyzer is in its attainable resolution. The observed values are generally below 500. In miniaturized versions, the resolution is frequently even lower. TOF mass spectrometers usually range between $20,000 and $45,000 in price.

4. Ion Detectors

Because of the stringent requirement for sensitivity, only a limited number of ion detection techniques can be used in mass spectrometry. In most spectrometers, the observed ion currents are of the order of 10^{-9} A and less, thus necessitating detectors of high sensitivity and inherently low noise. Three types are discussed below.

A. Electrical Detection

Because of the low ion currents involved, any electrical detector requires some form of current amplification. This necessitates amplifiers of low noise and high gain.

a. Faraday Cup Detector. This detector is, as the name implies, a simple plate or cup placed behind the detector slit to collect the incident ion beam. The ion current collected can be imposed on the grid of an electrometer tube, which may be further connected to one or more stages of amplification. The electrometer tube is extremely noise and vibration sensitive; and considerable care is taken to shock-mount the tube and place it, with careful shielding, as close as possible to the detector. An alternative approach is to use a vibrating-reed electrometer, which is much more rugged and reliable. The dc signal from the detector is applied across a metal reed vibrating close to a static anvil. The resulting ac signal is then amplified.

This method of detection has certain disadvantages. Because of the high current gains needed to provide a suitable signal to a recording device, the band-pass of the system is quite low. This presents difficulties in any work requiring a high band-pass, such as in a time-of-flight spectrometer or, for that matter, in any rapid-scanning instrument. For applications where the instrument sensitivity is high or when there is no requirement for rapid scanning, this detector is particularly suitable because the current measured is not a function of the incident ion beam energy. This dependence is observed if an electron multiplier detector is used. One particular application deserves mention at this time, and that is in the area of isotopic ratio measurements. For the determination of ratios such as $^{12}C/^{13}C$, $^{16}O/^{18}O$, and so on, detection is frequently carried out with a dual-collection system. Two parallel Faraday cup detectors with vibrating reed amplifiers are normally used to eliminate the effects of small variations in the ion beam. Even though the measured ion currents will vary in proportion to variations in beam intensity, the ratio remains constant, ensuring more accurate results.

b. Electron Multiplier Detectors. These detectors have proved to be excellent devices in practice. The ion beam is allowed to impinge upon the cathode of a multiplier, which may be simply the first of a string of dynodes. The geometry of the dynode string is arranged in such a manner that secondary electrons emitted from a dynode will strike the next dynode in the series. This dynode emits more electrons, and a cascade of electrons results. With dynode voltages of 100–200 V/stage in a 15–20 stage multiplier, gains of a factor of 10^6–10^7 can be achieved. The response time of the multiplier is extremely fast; and, because lower gains are required as compared to the Faraday cup, a high band-pass system results. The advantages of the multiplier are thus its high sensitivity and its ability to be used in any mass spectrometer system which requires rapid scanning of the spectra. The multiplier is capable of detecting single ions and is in fact used in systems for ion counting. Conventional multipliers have several disadvantages that should be mentioned. The emitting properties of the beryllium/copper dynodes change with time, resulting in a decrease in sensitivity. The number of secondary electrons emitted from the first dynode is a function of the ion energy. This means that, for those instruments employing voltage rather than magnetic scanning techniques, the sensitivity is lower for higher masses (less energy). The number of electrons emitted from the first dynode is also somewhat dependent on the composition of the impinging ion. This effect is, however, not important for most applications.

B. Photographic Detection

Ion detection on a photoplate was the first generally used method in early mass spectrometry. The Mattauch-Herzog double-focusing analyzer is capable of either photographic plate or electrical detection, but the photoplate is more widely used in practice. In modern instrumentation, only this geometry still uses the photoplate. The spectrum on the photoplate can be quantified by measurement of the band intensities with a high precision scanning microdensitometer.

The major advantage of the photoplate is its ability to integrate the signal over a considerable length of time. This property makes the photoplate an extremely sensitive detector because very weak signals can be determined by exposing the photoplate for sufficiently long periods. Because an instrument using this detector is run in the static mode, in other words, with a constant accelerating voltage and a constant magnetic field, all ions formed during the period of exposure fall on the photoplate. Ion beam fluctuations do not affect the results because all ions formed over the whole spectrum are simultaneously collected. The primary disadvantage is that, in spite of many years of effort in emulsion improvement, standardization, and calibration procedures, the photoplate is still not as good as electrical detectors for ion abundance measurements. There are other disadvantages, including the fact that the plate requires removal from the instrument and development before any data from it can be acquired. The lower accuracy in abundance measurements is primarily responsible for the gradual replacement of the photoplate by electrical detection methods.

5. Total Ion Monitor

An accessory which is especially useful for rapid calibration checks is the total ion monitor, which measures the intensity of the ion beam between the outlet of the ion source and the entrance to the mass analyzer section. In instruments with an electrostatic sector, the monitor is mounted at the exit of this tube. The device consists essentially of a disk with a rectangular slit in the middle which can also be used to collimate the ion beam. The amount of positive ions which do not pass through the slit but impinge on the plate is proportional to the total ion beam. The ion current between the plate and a negative ground is an indication of the total ion beam intensity.

Besides being very useful as a rapid check on whether the instrument is operating properly, the total ion monitor is also very useful in gas chromatography–mass spectrometry applications. The start of the elution of a compound from the chromatographic column is observed as an incipient deflection on the monitor. This signal can be used to trigger the mass scan of the spectrometer.

6. Output Devices

Rather than attempt to enumerate the various kinds of output devices, it would be more appropriate to mention the instruments that can be used for presenting data from different types of mass spectrometric experiments. For those measurements where single-ion beam intensities or isotopic ratios are being determined, a slow response output device is sufficient. This may be a reading from a simple galvanometer. Alternatively, the output can be connected to a low-speed, pen-and-ink potentiometric recorder. This type of output can be tedious to handle and is generally not very accurate. The most expensive approach to obviate these problems is observed in many modern isotope ratio instruments where a computer is employed to acquire and process the data. In general, however, any level of sophistication can be used.

For those experiments which involve the determination of complete mass spectra, the output device used depends on the required response time of the system. For low-resolution instruments, pen-and-ink recorders have been used extensively for slow scanning experiments. For higher scanning speeds, a galvanometer system with light-sensitive paper is frequently used. Again, digital computers are coming into greater use to relieve the tedious tasks of interpreting strip chart measurements. For high-resolution instruments, which even at moderate spectrum-scan speeds require high band-pass systems, or time-of-flight instruments, with their associated fast response times, different techniques must be used. Many time-of-flight instruments use photographic recording of oscilloscope traces for registering the spectrum. High-resolution instruments either record their output on high-speed analog tape or use a direct digital computer hook-up to accurately record the data.

There is also an alternative method of data recording for high-resolution instruments, a technique termed "peak matching." In this approach, ion beams are measured one at a time and compared to a beam of known mass. The reader is referred to a discussion by Beynon (4) for more information.

The point that should be stressed is that modern computer technology is making vast inroads in the field of mass spectrometry. The inherent high speed and accuracy of the computer makes it a particularly attractive choice for data acquisition and processing from all types of mass spectrometric experiments. Most instrument manufacturers are designing their new systems with computers in mind; and several companies provide complete mass spectrometer–computer packages, including all the necessary programming.

III. APPLICATIONS

1. Atomic Mass Measurement

The capability to measure atomic masses with increasing accuracy has been a permanent goal of mass spectrometry since its inception. The object of this continuing effort has provided much of the impetus for the development of improved instrumentation. In principle, with a given instrumentation of sufficient resolution to ensure separation of possible interfering ions of close m/e ratios, atomic masses can be measured by accurate determination of r, V, and B in Equation 10.5. In practice however, nonideal instrumental parameters preclude the direct determination of masses by this equation to the accuracies required in modern research. To correct for possible imperfections, the method of doublet measurement is used. Masses are determined by comparison to previously determined mass standards. The primary standard used to be ^{16}O but is now ^{12}C, which is defined to have a mass of 12.0000. Measurement of a mass relative to the primary standard will yield a secondary standard which can then be used in the determination of the other masses. In this way, one can obtain the necessary information to build up the values of the other nuclidic masses. These quantities are currently being determined with continually increasing accuracies.

This work is important in a number of ways. It forms the foundation for the development of modern nuclear physics by providing the prime data for the analysis of nuclear binding energies, nuclear shell structures, and the processes involved in the release of nuclear energy. The development and understanding of the nuclear packing fraction curve, for example, is a direct result of the ability to measure accurately atomic masses. A review of much of this earlier work and its implications has been presented by Nier (47).

An extension of this work to more recent development is in the field of high-resolution mass spectrometry for organic chemical applications. Determination of accurate masses can be employed to determine elemental compositions. This application is based on the small deviations from the nominal values in the atomic weight of the elements. This field will be covered in a subsequent discussion.

2. Electromagnetic Separations

Mass spectrometry has been extremely useful in the field of isotope separation and enrichment. Preparative-scale mass spectrometers are used to volatilize, ionize, separate, and collect the various isotopes of an element. This approach was one of those used in the enrichment of uranium isotopes crucial to the development of the atomic bomb. Although the gaseous diffusion process was eventually used for large-scale uranium isotope separations, the development of the mass spectrometric technique was largely instrumental in the successful separation and enrichment of these and other isotopes.

Enriched isotopes have played a major role in many areas of research, such as in the measurement of nuclear spin and magnetic moments, and in the correlation of these parameters with spectroscopic data. Enriched isotopes have made many mechanistic studies in synthetic organic and inorganic chemistry possible. In the fields of biochemistry and medical research, isotopes have been used extensively in metabolic studies and in the treatment of a variety of diseases. A review of these techniques, summarizing many of the basic principles and applications, has been published by Smith (53).

3. Nuclear Physics and Chemistry

The two previously discussed applications, those of atomic mass measurement and electromagnetic separation, provide an important foundation for studies in nuclear physics and chemistry. Another important advantage of the mass spectrometric technique in applications to nuclear studies is the ability to determine isotopic abundances accurately. One of the first applications was, of course, the detection of naturally occurring radioactive isotopes. In some cases, such as the discovery by Smythe (54) of ^{40}K as the potassium isotope responsible for potassium radioactivity, isotope enrichment via mass spectrometry was used for the analysis. Isotopic abundance techniques have been used to define more completely the periodic table of elements. Detailed measurements on both stable and radioactive isotopes have, for example, enabled fundamental studies of half-life determinations, neutron capture cross-sections, and yields of fission products to be carried out successfully.

The mass spectrometer has been used to determine half-lives of radioactive isotopes by monitoring the rate of growth of daughter nuclides or the rate of decay of parent nuclides. Absolute decay rates may also be determined by first measuring the disintegration rate by the use of radioactive counting techniques and then determining the amount of the nuclides present by mass spectrometry.

Isotopic cross-sections are important parameters in nuclear physics and chemistry. The determination of this property is facilitated if a mass spectrometer is available. Irradiation of an isotope by neutrons can result in the formation of new isotopes. Their rate of growth, from which the neutron absorption cross-section can be calculated, is conveniently monitored with a mass spectrometer by measuring the changes in isotopic abundances.

Knowledge of mass fission yields are important both in basic research on the behavior of the elements and in the development of the many uses of nuclear energy. Fission yields can be computed from accurate determinations of the mass and the isotopic distribution of fission products. From these data, one can also derive a measure of the energy released during fission. An excellent example of this application has been presented by Thode and Graham (57), who studied the rare gases resulting from the fission of uranium. Fission yield data on a number of radioactive elements have been presented by Katcoff (35).

4. Petroleum Chemistry

The requirement for improved qualitative and quantitative analyses of petroleum fractions has contributed significantly to the development of mass spectrometric instrumentation and techniques for analyzing organic compounds. Since petroleum chemistry is concerned primarily with hydrocarbons (49, 64), the first developments, in the early 1950s, focused on this particular area of chemistry. In later years, this technique has been expanded to cover all classes of organic compounds. This is the subject of a later section.

Mass spectrometry is, as already mentioned earlier, a technique which is eminently suitable for the analysis of single organic compounds. Each compound displays a characteristic and unique fragmentation pattern under electron impact ionization. This behavior can be used for identification purposes because the observed spectra can be compared with a file of known fragmentation patterns. This technique can also be extended to cover multicomponent mixtures. Because the mixture is introduced into the mass spectrometer through a molecular leak, the partial pressure of each component in the ion source is the same as its partial pressure in the sample reservoir. If the fragmentation pattern is known for each pure substance, a complete analysis of the mass spectrum of a complex mixture can be accomplished. It should be realized that each compound will yield many fragments, and overlap at many positions in the mass spectrum will occur if more than one compound is being analyzed. The spectrum becomes more complicated when an increasing number of components have to be considered. Since the necessary calculation increases geometrically with the number of components in the sample, a computer is generally necessary to unravel the separate contributions and to compute the composition of the original sample if more than two or three components are present. This is true even though calculations are not based on all peaks of the spectrum but only on a limited number of selected peaks. In practice, mixtures of 15–20 components can be analyzed quite accurately by this method.

An extension of this technique is observed in the so-called group analysis of mixtures for the estimation of the relative amounts of a particular chemical group, such as the *n*-paraffins, the cycloparaffins, the alkylbenzenes, and so on, in the sample. This approach is especially useful with complex mixtures where the large number of compounds preclude the simple determination of the separate components.

A number of identifying fragments are first determined by analyzing a large number of representative members of each group. By considering the relative abundances of these fragments in the mass spectrum of the mixture, one can then derive the relative amounts of compounds in each particular class.

Since mass spectrometry is a tool which works best with single compounds, the analysis of mixtures can be improved to a large extent by prior separation of the mixture components, such as by chromatography. Collection of individual components is then followed by mass spectrometric analysis (Chapter 9). Spectrometric analysis of multicomponent mixtures has, nevertheless, many applications in present day work, especially where a very small sample or a nonvolatile fraction is involved.

5. Studies on Compounds in the Solid State

Applications of mass spectrometry to studies of solids have been in three principle areas, namely, in trace element analysis, in the determination of heats of sublimation, and in the analysis of inorganic molecules. A major problem in these experiments is the difficulty in vaporization and ionization of the material. For this reason, one would tend to favor the use of spark or thermal ionization sources. Alternatively Knudsen cells may be used to create molecular beams of inorganic vapors which may then be ionized by electron impact.

Trace element analysis is the most important application in this field. This analysis is most frequently carried out with a spark source instrument with a double-focusing geometry (usually Mattauch-Herzog) to correct for the large ion beam energy spread from the ionization process. This method uses the photographic plate as a detector, and the limitation to sensitivity is determined by the signal-to-noise ratio of the photoplate. With this technique, trace metal or other solid material constituents can be measured at the 10–100 parts per billion level. Some of the most important applications are in the analysis of trace impurities in steel samples and in semiconductor materials such as silicon and germanium. Because the photoplate is used as a detector, much work has been done in emulsion studies, development techniques, and response curves for the various emulsions. Gelatin-free photoplates are the most promising type to come along in recent years (22, 31) due to their low background fog characteristics.

At present, mass spectrometry can already compete favorably with emission spectrometry for trace analytical applications. Although the accuracy of the mass spectrometric approach is not very high at this time, it has the advantage of a much greater ultimate sensitivity.

Studies on materials in the solid state involving heats of sublimation and equilibrium vaporization have proved extremely useful for obtaining data that are difficult to obtain by other techniques. For example, the mass spectrometric determination of the heat of sublimation of graphite by Chupka and Inghram (16) was an elegant study that necessitated recalculation of previously published bond dissociation energies. Sublimation studies on many elements such as copper and silver have led to the determination of the bond dissociation energies of their diatomic counterparts (Cu_2, Ag_2) (22). Recent advances include the use of laser-induced vaporization of solid materials (37), followed by mass spectrometric analysis of the resulting vapor.

6. Chemical Kinetics

Many aspects of chemical kinetics may be conveniently studied by mass spectrometry. Three areas in particular have received much attention, namely, that of free radicals, fast reactions, and equilibrium processes. The growth and decay of free-radical intermediates during pyrolysis or combustion reactions have been studied extensively. Pyrolysis of mixtures of gases such as deuterium with various organic molecules has been performed to determine the type and rate of atom exchange. These studies generally employ some specialized inlet systems, such as a reaction cell placed close to the ion source. Products containing free radicals are introduced to the source through a small leak. These studies yield information not only on the type and abundance of radical species formed but on the energetics of their formation. This approach can obviously also be used for the analysis of stable compounds formed by radical reactions.

Fast reactions can be followed by instruments capable of rapid spectrum determination, such as a time-of-flight mass spectrometer. For example, reactions occurring in shock tubes or in flash photolyses can be analyzed both for ultimate products and for short-lived product intermediates. The scan of a time-of-flight instrument is triggered on the shock or flash initiation. The growth and decay of various species in the reaction mixture can be measured if their lifetime is greater than the 20–50 μsec needed to scan a spectrum. Another technique in studying reactions is by the use of rapid flow tubes. By taking and analyzing gas samples at various points in the flow tube from the initial mixing point to the point of complete reaction, one can readily establish a profile of the reaction and the product intermediates.

Equilibrium and exchange processes have also been studied by mass spectrometry. An example involves the direct sampling of the gases in exchange reactions such as in the H–D exchange in ethylene over a nickel catalyst (59). Equilibrium reactions in solution have been studied by allowing the reaction to reach equilibrium in the reaction vessel, followed by isolation and analysis of the products.

7. Ionization and Fragmentation Processes

In spite of the somewhat explosive growth of mass spectrometry in recent years, the nature of ionization and fragmentation processes is not well understood. Much of the work in this area depends almost solely on observations of the empirical behavior of various classes of compounds under electron impact or other ionization methods.

The ionization potential of a molecule is the energy required to remove an electron from the highest occupied molecular orbital. Determination of ionization potentials is important to the understanding of the electronic structure of molecules. It is on this foundation that interpretations of ionization and fragmentation processes are based. Appearance potentials are the energies at which various fragmentation processes begin to occur. They are measured by increasing the ionization energy until a particular fragment ion appears in the spectrum. Correlations of these values with molecular structure are valuable in developing generalized fragmentation mechanisms. Reviews on ionization and fragmentation processes may be found in the book by Field and Franklin (25) and in a chapter by McDowell (39). The largest application of fragmentation patterns is in organic chemical studies. This is described below.

A. Organic Chemical Applications

This section will be devoted to a discussion of the type of information that can be obtained from mass spectra of organic compounds and some of the difficulties involved in interpreting such spectra. Most of the work in this area has been carried out with electron impact ionization.

To review terminology, the molecular ion is the ionized intact molecule. The mass spectrum is a representation of the peak intensity versus the mass of the different ions. The term fragmentation pattern refers to the mass spectrum and implies the distribution of peaks throughout the entire mass range below the mass of the parent peak.

Organic mass spectrometry is a highly empirical discipline. This is quite understandable because of the difficulties in dealing theoretically with the ionization and fragmentation processes in complex molecules. The features of mass spectra are best elucidated by simplified descriptions of the ionization and fragmentation processes which may take place. The transitions shown are based on current knowledge of the ground-state behavior of radicals and ions, especially in solution chemistry. The simplified ways in which these transitions are drawn are most valuable in the communication of ideas about fragmentation processes among scientists, and undoubtedly are *not* good descriptions of the actual physical processes involved in these highly energetic ions.

Ionization is generally pictured as the removal of an electron from some site in the molecule. If this molecule possesses a heteroatom, such as oxygen or nitrogen, with electrons of lower ionization potential, the ionization is depicted as resulting in a positive charge at the heteroatom. For example, in the case of a ketone or an amine, the following transitions take place:

$$R_1\text{—}\overset{\overset{\text{O:}}{\|}}{\text{C}}\text{—}R_2 \xrightarrow[\text{ionization}]{} R_1\text{—}\overset{\overset{\text{O}\cdot+}{\|}}{\text{C}}\text{—}R_2 \tag{10.8}$$

$$\mathrm{R_1CH_2\text{—}CH_2\text{—}\ddot{N}H_2} \xrightarrow[\text{ionization}]{} \mathrm{R_1CH_2\text{—}CH_2\overset{\cdot+}{N}H_2} \tag{10.9}$$

Once ionization has occurred, fragmentation of the molecular ion can take place. These processes are shown in terms of a localization of the charge in the molecular ion followed by bond cleavage as one may picture would also happen in solution chemistry. The site of the positive charge determines where the subsequent fragmentation will take place. This is shown below, again for a ketone and an amine:

$$R_1{-}\overset{\overset{\cdot +}{O}}{\overset{\|}{C}}{-}R_2 \xrightarrow{\text{loss of } R_1\cdot} \overset{+}{O}{\equiv}C{-}R_2 + R_1 \tag{10.10}$$

$$R_1{-}CH_2{-}CH_2{-}\overset{\cdot +}{N}H_2 \xrightarrow{\text{loss of } R_1CH_2\cdot} CH_2{=}\overset{+}{N}H_2 + R_1CH_2 \tag{10.11}$$

The ions thus formed are not necessarily stable and may fragment even further.

The mass spectrum that results can be extremely complex because many fragment ions are formed from both the molecular ion and from the intermediate fragments. Two things are apparent from a study of these mass spectra: (1) The spectrum obtained is not only characteristic of the molecule, it is *unique*. No two different organic compounds display identical spectra. (2) Closely related molecules yield spectra which are closely related in the sense that they display similar fragmentation patterns. These two characteristic properties of fragmentation patterns provide the basis for most of the applications in organic chemistry. Studies of various classes of organic compounds have led to certain generalizations on the various modes of fragmentation. In well-defined cases, such as ketones, these generalizations are sufficient to determine the structures of unknown compounds of the same type merely by an examination of the fragmentation pattern.

In addition to fragmentation, ions may undergo rearrangements, with further fragmentations occurring from the rearranged species. Most rearrangements are assumed to involve hydrogen atoms only, but there is a growing body of evidence that rearrangements of larger groups can take place quite frequently. These rearrangements are heavily dependent on molecular geometry because the hydrogen atom must be within a certain distance from the rearrangement site. This means that geometrical isomers may display very different fragmentation patterns due to a rearrangement in one isomer that is not possible in the other. Also, many organic compounds do not yield molecular ions under electron impact because of facile, rapid fragmentation.

This, of course, makes interpretation of the spectrum difficult. Another problem area of perhaps greater concern is that many compounds are extremely heat sensitive, and slight differences in instrument operating conditions can result in major differences in the spectra obtained. To some extent, this can be overcome by direct sample introduction techniques. Another major problem is that spectra of the same compound recorded on different instruments are, in general, not the same because there has been little effort at standardization of ion sources and mass analyzers.

For these and other reasons, much of the effort toward interpretation of mass spectra is now being directed at computer search techniques for matching an unknown spectrum with a large file of known spectra. This method has great potential for the near future but cannot be an ultimate answer to the problem of compound identification. The unknown may not be in the file. If it is there, there is still not an absolute certainty that it may be correctly identified. This is because most spectra have been obtained under nonstandard conditions and differences may be present with respect to the obtained spectra, which have not been taken into account by the computer program.

The author felt it necessary to discuss difficulties in spectral interpretation first so that the following discussion may be read and interpreted with these limitations in mind. The chemist possesses many techniques that are of great assistance in the interpretation of spectra of organic molecules. Important examples include the various methods of chemical alteration of the molecules.

For structural studies on unknown compounds, the mass spectrometric shift technique can be used. One such method is isotopic exchange, which is particularly helpful in elucidating fragmentation mechanisms. Exchange of labile hydrogens with deuterium results in peak shifts to higher mass for those fragments retaining the deuterium atom(s). An extension of this approach involves the preparation of derivatives with an increase in molecular weight such as the formation of methoxy derivatives ($—OCH_3$) from —OH groups. Those peaks containing the additional function are shifted to higher mass compared to the unreacted molecule, thus permitting localization of various functional groups within the molecule. For those compounds that are nonvolatile or heat sensitive, chemical alteration can be carried out to produce a more volatile or stable derivative. Some examples are the preparation of the methyl esters from acids, the conversion of alcohols into their trimethylsilyl ethers, or the esterification of alcohols with *o*-aminomethyl benzoic acid.

Many excellent books have been written covering in great detail sample handling techniques, fragmentation mechanisms, and chemical treatment of compounds. Such books are those by Biemann (6), and Budzikiewicz, Djerassi, and Williams (12). References cited therein provide detailed reference material on many aspects of organic mass spectrometry.

B. Interpretation of Mass Spectra

The mass spectrum of an inorganic sample is usually relatively simple. The spectrum of an organic molecule, or of an organic sample in general, can be quite complex because of the many fragment ions which can be formed from both the molecular ions and other intermediate fragments. The most important step is to recognize the peak formed by the parent ion because its position yields the exact molecular weight of the particular compound. Other peaks of interest are the parent peak plus one and the parent peak plus two peaks which are the result of isotope contributions. For organic samples, one would also determine the position of the other major peaks in the spectrum and the intervals between these peaks. From this information, one may deduce what fragments have been formed, what frragmentation processes have taken place, and possibly even what the structure of the original molecule could have been. Table 10.1 shows some common fragments observed in organic mass spectrometry.

Table 10.1. Some Common Fragments in Mass Spectrometry [adapted from (52)]

m/e	ION	*m/e*	Ion
14	CH_2	44	CO_2, $C_2H_4NH_2$
15	CH_3	45	C_2H_4OH, CH_2OCH_3
16	O	46	NO_2
17	OH	55	C_4H_7
18	H_2O, NH_4	56	C_4H_8
19	F	57	C_4H_9, $C_2H_5C{=}O$
20	HF	58	$C_3H_6NH_2$, CH_3COCH_3
26	CN	59	$(CH_3)_2COH$, $COOCH_3$
27	C_2H_3	60	CH_3COOH
28	C_2H_4, CO, N_2	69	C_5H_9, CF_3
29	C_2H_5, CHO	70	C_5H_{10}
30	CH_2NH_2, NO	71	C_5H_{11}, $C_3H_7C{=}O$
31	CH_2OH, OCH_3	73	$OCOC_2H_5$
33	SH	77	C_6H_5
34	H_2S	79	Br, $C_6H_5 + 2H$
35	Cl	83	C_6H_{11}
36	HCl	84	C_6H_{12}
41	C_3H_5	85	C_6H_{13}, $C_4H_9C{=}O$
42	C_3H_6	90	C_6H_5CH
43	C_3H_7, $CH_3C{=}O$	91	$C_6H_5CH_2$

For simple compounds, use can be made of existing tables to match the obtained spectrum. In general however, the spectra are more complex; and more background knowledge is necessary to extract the maximum information from the raw data. For further particulars on how to interpret mass spectra, reference is made to the excellent books by McLafferty (40), Biemann (6), and Silverstein and Bassler (52).

8. Geochemistry

Modern geochemistry is a field that encompasses many disciplines. This discussion will be broken down into two major sections, one dealing with the more convential aspects of geochemistry which I will choose to call "inorganic," the other dealing with a more recent branch of geochemical studies, that is, organic geochemistry. Mass spectrometric techniques are used quite extensively in both subdivisions.

A. Inorganic Geochemistry

This branch is concerned primarily with the study of past processes in the history of the earth based on current observations of the isotopic distribution of a number of elements. A wide variety of chemical, geological, and biological processes in nature alter the isotopic distribution of some elements. The mass spectrometer is an ideal tool in these studies because of its accuracy in determining isotopic ratios. Research of this nature can yield important information on the age of sediments, and on the temperature conditions and the amount of biological activity in certain periods in the past.

a. Age Determination. There are several methods which can be used in dating ancient material. Analysis of radiogenic lead is one of these. Three of the four stable isotopes of lead are the end products of the decay of uranium and thorium isotopes. With a knowledge of the half-lives of the radioactive nuclides, measurement of the relative amounts of the uranium and thorium parent nuclide in relation to the lead isotopes provides a measure of the age of the sample studied.

Many materials to be dated, however, are very low in uranium and lead. For some of these materials, potassium/argon dating may be used. The ^{40}K decays to ^{40}Ar with a half-life of approximately 1.4×10^9 yr. Measurement of the ^{40}Ar can be made from a sample of material heated to release the rare gases. The ^{40}K can be determined independently. This dating method gives the date of solidification of the material studied. It is assumed that at this time the amount of ^{40}Ar is zero and that subsequently formed and trapped ^{40}Ar in the material results from the decay of ^{40}K. An application of this method of particular interest to the scientific community was the dating of lunar material by Schaeffer and co-workers (50). It was found that the volcanic (crystalline) rocks are about 3.7×10^9 yr old, which is quite comparable in age to the oldest rocks found on earth.

Another method for dating extremely ancient materials is the rubidium/strontium method. ^{87}Sr is a radiogenic product of ^{87}Rb, and mass spectrometry is used to measure the desired isotopic ratios. This method has been used extensively in dating ancient sediments and meteorites. It has, for example, been applied by Wasserburg (65) to the dating of the Apollo 11 returned lunar samples. Results indicate an age of 4.5×10^9 yr for the lunar soil, which is quite close to the generally assumed age of the solar system. The birth of the solar system is defined to correspond with the time of condensation and solidification of the planets.

b. Isotope Thermometry. Equilibrium processes are known to have significant isotope effects. Of particular interest is the temperature effect on the equilibrium isotopic distribution in reactions occurring in nature. Urey (60) suggested that the carbonate–water exchange should be a fruitful basis for determining the temperature at the time of precipitation of limestone or calcium carbonate shell deposits. This material shows an enrichment of ^{18}O over that of the water from which it was deposited. The exact amount of enrichment depends on the temperature at the time of deposition. Isotope ratio measurements on the carbonate deposits are sufficiently accurate to determine the temperature of the water at the time of precipitation to $\pm 0.5°C$. By careful analysis of stratigraphic layers of limestone or shell deposits, a temperature history of the ocean can be determined. In principle, this method can also be used for phosphate–water, sulfate–water, and silicate–water systems. Studies of the carbonate–water systems have, however, been the most fruitful, especially in view of the widespread carbonate deposits in many parts of the world, which have been built up over many centuries.

c. Biological Applications. Biological systems also show significant isotope effects due to the equilibrium nature of most biological processes. Carbon isotope studies on carbon-containing material have been carried out for possible indications of past biological processes. This work is usually carried out by oxidizing the carbon in a sample to CO_2 and measuring the $^{13}C/^{12}C$ isotope ratios. Because biological systems discriminate against heavier isotopes, a depletion in ^{13}C is noted in comparison to, for example, a limestone of inorganic origin. The method is sufficiently sensitive to discriminate among limestone carbon, organic carbon of marine origin, and organic carbon of land-plant origin. The primary cause for the difference in $^{13}C/^{12}C$ ratios in biological systems as compared to this ratio in atmospheric CO_2 can be traced to the isotope effect, which, in turn, is due to the difference in zero point energies of the different isotopes. This isotope effect manifests itself in different transport properties in the environment and dissimilar assimilation rates of $^{13}CO_2$ versus $^{12}CO_2$ in living organisms.

Studies of sulfur isotopes have also been made in research on the origins of native and organic sulfur (58). The results of these studies with respect to indications of biological activity are, however, not as clear-cut as for those obtained with carbon isotopes.

B. Organic Geochemistry

Organic geochemistry is not a particularly well defined term. The field actually encompasses several disciplines, such as convential geochemistry, petroleum chemistry, and organic chemistry. An important area is those studies concerned with the origin of petroleum. Basically, these studies involve the isolation, identification, and the determination of the quantitative distribution of various organic constituents in a wide variety of petroleum samples. From these data, which cover both the hydrocarbons and other types of organic compounds, attempts are currently being made to elucidate the biogenic origins of these materials.

Another interesting area of research is concerned with the search for organic compounds in ancient sediments which would be indicative of past life processes. The search is aimed at isolating the identifying molecules that have survived the millions or billions of years that have elapsed since their formation. The isoprenoid hydrocarbons are examples of such molecules. Chlorophyll or chlorophyll-related molecules are definite indicators of past biological activity. These molecules possess a side chain consisting of an isoprenoid (e.g., phytyl) fragment. These fragments and their reduced derivatives, the hydrocarbons, are assumed to be the degradation products of more complex molecules associated with past life forms. The search for these so-called "biological markers" was originally undertaken to determine the point in the earth's past history when life first began. However these biological markers have been isolated and identified in the most ancient sediments found on earth, which are approximately 3.3×10^9 yr old. Although there is still considerable controversy over whether or not these compounds are indigenous to the sediments, there is no question that the time of the beginning of life on earth is considerably before that which has been previously assumed.

Several books are available that discuss the many aspects of geochemistry. From the books by Degens (18), Mason (41), Breger (10), and Eglinton and Murphy (23) and the references cited therein, the reader can get some idea of the broad nature of this field and the importance of mass spectrometry in these studies.

9. High-Resolution Mass Spectrometry

Although many mass spectrometers capable of performing high-resolution studies are currently commercially available, their acquisition is still limited to the well-funded laboratory. With these instruments, it is possible to separate closely spaced masses and to carry out accurate mass measurements of the separated ion beams. Historically, this application was of prime importance in determining accurate atomic masses. In modern chemical research however, the primary emphasis of high-resolution techniques is on studies of organic compounds. Although Nier (47) was the first to point out the potentials of high-resolution techniques for organic chemistry, Beynon (4) was the first to apply them to accurate mass determination. Because of the deviation of atomic masses from integral values, different elemental compositions that have the same nominal mass actually differ slightly in accurate mass. For example, considering only singly charged ions at *m/e* 28, there are four possible combinations of carbon, hydrogen, nitrogen, and oxygen, each differing slightly in mass. CO has a mass of 27.9949, the mass of N_2 is 28.0061, that of CH_2N is 28.0187, and that of C_2H_4 is 28.0313. Modern high-resolution instruments are capable of separating these species.

Figure 10.9 graphically demonstrates this example. This figure shows the *m/e* 28 region recorded on a Nier-Johnson double-focusing instrument at two resolutions, 1100 and 18,200. In the lower resolution spectrum, the peaks overlap one another. At higher resolution however, all the ion species mentioned above are well resolved. In addition, this resolution is even sufficient to resolve $C^{13}CH_3$ (the ^{13}C isotope peak from *m/e* 27) from C_2H_4.

The potentials of this technique are immediately obvious. If one can determine the masses of the molecular ion and all fragment ions to sufficient accuracy, then the elemental compositions of all ions can be determined. Thus the technique can provide not only the elemental composition of the intact molecule but also provide considerable information on the fragmentation pattern of the molecule and its relation to the molecular structure. This fact tends to eliminate many of the uncertainties of conventional low-resolution spectra.

Accurate masses can be determined in a number of ways. The first method used was that of Beynon (4) and is termed "peak matching." This technique employs a measurement of the accelerating voltages needed to focus a known mass and an unknown mass on the collector. The ratio of these accurately determined voltages is directly proportional to the mass ratio of the ions. This method suffers from the disadvantage of being able to measure only one peak at a time. This is a severe limitation when only small samples are available.

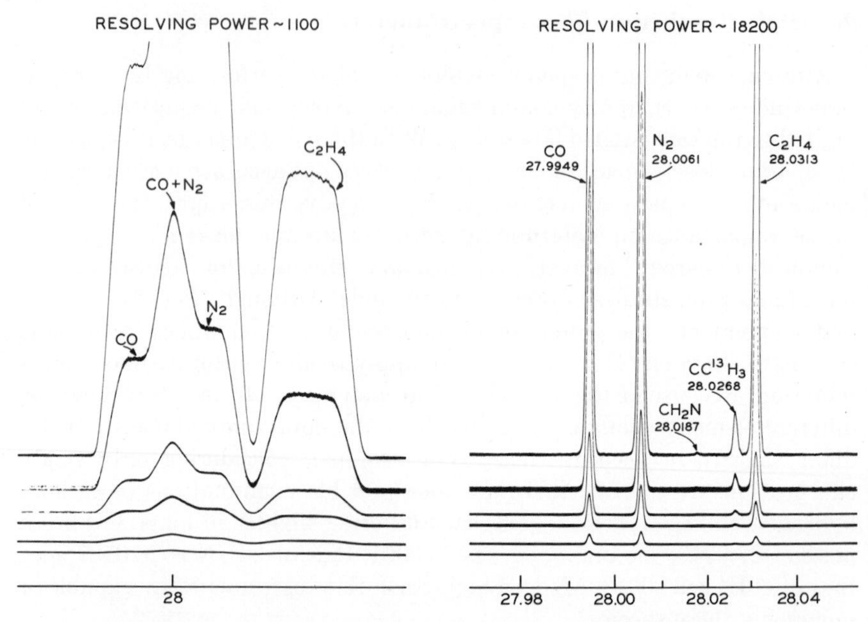

Figure 10.9. *M/e* 28 region recorded at resolving powers of 1100 and 18,200 on a high-resolution mass spectrometer (courtesy of Dr. R. M. Teeter, Chevron Research Co., Richmond, Cal.).

Because of the limitations of peak matching, other methods for determining complete high-resolution mass spectra have been developed. Biemann (7) pioneered in the development of systems designed to record complete spectra on photographic plates, using instruments of Mattauch-Herzog geometry. Line positions on the plate are measured with a high-precision microdensitometer. Masses, and ultimately compositions, are calculated with respect to an internal standard run with each sample. These procedures are carried out with a computer because of the large amount of calculations that must be done.

An alternative technique involves a scanning mass spectrometer. An internal standard is again used. The complete spectrum is recorded with the assistance of a computer, which is used to acquire the data, determine the peak positions, and calculate the masses and compositions. Discussions of the techniques involved and the accuracy of the data (masses accurate to <1 ppm can be determined) have been presented by Burlingame and co-workers (13, 14).

10. Mass Spectrometry in Space Research

Recent technological advances have resulted in the successful construction and operation of miniature mass spectrometers for various experiments in space probes. The problems of instrumentation in this application are obvious. A complete mass spectrometer including some device for sample introduction, ion optics, and ion detection system must be included in an instrument payload with severe power and weight restrictions. Important design considerations are therefore miniaturization of the ion optical system, utilization of light-weight metals or alloys in construction, and the use of solid-state electronics designed to dissipate the minimum amount of power. The instrument must also be very rugged to withstand the rigors of launch and flight.

Applications of these instruments may be divided for the purpose of this discussion into studies of the atmosphere of Earth and studies involving the atmosphere and surface of other planets. Mass spectrometers have been used to analyze the tellurian atmosphere for neutral and ionized gases, usually as a function of altitude. Instruments for these studies may be carried aloft by balloons for investigation of the lower atmosphere or by rockets or satellites for research at higher altitudes. Figure 10.10 is a photograph of a miniature double-focusing instrument which was part of an Aerobee rocket payload for studies of the upper atmosphere launched at Fort Churchill, Canada, in May, 1970.

From the information gained by these instruments one can determine the degree of mixing of the atmosphere and calculate the effects of ionizing radiation from the sun on the distribution of neutral and ionized or dissociated gases as a function of altitude. These subjects, together with the design and performance of instruments for atmospheric research have been reviewed by Von Zahn (62) and Spencer (56).

Mass spectrometers for use in planetary probes are now in the development stage. Although no mass spectrometer experiments are included in the Apollo Lunar Missions, two experiments involving miniature mass spectrometers are planned in the Viking Project for the 1975 dual orbiter/lander instrument packages to Mars.

The descent phase of the mission, from the orbiter to the surface, will include a mass spectrometer similar in concept to the one described in Figure 10.10. The experiment is designed to investigate the composition of the Martian atmosphere.

In addition, the lander portion of the vehicle will incorporate a second mass spectrometer for the search for and analysis of organic chemical compounds in the Martian soil (61). This instrument will be part of larger

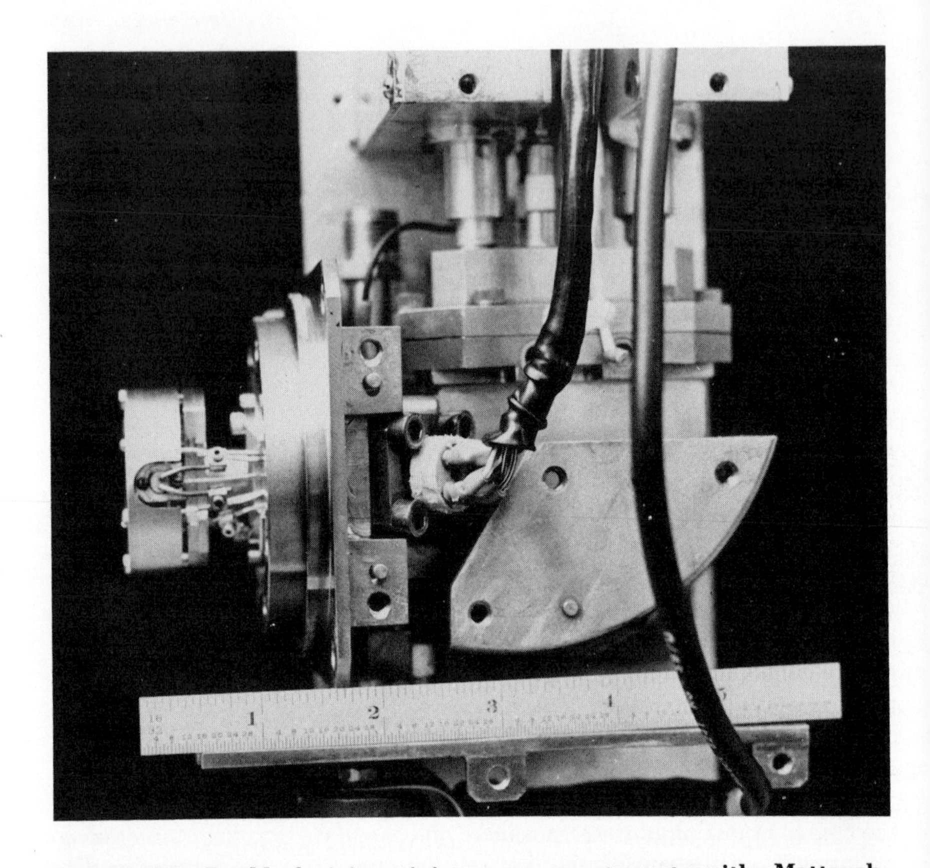

Figure 10.10. Double-focusing miniature mass spectrometer with a Mattauch-Herzog geometry used for studying the composition of the earth's upper atmosphere. Ions are produced in the ion source to the left and sent into the combination electric and magnetic analyzer in the center of the picture. The magnet appears just above right half of the scale. The electrometer amplifiers for measuring ion currents are just visible at the top of the picture. The remainder of the electronics appear as a separate unit. An instrument of this type, weighing under 10 lb, and consuming less than 5 W power, will be used to probe the Martian atmosphere in the proposed Mars Landing in 1975 (courtesy of Professor A. O. C. Nier, University of Minnesota, Minneapolis, Minn.).

package that includes a soil sampler and a combined pyrolysis unit/gas chromatograph for separation and subsequent mass spectral analysis of organic material.

IV. CONCLUSION

The author has attempted to summarize many aspects of the field of mass spectrometry. The discussions were not aimed at completeness; and indeed some facets of the technique, such as chemical ionization mass spectroscopy (44), metastable ions and ion kinetic energy spectra (2, 5), were not discussed. Sufficient references, however, have been included to permit the interested reader to obtain more information on the topics mentioned. In summary, in both instrumentation and applications, mass spectrometry is an extremely diversified method of analysis. The primary advantages of the technique are its sensitivity and accuracy, and its ability to analyze impure compounds or complex mixtures. The primary disadvantages are the complexity of instrumentation and the need for vaporization of the sample for most methods of ionization. These disadvantages can be overcome to some extent; the former by improved technology, and the latter by chemical transformation or degradation to more volatile materials. It is important to stress that the future of the technique will involve ever-increasing use of computers to relieve the drudgeries of routine analysis and to permit efficient application of techniques such as high-resolution mass spectroscopy to solution of chemical problems that can be handled by no other method.

REFERENCES

1. J. D. Baldeschweiler, Hanspeter Benz, and P. M. Llewellyn, "Ion Molecule Reactions in an Ion Cyclotron Resonance Mass Spectrometer," in *Advances in Mass Spectrometry*, Vol. 4, E. Kendrick, Ed., Institute of Petroleum, London, 1968, p. 113.
2. M. Barber and R. M. Elliott, "Comparison of Metastable Spectra from Single- and Double-Focusing Mass Spectrometers," *12th Annual Meeting ASTM Committee E-14 on Mass Spectrometry*, Montreal, Canada, June 7–14, 1964.
3. H. D. Beckey, H. Heising, H. Hey, and H. G. Metzinger, "The Use of Thin Wires in Field Ionization Mass Spectrometry," in *Advances in Mass Spectrometry*, Vol. 4, E. Kendrick, Ed., Institute of Petroleum, London, 1968, p. 817.
4. J. H. Beynon, *Nature* **174**, 735 (1954).
5. J. H. Beynon, *Anal. Chem.* **42** (1), 97A (1970).
6. K. Biemann, *Mass Spectrometry: Organic Chemical Applications*, McGraw-Hill, New York, 1962.

7. K. Biemann, P. Bommer, D. Desiderio, and W. J. McMurray, *Advances in Mass Spectrometry*, Vol. 3, W. L. Meade, Ed., Institute of Petroleum, London, 1966, p. 639.
8. W. Bleakney, *Phys. Rev.* **34**, 157 (1929).
9. Jochen Block, "Recent Developments in Field Ion Mass Spectrometry," in *Advances in Mass Spectrometry*, Vol. 4, E. Kendrick, Ed., The Institute of Petroleum, London, 1968, p. 791.
10. I. Breger, *Organic Geochemistry*, Pergamon Press, New York, 1963.
11. W. M. Brubaker, *Neuvieme Colloq. Spectroscopocicum Intern.*, Lyon, France, 1961.
12. H. Budzikiewicz, C. Djerassi, and D. H. Williams, *Mass Spectrometry of Organic Compounds*, Holden-Day, San Francisco, 1967.
13. A. L. Burlingame, D. H. Smith, and R. W. Olsen, *Anal. Chem.* **40,** 13 (1968).
14. A. L. Burlingame, D. H. Smith, T. O. Merren, and R. W. Olsen, "Real-Time High-Resolution Mass Spectrometry," in *Computers in Analytical Chemistry* (Vol. 4 in progress in Analytical Chemistry Series), C. H. Orr and J. Norris, Eds., Plenum Press, New York, 1969, pp. 17–38.
15. F. von Busch and W. Paul, *Z. Physik.* **164,** 581, 588 (1961).
16. W. A. Chupka and M. G. Inghram, *J. Chem. Phys.* **22,** 1472 (1954); **59,** 100 (1955).
17. E. M. Clarke, *Can. J. Phys.* **32,** 764 (1954).
18. E. Degens, *Geochemistry of Sediments* (*A Brief Survey*), Prentice-Hall, Englewood Cliffs, New Jersey, 1965.
19. A. J. Dempster, *Phys. Rev.* **11,** 316 (1918).
20. A. J. Dempster, *Proc. Amer. Phil. Soc.* **75,** 755 (1935).
21. A. J. Dempster, *Manhattan District Declassification Comm. Report* **370,** 1946.
22. J. Dowart and R. E. Honig, *J. Chem. Phys.* **25,** 581 (1956); **61,** 980 (1959).
23. G. Eglinton and M. T. J. Murphy, Eds., *Organic Geochemistry: Methods and Results*, Springer Verlag, Berlin, 1971.
24. R. M. Elliott, "Ion Sources," in *Mass Spectrometry*, C. A. McDowell, Ed., McGraw-Hill, New York, 1963, p. 84.
25. F. H. Field and J. L. Franklin, *Electron Impact Phenomena*, Academic Press, New York, 1958.
26. R. E. Fox, W. M. Hickam, T. Kjeldaas, Jr., and D. J. Grove, *Phys. Rev.* **84,** 859 (1951).
27. R. E. Fox, W. M. Hickam, D. J. Grove, and T. Kjeldaas, Jr., *Rev. Sci. Instr.* **12,** 1101 (1955).
28. R. F. Herzog, *Amer. Lab.*, May 1969, p. 15.
29. N. B. Hannay, *Rev. Sci. Instr.* **25,** 644 (1954).
30. H. Hintenberger, *Z. Naturforsch.* **3a,** 125 (1948).
31. R. E. Honig, J. R. Woolston, and D. A. Kramer, *Proc. of the 14th Ann. Conf. on Mass Spectrometry and Allied Topics*, May 22–27, 1966, Dallas, p. 481.
32. H. Hurzeler, M. G. Inghram, and J. D. Morrison, *J. Chem. Phys.* **27,** 313 (1957).
33. M. G. Inghram and W. A. Chupka, *Rev. Sci. Instr.* **24,** 518 (1953).

34. E. G. Johnson and A. O. Nier, *Phys. Rev.* **91,** 10 (1953).

35. S. Katcoff, *Nucleonics* **16,** 78 (1958).

36. L. Kerwin, *Rev. Sci. Instr.* **20,** 36 (1949).

37. B. E. Knox, *Advances in Mass Spectrometry,* Vol. 4, E. Kendrick, Ed., Institute of Petroleum, London, 1968, p. 491.

38. F. P. Lossing and I. Tanaka, *J. Chem. Phys.* **25,** 1031 (1956).

39. C. A. McDowell, "The Ionization and Dissociation of Molecules," in *Mass Spectrometry,* C. A. McDowell, Ed., McGraw-Hill, New York, 1963, p. 506.

40. F. W. McLafferty, *Interpretation of Mass Spectra,* W. A. Benjamin, Philadelphia, 1966.

41. B. Mason, *Principles of Geochemistry, 2nd ed.,* Wiley, New York, 1960.

42. J. Mattauch and R. Herzog, *Z. Physik.* **89,** 786 (1934).

43. J. Mattauch, *Phys. Rev.* **50,** 617 (1936).

44. M. S. B. Munson and F. H. Field, *J. Amer. Chem. Soc.* **88,** 2621 (1966).

45. A. O. Nier, *Rev. Sci. Instr.* **11,** 212 (1940).

46. A. O. Nier, *Rev. Sci. Instr.* **18,** 398 (1947).

47. A. O. Nier, *Science* **121,** 737 (1955).

48. W. Paul and M. Raether, *Z. Physik.* **140,** 262 (1965).

49. H. Powell and G. N. Ross, *Applied Mass Spectrometry,* The Institute of Petroleum, London, 1953, p. 6.

50. O. A. Schaeffer, J. Funkhouser, D. D. Bogard, and J. Zähringer, (members, Preliminary Examination Team), *Science* **165,** 1211 (1969).

51. P. Schulze, B. R. Simoneit, and A. L. Burlingame, *J. Mass Spectroscopy Ion Phys.* **2,** 183 (1969).

52. R. M. Silverstein and G. C. Bassler, *Spectrometric Identification of Organic Compounds,* Wiley, New York, 1963.

53. M. L. Smith, *Electromagnetically Enriched Isotopes and Mass Spectometry,* Butterworths, London, 1956.

54. W. R. Smythe and A. Hemmendinger, *Phys. Rev.* **51,** 178 (1937).

55. H. Sommer, H. A. Thomas, and J. A. Hipple, *Phys. Rev.* **82,** 697 (1951).

56. N. W. Spencer, "Upper Atmosphere Studies by Mass Spectrometry," in *Advances in Mass Spectrometry,* Vol. 5, The Institute of Petroleum, London, in press 1970.

57. H. G. Thode and R. L. Graham, *Can. J. Res.* **A25,** 1 (1947).

58. H. G. Thode, J. Macnamara, and W. H. Fleming, *Geochim. Cosmochim. Acta* **3,** 253 (1953).

59. J. Turkevich, F. Bonner, D. Schissler, and P. Irsa, *J. Phys. Colloid Chem.* **55,** 1078 (1951).

60. H. C. Urey, *J. Chem. Soc.* **1947,** 562.

61. *Viking Project Documents No. M73-101-5, Viking Mission Definition No. 2,* and *Viking Project Document No. M73-112-0,* Viking Lander Science Instrument Teams Report. Viking Project Office, Langley Research Center, National Aeronautics and Space Administration, 1969.

62. U. Von Zahn, "Space Mass Spectrometry," in *Advances in Mass Spectrometry,* Vol. 4, E. Kendrick, Ed., The Institute of Petroleum, London, 1968, p. 869.

63. J. D. Waldron, *Research* **9,** 306 (1956).
64. H. W. Washburn, *Physical Methods in Chemical Analysis,* Vol. 1, Academic Press, New York, 1950, p. 587.
65. G. J. Wasserburg et al., Conference on the Results of the Apollo 11 Sample Analysis, *Science,* **167,** 463 (1970).
66. W. C. Wiley and I. H. McLaren, *Rev. Sci. Instr.* **26,** 1150 (1955).

RECOMMENDED READING

J. H. Beynon, R. A. Saunders, and A. E. Williams, *The Mass Spectra of Organic Molecules,* Elsevier, Amsterdam, 1968.
H. Budzikiewicz, C. Djerassi, and D. H. Williams, *Mass Spectrometry of Organic Compounds,* Holden-Day, San Francisco, 1967.

D. D. GILBERT

Northern Arizona University
Flagstaff, Arizona

XI. Electroanalytical Methods

I. INTRODUCTION

1. Scope of Electroanalytical Methods

The scope of electroanalytical methods is much wider than is generally recognized. Essentially, these methods encompass the use of current–voltage relationships to determine the concentration of electroactive species in solution. Although the basic concept is quite simple, sufficient electroanalytical methods and forms have been developed over the years to cope with a wide variety of analytical problems, to provide the material for at least 100 or so textbooks on these and related techniques, to form the base for a torrential stream of papers, and to furnish the raison d'etre for programs in at least 50 graduate schools in the United States. A major reason for the upsurge in range and applicability of these methods can be traced to the increased sophistication of modern electronics which allow the control and accurate measurement of voltages and currents at very low levels. This, in turn, has led to significant advances in sensitivity and scope of these techniques.

Most of the electroanalytical methods used in practice are based on electron-transfer processes. These techniques are based on reactions which take place at the interface of a solution and a solid or liquid conductor. Other electroanalytical methods in use are based on conductance or on the dielectric properties of the material. Table 11.1 summarizes the applicability of various electroanalytical methods with respect to the determination of inorganic or organic species. Figure 11.1 shows the general concentration ranges over which these techniques have been used in a routine fashion.

Table 11.1. Applicability of Various Electroanalytical Methods

Electroanalytical Method	Applications
Titrimetry	Inorganic, organic
Zero-current potentiometry	Inorganic, organic
Constant-current potentiometry	Inorganic, organic
Null-point potentiometry	Inorganic
Amperometry	Inorganic, organic
Coulometry	Inorganic, organic
Electrogravimetry	Inorganic
Direct coulometry	Inorganic, organic
Direct potentiometry	Inorganic
Polarography	Inorganic, organic
Stripping analysis	Inorganic
Conductance	Inorganic, organic

In principle, electroanalytical techniques can be used for both qualitative and quantitative analysis. However qualitative electroanalysis is not commonly used because these techniques lack specificity. In quantitative work, species with concentrations from 10^{-1} to 10^{-9} M have been determined with accuracies from a few tenths of one percent to ten percent relative.

One of the most important applications of electroanalytical methods is in

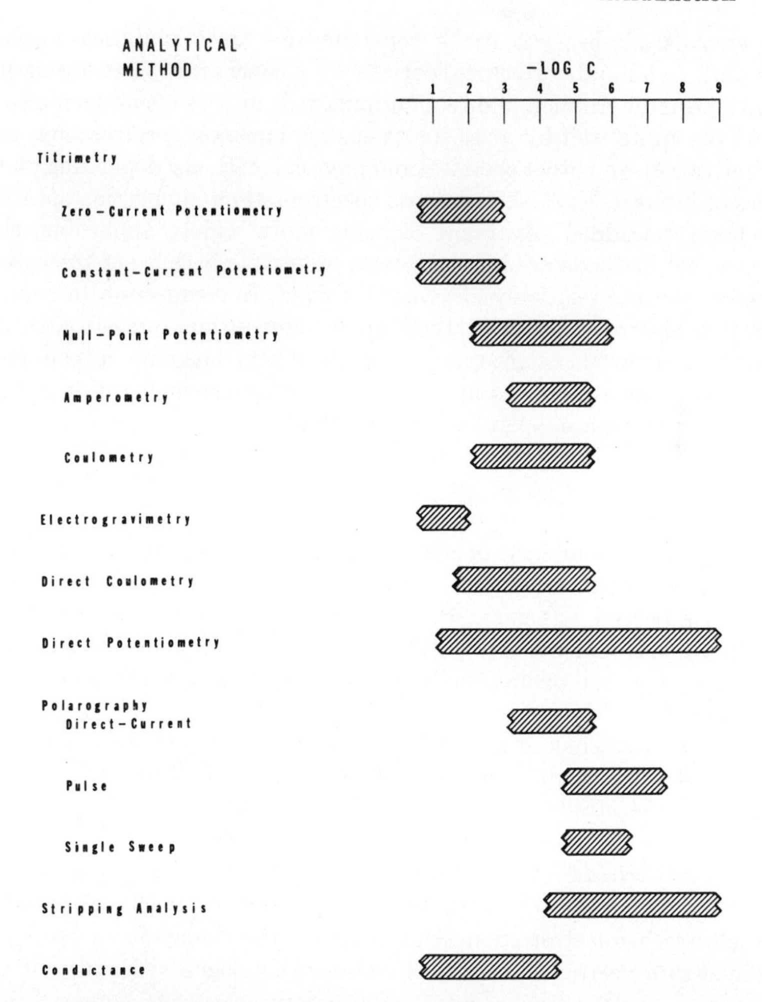

Figure 11.1. Approximate concentration ranges over which various electroanalytical methods are applicable.

its use as end-point detection systems for titrations in which visual indicators are either unavailable or unsuitable because of the color of the sample itself. Many laboratories use electrochemical end-point detection routinely as part of an automated titrimetric analytical system. The problem of maintaining a collection of colorimetric reagents, some of which may be unstable, is avoided. More than one species may often be determined from the data of one experiment with a single sample.

Electroanalysis is particularly well-suited for trace analytical applications (50). In a number of cases species with a concentration as low as 10^{-9} M have been determined. Trace determinations at these levels can also be carried out quite well by mass spectrometry, emission spectroscopy, neutron activation, and other chemical and physical methods, depending on the species of interest. Even though mass spectrometry and emission spectroscopy have the added advantage of being more widely applicable, electroanalytical methods are in many cases preferable because of the ease of operation and the relatively low costs involved. In comparison to neutron activation, electroanalytical methods show lower ultimate sensitivities but probably a somewhat larger range of applicability. The main advantage is that the equipment involved in the latter technique is much easier and less expensive to purchase, operate, and maintain.

2. Nomenclature and Conventions

There is a singular lack of agreement in nomenclature and sign conventions in the various electrochemical journals, texts, and reference works. Although a partial agreement was established by the IUPAC Stockholm Convention (11), older customs still persist in the literature. Various terms and conventions are operationally defined below as they are used in this chapter.

Electrode—a conductor at which the species of a redox system may be in equilibrium. Selected electrode systems are given in Table 11.2.

Anode—an electrode at which an oxidation occurs.

Cathode—an electrode at which a reduction occurs.

Indicator electrode—an electrode whose potential changes during an experiment because of a change in the concentration of a reactant or product of an electrochemical reaction. An example is the change in the potential of a silver wire electrode during the titration of a halogen with silver nitrate. The $Ag^+ - Ag(s)$ redox couple establishes the electrode potential. The electrode potential can be given by the Nernst equation. At 25°C, this is

$$E_{\text{electrode}} = E^\circ_{\text{electrode}} + 0.05916 \log [Ag^+] \tag{11.1}$$

Strictly speaking, the activity of Ag^+ should be used instead of the concentration, [ ]. E° is the standard electrode potential as given in Table 11.2. Unless further qualified, numerical values of electrode potentials are assumed to be those observed at the standard temperature of 25°C.

Electrode potential—the potential difference between one electrode system and another system used as a reference. Unless stated otherwise, the normal hydrogen electrode is assumed to be the reference.

Table 11.2. Selected Electrode Systems[a]

System	Standard Electrode Potential Standard Reduction EMF $E°$, V Versus NHE	Example Electroanalytical Application
$MnO_4^- + 8H^+ + 5e^- = Mn^{2+} + 4H_2O$	+1.51	Redox titration
$Br_2(aq) + 2e^- = 2Br^-$	+1.087	Coulometric titration
$VO_2^+ + 2H^+ + e^- = VO^{2+} + H_2O$	+1.000	Polarography
$Ag^+ + e^- = Ag(s)$	+0.7995	Precipitation titration
$Fe^{3+} + e^- = Fe^{2+}$	+0.771	Redox titration
$C_6H_4O_2(s) + 2H^+ + 2e^- = C_6H_4O_2H_2(s)$	+0.6994	Acid–base titration (quinhydrone electrode)
$I_2(aq) + 2e^- = 2I^-$	+0.620	Redox titration
$H_3AsO_4 + 2H^+ + 2e^- = HAsO_2 + 2H_2O$	+0.559	Direct coulometry
$UO_2^{2+} + 4H^+ + 2e^- = U^{4+} + 2H_2O$	+0.334	Direct coulometry
$Hg_2Cl_2(s) + 2e^- = 2Hg(l) + 2Cl^-$(sat'd)	+0.242	Reference electrode (SCE)
$AgCl(s) + e^- = Ag(s) + Cl^-$	+0.197	Reference electrode, precipitation titration
$H^+ + e^- = \frac{1}{2} H_2(g)$	0.000	Reference electrode (normal hydrogen electrode), acid–base titration
$Pb^{2+} + 2e^- = Pb(s)$	−0.126	Amperometric titration, polarography
$Ni^{2+} + 2e^- = Ni(s)$	−0.24	Polarography
$Cd^{2+} + 2e^- = Cd(s)$	−0.403	Anodic stripping
$Zn^{2+} + 2e^- = Zn(s)$	−0.763	Electrogravimetry
$Sn^{4+} + 2e^- = Sn^{2+}$	−1.90	Coulometric titration

[a] Data compiled from Lingane (27).

Reference electrode—an electrode system used as a reference point in electrode potential measurements. Some of the common reference electrode systems are noted in Table 11.2. Standard electrode potentials are given versus the normal hydrogen electrode (NHE). A very common reference electrode is the saturated calomel system, generally known as the saturated calomel electrode (SCE)

$$Hg_2Cl_2(s) + 2e^- = 2Hg(l) + 2Cl^- \tag{11.2}$$

for which the Nernst equation is:

$$E_{Hg_2Cl_2(s),Hg(l)} = E^{\circ}_{Hg_2Cl_2(s),Hg(l)} + 0.05916 \log \frac{1}{[Cl^-]^2} \tag{11.3}$$

If, during an isothermal experiment, the concentration of chloride ion in equilibrium with mercury metal and solid mercurous chloride (calomel) remains constant, the electrode potential does not change.

The monograph by Ives and Janz should be consulted if a very accurate reference electrode is required (22). The latter may be the case when electrode potential measurements are to be used in the calculation of thermodynamic or electrochemical kinetic data. In most electroanalytical analyses, temperature control of $\pm 0.25°C$ and the use of analytical reagents without further purification is satisfactory for preparing reference electrode systems.

Electromotive force (emf)—the driving force of a reaction expressed in electrical units. The emf of a cell is a measurable quantity, simply the difference in potential between two wires of the same material attached to the two electrodes. The emf of a half-reaction is a defined quantity.

Oxidation potential—the emf of a half-reaction written as an oxidation. This is the sense in which Latimer uses the term (25).

Reduction potential —the emf of a half-reaction written as a reduction.

There is considerable confusion in the terms and symbology of electrode potentials, and oxidation or reduction potentials. The potential of an actual electrode is an observed physical quantity; it is distinctly different from the latter two quantities, which are, in fact, defined quantities and are related to the thermodynamic Gibbs free energy function.

According to the IUPAC Stockholm Convention, an electrode potential is defined as the potential difference between the electrode in question and the normal hydrogen electrode. Although two totally different concepts are involved, the numerical value of the standard electrode potential is identical to that of the emf of the redox couple written as a reduction and is opposite in sign to the emf of the same couple written as an oxidation.

It would be better if there were a more generally accepted agreement on the usage of all these terms. Unfortunately, this is not the case. In fact the same symbol, E, is used for both an emf and for an electrode potential. For example, take the silver–silver ion system:

Oxidation half-reaction:

$$Ag(s) = Ag^+ + e^-$$

$$E^\circ_{Ag(s),Ag^+} = -0.7995 \text{ V}$$ This can be either a standard oxidation emf or a standard oxidation potential

Reduction half-reaction:

$$Ag^+ + e^- = Ag(s)$$

$$E^\circ_{Ag^+,Ag(s)} = +0.7995 \text{ V}$$ This can be either a standard reduction emf or a standard reduction potential

Standard electrode potential:

$$E^\circ_{Ag^+,Ag(s)} = +0.7995 \text{ V}$$

The symbol $E^\circ_{Ag^+,Ag(s)}$ is hence used for both the standard electrode potential and the standard reduction emf. Several authors have suggested that a symbol other than E be used to distinguish an electrode potential from an electromotive force. Unfortunately none of their proposals have been widely accepted.

An important aspect in an electrochemical reaction is related to the kinetics of the system. Frequently encountered terms are reversibility and irreversibility which are related to the kinetics of the electrochemical reaction (10, 14). Rapid kinetics lead to reversible behavior while slow electrochemical reactions are characterized as being irreversible. There is no sharp dividing line between the two. The consequences of slow electrochemical kinetics or irreversible electrochemical behavior will be pointed out in succeeding sections.

II. TITRIMETRY

1. End-Point Detection Systems

A. Zero-Current Potentiometry

A very important application of electroanalytical techniques is in their use as end-point detection systems for a wide variety of tritrations. Acid-base, precipitation, and complexometric titrations can usually all be carried out with an electroanalytical end-point detection system. In zero-current potentiometry the course of a titration is followed by measuring the change in potential of an indicator electrode at zero faradaic current.

Experimental conditions are arranged such that no significant current flows through the electrode system during the measurements. The magnitude of any current should be so low that it should not affect the measurement. This method has been applied to all types of titration systems including inorganic and organic acids and bases in aqueous and nonaqueous solvents. Precipitation titrations with silver, lead, and zinc often use this end-point detection method. Complexometric titrations including EDTA and redox titrations with cerium or iron solutions and many other titrants are well known (32). The principle of the titration apparatus is shown in Figure 11.2. The actual units are quite simple and commercially available.

The basis of the method depends on an indicator electrode whose potential changes during the course of a titration because of the change in concentration of either one of the reactants or one of the products of the electrode system in the titration reaction. If the electrochemical reaction is fast, the electrode potential may be expressed by the usual Nernst equation. The concentrations in the Nernst equation are those at the surface of the electrode. If potential measurements are made with no current flowing through the electrodes, the concentrations at the electrode surface will be the same as the concentrations of the species in the bulk of the solution. On the other hand, if current is flowing the concentrations at the electrode surface are changing and concentration gradients are created in the solution. Measurements at zero current are possible, if the voltage measuring device, V in Figure 11.2 draws no current during the measurement. A simple null-balance potentiometer serves satisfactorily for V, if the impedance of the electrode system is small. The V must be an electrometer (19), however, if the electrodes have an impedance of more than about 1 MΩ.

It has been implied that the electroactive species at the indicator electrode system are in equilibrium with each other at the time of potential measurement. This is certainly desirable, otherwise the electrode potential will drift. We must consider, however, the possibility of either a slow titration reaction in the bulk of the solution or a slow (irreversible) elec-

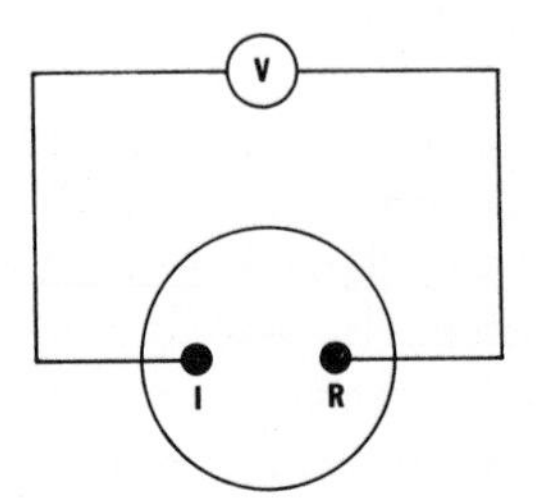

Figure 11.2. Zero-current potentiometric apparatus. I = indicator electrode, R = reference electrode, V = potentiometer or electrometer.

trochemical reaction at the surface of the electrode. For most accurate results it is necessary that the titrant is mixed rapidly and completely before a potential measurement is taken and that the titration is carried out manually point by point, rather than automatically.

Virtually every precipitation reaction equilibrates slowly near the equivalence point. If one is careless and does not allow the titration reaction to equilibrate, the indicator electrode's potential will appear to change less rapidly during the titration than it does in fact. An erroneous titration curve similar to that shown in Figure 11.3 will be the result. The observed curve is skewed and the break occurs at some distance beyond the true equivalence point. Sometimes the break is abnormally broad, in which case the equivalence point is difficult to determine.

Slow titration reactions do have profound implications for automatically recorded potentiometric titrations. There are many systems in which equilibrium is never achieved during the titration. Usually the displacement from equilibrium is the greatest near the equivalence point. Many automatic titrators reduce the speed of titration as the break in the curve is approached. This technique also reduces errors which arise from incomplete mixing of titrant and titrate at any given point in an automatic titration. The important point is learned in undergraduate analytical chemistry: standardize and analyze under identical conditions. If the analysis is to be done automatically, then the standardization should be done automatically also, with the same rate of titrant addition in both cases. If the titration is performed manually, both the analysis and standardization should be carried out with the same increments of titrant added and the same times allowed for the indicator electrode to reach equilibrium.

Once the titration curve has been obtained, one must then analyze it to determine the equivalence point of the titration. In most cases. it is assumed that the midway point of the break in the curve is the inflection point and

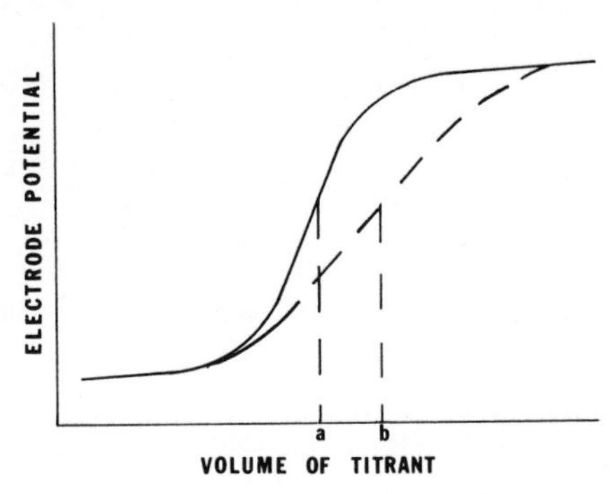

Figure 11.3. Zero-current potentiometric titration curves, *a* = equivalence point in which equilibrium is established during the titration, *b* = apparent equivalence point, if equilibrium is not established during the titration.

equivalence point of the titration. Actually this is true only for curves which are symmetrical about the equivalence point, in other words, if there is one mole of reactant to one mole of titrant. Another prerequisite is that the electrochemical electrode reaction is rapid.

Lingane discusses asymmetric titration curves in his monograph (27), and the reader is referred to it for more detailed information. In many laboratories the error introduced by taking the midpoint of the break as the equivalence point would probably be considered relatively small and can be ignored. For more accurate work however, $\pm 0.3\%$ or better, one should not *a priori* ignore this difference but check with accurately known standards whether or not the error is appreciable.

The inflection point of an ideally symmetrical titration curve, such as shown in Figure 11.4*a*, is rather easily determined graphically. The parallel foot and plateau of the curve are extrapolated so that a line perpendicular to

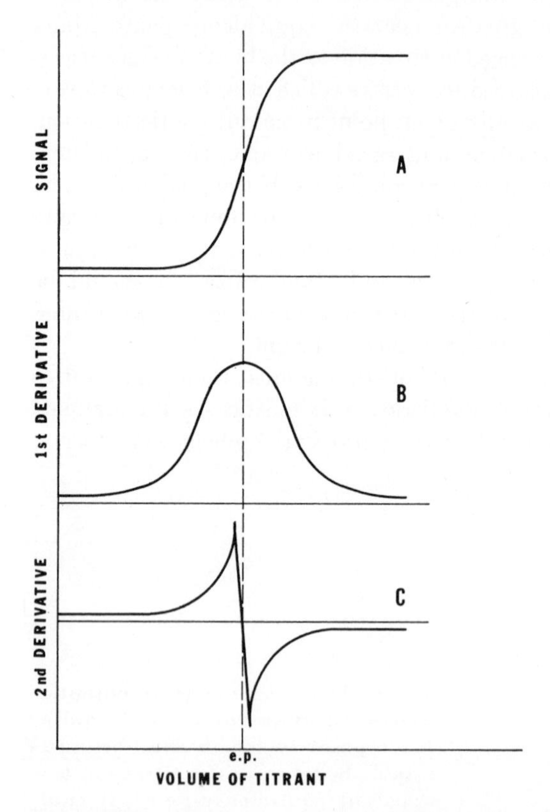

Figure 11.4. Zero-current potentiometric titration curves and first and second derivatives.

both can be drawn. The inflection point of a symmetric titration is located where this line is bisected by the titration curve. Frequently this method leaves something to be desired; the foot and plateau may not be parallel or the curve may be asymmetrical. In these cases, the use of the first or second derivatives of the titration curve may be more desirable. These are shown in Figures 11.4*b* and 11.4*c*. The derivatives can be obtained analytically (27) but this is rather laborious. There is little doubt that today it would be more economical for a laboratory with any significant amount of titration work to purchase a system which can record the first or second derivatives directly. To what extent one wants to automate his titrations will depend on the sample load, desired elapsed time for the report, and desired accuracy of the analysis.

The use of zero-current potentiometry is ultimately dependent on the quality of the curve one obtains. It is obvious that if one cannot obtain a reproducibly measured break, the method will not be applicable. Already it has been noted that slow equilibria will skew the curve and possibly reduce the usefulness of the data. We must also note that the magnitude of the potentiometric break is affected by the concentrations of reactant and titrant as well as the extent to which the titration reaction proceeds to completion. For very low concentrations of reactant or titrant and/or when the equilibrium constant of the titration reaction is small, the size of the break decreases until it can no longer be discerned and the equivalence-point determination becomes very inaccurate. Generally zero-current titrations are carried out with concentrations no lower than 0.01 N, if an accuracy of 0.1% or better is desired.

There are some applications described in the literature in which one titrates to a predetermined equivalence-point potential (27). While this technique offers some savings in both time and material it should be applied cautiously. Electrodes are not invariant objects; their surfaces can and often do change with time. This can lead to an apparent change in the equivalence-point potential. If one uses a titration to a predetermined potential, the system should be checked regularly to ascertain that the end-point potential has not changed with time.

It should not be implied that the electrode reaction must be rapid, if zero-current potentiometry is to be used. In some cases it is still possible to use this technique with a slow electrode reaction, for example, in the permanganate titration of iron(II), but this system has definite disadvantages. In general, with a slow electrode system zero-current potentiometry may not be useful, usually because inflection points become difficult to determine.

The *Handbook of Analytical Chemistry* (32) has a relatively recent compilation of potentiometric titration references. Potentiometric titration biennial reviews in *Analytical Chemistry* provide many references along with the *Collective Indices of Analytical Chemistry* (1) and various series which have reviews, for example, Part II of the *Treatise on Analytical Chemistry* (24).

B. Constant-Current Potentiometry

Another widely applicable end-point detection technique in many titrations is called constant-current potentiometry. In this technique, a small but constant current, about 5 μA or so, is forced to flow through the electrode system during the titration. The potential change during the titration is monitored as usual (Figure 11.5). This technique is used in preference to zero-current potentiometry when the electrode system is so slow that a break in the zero-current potentiometric curve is not readily obtained or when it is more advantageous to use a second metallic electrode in place of a conventional reference electrode such as the SCE. This configuration is desirable in many nonaqueous systems. With the SCE the potassium chloride electrolyte may be precipitated at the electrode/solution interface, effectively causing a break in the circuit or too much aqueous solution may leak into the titration system and interfere with the titration. By using metallic electrodes for both cathode and anode these potential problems are obviated. The titration involving coulometrically generated bromine in a nonaqueous system is a good example of a constant-current potentiometric end-point detection system with some of the problems which can be encountered (6). Figure 11.5 schematically shows the typical constant-current potentiometry apparatus. Two electrodes are immersed in a solution of the titrate. The electrodes are generally of the same size and composition, for example, platinum–platinum, but there are applications in which dissimilar pairs, for example, platinum–tungsten, have been used.

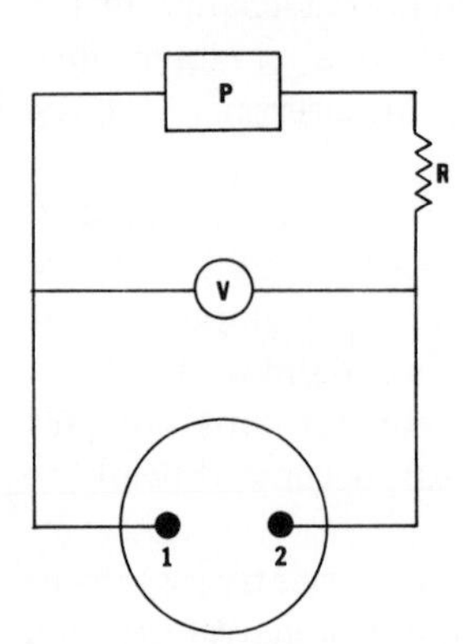

Figure 11.5. Constant-current potentiometric apparatus. *P* = dc power source, ~45 V; *R* = resistance, ~10^7 ohms; *V* = voltage-measuring device; electrodes 1 and 2 may be identical or dissimilar.

A relatively large dc voltage is applied across the electrodes. A large resistance, in series with the electrodes, is chosen to produce a current of 2–10 μA. Smaller indicator currents lead to sharper titration curve changes at the equivalence-point.

Because of the applied voltage, some electrolysis takes place at the electrodes. The error caused by this phenomenon is, however, insignificant.

The potential difference between the electrodes is determined by the nature and concentration of the electroactive species in the solution. Since potential difference measurements are not dependent on an equilibrium condition at the electrode surface, this technique is applicable to electrochemically slow reactions, in other words, where the rate of reaction at the electrode is determined by other factors than diffusion.

Charlot (10), and Lingane (27) discuss the shapes of constant-current potentiometric titrations curves, based on a set of idealized current-potential curves taken at various stages of the titration. The curves are obtained essentially the same way as a polarogram (Figures 11.11 and 11.12), except the dropping mercury electrode is replaced by one of the electrodes used in the titration apparatus. Depending on the electrochemical kinetics of the titrate and titrant couples, various titration curve shapes are possible, as shown in Figure 11.6. Titration to a preselected end-point difference is particularly hazardous in constant-current applications. Platinum oxide may be formed on the surface of the more anodic platinum electrode which can lead to anomalous titrations (6). Commercial apparatuses for constant-current potentiometry are readily available. Most pH meters have this option available and some automatic titrators have this capability. The concentration range over which constant-current potentiometry has been used is very similar to that of zero-current methods. A 10^{-3} normal solution can be titrated with an accuracy on the order of 1%. Meites' *Handbook of Analytical Chemistry* (32) and the biennial *Fundamental Reviews in Analytical Chemistry* provide readily accessible references for this method.

C. Null-Point Potentiometry

This technique (Figure 11.7) was developed primarily for trace applications, in other words, to determine concentrations at which the zero-current method is no longer applicable (30). The process is based on adding titrant to the unknown sample in order to bring the concentration of the species of interest to the same concentration of the same species in the other half of the cell system. Identical indicator electrodes are used in both sides of this two-chamber cell. The potential difference between the two electrodes is zero when the concentration of the investigated substance is the same in

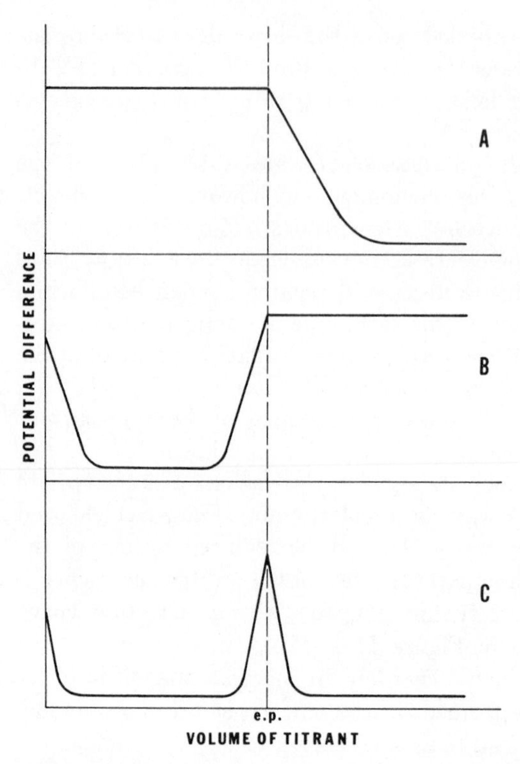

Figure 11.6. Constant-current potentiometric curves. The various shapes result from the reversibility or irreversibility of the titrant and titrate couples.

both half-cells. It is particularly important that the ionic strength of both half-cells is the same and rather concentrated titrants, of a 1–3 *M* range, be used to minimize dilution effects. Clearly, one must have some prior knowledge about the nature of the unknown sample if ionic strength effects are to be minimized. The technique has been applied to chloride determinations at the part per billion level (18). Null-point methods are included in the potentiometry biennial review in *Analytical Chemistry*.

D. *Amperometry*

An electrochemical equivalence point detection technique which is particularly well-suited to trace analysis is amperometry. In amperometric titrations the concentration of an electroactive substance is measured by the current which is generated from the reaction of these species at an electrode. Concentration changes during a titration are reflected in a

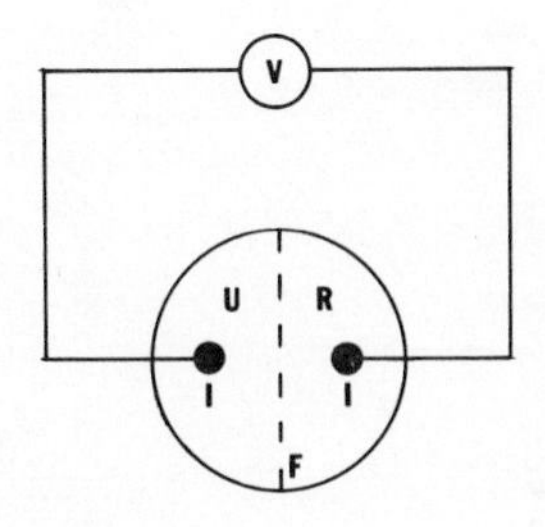

Figure 11.7. Null-point potentiometric apparatus. *F* = a bridge between two half-cells, *U* = solution with unknown concentration of analyte, *R* = solution with known concentration of analyte, *I* = identical indicator electrodes, *V* = voltage-measuring or null-balance device.

change in the current. Solutions of 10^{-3} *M* and less are titrated in 5–10 min unless the titration reaction itself is rather slow. Amperometric titrations involving silver, mercury, acids, bases, lead, ferro and ferricyanide, and many others are well documented (48). Methods involving organic compounds such as EDTA, heterocyclic nitrogen compounds, and sulfur compounds have also been described. The technique may be used with either one or two indicator electrodes and is depicted schematically in Figure 11.8. The electrodes are usually relatively small with surface areas of a few tenths of a square centimeter. In the single-indicator electrode case one applies a potential to the indicator electrode such that one of the species in the titration reaction is electroactive. The electrode potential is held at a constant value during the titration and the current which flows through the system after a certain period is the so-called limiting current (see Section VI). The second electrode is a reference electrode. Depending on whether the electroactive species is the titrate, titrant, or a product of the titration reaction, the amperometric curves may take on a variety of forms, as shown in Figure 11.9.

In dual indicator electrode systems, the potential of both electrodes vary during the titration, but the potential difference between the electrodes remains constant. The derivation of the titration curves is most easily done

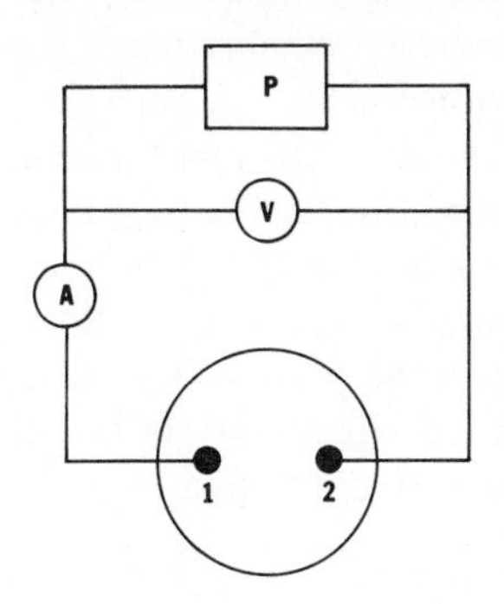

Figure 11.8. Amperometric apparatus. Single indicator electrode: *P* = dc power supply, 1 is held at a constant potential versus 2. Dual indicator electrodes: 1 and 2 are identical electrodes, held at a small, constant, potential difference.

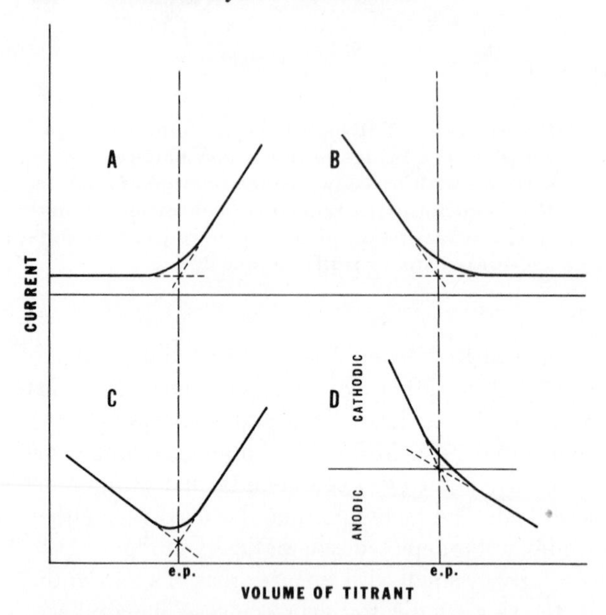

Figure 11.9. Typical single indicator electrode amperometric titration curves. Shape is dependent on the potential of the indicator electrode and the electroactive species at that potential.

from a set of current-potential curves of the titration system at various points in the titration (10). The applied potential difference is relatively small, 10–15 mV, and the current is generally not as large as in single-indicator electrode amperometry since limiting currents are not being measured.

Amperometry may be applied with either a stationary or rotating electrode system. Alternatively, the solution can be stirred—if the electrode is mechanically stable—in other words, it is not a dropping mercury electrode. A significant increase in sensitivity is realized with a rotating electrode system (or stirred solution) since more electroactive species reach the electrode per unit time than if the solution were quiescent.

In the single-electrode system, limiting currents are usually measured. They are dependent on the solution temperature and the rate of supply of the electroactive species to the electrode. Since in a quiescent solution the limiting current is diffusion controlled the experimental parameters affecting mass transport should be kept constant. When the system is stirred, the rate of mass transfer of electroactive species to the electrode is not diffusion-controlled any more. For best results the rate of stirring must be constant.

In the dual electrode system, limiting currents are generally not being measured and the control of mass transport parameters is not so important. Since limiting currents are being measured in the single-indicator electrode system and are larger than the nonlimiting currents measured in the dual-indicator electrode system, the single-electrode systems can be used in more dilute solutions; for example, in the 10^{-5}–10^{-6} M range.

Amperometric titrations are usually carried out with a relatively concentrated titrant so the total volume of the system will not change significantly during the titration. This is particularly important in single indicator electrode applications when limiting currents are being measured. In these cases an increase in volume will lower the observed current noticeably. To correct for this volume effect, the observed current can be multiplied by the factor $(V + v)/V$ where v is the titrant volume added to the original volume, V, of the system.

Another noteworthy advantage of amperometric techniques is that one does not have to obtain a complete titration curve. As seen in Figure 11.9 the curves just prior to and after the equivalence-points are linear. Hence one determines four or five experimental points on either side of the equivalence-point and extrapolates these lines until they intersect at the equivalence-point. The curvature in the actual titration curve at the equivalence-point is due to a nonquantitative titration reaction. Those two indicator electrode amperometric applications in which the current at and after the equivalence point is zero or close to zero were first introduced by Foulk and Bawden as "dead-stop" techniques (18*a*). In a "dead-stop" technique one can add the titrant dropwise as the endpoint is approached while observing the change in indicator current without bothering to plot the exact values of the current. Stock treats amperometric titrations thoroughly in his monograph (48).

2. Titrant Generation: Coulometry

In the previous section on titrimetry, titrant is added from a previously prepared source to the solution under study. In coulometry, electrolysis is used to generate the titrant *in situ* or externally immediately before use. This technique is used most often in two experimental situations: the titrant is unstable or the amount of titrant to be added is relatively small, in other words, in trace analysis. The method is based on Faraday's law—one equivalent of a substance is produced by 96,491 coulombs or A-sec of electricity.

In general, a titrant is generated with a constant current, i, for a time, t, until the equivalence-point of the titration is reached. Thus the number of equivalents of titrant generated is given by $it/96491$ where i and t are in units of amperes and seconds respectively. It is important that the generating process is either 100% efficient or that the efficiency is known and that the other substances generated at the electrodes in no way interfere with the determination—such as by the consumption of the titrate. Titrants including I_3^-, H^+, OH^-, Br_2, Sn^{2+}, BrO^-, Cr^{2+}, and alkali metals have been generated. Compounds such as water, phenol, unsaturated hydrocarbons, ammonia, oxygen, and mercaptans have been determined by coulometric titration. Considerable data are available from DeFord and Miller (13) and Milner and Phillips (35).

A convenient arrangement, if possible, is to place the generating electrodes in the solution which is to be titrated. The solvent for the titrate is added and the titrant is generated *in situ*. With this system the reaction rate between the titrate and titrant must be greater than the rate of generation of the titrant. Otherwise, whatever detection system is used for the end-point will register an early signal and the titration will be stopped. In some cases, the conditions necessary to generate the titrant also cause an adverse electrochemical reaction in the titrate solution. In this case, the titrant is best generated externally and added to the solution immediately.

The end-points may be detected visually, spectrophotometrically, or electroanalytically. Several apparatuses have been constructed to detect end-points potentiometrically or amperometrically and some will automatically switch off the titration at a preselected end-point potential or current. Whenever electroanalytical end-point detection is used in coulometric titrant generation one should take care to isolate the two electrical systems from one another. Otherwise, electronic cross-talk may cause spurious end-point signals.

III. ELECTROGRAVIMETRY

This technique rates as the oldest of the quantitative electroanalysis methods. In electrogravimetry the solution is exhaustively electrolyzed, in other words, all of the electroactive material is plated onto a cathode which is washed, dried, and weighed. Obviously the process is restricted to species which can be reduced to metals. Bismuth, cadmium, cobalt, copper, lead, tin, and zinc are some of the metals which have been determined electrogravimetrically. The three most widely used electrogravimetric methods are described below.

1. Constant Applied Potential Difference Electrogravimetry

A constant potential difference is applied across two platinum gauze electrodes. Although the potential difference between the two electrodes remains constant, the potential of each electrode will depend on what electroactive species are present in the solution. Furthermore, the cathode's potential drifts to more negative values as the concentration of the oxidized form of the electroactive species is decreased during the experiment. If a second metal is present, it may also be plated onto the cathode as the electrode potential becomes more negative. As these electroactive species near depletion, hydrogen evolution can take place at the cathode. It is a good practice to apply a relatively small potential difference across the electrodes, about 1–2 V, to minimize the chance of simultaneous reduction of two or more species.

With relatively pure materials, one does not encounter the problem of simultaneous reduction of two electroactive species. In such cases, exhaustive electrolysis with a constant applied potential can yield very accurate results, in the order of 2–5 parts per thousand. Evolution of hydrogen at the cathode is likely to cause a mechanically less stable plate and lower the accuracy of the determination because of possible loss of some material. Evolution of oxygen at the other, counter electrode is also undesirable since there may be a subsequent chemical reaction with the species being plated. A cathodic depolarizer, for example, nitric acid, is sometimes added to prevent the discharge of hydrogen gas. The depolarizers themselves are reduced at less negative potentials than hydrogen ion and do not interfere with the production of a stable plate. They are added in sufficient concentration to hold the cathode's potential at sufficiently positive levels that hydrogen will not be evolved during the electrolysis. Anodic depolarizers, for example, hydrazine or hydroxylamine are added whenever the evolution of oxygen at the anode is undesirable.

A trade-off arises in constant applied potential electrolysis between the selectivity and the time for completion of the electrolysis. The smaller the applied potential difference between the electrodes, the greater the selectivity of the electrolysis reaction. This is very desirable, if more than one electroactive metal is present. On the other hand, as the applied potential difference is reduced, the rate of the electrolysis reaction is also decreased. This is reflected in a smaller electrolysis current and a longer elapsed time for the experiment.

2. Constant-Current Electrogravimetry

The slow rate of deposition and long electrolysis times frequently encountered in constant applied potential applications is obviated by maintaining a relatively large constant current through the electrodes, for example, of the order of milliamperes or amperes. The potential difference between the electrodes as well as the individual electrode potentials change slowly during this process. The same problems with evolving gases or other platable impurities as in the constant applied potential case are also observed. Nevertheless if one deals with relatively pure samples or samples which have no other electroactive species except the one to be determined, this technique is rapid, simple, and accurate. Accuracies on the order of 5 part per thousand or less are readily attained.

3. Controlled-Potential Electrogravimetry

In this case the potential of the working electrode (cathode) is carefully controlled at a preselected potential. The potential is determined, usually by obtaining a prior current–potential curve in the system, such that only the desired electroactive species is reduced. A three-electrode system, in contrast to a two-electrode system, is used. The two systems are shown schematically in Figure 11.10. The potential of the working electrode is constantly monitored using some reference electrode. A potentiostat, P, adjusts the applied potential difference between the working electrode and counter electrode to maintain the working electrode's potential constant with respect to the reference electrode. This technique has obvious advantages in those samples which contain several electroactive species. The principal drawback is the decrease in electrolysis current and concomittant decrease in the rate of deposition of metal as the electrolysis proceeds.

With the advent of readily available modular analog electronic components one can build any one of these three types of electrolysis apparatuses at relatively low cost. Commercially available potentiostats are also rather inexpensive today. In addition to Meites' *Handbook* (32), Tanaka has written another good reference work for electrogravimetric techniques (49).

IV. DIRECT COULOMETRY

Nonmetal electroactive species are not subject to gravimetric electrolysis methods; hence other methods are necessary to determine the concentration of these species in solution. One good method is direct coulometry where electrons are used as the titrant. If the current efficiency for a particular electrochemical process is 100%, one may use Faraday's law to determine the amount of the substance of interest. The completion of the electrolysis may be noted by an electrochemical or visual indicator, much in the same manner as a conventional titration end-point is detected. The electrolysis can also be stopped at a predetermined residual current value.

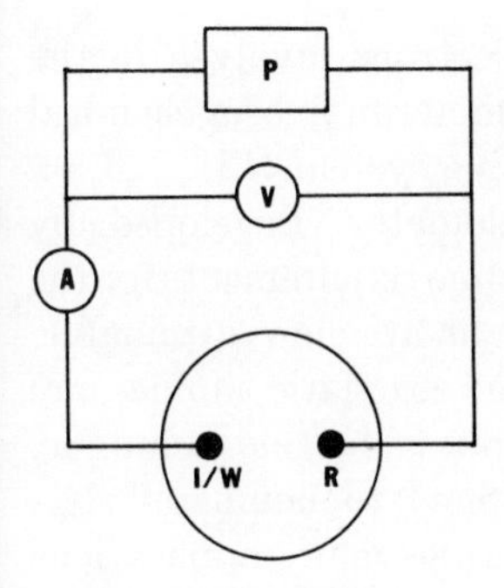

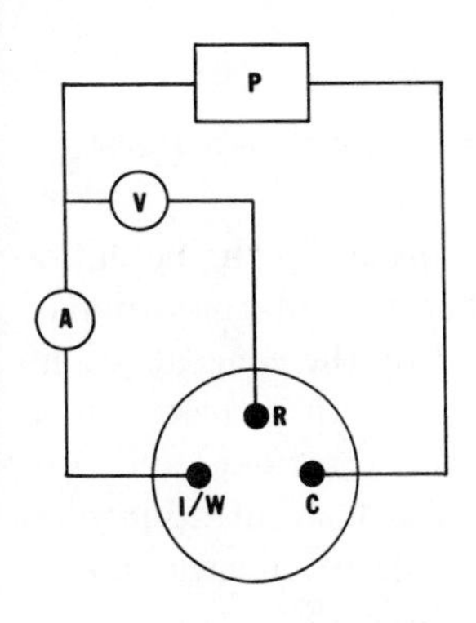

Figure 11.10. Two- and three-electrode systems. P = voltage source, fixed or variable; V = voltage-measuring device; A = current-measuring device; I/W = indicator or working electrode; R = reference electrode; and C = counter electrode.

It is imperative that the current efficiency be 100%. Therefore, the potential of the working electrode must be maintained at a value such that only the desired electrochemical reaction occurs. The apparatus and technique is similar to a controlled-potential electrolysis experiment in which a potentiostat controls the working electrode's potential. A coulometer must be used to determine the total number of equivalents of electricity which flowed during the experiment. The current decreases as the concentration of electroactive substance is reduced during the electrolysis. One must integrate the total current which flows during time, t, of the experiment to determine the mass of the electroactive material. The mass, in grams, is given by

$$\text{mass} = \frac{itM}{nF} \tag{11.4}$$

where i is in amperes, t is in seconds, M is the formula weight of the electroactive species in grams, n is the number of electrons involved in the electrode reaction, and F is the faraday. The coulometer may be a chemical system, an electromechanical device, or an electronic system (13).

Techniques of rapid controlled-potential coulometry, developed by Bard (2) ,have substantially improved the elapsed time requirements for the technique. Analysis times of 20 min instead of 2 hr are now attainable. Aliphatic amides, sulfur, nitro, nitroso, and some aromatic amines are examples of organic compounds which can be determined by this technique.

Direct coulometry is, of course, not only confined to nonmetal electroactive species. Rechnitz (42), for example, describes many applications to metals, including arsenic, chromium, neptunium, nickel and uranium. The method is particularly well-suited for trace analytical problems. Accuracies attainable in many systems are in the order of 5 parts per thousand or less.

V. DIRECT POTENTIOMETRY; ION-SELECTIVE ELECTRODES

The quantitative determination of electroactive species by the potentiometric techniques discussed thus far has depended on the interpretation of changes of electrode potential. The potential is fixed by the concentrations of the species involved in an electrochemical reaction. In principle, it is possible to correlate electrode potentials to concentrations and use potential measurements themselves to determine the concentration of a given electroactive species. This technique is used for pH measurements with glass electrodes and for a variety of other species with so-called specific ion electrodes including K^+, Na^+, Ca^{2+}, Mg^{2+}, Cl^-, and F^-. This type of application, however, is the only one in which concentrations are determined directly from potential measurements.

The drawback to most electrode systems for this type of application is the lack of reproducibility of the electrode potential. The electrode may respond to more than one electroactive species simultaneously and yield a mixed potential. A variable surface condition, such as the presence of an oxide film, may inhibit a particular electrochemical reaction and make the electrode less responsive to a particular species. Specific-ion electrodes are outstanding exceptions. The development of novel ion-selective electrodes has been called the single most important factor in the revitalization of analytical potentiometry. These electrodes, which are quite rugged, portable, easy to use and relatively inexpensive measure directly the activity (rather than the concentration) of a specific ion in solution. The electrodes are sometimes subject to surface conditions which can cause a malfunctioning of the electrode. By following the usual standardization

Table 11.3. Commercially Available Ion-Selective Electrodes, 1969 (26b)

Electrode Designated For	Class	pH Range	Principal Interferences
Bromide	Solid	0–14	CN^-, I^-, S^{2-}
Cadmium	Solid	1–14	Ag^+, Hg^{2+}, Cu^{2+}, Fe^{2+}, Pb^{2+}
Calcium	Liquid ion exchange	5.5–11	Zn^{2+}, Fe^{2+}, Pb^{2+}, Cu^{2+}, Ni^{2+}
Chloride	Solid	0–14	Br^-, I^-, S^{2-}, CN^-, SCN^-, NH_3
Chloride	Liquid ion exchange	2–10	ClO_4^-, I^-, NO_3^-, Br^-, OH^-, OAc^-, HCO_3^-, SO_4^{2-}, F^-
Cyanide	Solid	0–14	S^{2-}, I^-
Cupric	Solid	0–14	Ag^+, Hg^{2+}, Fe^{3+}
Fluoride	Solid	0–8.5	OH^-
Fluoroborate	Liquid ion exchange	2–12	I^-, NO_3^-, Br^-, OAc^-, HCO_3^-
Iodide	Solid	0–14	S^{2-}, CN^-
Lead	Solid	2–14	Ag^+, Hg^{2+}, Cu^{2+}, Cd^{2+}, Fe^{2+}
Lead	Liquid ion exchange	3.5–7.5	Cu^{2+}, Fe^{2+}, Zn^{2+}, Ca^{2+}, Ni^{2+}, Mg^{2+}
Nitrate	Liquid ion exchange	2–12	ClO_4^-, I^-, ClO_3^-, Br^-, S^{2-}, NO_2^-, CN^-, HCO_3^-, Cl^-, OAc^-, CO_3^{2-}, $S_2O_3^{2-}$, SO_3^{2-}
Perchlorate	Liquid ion exchange	4–10	OH^-, I^-, NO_3^-
Hydrogen (pH)	Glass	0–14	
ORP (OX-RED POT.)	Solid	0–14	All redox systems
Potassium	Glass	7–13	H^+, Ag^+, NH_4^+, Na^+, Li^+
Silver	Solid	0–14	Hg^{2+}
Silver	Glass	4–8	H^+
Sodium	Glass	3–12	Ag^+, H^+, Li^+, K^+
Sulfide	Solid	0–14	
Thiocyanate	Solid	0–14	I^-, Br^-, Cl^-, $S_2O_3^{2-}$, NH_3
Water Hardness	Liquid ion exchange	5.5–11	Zn^{2+}, Fe^{2+}, Cu^{2+}, Ni^{2+}, Ba^{2+}, Sr^{2+}

procedures however, problems are readily identified. Manufacturers give ample literature describing precautions to be observed in the use of such electrodes.

Some of these systems have outstanding sensitivities, for example, concentrations as low as 10^{-12} *M* hydrogen ion can be detected and measured with the glass electrode. At these low levels some accuracy is sacrificed. Results which are accurate to 5% are common, but for a large number of applications this is quite acceptable. The systems are particularly well-suited for continuous monitoring applications for example, in water analysis, oceanographic work, and routine laboratory applications. A variety of equipment is available commercially. Table 11.3 lists a number of available ion-selective electrodes. Most of these electrodes can be obtained from more than a dozen companies. The *Annual Laboratory Guide* published by the American Chemical Society as a supplement to *Analytical Chemistry* gives an up to date list of suppliers.

Bates' monograph on the determination of pH is particularly useful (5). The monograph edited by Eisenman discusses the glass electrode for the determination of hydrogen as well as other ions (17). The discussions of a symposium on various types of ion-selective electrodes, including electrodes other than glass, are summarized by Durst (16). Rechnitz (43) reviewed this field in 1967.

VI. POLAROGRAPHY

1. Direct-Current Polarography

Direct-current polarography is used to determine the concentration of electroactive species in solution by changing the oxidation state of these elements or compounds of interest. Hence dc polarography is only applicable to the determination of electroactive species with two or more relatively stable oxidation states. The literature is, however, replete with dc polarographic work, covering both organic and inorganic applications. Some form of nearly every stable element in the periodic table has been investigated polarographically. The method is not confined to inorganic elements. Polarographically active organic species include aromatic hydrocarbons, aromatic and unsaturated aliphatic carboxylic acids, thiols, carbonyl compounds, amines, proteins, and amino acids (see Table 11.4).

The method can detect elements in concentrations as low as a few ppm. The sensitivity is proportional to the number of electrons transferred in the reaction. Organic compounds, where six or more electrons can be transferred during a reduction or oxidation reaction, can be determined to much lower levels than those elements showing only a one-step change in oxidation level. At higher concentrations of the substance of interest, polarography is not always the best approach, especially if only one species is of interest. This is due to the inherently low accuracy and precision of the method.

Polarography is especially applicable when different oxidation states of an element are found simultaneously in one solution and one wishes to determine the relative concentrations of this element. Polarography is also a very good technique when an element and an organic portion of a metallo-organic compound have to be determined simultaneously in solution.

In a standard dc polarograph a linearly increasing dc potential is applied between a dropping mercury electrode (DME) and a reference electrode, usually a saturated calomel electrode, and the resulting current is plotted as a function of the applied potential voltage. A schematic diagram of a polarographic cell is shown in Figure 11.11. The configuration in modern instruments is essentially the same, except that the contact across the calibrated slide-wire is driven by a synchronous motor and the output current is measured by an automatic recorder.

In polarography no titrant is added from an external source or generated internally; the observed current is due to either an oxidation or a reduction process which takes place in the sample solution. The observed current

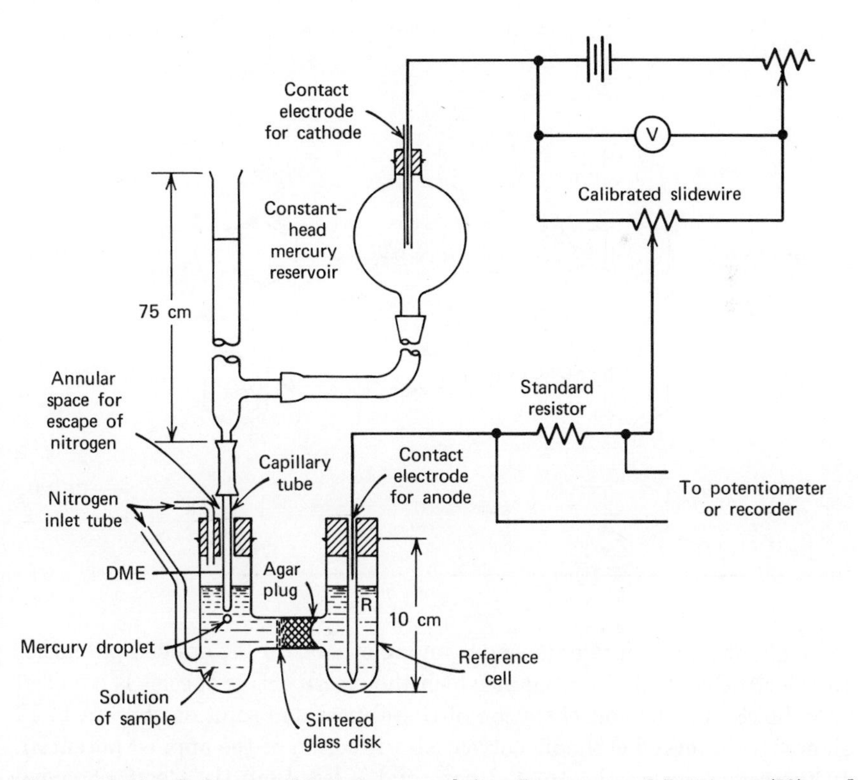

Figure 11.11. Polarographic apparatus [after Lingane and Laitinen (28) and Pecsok and Shields (36)]. DME = dropping mercury electrode; *R* = reference electrode, usually a saturated calomel electrode.

Table 11.4. Simplified Organic Reduction Processes in Polarography (52)

Bond Grouping	Example
C—C	p-$NC-C_6H_4-COOR + 2e + 2H^+ \longrightarrow C_6H_5-COOR + CN^-$
C—N	$R-CO-CH_2-\overset{(+)}{N}R_3 + 2e + H^+ \longrightarrow RCOCH_3 + NR_3$
C—X	$R-CH_2-Br + 2e + H^+ \longrightarrow R-CH_3 + Br^-$
	$R-CH(Br)-CH(Br)-R + 2e \longrightarrow R'\ CH{=}CHR + 2Br^-$
C=C	$C_6H_5-CH{=}CH_2 + 2e + 2H^+ \longrightarrow C_6H_5-CH_2CH_3$
C=N	$R'-C(={\overset{(+)}{HNR}})-R'' + 2e + 2H^+ \longrightarrow R'-CH(HNR)-R''$
C=O	$C_6H_5-COR + e + H^+ \longrightarrow C_6H_5-\dot{C}(OH)-R \longrightarrow$ Dimer
C≡C	$C_6H_5-C{\equiv}C-CHO + 2e + 2H^+ \longrightarrow C_6H_5-CH{=}CHCHO$
C≡N	$CH_3CO-C_6H_4-CN + 4e + 4H^+ \longrightarrow CH_3CO-C_6H_4-CH_2NH_2$

throughout a polarogram depends, among others, on the applied potential, the electrochemical potential at which the electroactive species is oxidized or reduced and the concentration of these species in solution. Figure 11.12 shows the observed cathodic current as a function of the applied potential. In the region where the applied potential is less than the electrochemical potential at which an electrochemical reaction occurs no appreciable current is observed. At higher applied potentials a current is observed. The

Table 11.4.—Continued

Bond Grouping	Example
N—O	R—N—CH (ring: N(+)—C—O(−), N—O—C) + 4e + 4H$^+$ ⟶ R—N(NH$_2$)—CH$_2$—COOH
	R—C═N(→O)—R′ + 2e + H$^+$ ⟶ R—C═N—R′ + OH$^-$
	R—NO + 2e + 2H$^+$ ⟶ R—NHOH
N═N	C$_6$H$_5$—N═N—C$_6$H$_5$ + 2e + 2H$^+$ ⟶ C$_6$H$_5$—NHNH—C$_6$H$_5$
N═O	R—NO$_2$ + 4e + 3H$^+$ ⟶ R—NHOH + OH$^-$
O—O	R—O—O—R + 2e + 2H$^+$ ⟶ 2R—OH
S═O	R—SO—R′ + 2e + H$^+$ ⟶ R—S—R′ + OH$^-$
S—S	R—S—S—R + 2e + 2H$^+$ ⟶ 2RSH
	Naphthalene + 2e + 2H$^+$ ⟶ 1,4-dihydronaphthalene (or) 1,2-dihydronaphthalene
	Tropylium ion (+) + e ⟶ cycloheptatrienyl radical (•) ⟶ Dimer
	N-alkylpyridinium, R (+) + e ⟶ pyridinyl radical (•), N—R ⟶ Dimer

half-wave potential, $E_{1/2}$, is characteristic for each oxidation state of each electroactive species. Unfortunately, this value is not sufficiently constant to be used for qualitative purposes because $E_{1/2}$ is very much influenced by the pH, complexing agents, ionic strength, the presence of organic impurities, and other factors.

The magnitude of the limiting current is determined by the rate at which the electroactive species can reach the surface of the electrode. With a

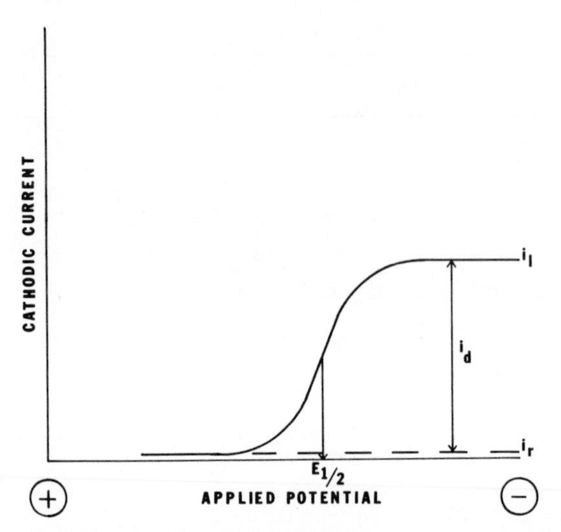

Figure 11.12. Current-potential curve. $E_{1/2}$ = half-wave potential; i_l = limiting current; i_d = diffusion current; i_r = residual current.

small electrode, such as the DME, this rate is frequently diffusion controlled The electrode is thus said to be "polarized," from which the term polarography is derived. Since the magnitude of the current is a function of the concentration of the electroactive species in solution, under carefully controlled conditions this concentration can be derived from the size of the current. It is noteworthy that polarography is the only electrochemical method in which concentrations are determined directly from a current measurement.

One of the most important components in a dc polarograph is the dropping mercury electrode. Mercury flows from a reservoir through a 0.06–0.08-mm ID capillary to form droplets at the end of the tube in the solution of interest before dropping off to the bottom of the sample container. At any applied potential the current increases as the mercury drop grows in size; it drops off sharply as the drop detaches itself from the tip of the capillary. An actual polarogram is shown in Figure 11.13, from which the idealized polarogram of Figure 11.12 can be derived. The main advantage of a DME is that one always has a clean and reproducible electrode surface, even though metals which are reduced at this electrode are dissolved in the Hg by amalgamation.

Electroactive species at 10^{-3}-10^{-6} M can be detected and determined by dc polarography. Such low concentrations can be analyzed because of the excellent reproducibility of successive mercury drops. Another advantage

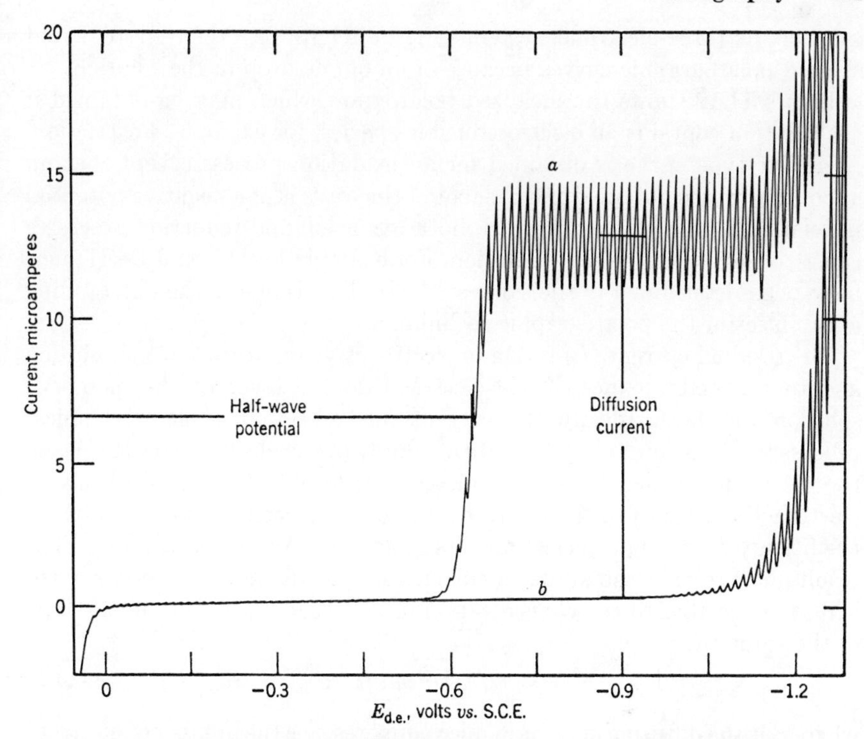

Figure 11.13. (*a*) Polarogram of 1 *M* HCl containing 0.5 m*M* Cd(II). The wave is due to the reduction of Cd(II) to Cd amalgam; the plateau, which begins at about −0.75 V reflects the practically complete reduction of all Cd(II) ions reaching the DME. (*b*) Polarogram of 1 *M* HCl alone [after Meites (34)].

of mercury is its rather high hydrogen over-voltage so one can reach more negative potentials and observe a greater variety of electroreductions than with platinum. The mercury electrode is unsuitable at more positive potentials due to the oxidation of the mercury metal. In such a situation, a platinum electrode is used. Whenever platinum electrodes are used, the technique is called voltammetry. As mentioned earlier, the platinum electrode is not very reproducible and is not used in applications in which concentrations are determined from the magnitude of the current flowing through the electrode.

Direct current polarographic instrumentation is fairly simple and relatively inexpensive, on the order of $2000 per unit. In the standard configuration a dropping mercury electrode (DME) and a reference electrode, usually a saturated calomel electrode is used. If the solution containing the electroactive species has a high impedance, for example, in a nonaqueous solution, a three-electrode system (Figure 11.10) is preferred to avoid skewed polarographic curves because of an ohmic drop in the solution.

Figure 11.12 shows the idealized polarogram which may be obtained if the solution contains an electroreducible species, for example, Fe(III) ion. A similar type curve is obtained for an oxidation process except that an anodic current is obtained and the foot of the wave is at a negative potential in relation to the plateau. Indeed, both oxidation and reduction processes can be observed in the same solution. For example Fe(III) and Fe(II) ions have been determined in the same solution. This is one of the outstanding capabilities of the polarographic technique.

The residual current (i_r) is due to electroactive impurities in the solution and to the maintenance of the electrical double layer at the mercury–solution interface. Ideally the current for a particular electrochemical process becomes limited by the rate of supply of the electroactive species to the electrode surface (i_l). In a quiescent solution, the rate of supply of electroactive species to the electrode surface is dependent on the diffusion coefficient, D, of the species, and its concentration in the solution. The resultant current is the diffusion current, i_d. The average diffusion current is directly related to the concentration of electroactive species in the bulk of the solution:

$$i_d = 607nD^{1/2}m^{2/3}t^{1/6}C \qquad (11.5)$$

where i_d is the diffusion current in microamperes, n is the number of electrons exchanged in the electrode process, D is the diffusion coefficient of the electroactive species (cm²/sec), m is the rate of mercury flow through the capillary (mg/sec), t is the lifetime of a drop of mercury (sec), and C is the millimolar concentration of the electroactive species. Equation 11.5 is known as the Ilkovic equation, and m and t are the "capillary characteristics" of the particular capillary tubing used. Although the Ilkovic equation is given for a rapid electrochemical process, polarography can also be used for a number of slow processes, such as for the detection of Cr(VI) and V(V). In general, one must be more careful that experimental conditions are held constant when dealing with slower electrode processes.

The general limit of detection of a species which undergoes a rapid electrode reaction is about 10^{-6} M. This limit is determined by the residual current which results from the maintenance of an electrical double-layer at the electrode surface (capacitance current). While this 10^{-6} M limit can occasionally be attained, routine polarography applications might better consider the 10^{-5} M level as a good practical limit because of the experimental effort required to decrease the limit another order of magnitude.

Rarely does one compute the concentration of a species from the Ilkovic equation. A variety of techniques are used to correlate the experimental results to the concentration of electroactive species in solution. Calibration or working curves similar to those obtained for spectrophotometric measurements are used. A series of standards are prepared from which polarographic curves are obtained. The diffusion current, preferably measured at a given potential, is determined for each solution and the results plotted on linear graph paper with current versus concentration. If working curves are used, a revision must be made if the capillary characteristics, m and t change. One technique to avoid a dependence on the capillary is to use an internal standard, similar to what is used in emission spectroscopy. One adds a known amount of a second electroactive species. A working curve is established by using solutions with a range of concentrations of the electroactive species to be determined but always with the same concentration of internal standard. The ordinate is the current ratio from the species being determined to that from the internal standard. The abscissa is the concentration of the species being determined. For relatively large numbers of samples, or at least where one can expect to repeat the analysis over several months, the internal standard approach is an excellent one.

The use of a standard addition technique, in other words, spiking a sample with a known concentration of the material to be determined, is particularly efficient if an analysis is done only occasionally. If an appropriate internal standard cannot be found for a particular case, the standard addition technique can be used to advantage. The method is particularly easy in polarography. A polarogram of the solution containing the species of interest is obtained and the diffusion current noted. A known amount of the same species is added from a rather concentrated solution so that the total volume of solution in the cell is changed by less than 1 or 2%. A second polarogram is obtained and the new diffusion current noted. The difference between the two diffusion currents i_x is directly proportional to the added concentration of x:

$$i_x = kC_x \tag{11.6}$$

and

$$\frac{i_u}{i_x} = \frac{kC_u}{kC_x} \tag{11.7}$$

so that the original concentration of the species C_u can be calculated from

$$C_u = \frac{C_x i_u}{i_x} \tag{11.8}$$

where C_x is the increased concentration of x, i_u is the original diffusion current, and i_x is the diffusion current increase due to the increase in concentration of x. Great care must be taken that all experimental conditions for the two polarograms are the same.

A variety of problems can make polarographic measurements difficult. Dissolved oxygen is polarographically active and must be removed and then excluded from the solution. This is accomplished by bubbling oxygen-free nitrogen gas through the solution for 5–10 min prior to obtaining the polarogram. Longer deoxygenation times are needed as the concentration of species being determined approaches 10^{-6} *M*. The presence of maxima, or humps, on the plateau of the polarogram can sometimes distort the plateau to make it virtually unusable. The maxima frequently are due to a potential dependent adsorption phenomenon on the DME which can be eliminated by adding 10^{-4}% of a surface-active agent such as gelatin, or a nonionic detergent, such as TRITON X-100.*

A singular advantage of polarographic methods is their capability to allow simultaneous determination of two or more species in the same solution. However if the current due to the species at more positive potentials is more than 10 times that due to the electroactive species at more negative potentials, the determination of the latter becomes very difficult. In general, one can resolve two electrochemical processes which yield the same current, if their half-wave potentials ($E_{1/2}$, Figure 11.12) are separated by at least 120 mV. Generally one may tolerate rather large relative amounts of the second electroactive species at more negative potentials, if reductions are involved or at more positive potentials, if oxidations are involved. However the foot of the second process will eventually contribute significantly to the value of the plateau of the first so that good results are hard to obtain. Accuracies of 3% of the amount present are considered satisfactory for routine polarographic analysis.

One can make use of electrode reactions which are other than diffusion controlled. For example in the case of many nitrogen base compounds a set of reactions is proposed:

$$B + H^+ \rightarrow BH^+ \tag{11.9}$$

$$BH^+ + e^- \rightarrow BH\cdot \tag{11.10}$$

$$2\,BH\cdot \rightarrow 2\,B + H_2 \tag{11.11}$$

where all reactions take place at the electrode surface. The nitrogen base (10^{-6}–10^{-7} *M*) is protonated by a proton donor which is in ample supply in the bulk of the solution (10^{-1} *M*). An electrochemical reaction takes place which generates a radical and is followed by a bimolecular collision of two radicals yielding the original base compound and a hydrogen molecule. The original base compound can enter the reaction sequence again since it is already at the electrode surface. The observed wave is called

* Trademark of the Rohm and Haas Company.

a catalytic hydrogen wave since molecular hydrogen is discharged at potentials more positive than found in the absence of the catalyst, B. Sulfur compounds also give rise to catalytic hydrogen waves which make polarography very useful in biochemical systems (8, 51, 52).

A wide range of polarographic techniques have been developed to increase the sensitivity of dc methods beyond 10^{-5} M or so and to resolve successive processes which are within 120 mV of each other and to separate two species where the current generated by the second component is less than one-tenth the first. Schmidt and von Stackelberg (44) have presented the basis of the techniques very succinctly.

In the author's opinion, dc polarography can solve perhaps 90% of the analytical problems for which polarography is a satisfactory approach. In many cases one can concentrate a sample, if polarographic sensitivity is not adequate. A chromatographic separation can be carried out, if two species are unresolved.

The three most important improvements in dc polarography, when cost and experimental methods are kept in mind, are single-sweep or cathode-ray methods, pulse polarography, and stripping analysis. These three techniques are now discussed.

2. Single-Sweep Polarography

Single-sweep polarography is carried out by the application of a dc sweep which is applied during a selected period of the lifetime of a single drop of mercury. The scan period usually ranges from 0.1 to 10 sec. Analysis times are, therefore, a magnitude or more smaller than in regular dc polarography. This is commonly stated as an advantage over dc polarography. One must remember, though, that in whatever technique is used, sample preparation time is usually the governing factor for the elapsed time of the analysis. The output is usually observed on an x–y recorder or oscilloscope. The generated current peaks are somewhat reminiscent of a gas chromatography trace with different peaks for different species. The size of each peak is a function of the concentration of the corresponding compound. The rate of change of the sweep is most important as well as the choice of the time in the life of the mercury drop at which the sweep begins. It is much easier to resolve two electroactive species by single sweep than by regular dc polarography because in the former technique there are no oscillations from the mercury drop during the experiment. In the interpretation of the polarogram, the separate peaks for the different species are more easily distinguished from each other than the different current levels in regular dc polarography, also because these levels are somewhat obscured by the oscillations due to the DME. On the other hand, the small size of an oscillosocope read-out can present problems in data reduction. Two peaks, rather close to one another, are readily detected but the extrapolation of the curve for the preceding process to establish a baseline for the second process can be difficult. Some workers use the derivative of the original signal to decrease this problem. Peaks separated by 40–60 mV have been used analytically in single-sweep polarography. In general, a one-order-of-magnitude increase in sensitivity is obtainable by single-sweep techniques over dc methods; 10^{-6} *M* solutions are typically determined. Very fast recording devices must be used in this technique, hence the names "cathode-ray" or "oscillopolarography." Depending on the experimental situation, a fast x–y recorder can also be used. The method has found wide acceptability in biocide residue analyses (20, 21, 31).

3. Pulse Polarography

There are two forms of pulse polarography. The normal type is obtained by the application of increasingly larger voltage pulses to successive drops of mercury. The recording resembles a dc polarographic wave. The other form is called the derivative or constant-amplitude mode and is obtained by superimposing pulses of constant-amplitude on an increasing dc voltage ramp or by taking the difference in current resulting from two successively larger pulses (9). In both cases, the pulse is applied only once in the lifetime of a mercury drop and the current is measured during a predetermined portion of the pulse. The measurement is taken in the latter part of the pulse. By this arrangement the capacitance current is nearly zero at the time of current measurement and only the faradaic current due to the electrochemical process, is measured. Thus the residual current due to the charging of the double layer, which limits dc measurements, is significantly reduced in pulse polarography. Normal pulse polarography is somewhat more sensitive for rapid electrode reactions but the derivative mode is more advantageous when dealing with mixtures of electroactive species. The reason is that the latter yields peaks rather than waves. One can resolve two peaks of equal height separated by 90 mV, if both peaks are due to one-electron processes. The resolution will not be complete, but is adequate to obtain results for each species within 10% of the amount present.

Typically, species present in a concentration as low as 10^{-7} *M* may be determined by pulse polarography. Several workers have noted that one may have ratios in the order of $1:10^3$ or $1:10^4$ of successive electroactive species and still separate the peaks usefully in the derivative pulse mode. Naturally the limit with which one may separate peaks and determine the species quantitatively is largely dependent on the individual species involved. These figures are only general guidelines.

Both single-sweep and pulse polarography require more sophisticated instrumentation than dc polarography. Several manufacturers have recently introduced all purpose electrochemical units which have the capability of all these techniques as well as others. The prices quoted are quite reasonable and well within the reach of a laboratory which has any substantial need for polarographic instrumentation. The references by Meites (33), and Kolthoff and Lingane (23) are particularly useful when starting any polarographic work.

VII. STRIPPING ANALYSIS

A very useful adjunct to single-sweep polarography is anodic stripping. It is applied frequently to the analysis of very dilute solutions of mercury-amalgam forming metals. Cadmium, copper, gallium, indium, lead, and zinc are among the metals which have been determined by this technique.

Stripping analysis is a two stage technique. In the first stage an electrode usually consisting of a single mercury drop is held at a sufficiently negative potential so that the electroactive species of interest are reduced. If the species form metals upon reduction, the metal forms an amalgam and diffuses into the mercury drop. This concentrates the electroactive material, as the metal-amalgam, in the electrode. This cathodic electrolysis is continued until the concentrations of the various amalgams are sufficiently high.

In the second stage of the technique the cathode potential is increased. In this process of sweeping to more positive potentials an anodic current peak is generated at those potentials when the dissolved metals are oxidized. Because of the preconcentration step these peaks are large enough to be easily measured. More than one substance can often be determined in a single experiment because successive anodic current peaks corresponding to the different metals in the amalgam are observed as the potential of the electrode is increased. Dilute solutions of 10^{-6}–10^{-9} M have been analyzed with cathodic electrolysis times of 5–60 min and resultant accuracies of 0.1–4%. Besides mercury, an inert solid electrode, such as platinum or carbon, has been used (3, 45).

VIII. CONDUCTOMETRIC ANALYSIS

Techniques where the conductance of a solution is measured to determine the concentration of ionic species in solution rate among the oldest electroanalytical methods. The chief distinguishing feature of conductometric analysis from other electroanalytical methods is its lack of specificity.

The conductance of a solution is the sum of the contributions of all the ionic species present, and is hence not only dependent on the species for which one is analyzing. If, however, the concentration of all the other ionic species are maintained at a constant level, the observed conductance can then be related to the concentration of the species of interest. Conductance measurements have some marked advantages. In many cases they are quite accurate, and results to within 0.1% or better are easily obtained. Parameters other than the conductance itself may be the limiting factor of accuracy in a particular experiment, for example, the accuracy with which titration volumes or the strength of a standard titrant can be determined.

Generally a conductance cell is made as one arm of a Wheatstone bridge. An alternating current of 10^3–10^4 Hz at an applied voltage of approximately 5 V is used to avoid a change in solution composition created by electrode reactions. Direct-current measurements have been made but then very small currents are used so that the change in the solution composition is minimal. The design of the experimental cell is dependent on the conductance of the solution being measured. Temperature constancy in conductance work is somewhat more important than in most other electroanalytical methods. Conductivity temperature coefficients range from 2.0 to 2.5%/degree except for the hydrogen and hydroxyl ions which are respectively 1.5 and 1.8%/degree (40). In conductance titrations, the temperature should be held constant during the experiment, preferably to within ±0.1°C.

When conductance is used as an end-point detection method, it has the advantage that one can obtain data points before and after the equivalence point. Extrapolation of the linear portions of these curves to a point of intersection yields the end-point of the titration. Conductometry is similar in this respect to amperometric or spectrophotometric end-point detection. End-point detection by conductometry has been applied in acid–base, precipitation, and redox titrations.

High-frequency conductometry, at several megacycles, is particularly advantageous in those cases in which electrode poisoning might be a problem, for example, in precipitation titrations, or when trace sulfur impurities are present. In high-frequency applications, electrodes are external to the system (29, 40).

IX. AUTOMATED SYSTEMS

Automation in titration systems has been widely used in analytical laboratories for over 25 years (39). Most systems use a zero-current potentiometric end-point detection system. The titrant is added at a constant rate or is generated coulometrically. Any already existing titration technique can be carried out on an automatic titrator, if desired. It is interesting to note that only the actual titration system has been automated to a high degree. Automatic sampling, data acquisition and data processing are not often incorporated in a complete system. With the advent of the small digital computer, this is likely to change rather rapidly. There is no reason why a small digital computer cannot be used more frequently to control the titration experiment as well as to acquire and to process the data. Automatic titrators with constant-current potentiometric and amperometric end-point detection systems can now be purchased. Another useful system, which can now be obtained commercially at reasonable cost is the so-called pH stat, which is a system to control the pH.

Real-time computer applications in single-sweep polarography have been described by Perone and co-workers (37, 38). Smith has developed real-time applications for alternating-current polarography (47). Both systems use computers to control the experimental conditions as well as to acquire and process the generated data. Lauer and Osteryoung (26), and Ramaley and Wilson (41) have described general-purpose systems for computer data acquisition and processing, and for control of electrochemical experimental apparatus. In the next few years considerable developments of technique enhancement, data acquisition, and processing will be seen in electroanalytical methodology.

X. MISCELLANEOUS METHODS

There are a variety of electroanalytical methods which a reader may find missing in this chapter. They include chronoamperometry, chronopotentiometry (12, 15), alternating-current polarography (7, 46), and square-wave polarography (4). All of these techniques have their particular usefulness and in some analytical problems one of these methods may very well be the most appropriate. In general, they are applied to rather specific analytical problems, for example, in the study of thin films or the determination of intermediate reaction products. These techniques are therefore more often used in fundamental studies in electrochemistry and less often in applied analytical problems.

REFERENCES

1. American Chemical Society, 1155 Sixteenth St., N.W., Washington, D. C. 20036, *Collective Indices of Analytical Chemistry*, available for the years 1929–1968.
2. A. J. Bard, *Anal. Chem.* **35,** 1125 (1963).

3. E. Barendrecht, in *Electroanalytical Chemistry*, Vol. 2, Ed., A. J. Bard, Marcel Dekker, New York, 1967, p. 53.
4. G. C. Barker, *Anal. Chim. Acta* **18,** 118 (1958).
5. R. G. Bates, *Determination of pH Theory and Practice*, Wiley, New York, 1964.
6. F. Baumann and D. D. Gilbert, *Anal. Chem.* **35,** 1113 (1963).
7. B. Breyer and H. H. Bauer, *Alternating Current Polarography and Tensammetry*, Interscience, New York, 1963.
8. M. Brezina and P. Zuman, *Polarography in Medicine, Biochemistry, and Pharmacy*, Interscience, New York, 1958.
9. D. E. Burge, *J. Chem. Educ.* **47,** A81 (1970).
10. G. Charlot, J. Badoz-Lambling, and B. Tremillon, *Electrochemical Reactions*, Elsevier, New York, 1962.
11. J. A. Christiansen and M. Pourbaix, *Proc. 17th Conf. IUPAC, Stockholm*, Butterworths, London, 1953, pp. 82–5. Reprinted by J. A. Christiansen, *J. Amer. Chem. Soc.* **82,** 5517 (1960).
12. D. G. Davis, in *Electroanalytical Chemistry*, Vol. 1, Ed., A. J. Bard, Marcel Dekker, New York, 1966, p. 157.
13. D. D. Deford and J. W. Miller, in *Treatise on Analytical Chemistry*, Part 1, Vol. 4, I. M. Kolthoff and P. J. Elving, Eds., Interscience, New York, 1963, Chapter 49.
14. P. Delahay, *New Instrumental Methods in Electrochemistry*, Interscience, New York, 1954.
15. P. Delahay, in *Treatise on Analytical Chemistry*, Part 1, Vol. 4, I. M. Kolthoff and P. J. Elving, Eds., Interscience, New York, 1963, Chapter 44.
16. R. A. Durst, Ed., "Ion-Selective Electrodes," *U.S. National Bureau of Standards Special Publication No. 314*, U.S. Govt. Printing Office, Div. Public Documents, Washington, D. C., 1969.
17. G. Eisenman, *Glass Electrodes for Hydrogen and Other Cations, Principles and Practice*, Marcel Dekker, New York, 1967.
18. P. Emmott, *Analyst* **90,** 482 (1965).
18a. C. W. Foulk and A. T. Bawden, *J. Amer. Chem. Soc.* **48,** 2045 (1926).
19. N. H. Furman, in *Treatise on Analytical Chemistry*, Part 1, Vol. 4, I. M. Kolthoff and P. J. Elving, Eds., Interscience, New York, 1963, Chapter 45.
20. R. J. Gajan, *Residue Rev.* **5,** 80 (1964).
21. R. J. Gajan, *Residue Rev.* **6,** 75 (1964).
22. D. J. G. Ives and G. J. Janz, *Reference Electrodes, Theory and Practice*, Academic Press, New York, 1961.
23. I. M. Kolthoff and J. J. Lingane, *Polarography*, 2nd ed., Interscience, New York, 1952.
24. I. M. Kolthoff and P. J. Elving, Eds., 1961, et seq., *Treatise on Analytical Chemistry*, Interscience, New York, Part II, in several volumes. Analytical methods for individual elements are reviewed.
25. W. W. Latimer, *Oxidation Potentials*, 2nd ed., Prentice-Hall, New York, 1952.
26. G. Lauer and R. A. Osteryoung, *Anal. Chem.* **40,** 30A (1968).

26b. T. S. Light, in "Ion Selective Electrodes," R. A. Durst, Ed., *U.S. National Bureau of Standards, Special Publication 314*, U. S. Govt. Printing Office, Div. Public Documents, Washington, D. C., 1969, p. 350.

27. J. J. Lingane, *Electroanalytical Chemistry*, 2nd ed., Interscience Publishers, New York, 1958.

28. J. J. Lingane and H. A. Laitinen, *Ind. Eng. Chem. Anal. Ed.* **11,** 504 (1939).

29. J. W. Loveland, in *Treatise on Analytical Chemistry*, Part 1, Vol. 4, I. M. Kolthoff and P. J. Elving, Eds., Interscience, New York, 1963, Chapter 51.

30. H. V. Malmstadt and J. D. Winefordner, *Anal. Chim. Acta* **20,** 283 (1959).

31. P. H. Martens and P. Nangniot, *Residue Rev.* **2,** 26 (1963).

32. L. Meites, Ed., *Handbook of Analytical Chemistry*, McGraw-Hill, New York, 1963.

33. L. Meites, *Polarographic Techniques*, 2nd ed., Interscience, New York, 1965.

34. L. Meites, in *Treatise on Analytical Chemistry*, Part 1, Vol. 4, I. M. Kolthoff and P. J. Elving, Eds., Interscience Publishers, New York, 1963, Chapter 46.

35. G. W. C. Milner and G. Phillips, *Coulometry in Analytical Chemistry*, Pergamon, New York, 1967.

36. R. L. Pecsok and L. Donald Shields, *Modern Methods of Chemical Analysis*, Wiley, New York, 1963, p. 339.

37. S. P. Perone, J. E. Harrar, F. B. Stephens, and R. E. Anderson, *Anal. Chem.* **40,** 899 (1968).

38. S. P. Perone, D. Jones, and W. F. Gutknecht, *Anal. Chem.* **41,** 1154 (1969).

39. J. P. Phillips, *Automatic Titrators*, Academic Press, New York, 1959.

40. E. Pungor, *Oscillometry and Conductometry*, Pergamon, New York, 1965.

41. L. Ramaley and G. S. Wilson, *Anal. Chem.* **42,** 606 (1970).

42. G. A. Rechnitz, *Controlled-Potential Analysis*, Macmillan, New York, 1963.

43. G. A. Rechnitz, *Chem. Eng. News*, June 12, 1967, pp. 146–58.

44. H. Schmidt and M. von Stackelberg, *Modern Polagraphic Methods*, Academic Press, New York, 1963.

45. I. Shain, in *Treatise on Analytical Chemistry*, Part 1, Vol. 4, I. M. Kolthoff and P. J. Elving, Eds., Interscience, New York, 1963, Chapter 50.

46. D. E. Smith, in *Electroanalytical Chemistry*, Vol. 1, A. J. Bard, Ed., Marcel Dekker, New York, 1966, p. 1.

47. D. E. Smith, 1970, Department of Chemistry, Northwestern University, Evanston, Ill., private communication.

48. J. T. Stock, *Amperometric Titrations*, Interscience, New York, 1965.

49. N. Tanaka, in *Treatise on Analytical Chemistry*, Part 1, Vol. 4, I. M. Kolthoff and P. J. Elving, Eds., Interscience, New York, 1963, Chapter 48.

50. J. J. Taylor, E. J. Maienthal, and G. Marienko, in *Trace Analysis: Physical Methods*, G. H. Morrison, Ed., Interscience, New York, 1965, Chapter 10.

51. P. Zuman, *Organic Polarographic Analysis*, Macmillan, New York, 1964.

52. P. Zuman, *Chem. Eng. News.* **46** (13), 94 (1968).

EDWARD M. BARRALL II

IBM Research Laboratory
San Jose, California

XII. Differential Thermal and Thermogravimetric Analysis

I. DEFINITION AND THEORY OF OPERATION

1. Differential Thermal Analysis (DTA)

Historically, DTA is a relatively old instrumental method. The first reported use was by Le Chatelier in 1887. The technique was essentially perfected by Roberts-Austen in 1899 (42).

The DTA technique involves heating or cooling a sample and a reference material in close proximity at some linear heating rate in a furnace. The temperature of the furnace, and the difference in temperature between the sample and reference are monitored and recorded. Thus a differential thermogram consists of a record of the difference in sample and reference temperature (differential temperature, ΔT) plotted as a function of time (t), sample temperature (T_s), reference temperature (T_r) or furnace temperature (T_f). Differential thermograms may also be obtained isothermally (ΔT is then plotted as a function of t) as will be discussed in Section II.4.C.

A technique which is closely related to DTA is differential scanning calorimetry (DSC) (41, 52). This technique although roughly analogous to DTA in final results, works on a different principle. The sample and reference (usually an empty sample pan) are mounted on two separate small heaters. The temperatures of the two cells are monitored by platinum resistance thermometers. The two heater windings are supplied with current so that both windings heat or cool at the same rate. No temperature difference is allowed to appear. The difference in power requirements for the two heaters is measured (ΔP) and recorded as a function of the indicated program temperature (T_p). The process is more closely analogous to classical calorimetry than is DTA.

2. Thermogravimetric Analysis (TGA)

The roots of thermogravimetric analysis also lie at the beginning of the 20th century (30). TGA involves the measurement of the weight of a sample either as a function of time at constant temperature or as a function of some parametric temperature as the system temperature changes. The sample may either lose weight to the atmosphere or gain weight by reaction with the atmosphere. The record is usually in the form of an integral curve with absolute weight (w) as the y-axis and time (t) or temperature (T) as the x-axis. However systems have been developed which record change in weight as a function of t or T. This produces a rate differential curve dw/dt, or better, dw/dT.

II. APPLICATIONS OF DIFFERENTIAL THERMAL ANALYSIS

DTA first found broad application in the analysis of inorganic materials—clays and microcrystalline materials (34). The thermal transitions of these materials are typically very large and easy to detect with the simplest equipment. DTA found very broad application in mineralogy before the wide-spread availability of good x-ray diffraction equipment. The techniques of inorganic DTA prior to 1958 are given in detail by MacKenzie (34) and many valuable references to applications are given by Smothers and Chiang (45).

In the late 1950s interest increased in applying DTA to the study of phase transitions in organic and polymeric materials. This new application required instrumentation several orders of magnitude more sensitive than that required for inorganic studies. With steadily increasing sensitivity and stability the techniques of DTA have been applied to a broad spectrum of organic materials for a large number of thermodynamic studies. In theory any transition or reaction which is accompanied by an absorption or emission of heat or by a change in heat capacity can be examined by modern DTA techniques.

1. Instrumentation

A functional differential thermograph requires the following components:

1. Sample cell
2. Temperature programmer
3. Temperature detectors
4. ΔT amplifier
5. T amplifier
6. Recorder
7. Atmosphere control
8. Cooling system

A workable group of DTA arrangements are shown schematically in Figure 12.1. The items above are numbered in the drawing. The type of equipment chosen depends greatly on the materials to be studied. Modular equipment, such as the DuPont thermograph, has the convenience that both high-temperature and low-temperature functions can be carried out by simply changing the sample cell.

A. Sample Cell

Dozens of cell designs have been proposed since the first DTA equipment was constructed. Most early types have been described by MacKenzie. (34)

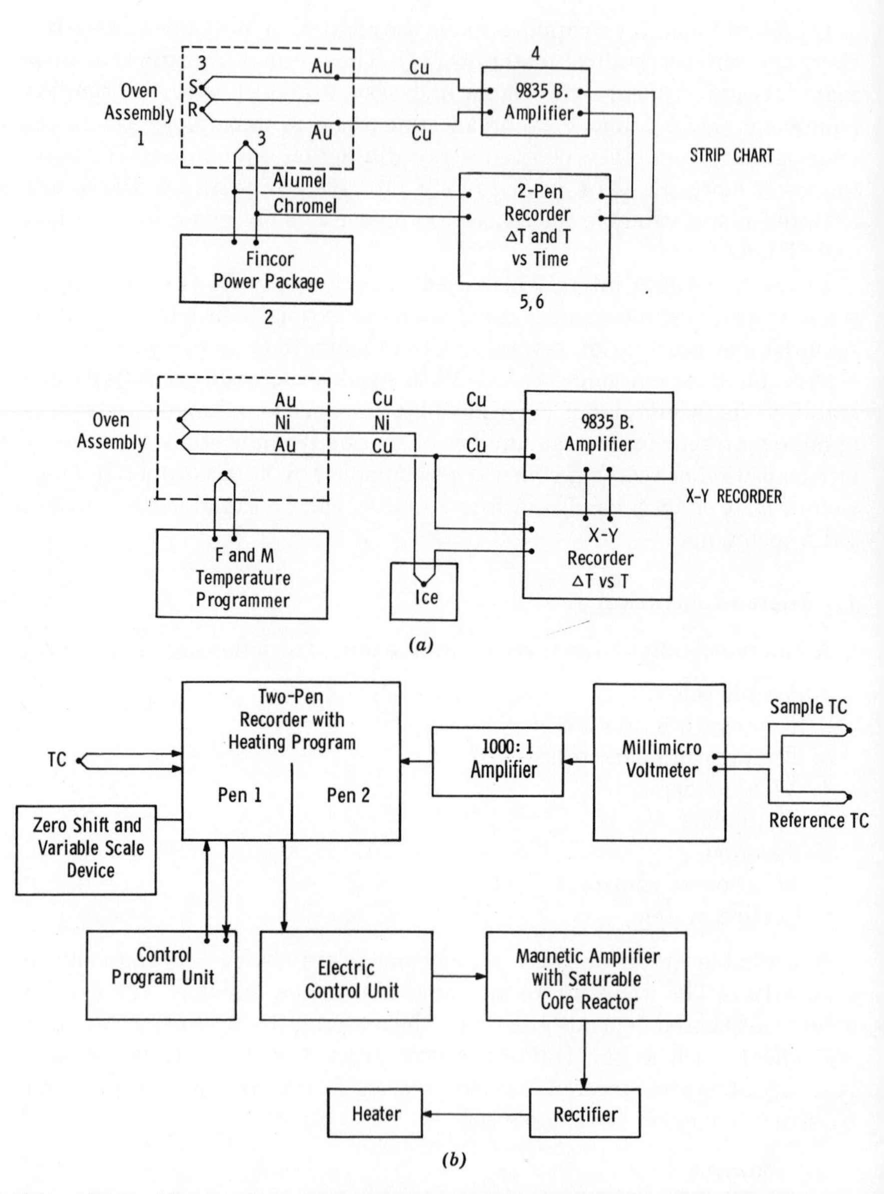

Figure 12.1. Three schematic diagrams of DTA instruments (31, 46).

All of the designs are based on essentially three more or less unique configurations. These are shown in Figure 12.2. Cell-type *a* is suitable for temperature measurements. Some work has been expended making this cell suitable for heat of transition measurement, but the design is poor for such application. The centrally located temperature sensing element responds to a number of sample parameters; a few of which are: thermal conductivity, thermal gradients, and exact temperature sensor location. The heat capacity of the sample and the enthalpy change (ΔH) at the

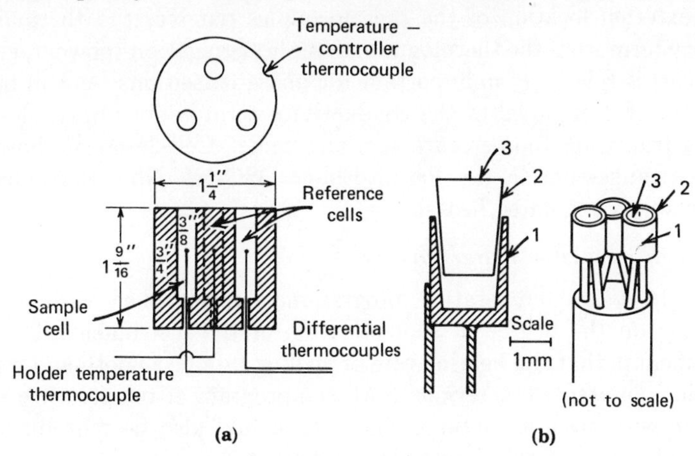

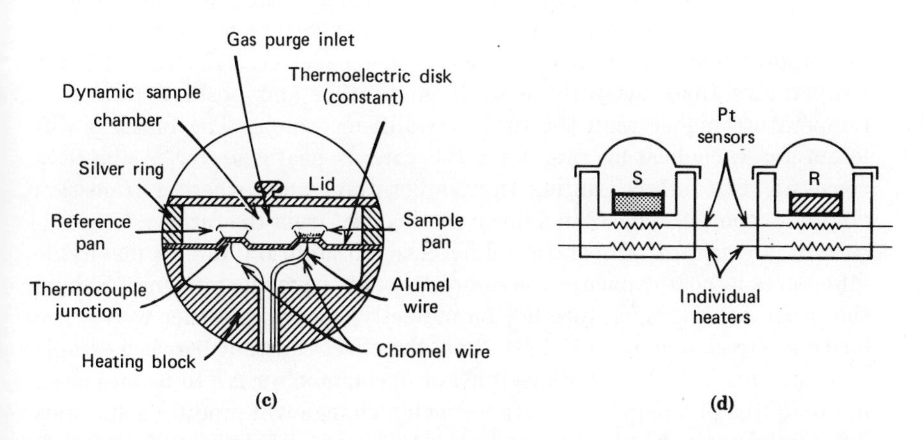

Figure 12.2. Cell designs for differential thermal analysis and differential scanning calorimetry (11, 38, 43, 47). (*a*) With central thermocouple; (*b*) with external thermocouple; (*c*) thermoelectric plate; and (*d*) Perkin-Elmer DSC, platinum resistance and sample (*S*) and reference (*R*) heaters.

transition are sometimes smaller than the effects caused by thermal conductivity alone. Thus the resulting thermogram of a transition while quite precise as to temperature may not reflect the thermal magnitude of the transition, ΔT, with any accuracy.

Cell-design *b* is satisfactory for both T and ΔH measurements within limits as is design *c*. Design *c* is a relatively new innovation and has several points recommending it: sample encapsulation is simple, sample placement is unrestricted but accurate, and the cell is not overly fragile. In designs *b* and *c*, external location of the thermocouples removes the thermal conductivity term from the thermogram. Sample aggregation (powder, crystal, thin layer) is relatively unimportant for phase transitions. As will be seen in Section II.2, removal of the conductivity term is not always desirable in glass transition temperature measurements. Cell-design *d* shows the sample arrangement in the Perkin-Elmer DSC-1B. Note the individual and separately regulated heaters.

B. Temperature Programmer

Over the years temperature programmers have taken various forms depending on the degree of sophistication of the instrument. The main requirement is that the heating rate or cooling rate be smooth and without oscillations or steps. For serious work the program of temperature should be linear with respect to time. The rate should also be reproducible to $\pm 0.1°C$ or better, if calorimetry is contemplated.

The reasons for the restrictions above are best understood by examining Figures 12.3 and 12.4. In Figure 12.3 the heating of the reference is linear. The reference is offset from the furnace by a temperature gradient which to a first approximation is constant. The offset is negative (reference at lower temperature than the furnace wall) on heating and positive (reference temperature higher than the furnace wall) on cooling. The offset is different for each heating rate since the rate of heat transfer is almost a constant. The sample, during the temperature range where a transition does not occur, also heats in a linear mode offset from the furnace temperature by a constant temperature difference. In addition, due to inevitable differences in heat capacity, the sample is offset from the reference. Should the program of temperature not be perfectly smooth, in other words, the heating rate change from time to time, the offset constant between sample and reference will change. Since it is not uncommon for ΔT to be measured in hundredths of a degree centigrade, such a change will produce a spurious differential peak or at best a wandering ΔT baseline. Interpretation of such records is all but impossible, since real events in the enthalpy curve of the sample can cause identical effects. The question of whether this a real effect or a program artifact is most troublesome. Sample dilution techniques

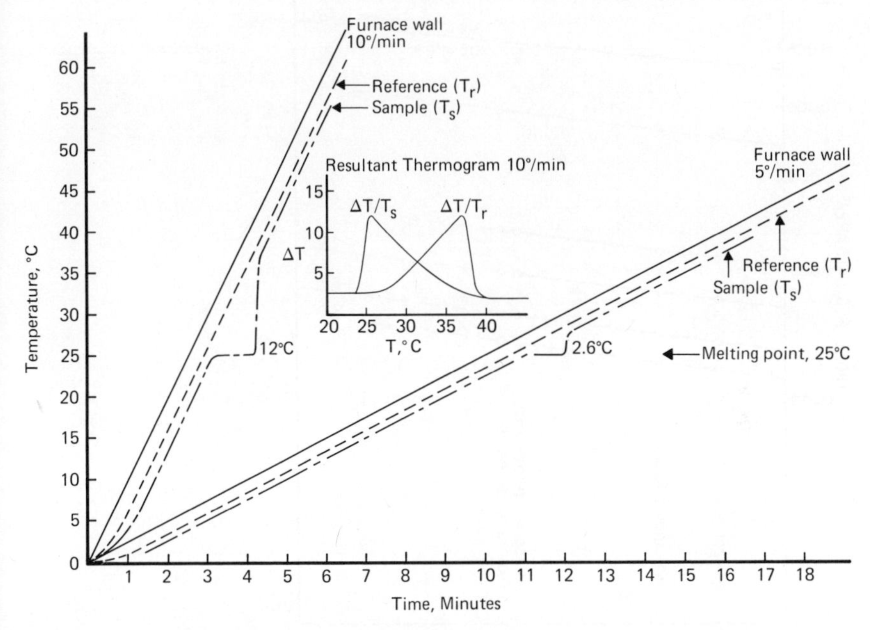

Figure 12.3. Temperature profiles in the DTA cell.

can minimize such effects by making the heat capacity of sample and reference more nearly equal at the expense of sensitivity.

Heating rate should be reproducible for calorimetry, if the plot of ΔT versus T_s is used. The calibration constant (calorie equivalent per unit area of thermal record) is very sensitive to the heating rate. This is shown in the experimental results (Figure 12.4). If a record of ΔT versus T_r is used with cell *c* of Figure 12.2, the response/heating rate dependence is somewhat reduced (11).

The simplest temperature programmer is a variable transformer. This is set to some fixed voltage and the furnace heats up along a smooth but nonlinear profile. This is so-called ballistic programming. The shape of the T versus t curve resembles a ballistic profile. Various programmers consisting of motors attached to a Variac shaft have been described (51). It is possible to get a linear program over a short range with this arrangement. Proportional band-width temperature controllers offer an additional level of sophistication. The signal from a thermocouple in the furnace is compared electronically against a reference potential which can be programmed to correspond to a variety of heating modes and heating rates. This kind of programmers are the best and most applicable for serious work. Modern

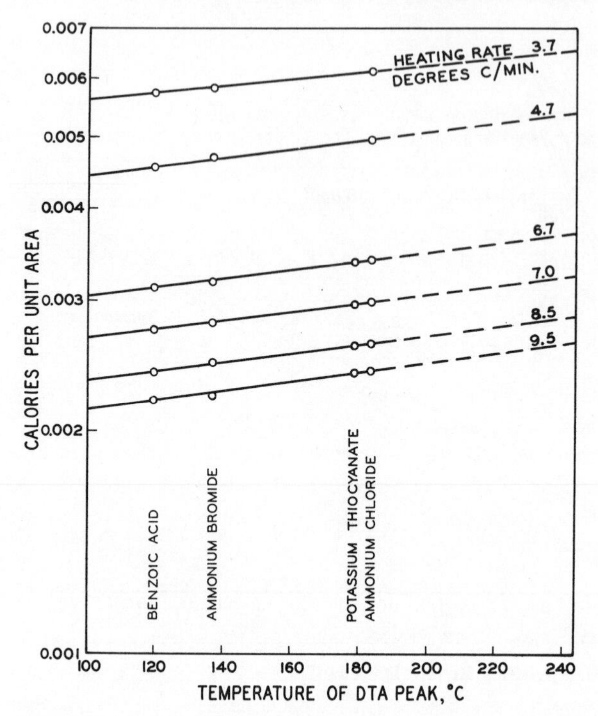

Figure 12.4. Variation of calories per unit area with temperature and heating rate (8). Differential temperature plotted as a function of sample temperature of an *x*-*y*-recorder.

solid state silicon triodes can control directly up to 50-A currents. This is more than adequate. Originally cam follower potentiometers were used to generate the reference potential (4). However various solid-state devices are now employed in modern research equipment.

C. Temperature Detectors

Almost every known bimetallic pair has been employed at one time in DTA for thermocouple construction. For high-temperature work platinum/platinum–10% rhodium thermocouples are the most frequently used. For very high-temperature work, up to 3000°C, recording optical pyrometers have been described. Low temperature ranges (−100 to 600°C) are usually measured with the higher output pairs: copper/constantan, iron/constantan, gold/nickel, and chromel/alumel. Copper/constantan has the advantage of requiring no special contact thermostating or temperature equalization circuits to connect with the rest of the instrumental electronics. Obviously the other pairs would form a second and unwanted set of thermocouples at the point where they are connected to the copper conductors. Lead resistance is not a problem with copper/constantan.

Others detectors such as thermistors and platinum resistance thermometers have been employed. Both the Perkin-Elmer DSC-1 and DSC-1B use platinum resistance circuits. In this case the resistance/temperature relation is easier to adapt to the power measuring function of the instrumental electronics.

D. ΔT Amplifier

A large number of good dc amplifiers are commercially available. It is important that the amplifier does not draw a measurable current from the ΔT thermocouple. The voltage/temperature relationship of a thermocouple pair becomes very nonlinear and unpredictable if a current flows in the circuit. The amplifier should be capable of factors as high as 1000. Input impedance should be at least 5000Ω or higher and the time constant should be less than 1 sec.

E. T Amplifier

The remarks given in Section D also apply here. The amplification factor needs only be about 10 and the time constant can be as long as 2–5 sec. The voltages usually handled are of the order of millivolts for T whereas the ΔT signal is usually in microvolts.

F. Recorder

The recorder is to record with as little distortion as possible the ΔT and T signals. Systems using two strip chart recorders or a single dual pen strip chart have been described. To get ΔT as a function of T it is necessary to cross plot the two records. Several authors have suggested recording on an x–y recorder with ΔT as the y-axis and T as the x-axis (4, 34). This eliminates the need of a cross plot. In either case, the recorder should have a sensitivity of 1 to 0.1 mV/in. and a time constant of 1 sec or less. The x–y recording system has one drawback; some form of electrical null system is required at reasonably high temperatures (above 100°C) if T-axis sensitivity is to be preserved. That is, part of the T signal must be subtracted if the system is to be operated at high sensitivity over a wide range. Otherwise the recorder range will be exceeded (around 120°C for copper/constantan operating a 12-in. bed, 1 mV/in. sensitivity, ~10°C/x-axis in.). Usually a very stable potentiometer is employed to null out a certain fixed number of millivolts (4). All recording systems given above have certain advantages in individual cases: for calorimetry ΔT versus t is useful (11), for kinetics ΔT versus t with a record of T is convenient, for transition temperature measurements ΔT versus T is required and for isothermal studies ΔT versus t is the only method.

The ultimate method of recording ΔT, t, and T is by direct conversion of the analog to digital data. This can be done with a pair of printing digital voltmeters or by direct connection to the analog to digital converter of an IBM 1800 or comparable computer. A 10–50 mV signal is desirable. This direct conversion removes the data reduction step from recorder traces. Also, the noise and range limitations of recorders *per se* are removed.

G. Atmosphere Control

Some form of atmosphere regulation must be provided for the operation of a differential thermograph, if only for the protection of the sample from atmospheric oxygen. The systems which are in current use are of four general types: inert gas purge for the protection of the unencapsulated sample; inert gas covering within the encapsulation; inert or reactive gas at different pressures for thermodynamic and reaction kinetic studies; and flowing inert or reactive gas for reaction kinetic studies.

H. Cooling System

The cooling system is considered separate from the temperature programmer, since in most commercial instruments and those described in the literature cooling is completely independent from heating. The simplest and probably the most satisfactory arrangement is cooling with a gas as the heat exchange medium. The temperature program is carried out against a constant temperature thermal reservoir. Some form of quench cooling is extremely useful in instruments not expected to function below ambient temperature. The cool-down time from a high-temperature run can be inconveniently long if one relies solely on radiation to dissipate heat from the system. In addition, most temperature programmers do not function efficiently unless a thermal reservoir at least 30°C below the program temperature is available. No system using thermoelectric cooling has been described. This should offer the ultimate in low-temperature dynamic control. A simple, workable cooling system which is satisfactory for cooling most commercial instruments is shown in Figure 12.5.

2. Melting or Transition Point Determination

Prior to a consideration of methods it is necessary to understand the significance of the various parts of the differential curve. This is turn requires a consideration of how and in what part of the system ΔT and T were measured.

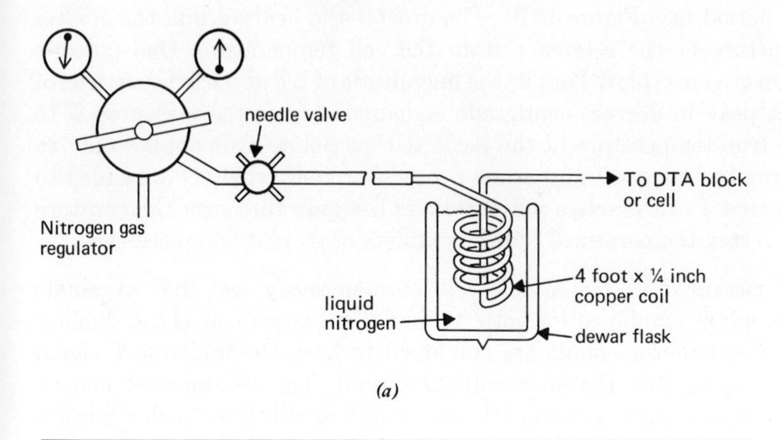

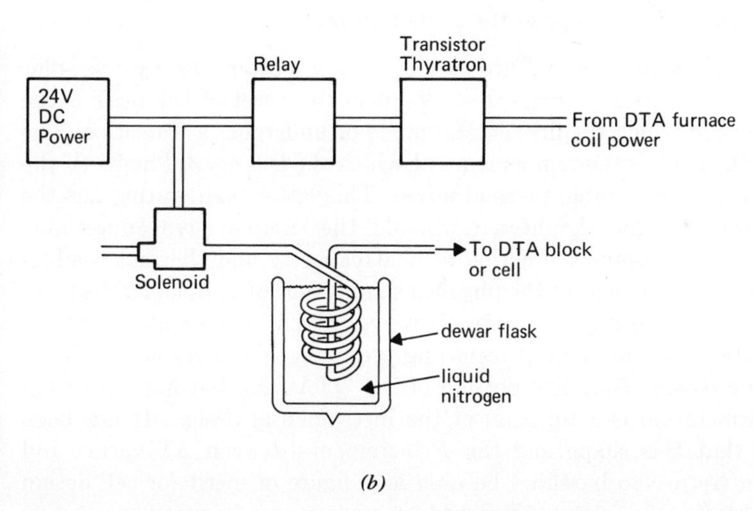

Figure 12.5. Simple cooling system for DTA. Automatic system using a thyratron-relay to turn off gas while current is applied to the heater windings.

If the cell-type *a*, Figure 12.2, is employed the following conditions prevail.

1. *T measured in the furnace or the reference cell.* This arrangement is usually unsatisfactory for precise temperature measurement as the location of the onset, vertex, and conclusion of the DTA transition are heating rate dependent. By definition the sample temperature must be nearly isothermal

during the transition. The cell and reference heat linearly during the isothermal period (see Figure 12.3). The greater the heating rate the greater the departure of the reference from the cell temperature. One possible correction is a cross plot. That is, the magnitude of ΔT at various portions of the DTA peak in degrees centigrade is deducted from the measured T to give the true temperature of the peak at that point. This applies only to endothermal processes. Exothermal processes require that ΔT be added to the indicated T. In practice most workers use only the onset temperature and the vertex temperature; thus a complete cross plot is unnecessary.

2. *T measured in the sample cell simultaneously with ΔT.* A single thermocouple is employed for both ΔT and T measurement (4). A dummy resistance and thermocouple are employed to keep the millivolt T signal from backing up into the microvolt ΔT circuit. The resistance should be equal to that of the amplifier and thermocouple circuit. Exact balance is not critical. An electronic compensation circuit or very high impedance dc amplifiers can also accomplish the same purpose.

In the case where ΔT and T are plotted on x–y coordinates (by cross-plot from dual recorders or a single x–y recorder) the onset of the peak is the temperature at which the first crystal melts or undergoes a transition. The vertex of the peak is the temperature at which the last crystal melts (if the sample has no appreciable thermal mass). This latter temperature has the most thermodynamic significance. Should the sample have appreciable thermal mass (a complex function of heat capacity and thermal conductivity), exact assignment of the physical significance of various ΔT features is not possible with any certainty. However given a reasonably well designed system, this method of recording produces thermograms which are *heating rate independent.* The portion of the DTA peak beyond the vertex towards conclusion is a function of the instrumental design. It has been proposed that this shape and the T increment between ΔT vertex and conclusion (return to baseline) be used as a figure of merit for cell design (41, 52). Ideally the increment should be as near zero as possible.

If cell-types b or c, Figure 12.2, are used the following conditions are significant:

1. *T Measured in the Furnace or Reference Cell* (under the cell in case c). When this configuration is used, the temperature records are as complex as in case 1 above with the addition of a thermal resistance factor. Such records are not suggested for use in transition temperature determinations.

2. *T Measured In or Under the Sample Cell.* Since the thermocouple is not in direct contact with the sample but separated by several layers of metal, a thermal resistance factor must be considered. This also holds for

the Perkin-Elmer DSC-1 and DSC-1B (see Figure 12.2 cell *d*). This thermal resistance is equivalent to an electrical resistance with the temperature difference between the sample and thermocouple equivalent to a voltage difference. This type of system has been considered in detail by Watson et al. (52), and O'Neill (41). The result of the thermal resistance term is to delay the response of the sample temperature detector by a constant slope function. *This function is heating rate dependent.* It is possible to determine precise temperatures if the melting curve of a very pure material (at least 99.999%) made at the sample heating rate and melting within 50°C of the sample is available.

The method for determining the temperature at any value of ΔT is shown in Figure 12.6. The melting of a 100% pure material should be a step function (enthalpy vs T) with a first derivative ($\Delta H/\Delta T$ vs T) consisting of a

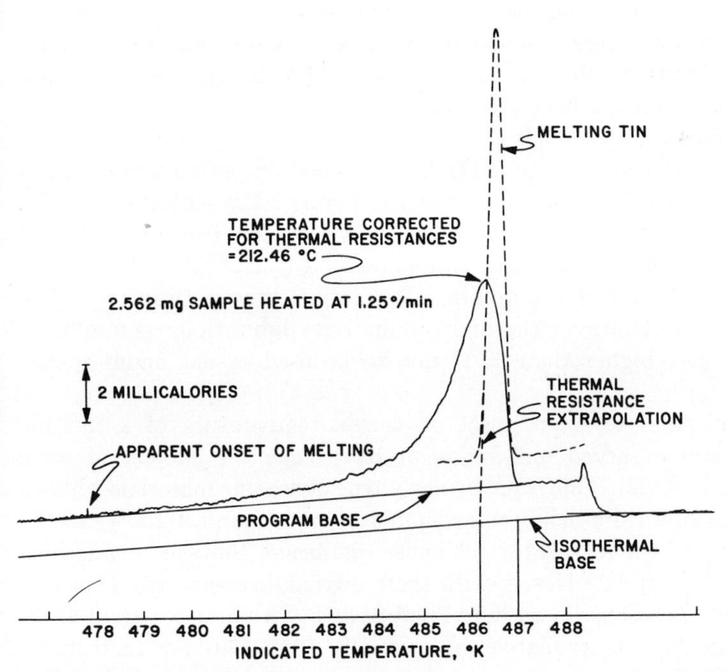

Figure 12.6. Correction of a DSC curve for thermal resistance (3). Sample: 9,10-Dichloroanthracene containing 1.10 mole % anthracene, 2.562 mg working sample heated at 1.25°C/min.

vertical line of no T width and vertex at $\Delta H/\Delta T = \infty$. Any departure from this shape is due to impurity and thermal resistance. By the use of very pure standards the effects of thermal resistance can be computed. The resistance in a well designed system should not amount to a correction of more than 2°C. To a very rough approximation, the T at onset of the differential peak corresponds to the thermodynamic transition temperature (T vertex of cell d, Figure 12.2).

Temperature calibration of a DSC is somewhat more difficult. The electronic components of the Perkin-Elmer DSC-1B cause the temperature error curve as a function of system temperature to assume the shape of a parabola (48). This requires careful calibration with many points. The error may be very large, ± 18°C, at the extremes of the curve. The shape and magnitude of the error curve is heating rate and encapsulation sensitive. In addition, it is necessary to recalibrate if the temperature average and differential potentiometers are changed in any way. Each cell must be calibrated individually. The error is also sensitive to purge gas flow rate. A new error curve is obtained when the instrument is switched from the -100 to $500°C$ range to the 0 to $600°C$ range. It is possible by the adjustment of the slope and average controls to minimize the error over a limited working range. Obviously the DSC requires more careful temperature calibration than a conventional DTA using thermocouples. However this author has been able to measure temperatures with an accuracy of $\pm$ 0.05°C on a DSC-1B (3).

The most precise and valuable DTA determined transition temperatures are obtained with cells of the type shown in Figure 12.2a with ΔT recorded as a function of sample temperature on an x–y recorder. The system should consist of the smallest amount of sample possible consistent with good ΔT recording. The height of the differential curve is directly proportional to the heating rate. However thermal conductivity limitations generally do not permit rates higher than 20°C/min to be used except under special conditions.

The correlations between onset or vertex temperature of a melting endotherm and observed melting point have been the subject of some literature debate (22). This is particularly true of organic materials where a large body of literature has been accumulated over the past 150 years. In general, many of the features which cause significant "optical" changes do not necessarily correlate closely with thermodynamic events (5). One way to settle such questions on an individual basis is with a transparent DTA cell. Such an item is available commercially from DuPont Instrument Products Division or may be constructed from an aluminum bolt and a large slit glass tube. The sample may be watched without disturbing the DTA system. Open cells may also be used on the Perkin-Elmer DSC-1B.

Although remarkably standard thermocouples may be produced from Leeds and Northrup duplex thermocouple wire, calibration at a few points is necessary. Table 12.1 lists a group of standards which may be helpful. For the cell designs given in Figure 12.2*b*, *c*, and *d* the fusion of ultrapure metals is highly recommended for calibration. Suitable materials up to the high-temperature limit of most equipment are readily available.

Table 12.1. Materials Used for the Calibration of a Differential Scanning Calorimeter (6)

Temperature, °C	Material	Melting (M) or Inversion (I)	Enthalpy Change, cal/g
−38.9	Mercury	M	2.74
29.8	Gallium	M	19.9
69.4	Stearic acid	M	47.6
125.2	Ammonium nitrate	I	12.6
169.6	Ammonium nitrate	M	16.2
122.4	Benzoic acid	M	33.9
156.6	Indium	M	6.79
231.9	Tin	M	14.2
264	Lithium nitrate [a]	M	88.4
273	Sodium nitrate	I	9.5
327.4	Lead	M	5.89
419.5	Zinc [a]	M	27.0
588	Potassium sulfate	I	12.3

[a] Place between mica strips prior to encapsulation.

DTA is perhaps the ideal way to rapidly study phase transitions. The thermograms contain not only the range of melting and the temperature of maximum melting, but also furnish a permanent record of a material's thermal behavior without a personal element.

In addition to solid-to-solid and solid-to-liquid transitions, DTA and DSC have been successfully applied to the determination of boiling point (7, 49, 32, 50). Using cell *a* (Figure 12.2), a few micrograms of liquid placed on a nonchemisorbing solid support (such as calcined carborundum, 500 mesh) will furnish a sharp endotherm on boiling. The boiling point is taken as the extrapolated onset temperature, the intersection of a straight line through the nontransition base and a straight line through the most linear part of the endotherm. With cells *b*, *c*, and *d*, (Figure 12.2) a small hole is made in the encapsulation lid. Samples are *not* run in open cells since one plate of equilibration is required for a boiling point. If the system pressure is altered and a series of runs made, the heat of vaporization may be determined. A plot of log pressure (in atm.) versus $1/T$ (in °K) has the slope $-E/R$ where R is the gas constant. The most convenient method of data collection is by x–y recording. The sample size required is so small that satisfactory endotherms have been obtained from peaks trapped from an analytical gas chromatograph on carborundum.

3. Transition Heat Determination

The measurement of the number of calories involved in a transition poses new problems over those encountered in simple temperature measurement by DTA. It is necessary to convert a temperature difference, $\Delta T/T$ or $\Delta T/t$, into an energy difference. This, in theory, can be done by considering the heat capacity of the sample and instrument. In practice many more variables enter the picture. It is necessary that the ΔT detector note *all* heat entering or leaving the sample—or at least a reproducible fraction. For most purposes the calorimetric measurements which can be obtained with cell *a* (Figure 12.2) are very rough. Thermal conductivity changes in the sample are far too important to permit accurate calorimetry. For that reason and others, the centrally located thermocouple cell will not be considered in connection with calorimetry. Instead, the discussion will be limited to cells *b*, *c*, and *d*.

For precise phase-change calorimetry the sample must be accurately weighed and encapsulated in an inert environment in a reproducible manner. Since the heat capacity of the container and sample are to be minimized, the sample weight should be under 20 mg. For most purposes a record of ΔT versus t is satisfactory. For determination of ΔH it is not necessary to know T accurately. The heating rate for the reasons given in Section II.1 should be very linear; knowing the starting temperature, rate, and elapsed time is sufficient. The heating rate should not exceed 5°C/min if the detectors are to follow the sample temperature closely.

The area under the transition peak is related to the total heat of transition by a complex factor. This factor can be evaluated by running known materials. Since the instrument heat capacity is part of the factor and the heat capacity increases with increasing temperature, it is necessary to calibrate at several temperatures. Generally the instrument is more sensitive at lower temperatures than at higher temperatures (see Figure 12.4). If ΔT is recorded as a function of t or T_r heating rate will not affect the conversion factor. Heating rate does alter the factor if ΔT is recorded as a function of T_s. With either system the conversion factor is expressed in cal/unit area. The area units depend on the means of integration, in other words, planimeter, weight of cutout peak, and so on. Any technique satisfactory for the determination of the area under a gas chromatographic peak may be applied to DTA.

Calibration of a DSC is somewhat simplified over DTA, since the record is of differential power as a function of time versus temperature. The calibration constant is obtained in the same way as for DTA. It is much less sensitive to heating rate and transition temperature, although calibration must be carried out for precise work at different heating rates.

As previously mentioned, the heat of transition is proportional to the area under the transition peak. However it is necessary to construct a baseline to enclose the area for integration (Figure 12.6). This is not a straightforward process. In practice, the temperature at which ΔT first departs from the program line is joined to the temperature at which ΔT returns to the program line with a straight line. This completely ignores the fact that the heat capacity of the system probably changes in a nonlinear manner during the transition. Baseline construction has been the subject of several studies (see Figure 12.7*a*). The method used in Figure 12.7*a* is very useful if polymers are being studied. Due to supercooling of the melt, amorphous material for use in the reference cell is easy to obtain. Most authors agree that for sharp transitions the unwanted heat capacity included on one side of the transition is lost by the baseline assumption on the other side of the transition. For broad transitions with an unsteady program line the detection of onset and conclusion can be impossible. This is the usual case in DTA of semicrystalline polymers. Gray has given a very useful method for partial solution of such problems when encountered in DSC (26). These polymers at any temperature below the fusion temperature consist of a mixture of amorphous and crystalline material.

The reasoning towards a solution considers the change in enthalpy the crystalline and amorphous polymer experience in being heated from T_1 to T_2. The term T_1 is below the fusion point of the crystalline phase and T_2 is above the fusion point. At the initial temperature below the fusion point the total enthalpy, H_1, may be expressed as:

$$H_1 = X_1 H_{c1} + (1 - X_1)\mathrm{H}_{a1} \tag{12.1}$$

where X_1 = the fraction of crystalline polymer present at T_1

H_{c1} = the enthalpy of the crystalline polymer at T_1

H_{a1} = the enthalpy of the amorphous polymer at T_1.

At temperature T_2, above the crystalline melting point, the total enthalpy is

$$H_2 = H_{a2} \tag{12.2}$$

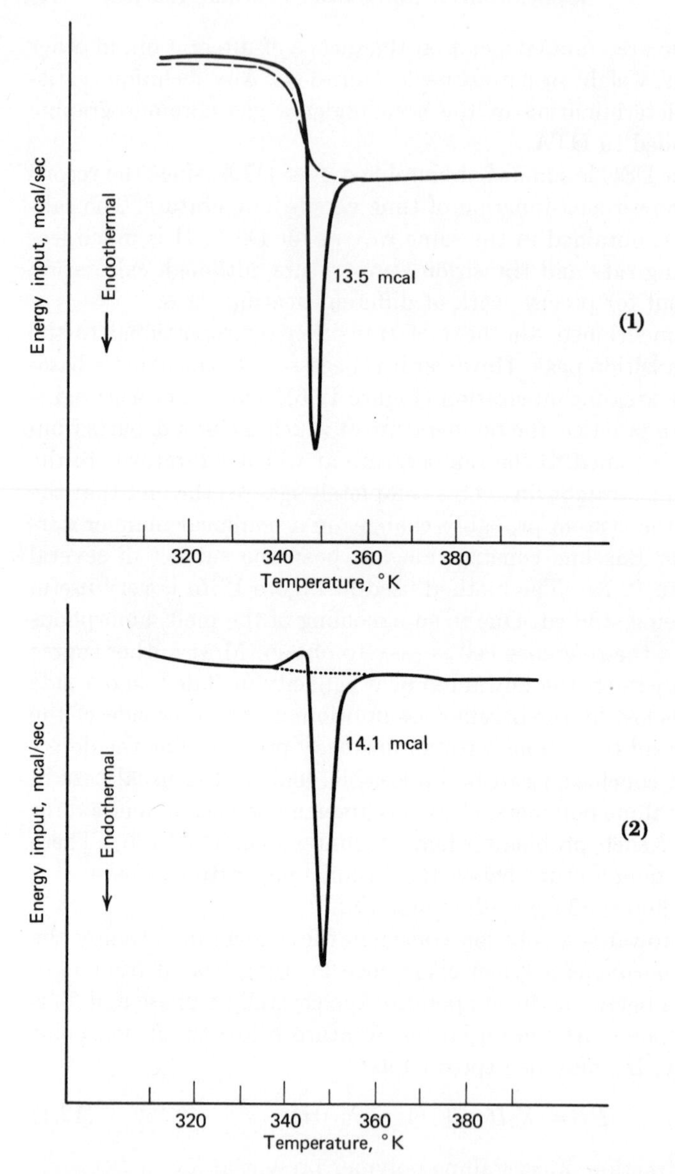

Figure 12.7a. Baseline construction using an amorphous polymer as reference (25). (*1*) Dashed line represents a quenched amorphous polymer, solid line is the thermogram of a well crystallized sample of the same material. (*2*) Thermogram obtained when the crystallized and amorphous polymers are run versus one another as sample and reference respectively. Dashed line is the integration baseline for heat estimate.

The total enthalpy change in going from T_1 to T_2, $\Delta H_{2,1}$ is the difference between Equations 12.1 and 12.2.

$$\Delta H_{2,1} = H_2 - H_1 = (H_{a2} - H_{a1}) + (H_{a1} - H_{c1})X_1 \qquad (12.3)$$

This reduces to

$$\Delta H_{2,1} = \Delta H_{a(2,1)} + \Delta H^{\circ}_{F_1} X_1 \qquad (12.4)$$

where $\Delta H_{a(2,1)}$ is the enthalpy change experienced by the amorphous material in going from T_1 to T_2. The $\Delta H^{\circ}_{F_1}$ is the heat of fusion of the pure crystalline material. Since we desire to know X_1, the fraction of crystalline material at T_1, Equation 12.4 is rearranged to

$$X_1 = \frac{\Delta H_{2,1} - \Delta H_{a(2,1)}}{\Delta H^{\circ}_{F_1}} \qquad (12.5)$$

These parameters can be equated to features of a DSC curve (Figure 12.7*b*) $\Delta H_{2,1}$ is the area *ACDEF*. The $\Delta H_{a(2,1)}$ is the area *ABEF* and the difference is *BCDG*. This last point is important, since it indicates that the correct method for the construction of the baseline is by extrapolation of the liquid and not the solid baseline as is usually done.

Unfortunately for direct resolution of the problem there are several assumptions and experimental problems. The first and most serious assumption is that the true baseline between *G* and *E* (Figure 12.7*b*) may not be extrapolatable, since the enthalpy curve for the liquid or amorphous phase is probably not linear. Second, the term $\Delta H^{\circ}_{F_1}$, the heat of fusion of the 100% crystal form, is very difficult to determine experimentally. For example, over twenty values for the ΔH°_{F} of polypropylene have been reported. These range from 15 to 60 cal/g. This is because the crystallinity of the polymer changes continuously from below −20°C to the ~168°C

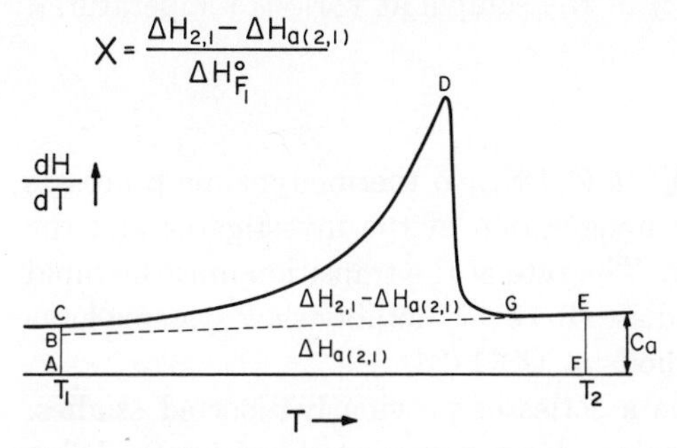

Figure 12.7b. Baseline independent method for integrating fusion endotherms (26).

"melting point." Therefore it is impossible to unambiguously define the temperature at the onset of melting from which ΔH_F° must be calculated.

Semicrystalline polymers offer difficulties not encountered with lower-molecular-weight pure crystalline materials. A systematic evaluation of all necessary parameters may yield a solution to the problem, but errors and assumptions would void the results of any analytical usefulness.

In consideration of the situation above a relative method may be the best and only solution. The following method is suggested. The first DSC scan is made with no sample from T_1 to T_2 with isothermal baselines recorded at T_1 and T_2. The sample is then placed in the pan and the above operation is repeated. The isothermal lines at T_1 and T_2 must be identical for the "no sample" and sample runs. A scan rate of 10°C/min on a 10-mg sample is usually satisfactory. The "no sample" line is deducted from the sample scan by superpositional or mathematical techniques. The area thus defined is integrated from T_1 to T_2. This area can then be compared with the similarly generated area from another sample and a relative crystallinity can be estimated. If several samples can be examined by alternate methods (x-ray diffraction, IR, etc.) a relative crystallinity index can be established.

The DTA and DSC trace also contains sufficient information to calculate the heat capacity of the sample in the nontransition temperature range. To carry out this calculation it is necessary to have three scans: empty encapsulation; encapsulation plus standard; and encapsulation plus sample. The operations are detailed in Figure 12.8. Amorphous alumina (optical sapphire) is the commonly used standard. The specific heat of sapphire has been determined to a high degree of accuracy from a few °C to the melting point. For statistical reasons it is necessary that the sample deflection be at least one half of the standard deflection. A simple ratio (taking into account the sapphire and sample weights) is used to calculate the heat capacity (more correctly specific heat) of the sample at various temperatures.

4. Applications

The general application of DTA and DSC to thermodynamic processes is generally limited only by the imagination of the investigator and the sensitivity of the instrumentation. The rate of the transition must be rapid and heat dependent for most studies. However, notably slow systems have also been studied by special methods of DSC (24).

The following text will outline a series of previously reported studies. The list is not intended to be complete. Its purpose is to furnish a guideline and a basis for thinking.

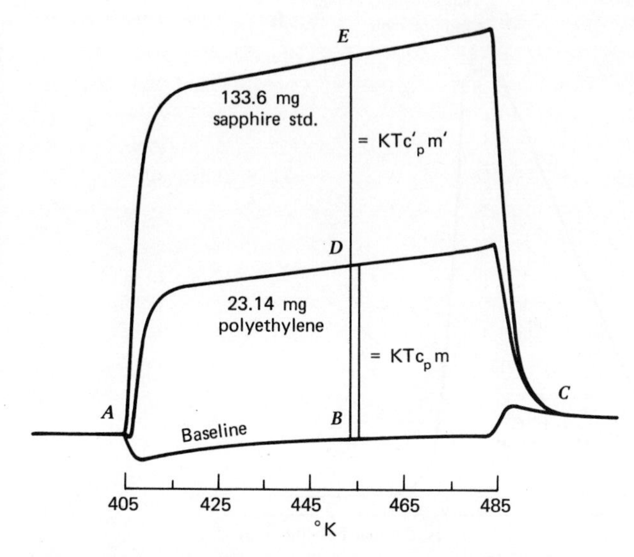

Figure 12.8. Typical specific heat determination, molten linear polyethylene in the range 405 to 485°K (27). Calculation: $C_s = H_s W_a C_a / H_a W_s$ where C_s = specific heat of the sample at T_1; H_s = amplitude of the sample at T_1; H_a = amplitude of the sapphire at T_1; W_a = weight of the sapphire; W_s = weight of the sample; and C_a = specific heat of the sapphire at T_1.

A. Analysis of Mixtures

If a mixture undergoes a phase transition, preferably melting, DTA or DSC may be of some aid in analysis. Mixtures encountered are normally of two types: principal component with traces of impurity; and two or more components in roughly equivalent concentrations. In the first group, the impurity may obey the van't Hoff relationship on melting, in other words, it is completely insoluble in the solid phase and completely soluble in the mixture melt. This is not always the case. In the second case each component may crystallize out separately or in various complex co-crystals. In addition, one or more component may have no transitions in the thermal range of interest (a filler in a crystalline polymer, for example).

In cases where materials are pure (97% to 100%) and the van't Hoff equation can be verified as being applicable, it is necessary to obtain only a single endotherm of the melting process by DSC. The curve is divided up as shown in Figure 12.9. The curve is integrated to the vertex by parts and the temperatures are tabulated with the partial areas *up to* that temperature. The temperatures are corrected for thermal resistance by using the melting curve of a very pure material melting within 25°C of the transition

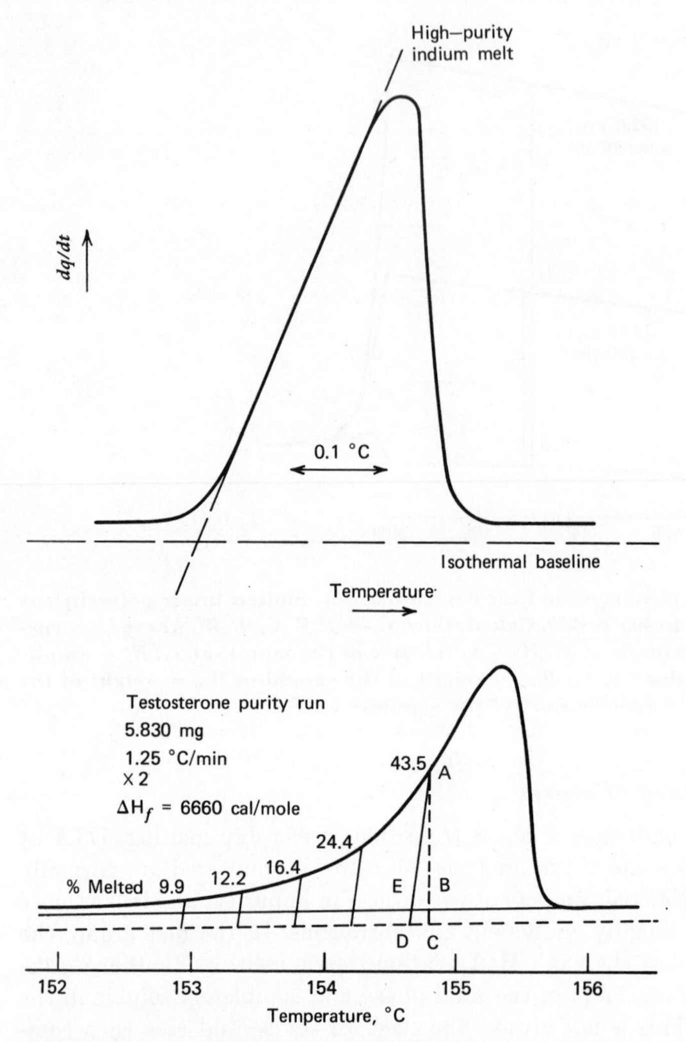

Figure 12.9. Differential scanning calorimeter purity determination, testosterone (48). Solid baseline is constructed between apparent onset and conclusion temperatures. Dashed baseline is constructed between isothermal points (program off but instrument on) at the beginning and end of the run. Sloped lines are obtained from the above indium scan to correct for thermal lag in the temperature axis.

in question (see Figure 12.6). A plot of temperature as a function of the reciprocal of the fraction melting ($1/F$) to that temperature is made. The fraction melted (F) is the area to a given temperature divided by the *total* area of the endotherm. Given the van't Hoff equation the following relationship may be developed:

$$T_s = T_0 - \frac{RT_0^2X}{\Delta H}\left(\frac{1}{F}\right) \tag{12.6}$$

where T_s is the instantaneous temperature of the sample in °K; T_0 is the melting point of the infinitely pure sample (solvent) in °K; R is the gas constant (1.987 cal/mole. °K); ΔH is the heat of fusion of the sample (solvent) in cal/mole; X is the mole fraction of impurity; and F is the fraction of the total sample melted at T_s. Thus a plot of T_s versus $1/F$ should give a straight line of slope $-(RT_0^2X/\Delta H)$ with an intercept of T_0. ΔH is obtainable from the DSC curve.

The equation assumes the following.

1. The impurity is *insoluble* in the predominantly solid phase.
2. The impurity is completely soluble, in other words, it forms an ideal solution in the molten or liquid phase.

These assumptions make the equation inapplicable in the following cases: co-crystals are formed where the impurity can crystallize in the host lattice; the impurity forms a nonideal solution in the melt by association or chemical reaction; and the impurity is totally insoluble in the liquid phase. A simple case of the third item is sand in water. Essentially, the van't Hoff equation describes the molar entropy of mixing and solution of the impurity in the melt. In this respect we refer to the total melt, not to just a small fraction (3).

Unfortunately, for this simple application two corrections are required to account for instrumental variables: (1) thermal resistance of the instrument; and (2) undetected melting. Correction for thermal resistance has been covered. The undetected melting can be corrected by the following calculation:

$$\text{true } F_1 = \frac{a_1 + x}{A + x}, \qquad \text{true } F_2 = \frac{a_1 + a_2 + x}{A + x} \qquad \text{and so on} \tag{12.7}$$

where a_1 and a_2 are the partial areas; x is the small area missed in integration at low temperatures; and A is the total area under the curve.

Plots are shown in Figure 12.10. The x has a large effect at small fractions melted. Overcompensation results in a curve which bows downward. The method is not arbitrary since only one value of x will result in a linear plot.

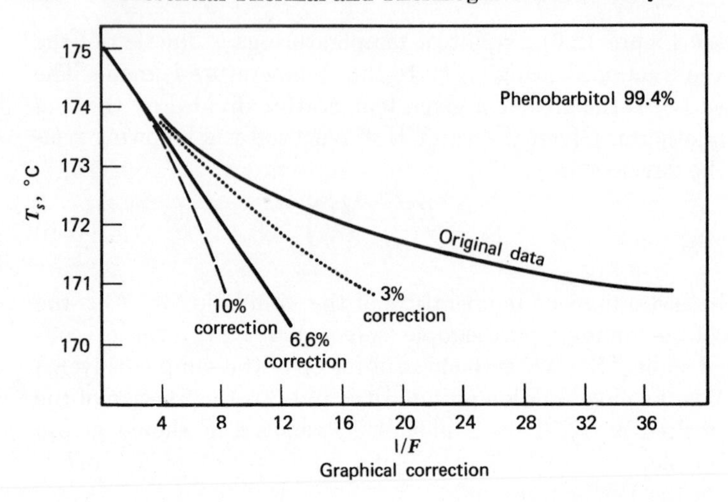

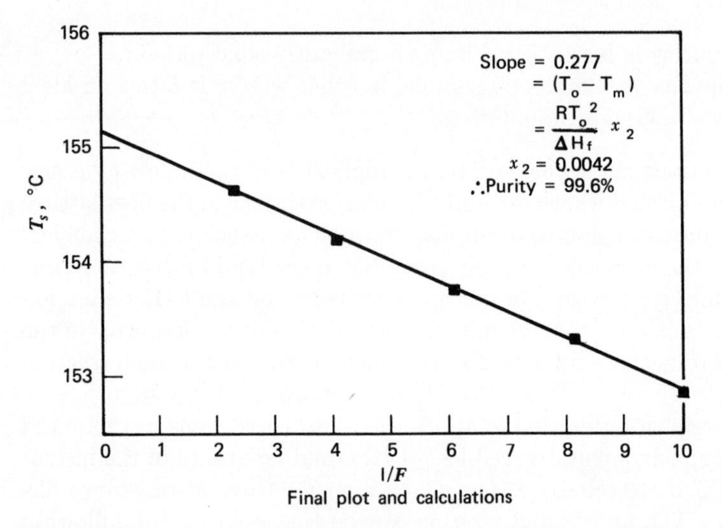

Figure 12.10. Graphical linearization of a 1/*F* plot and final corrected form (48).

The assumptions of the van't Hoff equation require that the plot of T_s versus $1/F$ be linear.

In applicable cases the following precautions must be taken for accurate results (3).

1. The sample size should be less than 3 mg.
2. The heating rate should be $\leq$ 1.25°C/min.

3. Due to the thermal conductivity and the finite vapor pressure of most organic materials encapsulation in a "volatile sample sealer" modified to maintain good thermal contact is necessary.

4. The temperature axis should be precisely calibrated.

5. The area considered for $1/F$ calculation must start at the first detectable melting and finish with a point at the endothermal minimum and contain at least six points.

6. Heating rate and sample size must be adjusted so that the slope of the endotherm never exceeds the slope of the pure standard at $\frac{1}{2}$ peak height of the standard.

7. The thermal lag must be measured with a standard which melts near the sample.

These seven steps are all necessary because of the inherent thermal lag and response time of the calorimeter and the requirements of a close approach to thermal equilibrium.

In cases where the van't Hoff equation does not apply, thermal methods may still be of help. A co-crystal melting point is usually very sensitive to composition (12). Calibration with standards is necessary. The amorphous polymer fraction in crystalline polymers is customarily determined by DTA or DSC— and these systems are far from ideal. It is not uncommon in mixtures and copolymers of crystalline polymers that chain types will segregate. Then the two endotherms of melting may be used for a relative analysis (9). Clay and other mineral mixtures have been successfully analyzed on a quantative and qualitative basis by DTA. It is not unusual that mixture analysis by DTA or DSC is simpler and more accurate than chromatographic or spectrophotometric methods, for solution, separation and reshaping of the sample are not required in most thermal methods.

B. Phase Diagrams

DTA has been suggested as an ideal means of rapidly obtaining phase diagrams of two and three component systems. All that is required is that phase transitions, eutectic formation, and so on, be rapid. This is usually the case. A small body of literature is available (55), but broader application is possible. Frequently it is possible to detect eutectic melting from only the shape of the DTA trace. Eutectic melting produces a symmetrical endotherm, while impure materials produce skewed endotherms. Glasses or supercooled liquids can be detected from heat capacity data. In addition many glasses nucleate and recrystallize near the nominal melting point. This produces a DTA trace of an exotherm (recrystallization) followed by an endotherm (melting). Metastable solids produce the same effect but lack unique heat capacity changes.

C. Kinetic Parameters

Numerous studies of the types

Solid I → vapor
Solid I → solid II + gas
Solid I → solid II

and variations thereof have been reported. Several mathematical treatments of kinetic data have been published. All make use of the Arrhenius equation:

$$K = Ae^{-E/RT} \tag{12.8}$$

where K is the rate of reaction; A is the "frequency" factor; E is the energy of activation; R is the gas constant; and T is the temperature at which the rate $= K$.

If logs are taken of both sides of the equation,

$$\log K = \log A - \frac{E}{2.303RT} \tag{12.9}$$

A plot of $\log K$ as a function of $1/T$ should be linear with a slope of $-E/2.303R$. If K is known at two temperatures:

$$\log \frac{K_2}{K_1} = \frac{E}{2.303R}\left(\frac{T_2 - T_1}{T_1T_2}\right) \tag{12.10}$$

A number of approaches have been developed to use the above relations for first order reactions. Higher reaction orders have been explored, but usually do not give results in agreement with other methods.

The value of K may be estimated in a number of ways.

1. The instantaneous slope of a single thermogram may be determined at several temperatures. This is related to cal/deg or cal/sec which can be converted to grams/sec if we know the total heat capacity.

2. The partial area to various temperatures may be determined. Divided by the total area this is proportional to K.

3. From a series of *isothermal* runs at different temperatures it is possible to measure the rate by determining the area fraction converted at a fixed time or time at a fixed fractional area. This assumes that a series of temperatures exist where the reaction is slow enough to permit the thermograph to balance before reaction starts (24).

4. From a series of runs at different heating rates and the peak temperature enough data are obtained to use Equation 12.10. This assumes that the rate is infinite at the vertex temperature. This is experimentally true in many cases.

From the above, it is obvious that relative values of E may be obtained if certain rather restrictive assumptions have been made. Fava has developed a technique which overcomes some of the above problems (24). The sample is rapidly heated to some fixed temperature (at which the reaction goes slowly) for a known amount of time and then quenched. The quenched sample is reheated at about 10°C/min in a DSC or calorimetric DTA and the residual (unreacted) material allowed to react—producing an exotherm. The area of this exotherm is compared with the exotherm of a sample which has received no heat treatment. Thus K may be obtained from the ratio (fraction) and the time of heat treatment. Several experiments at different isothermal temperatures are made and Equation 12.9 or 12.10 is employed. This approach assumes only that Equation 12.8 is applicable.

In all rate measurements it is absolutely essential that all products be kept within the sensitive area of the DSC or DTA. If products are allowed to escape, the rules of equilibrium reaction cannot be employed. This removes the study from the realm of classical kinetics and, concurrently, from the realm of the interpretable. In addition, the heat of vaporization (often very large) is added to the heat of reaction (opposite in sign). Thus an exothermal reaction may be noted by an endotherm if the high heat content products are allowed to escape.

The kinetics of the above measurements are taken at constant volume. This is unfortunate since the large body of physical chemical literature is based on observations obtained at constant pressure. Various attempts at obtaining isobaric data have been made, but these have met with limited success.

The kinetics of explosives have received some attention by DTA. The actual calculations are no different from conventional measurements. However encapsulation and cell design are specialized to withstand the rapid rise in gas pressure at detonation (14). Although sample sizes are kept in the milligram range, the final pressure can amount to several hundred atmospheres. DTA scans can often be used to separate a detonation from a simple decomposition on the basis of rate alone.

Crystallization processes in minerals, crystalline polymers and other materials are amenable to rate studies by DTA and DSC. The previously given Arrhenius equation treatments and the Avrami equation (12.11) may be used.

$$\ln (1 - \theta) = Kt^n \tag{12.11}$$

where θ = degree of crystallinity
t = time to reach a given crystal size
K = constant containing the nucleation and growth parameters
n = an integer whose value depends on the nucleation mechanism

Morgan (40) has developed a large body of data concerning the significance of n. DTA is used to evaluate θ. A plot of $-\ln(1 - \theta)$ versus $\ln t$ produces a straight line of slope n if the system fits within the limits of nucleation assumptions. Nonintegral values of n are obtained if more than one growth mechanism is at work in transforming the liquid into a solid.

D. *Thermal Stability Studies*

The measurement of thermal stability is a difficult process. Frequently physical transformations and reactions which require only a few millicalories have profound effects on the physical properties of a material. Determining such a small amount of heat is very difficult. The DTA can be used to heat-treat samples for other, more sensitive methods (see Section II.5.D). In cases where applicable—and these are far fewer than the general literature indicates—the problem is a kinetic study. Frequently the sample atmosphere must be dynamic (flowing) and controlled. This introduces a group of problems which have been partially solved in specialized studies. In any given case it is impossible to state if DTA or DSC is applicable with any degree of certainty in the absence of several experimental runs. Any correlation between "in use" thermal stability and DTA endotherms or exotherms is highly compound and system specific. DTA is more likely to be a useful technique if the decomposition is auto-catalytic.

E. *Combined Methods*

DTA and DSC reach their maximum usefulness when combined with other instrumental methods. These applications may be either serial or simultaneous. Serial techniques are recommended when it is desirable to use two techniques under optimum conditions for each technique. Simultaneous methods usually involve sacrificing one or more optimum conditions for the sake of speed or experimental expediency. A case point is a study by Wendlandt (53). DTA and gas chromatography were used to study the decomposition of complex salts and chelates. The DTA required that the injection volume to the chromatograph be large and the chromatograph required that the DTA operate in a flowing atmosphere with free diffusion of the products. These two requirements do not allow either of the two techniques to be operated under optimum conditions. However, because of careful experimental design and a good understanding of limitations, the identity of each gaseous component given off at each endotherm or exotherm was positively established and the original problem could be solved.

There are a number of commercial machines which combine two or more methods. Instruments which combine as many as four methods have been described (16). Reasonably compatible sets are: DTA/thermogravimetry,

DTA/gas chromatography, DTA/microscopy, DTA/conductivity, DTA/x-ray diffraction, DTA/mass spectroscopy, and DTA/reflectance spectroscopy.

Two instrumental solutions for simultaneous operation are possible: (1) use one sample and one furnace, and (2) use two samples and one furnace. In the first case the thermocouples must be mounted in the sample in such a manner that they do not interfere with the other measurement. This offers some geometrical problems. In the second case the assumption must be made that both samples are acting synchronously. This is not always the case in kinetic processes due to furnace thermal gradients or samples of greatly different masses.

Great caution must be exercised in the analysis of data obtained by either simultaneous or consecutive techniques. Many materials have unexpected polymorphic and metastable forms. However because of the great increase in usefulness of the thermal information, when combined with other methods, these techniques are recommended for the solution of many problems.

5. Useful Auxiliary Methods

In many cases it is impossible to assign a specific phase transition to a given endotherm or exotherm. There is no rule of thumb for separating a solid-to-solid transition from a solid-to-liquid transition on the basis of DTA curves alone. Many transitions are thermally very small. There is great opportunity to overlook important phase changes due to operation at the wrong sensitivity. This is very true of mesophase forming materials. DTA is incapable of either identifying the gases evolved during the decomposition of a complex salt or separating phase changes from decomposition endotherms. In short, auxiliary methods are required with both DSC and DTA in many cases. A few methods which the present author has found useful are given below.

A. Microscopy

A simple hot-stage, low-power microscope equipped with a substage polarizer and an eyepiece analyzer is a great aid in both DTA and DSC. With this simple equipment transitions may be located rapidly, the nature of phase changes determined and the sample examined for evidence of polymorphy. A somewhat more sophisticated system has been reported in which the light translated from the polarizer plane to the analyzer plane is monitored by a photocell (5). A record of the light received at the photocell versus temperature of the stage is very useful. The method has been called depolarized light intensity analysis (DLI). Since DLI is sensitive to optical anisotropy and not to heat uptake, the records are somewhat dif-

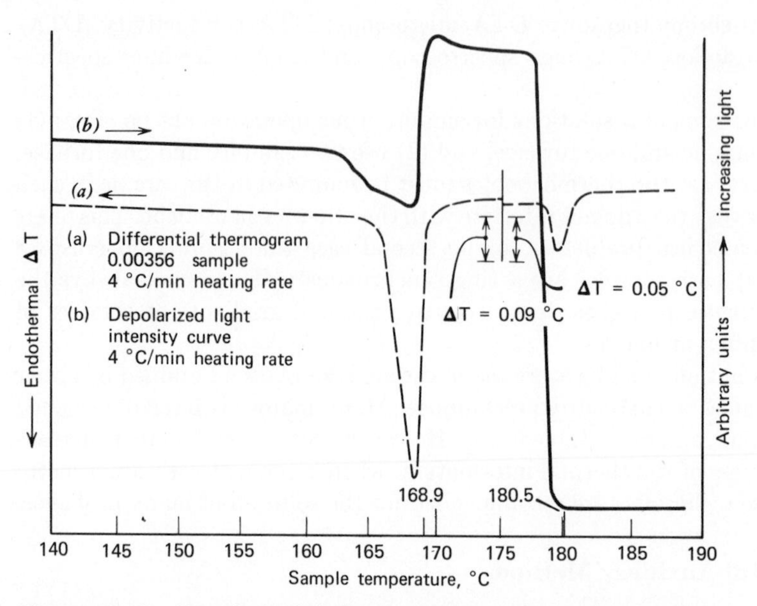

Figure 12.11. Depolarized light intensity analysis of anisaldazine compared with differential thermal analysis on the same sample (5).

ferent from thermograms. Since large changes in anisotropy can occur with only a minor change in enthalpy, the records are useful in determining if all transitions have been accounted for by DTA. Figure 12.11 is an example of such behavior. Several authors have used DLI in conjunction with DTA and/or x-ray diffraction in the study of organic and polymeric systems (37).

B. Gas Chromatography

The identification of gaseous products from DTA has been the subject of several papers (53). Gas analysis without a chromatographic column, also called effluent gas analysis (EGA), has been used by Gray in a very excellent study of the decomposition of lithium aluminum hydride (13). Several commercial instruments have simultaneous EGA as an optional feature. For materials which are gaseous at or near room temperature the technique should be very helpful in the identification of individual processes responsible for endotherms and exotherms.

C. Mass Spectrometry

This technique which is capable of dealing with the analysis of relatively complex molecules should be very useful in DTA. Application has not been frequent probably due to the large instrumental cost (33). Usually this method is used in conjunction with thermogravimetry (TGA) of polymeric and other types of organic materials. The analysis of sorbed gases should be greatly facilitated by this method and DTA. However this has usually been studied by TGA/MS.

D. Sample Preparation for Other Methods

DTA and DSC are excellent preparative methods for x-ray and other physical techniques. A sample may be prepared in a known crystal form and quenched to preserve the final form before subjecting it to other analyses. The DTA record furnishes proof that the required thermal transformations have taken place. Heat treating on a small, precise scale can usually be done more elegantly with a DTA than by any other method. The release of strains in metal and polymer samples can be verified prior to running the prepared sample in an Instron Rheometer, for example. Heat treatment cycles for catalyst activation and rejuvenation are most conveniently established by DTA.

III. APPLICATIONS OF THERMOGRAVIMETRIC ANALYSIS

Procedures of classical gravimetric analysis, long discarded because of difficult weighing forms and tedious oven dryings, can be profitably revived by the use of TGA. Continuous recording of weight as a function of temperature to a large extent eliminates the necessity of "constant weight" as conceived in classical gravimetry. Duval has composed an encyclopedic text of TGA applied to gravimetric analyses (19). Many analyses which require difficult separations and expensive spectrophotometric apparatus can be done by relatively inexpensive TGA equipment in a single operation.

1. Instrumentation

A. Types of TGA

Modern thermobalances fall into four working categories: beam and fulcrum, spring, cantilever, and torsion arm. These have found broad application and a number of satisfactory commercial instruments are available. Figure 12.12 shows the four types. No division can be made depending on the organic or inorganic nature of the samples.

B. Heating

Temperature programming is conducted in the same way as in DTA. Many commercial DTA and DSC instruments also accommodate thermo-

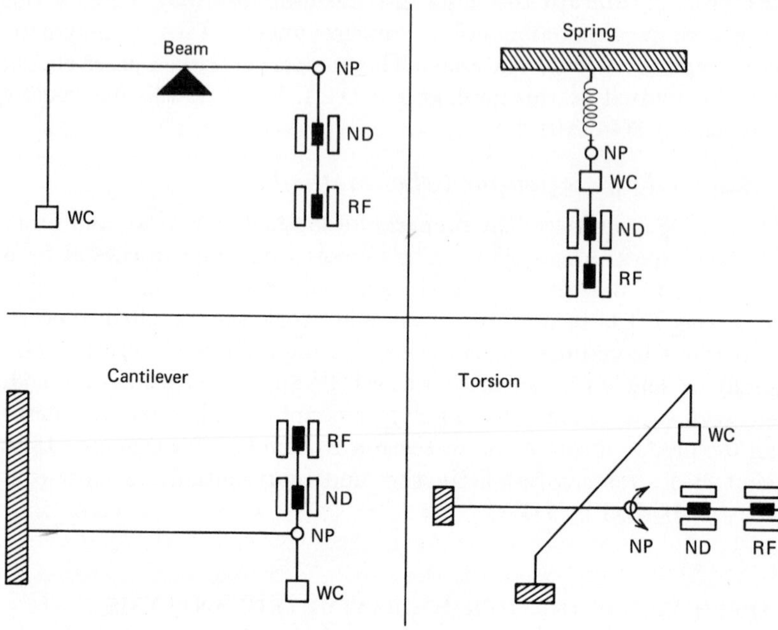

Figure 12.12. Basic types of recording balances (2). *WC* = weight change; *NP* = null point; *ND* = null detector; *RF* = restoring force.

balances. The furnace of the Perkin-Elmer thermobalance is unique since the furnace winding resistance is used to measure the furnace temperature. Other equipment uses conventional thermocouples or resistance thermometers. Furnaces, because of the generally higher temperature ranges (even for organic TGA), are of heavier design than in DTA. Platinum or tungsten windings are commonly used. The upper temperature of most instruments is ~ 1200°C. At least one commercial instrument is capable of reaching 1600°C. At higher temperatures an inert gas blanket is usually required. With the rapid heating and cooling commonly employed the thermal shocks experienced by the furnace structure are severe. Quartz and magnesium oxide are commonly employed as well as other refractory materials.

C. *Weight Measurement*

If very large samples are to avoided, very sensitive mechanisms must be employed to detect the sample weight change. In addition, these devices must be very stable in an absolute sense. It is not uncommon in certain kinds of research for single samples to be weighed continuously for a matter of days. Although TGA runs can be rapid and on the time scale of DTA and DSC, studies at isothermal conditions may go for very long time periods. The long term requirement for an absolute measurement of weight pushes electronic technology to near its limit.

As a result, almost every conceivable system has been employed to follow weight change. The early thermobalances used a mirror attached to the beams of a conventional balance assembly. The mirror reflected a ray of light onto a slowly rotating sheet of photosensitive paper for recording. Servomotors have been used to drive conventional chain assemblies, with null being detected by split photocells. Electrodeposition has been employed. The out of balance signal from a pair of null photocells caused an emf to flow in a plating circuit. This resulted in copper or silver being plated onto or stripped from an electrode on the reference pan. Telescope systems have been employed in long term experiments to measure beam or spring movement with an eyepiece micrometer. Movable core transformers have been used. The signal in the secondary coil is proportional to the location of the linear core. This system is employed in some modern commercial instruments. A beam of light obstructed by a vane attached to the end of a quartz spring has been used in quartz spring balances. Photoelectric detection was used in that case. Several modern systems make use of a torque motor. This is essentially a galvanometer with the sample attached to the needle. The weight of the sample is proportional to the current required to restore the needle to some null position. Null is detected by a pair of split photocells behind a vane in a light beam. A simple differential thermo-balance has been constructed in which the rate of beam or spring movement is detected. A magnet suspended in a coil of wire acts as a linear generator. The faster the weight change the greater the linear velocity of the magnet and the greater the voltage generated in the coil. The polarity of the signal is determined by the direction of movement. Wendlandt's text gives many useful examples of balance design (54).

Various damping systems have been employed. The modern trend is away from oil dash pots or magnetic damping with a metal plate and large magnets. The torque motor is self-damping. Magnetic reactance damping is used in some nontorque motor systems. Most modern systems seek to remove the necessity for damping by using a very small moving mass. This requires that all components of the balance be as low in mass as is consistent with structural integrity. The low mass requirement excludes the use of heavy magnets and transformer cores in weight detection. In quartz spring balances a certain degree of damping is obtained from the torque placed on the individual quartz wires by the load. To make optimum use of this structural damping it is important to keep the period of the vertical oscillations as short as possible. This requires a very low mass sample weight detector. Ideally the largest moving mass in the system should be the sample.

Various suspension systems have been described (2, 21). The torque motor systems usually employ a ribbon suspension to carry the moving coil. Since the motor operates only in the null position, the potential torque applied by the deformed ribbon is never realized and is thus not a source of error. Beam balances, especially those without a continuous null system, employ conventional agate plate and knife edge suspension. The suspension in quartz spring balances is the spring itself.

D. Temperature Measurement

Precise measurement of the sample temperature is one of the most difficult problems in TGA. It is also the chief source of error. During reaction of a sample with the balance atmosphere, the important quantity is the temperature of the surface of the sample. The temperature of the gas on and over the surface is usually different and not important to the reaction kinetics. Self-heating during oxidation is a very serious problem. During autodecomposition the temperature of the interior of the sample is important. No generally applicable solution to these measurement problems compatible with the rest of the balance systems has been found. Thus a series of compromises has been worked out. It is important to remember that the measured temperature can be different by as much as 200°C from the rate controlling temperature.

Some balances of the beam type permit the location of a thermocouple within the sample. This is a physical impossibility in many other systems. Other instruments have a movable thermocouple which can be placed close to the sample cup or sample surface (but *not* in contact). A limited application of bolometers and optical pyrometers has been made. Since almost any system is subject to more or less serious error, it is important to develop a procedure and adhere to it closely for consistency. The following steps are suggested for work with thermocouples placed near to but not in contact with the sample:

1. Use the smallest weight of sample consistent with balance sensitivity.
2. Maintain a constant sample mass.
3. Insofar as possible maintain a constant sample geometry (particle size, shape or surface area).
4. Use the same heating rate for all samples. In some cases, for kinetic studies different heating rates are desired. These should not exceed 10°C/min.
5. Maintain the same constant flow rate in the balance atmosphere during all experiments.

6. Do not alter the thermocouple location.
7. Unless required by the experiment, the thermocouple should be located so that evolving gases do not cross it. These gases are generally at a temperature somewhat removed from that of the sample interior or surface.

E. Atmosphere Control

Since weight change in the sample is due to the formation of a gaseous product or the reaction of the sample with the balance atmosphere, regulation of this quantity is extremely important in TGA. Atmosphere control can be of two types: in the sample, or in the balance. Regulation of the atmosphere within the sample is usually effected by simply covering the sample container with a loose-fitting lid. Garn has described several such arrangements (29). If the overhead of the sample cup is large and the weight of the evolved gas is small, it is possible to maintain a sample generated atmosphere during the whole of a process. This is extremely important in certain kinds of kinetic studies. Gas producing reactions are very sensitive to product concentration (Le Chatelier's principle). Back reaction and side reactions shape the reaction path and kinetics to a very great extent. It is particularly important if TGA is being used as an analog of some real process (example, industrial scale calcination) that atmosphere conditions closely follow those present in the real process. This requirement and the lack of its observation accounts for many varied results reported by TGA. The weight loss curve of a simple hydrate or carbonate is greatly affected by the way in which the sample atmosphere is generated.

It is very important to remember that the rate of enclosed reactions can be almost totally controlled by the diffusion rate of the products away from the sample. That is, temperature may become a secondary term in controlling the rate. This is particularly true of dehydrations and carbonate decomposition.

As an alternative to a self-generated atmosphere the whole balance may be filled with a given gas. This has led to the development of complex gas control devices. In precise decomposition studies it is not uncommon that the greater portion of the TGA equipment is devoted to gas regulation. Atmosphere control is *not* a secondary function in TGA.

A few general statements may be made concerning atmospheres. "Inert" atmospheres of nitrogen, argon, carbon dioxide, hydrogen, and helium have been used. These gases in general use permit TGA at known pressures without back or side atmospheric reactions.

The author recommends the use of helium as inert atmosphere for the study of polymer decompositions. Other than the inertness helium has a very low density and very high thermal conductivity. Low density is important to avoid "buffeting" of the balance mechanism at high flow rates. Polymers generally decompose to yield relatively high molecular weight materials. Since back reaction is not important, once a fragment has formed it must be removed as efficiently as possible to avoid condensation elsewhere in the weighing mechanism. Helium flow rates of 50 ml/min through most balances are possible without introducing noise into the weight curve. This is not possible with denser gases.

Liquids are also used in TGA. Corrosion studies lasting several months are possible. The corrosion plate is suspended dipping into a pool of the liquid under study. Manual recording techniques are frequently used.

For serious work in the study of decompositions and atmosphere reactions it is necessary to regulate gas pressure as well as composition. Regulation should be from 10^{-6} torr to at least two atmospheres. Higher pressures have been used by several workers to good advantage. Studies involving hydrogenation usually require pressure.

In many studies it is desirable to have the reactive gas diluted to different concentrations with an inert gas. This simplifies kinetic considerations by keeping the total pressure constant. Many reactive gases can be bought premixed. This simplifies metering problems.

It is extremely important not to expose the metal parts (if any) of the balance mechanism to a corrosive atmosphere introduced into the balance or evolving from the sample. Many commercial TGA machines have provisions for blanketing the delicate balance parts with dry inert atmosphere. It is good practice to do this even when no corrosion is expected. This also prevents the deposit of high-boiling liquids in the mechanism when organic materials are studied.

F. Sample Cups

Sample cups are of four basic designs: shallow pan, deep crucible, covered and retort.

Shallow pans are used where it is desired to eliminate diffusion as the rate controlling step. As volatile material is produced throughout the sample mass it must diffuse to the surface to escape and be registered as a weight loss. Side reactions may occur. For polymer decomposition studies this effect is very undesirable; thus the sample is arranged in as thin a layer as possible. As soon as a volatile fragment is formed it is free to leave. Therefore bond breakage is noted as a prompt weight loss at the appropriate temperature.

Deep crucibles are used where side reaction and/or partial equilibration is desired. Such is the case where industrial scale calcinations are studied. These crucibles are also useful in surface area studies which will be discussed later (see Section III.4.B).

The usefulness of loosely covered cups in self-generated atmosphere studies has been mentioned earlier. Since many of the studies reported with this equipment have been done isothermally, rate of weight loss and not exact temperature is important. Delays due to diffusion are not important at the steady state.

Retort cups are a special class of covered cup. These usually resemble an alchemist's retort suspended so that the lip points up. Such cups with a controlled orifice size are useful in boiling point studies. The retort furnishes the single plate of reflux essential for a simple boiling point determination.

There are many other designs reported and available commercially, but the majority will fit into one or more of the above classes. Some unique shapes have been evolved for use in simultaneous DTA and TGA studies. The literature describes these (28).

2. Qualitative and Quantitative Analyses

A. Inorganic Mixtures

These mixtures are of two types—natural and synthetic. An example of a natural mixture would be a complex ore or mineral. The stepwise decomposition of each component gives a characteristic weight loss. If the identity of component lost can be determined or is known, a very good analysis is possible. A synthetic mixture would be the result of a precipitation of a group of metals (for example) by a known agent. The present author has found the classical qualitative analysis sulfide scheme very useful in rapid and accurate metal analysis. The problems of sample preparation are much simplified over spectroscopy and spectrophotometry.

The following program is suggested:

1. Heat the mixture in flowing nitrogen and follow the weight changes. Identify the effluent gases (these are usually limited to water, carbon monoxide and dioxide, sulfur dioxide, and oxygen). Some metal oxides are also volatile.
2. Admit oxygen to the system carefully and note the weight increase.
3. Cool and reheat in flowing hydrogen. The reduction of many materials is highly specific. Given the temperature-weight loss relationships reasonably good analyses may be obtained. Duval's book is invaluable in this work (19).

B. Organic Mixtures

Use is made here of the different volatilities and reactivities of organic materials. The techniques are somewhat sensitive to the atmosphere within the balance. If the material is a high-boiling liquid or polymer, the study should first be carried out in an inert atmosphere. The method is very useful for the analysis of small amounts of volatile or unstable material in the presence of nonvolatile or thermally stable material. In addition, a char frequently forms which can be measured by burning off in a second run in oxygen. Filled polymers, elastomers, and many natural products may be studied in this way. TGA is one of the few dependable methods for the determination of carbon black (a common filler) and zinc oxide in rubbers. The thermogram of a polymer mixture is frequently useful in determining the number of components, the type of polymer, and the presence of residual fillers and metal catalysts. For example, polyvinyl chloride decomposes in two sharp steps due to the dehydrohalogenation at low temperatures followed by the high-temperature decomposition of the polyene structure. Nitrile polymers cyclisize to form high temperature char products (styrene–acrylonitrile mixtures and copolymers are easily treated by TGA). Residual Ziegler catalysts may be identified by the metal residue. Plasticizing agents generally vaporize from thin polymer films prior to polymer decomposition. Cross-linked polymers such as urethanes and phenolics may be identified and categorized by TGA.

3. Thermal Stability Studies

TGA is suitable only for *gross* thermal stability determinations. Changes which affect the physical properties of many materials (especially polymers) occur at temperatures much lower than the first significant weight loss. The onset of weight loss is a reliable measure of the temperature at which the physical properties of a material are surely changing. Thus most TGA data set an upper limit on expected thermal stability.

A. Kinetics

In theory the kinetics of any reaction of the types:

Solid I → gas
Solid I → solid II + gas
Solid I + gas → solid II

may be determined by TGA. An excellent discussion of dynamic and isothermal methods has been given by Doyle (21). A detailed description of the methods will not be given here. It is sufficient to say that all current methods are based on the Arrhenius equation discussed in Section II.4.C. This equation is self-limiting in many respects. In addition all methods suggested to date suffer more or less seriously from procedural and experimental errors. Energy of activation and reaction order from programmed

TGA evidence should always be held in doubt until confirmed by other techniques. In many cases TGA data are the only experimental facts available and must be used. To facilitate such calculation the following three methods are given. These methods have the advantage of simplicity of manipulation and relative freedom from error. In addition if error is present the results will be complete and obvious nonsense.

The first method given here was reported by Broido (15) and in the form developed below, applies only to first order reactions. This includes the majority of simple pyrolyses. This equation is derived in detail in the above cited paper *only* for situations where the simple Arrhenius equation is applicable. The weight loss at any given temperature is defined as

$$y = \frac{(W_t - W_\infty)}{(W_0 - W_\infty)} \tag{12.12}$$

where W_t is the indicated weight at time or temperature, t; W_0 is the initial sample weight; and W_∞ is the residue weight or weight at the end of the process under study.

In practice a plot of $\ln \ln(1/y)$ as the y-axis versus $1000/T$ (where T is the temperature at weight loss t in °K) is made. This must yield a straight line if the data are accurate and the reaction is first order. If a straight line is obtained the energy of activation, $-E$, is R times the slope of the line. The method is very sensitive to weighing errors, especially in W_0 and W_∞. Proof has been offered that if the plot is linear over the total range of decomposition the reaction must be first order.

Coats and Redfern have proposed a method which will give reasonable values of E only if the assumed kinetics are correct (17). In practice a plot of $\ln 2[1 - (1 - \alpha)^{1/2}]/T^2$ versus $1000/T$ is made as above. The term α in this case is y of the previous notation. All temperatures are in °K. The slope of this linear plot is equal to E/R for first-order reactions.

Achar et al. (1) have proposed a differential method for the calculation of E. The method involves taking the instantaneous slope of the thermogram, $d\alpha/dT$. Thus the method cannot be recommended if a conventional TGA plot is obtained. Instantaneous slopes are difficult to measure and subject to great error in the important regions of the decomposition. If a plot of the differential thermogravimetric trace is available (generated electronically), then the method is worth at try. The height of the differential plot at any temperature is proportional to $d\alpha/dT$ and is easily measured. In practice a plot of $\ln [(d\alpha/dT)/(1 - \alpha)^{1/2}]$ is made versus $1/T$. If the first-order kinetics are correct, a straight line will be obtained with a slope of $-E/R$ and a y-intercept of A, the Arrhenius frequency factor.

Caution is urged in the application of the above methods for the reasons given previously. However if isothermal TGA data are available the chances for determining a real value for E are somewhat better. In isothermal TGA the sample is rapidly heated to some predetermined temperature and the weight loss is recorded as a function of time. A family of such plots is obtained from isothermal runs at different temperatures. The slopes of the lines obtained at different temperatures (a rate term) are plotted as a function of weight percent of sample volatilized. Depending on the region of interest, a fixed percentage of decomposition is chosen and the log of the rate is plotted as a function of the reciprocal temperature for that curve. The slope of the curve is $-2.303E/R$. With this technique the activation energy for various stages of the reaction may be readily calculated (36).

B. *Relative Stabilities*

The determination of relative stability is not so much concerned with actual computation of values of E as with the determination of critical parameters which will permit intercomparison of several materials. The simplest notion rests on selecting critical temperatures which will describe the thermogram. Figure 12.13 shows a TGA trace with several critical temperatures indicated. The onset temperature is taken at the first sensible change in weight. This is a very variable number and not respresentative of any particular function—weight must have been lost before the balance

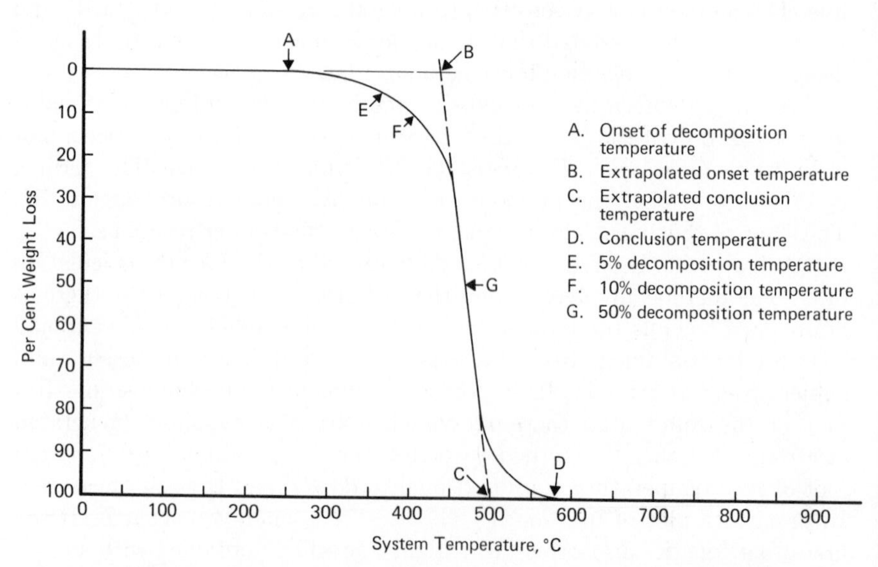

Figure 12.13. Temperatures frequently cited in thermogravimetry.

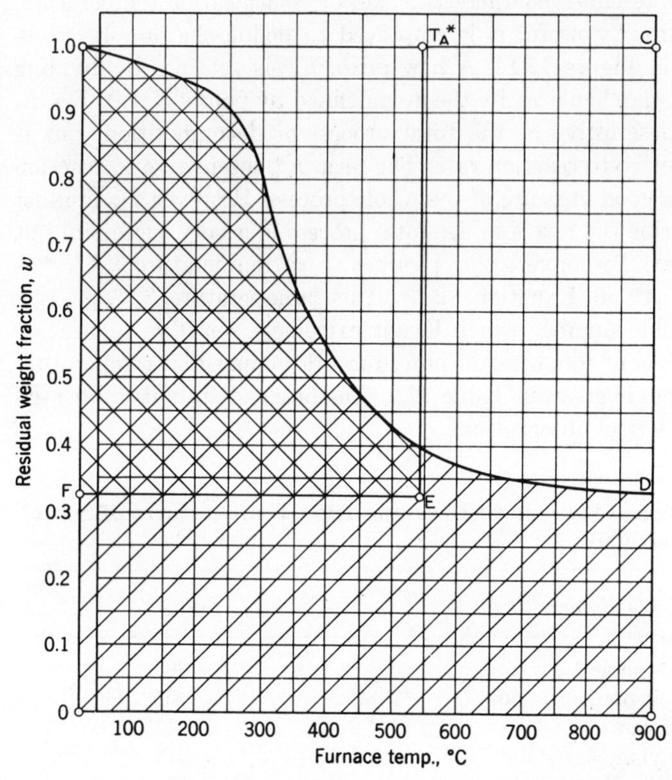

Figure 12.14. Thermogravimetric trace showing the method of integration for calculation of integral procedural decomposition temperature (IPDT) (20).

detected a change. The extrapolated break temperature is the intersection of two lines constructed as shown in Figure 12.13. This temperature is fairly reproducible and a useful point to know in estimating thermal stability. The temperatures at 5, 10, and 50% weight loss are even better measures of stability. These temperatures give an indication of the rate of decomposition.

In practice the curve in Figure 12.14 extends over the total range of the determination. The total area of the rectangle defined by points 0, 1.0, C, 900 is divided into the total cross hatched area defined by points 0, 1.0, D, 900. This produces the normalized fraction, A^*. This A^* is normalized not only with respect to weight, but also with respect to temperature. The term A^* is converted to a "temperature," T_{A^*} in Equation 12.13.

$$T_{A^*} = 875A^* + 25 \tag{12.13}$$

The term T_{A^*} represents a characteristic end of volatilization temperature. Although not directly useful it is employed to define the doubly cross-hatched region in Figure 12.14. A new ratio, K^*, is obtained by dividing the doubly cross-hatched area by the area defined by points F, 1.0, T_{A^*}, E.

Now, A^* is descriptive of the total process of decomposition, but is relatively insensitive to reaction rate. The term K^* is sensitive to reaction rate, but is not a good measure of the whole process. However the product A^*K^* does describe the rate and the total process in a very successful but empirical manner. To convert this product to a "temperature," A^*K^* is substituted for A^* in Equation 12.13. This "temperature," $T_{A^*K^*}$, is a highly reproducible number which is not extremely sensitive to heating rate. A comparison of this integral procedural decomposition temperature for eight polymers is given in Table 12.2. The order of stability indicated fits well with practical observations.

Table 12.2. Integral Procedural Decomposition Temperatures of Some Familiar Polymers (20)

Polymer	ipdt, °C
Polystyrene	395
Maleic-hardened epoxy	405
Plexiglass	345
66 Nylon	419
Teflon	555
Kel-F	410
Viton	460
Silicone resin	505

Although there are many complicating factors in the calculation of $T_{A^*K^*}$ the numbers generated furnish an index for the evaluation of thermal stability. In many cases this index defines a structure and a starting point for more sophisticated studies.

4. Special Applications

The usefulness of a modern thermobalance extends beyond the study of gas forming or gas consuming reactions. A short description of some more important non-kinetic techniques is given to show TGA flexibility.

A. Curie Point Determination

If a ferromagnetic material is placed on a thermobalance with the pole of a magnet above the sample, the indicated weight will be less than the real weight or the indicated weight without the magnet. If the sample is heated above the Curie point, the ferromagnetic properties will vanish causing an increase in indicated weight. A thermogram carried out in the presence of a magnetic field will indicate weight increases as each ferromagnetic component in the sample passes above its Curie temperature. Transitions from one ferromagnetic form to another may also be followed. The reverse case is obtained if the magnet is placed below the sample. Gray (26) has suggested using a series of wires of differing Curie temperature to calibrate a thermobalance.

With a more sophisticated electromagnet and shaded pole pieces advanced magnetic studies may be carried out (44). The technique is ideal for studying thermal effects on induced magnetism and evaluating the effects of heat on ferromagnetic materials.

B. Surface Area Measurement

A gas adsorbed on the surface of a solid obviously contributes to its weight. This effect can be used to determine nitrogen and other surface areas. In the conventional BET surface area measurement the volume or pressure change observed in a chamber with and without sample is determined. The difference is the amount of gas held on the surface of the sample. Many interesting gases for surface area studies are sorbed more by the walls of the apparatus than by the sample. This gives rise to a large blank and poor statistics in volumetric measurements. If the sample is weighed this problem is no longer important. An all-quartz balance can handle almost any vapor with the exception of wet hydrogen fluoride. The equipment is not much different from a well designed conventional TGA. Needed parts are as follows.

1. Thermobalance constructed of inert materials (quartz is preferable).
2. A good vacuum pump (fore pump and oil diffusion pump).
3. Low pressure gages.
4. Gas manifold.
5. Thermostat.
6. Stable time base recorder for weight change.

The sample surface is cleaned in a reasonably good vacuum ($\sim 10^{-6}$ Torr) using the TGA furnace to heat and the balance mechanism to record completeness of weight loss. The sample is then thermostated at the desired

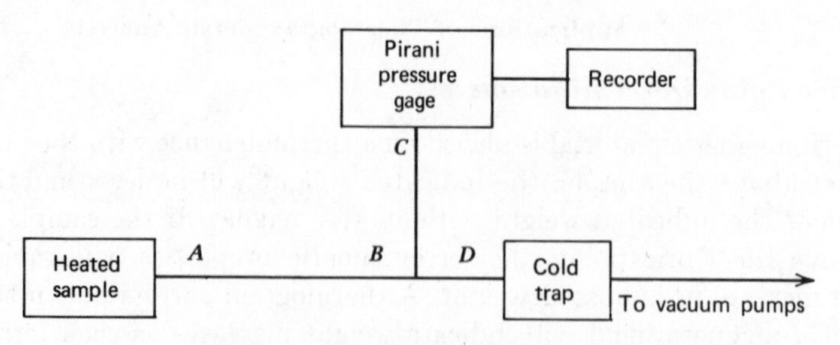

Basic layout for thermal volatilization analysis.

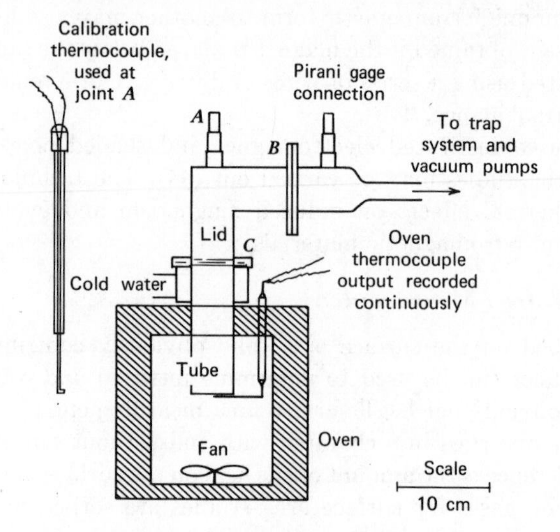

Arrangement of oven, sample tube, lid assembly, etc. for TVA.

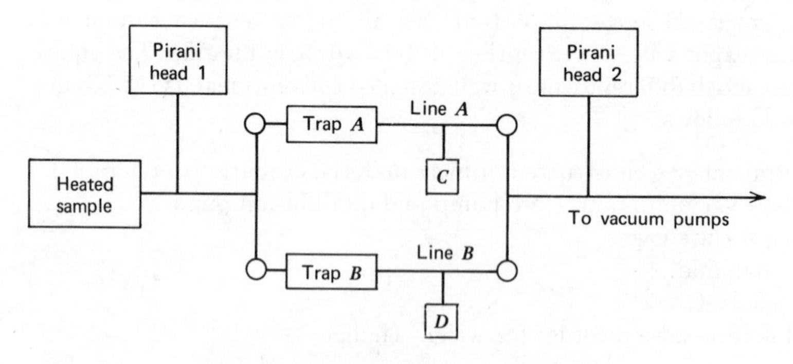

Parallel line TVA arrangement, to facilitate product fractionation according to temperature range.

Figure 12.15. Instrumentation and block diagram of a thermal volatilization apparatus (TVA) (39).

temperature and a known pressure of gas to be sorbed is set up. The balance mechanism and recorder indicate and follow the weight gain with time. After equilibration the gas pressure is increased to the next desired level. This method in the author's laboratory has produced a complete adsorption isotherm in sulfur hexafluoride as well as in nitrogen in 4 hr. The method should be applicable to the vapor phase swelling of polymers as well.

C. Sample Preparation for Other Methods

The thermobalance is an ideal reactor for reacting a solid with an inorganic and/or organic gas. The weight gain is an excellent indicator for rate and completeness of reaction. The formation of complex sulfides, carbonates, and oxides is easily observed and heating cycles can be regulated accordingly. Polymeric materials in various stages of degradation for physical testing can be made with some degree of facility.

5. Useful Auxiliary Methods

The effluent gas from a thermobalance during a complex decomposition has been analyzed by mass spectrometry and by various chromatographic methods. At present there is an increasing tendency to use TGA as the first stage in a multi-instrument approach for the analysis of such varied materials as polymers, oil shales, moon rocks, and catalysts (33). Combined with computer acquisition of data such arrangements as TGA–mass spectrometry or TGA–gas chromatography–mass spectrometry are among the most powerful general analytical approaches. Once again these methods are applicable only when the process of interest produces volatile fragments.

An outgrowth of TGA is a method which measures not weight change but pressure change over the sample. The instrument is shown in Figure 12.15. The pressure change noted is proportional to the pumping rate, evolution rate, and temperature of the trap. Volatile products which are formed at low temperatures and in very small yield may be detected. Detection somewhat below the sensitivity limits of a conventional TGA are possible. This instrument is especially useful in the study of the early critical phases of polymer decomposition (39). Alternatively flame ionization has been used to detect evolving gases (23). The apparatus is shown in Figure 12.16, and is commerically available from Carle Instruments, Fullerton, California.

IV. CONCLUSION

It is hoped that this brief chapter on DTA and TGA will furnish a guide to workers starting in these fields and as a reference of thermal methods to

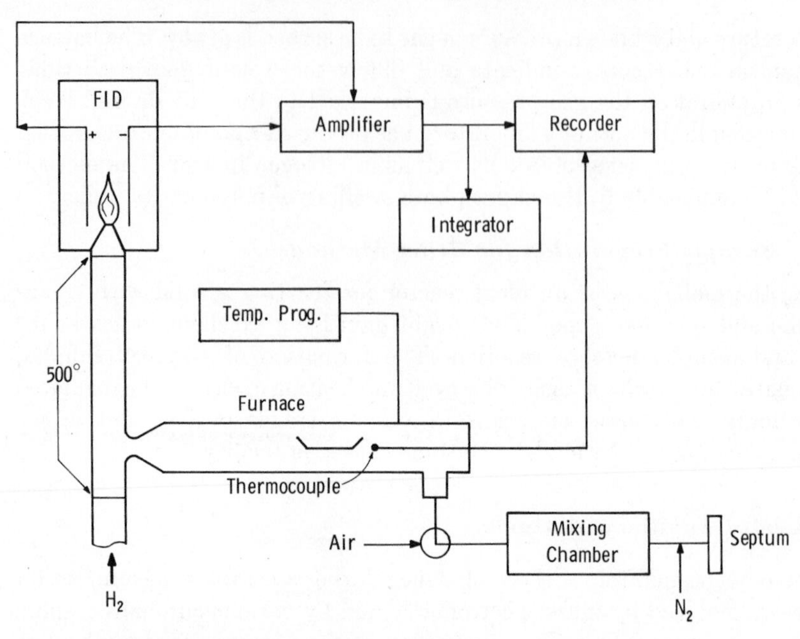

Figure 12.16. Diagram of a flame ionization apparatus for the detection of decomposition products (23).

analysts in other fields. Although the bibliography is not extensive, the references will furnish a key to the large body of the literature. The fields of DTA and TGA are currently growing so rapidly that any "authoritative" source retains its currency for but a short time. Recently *Chemical Abstracts* has included subject headings appropriate to various areas in DTA and TGA. However these headings function only when the authors of a paper have had the foresight to include certain key words in title and abstract. The listing is becoming more complete with time. The year 1970 saw the appearance of two new journals in the field of thermal analysis, *Journal of Thermal Analysis* and *Thermochimica Acta*. It is hoped that these two journals will do for thermal analysis what *Journal of Polymer Science* and *Makromolekulare Chemie* did for polymer chemistry. Heretofore thermal analysis papers have been spread over 50 or more journals. Excellent bibliographies for inorganic and organic TGA and DTA have been compiled by experts in the fields (17, 28, 35, 54).

The North American Thermal Analysis Society is beginning to exert some influence in areas of standardization and terminology. The American Chemical Society has sponsored three thermal analysis symposia in the past three years. The papers presented have been published as symposia volumes (see references). Current indications are that finding specific papers on these subjects will become easier in the future.

ACKNOWLEDGMENTS

The author wishes to thank Dr. Roy J. Gritter of IBM Research for his many helpful suggestions and proof reading of the manuscript and Mrs. Margene Yeaton for typing the manuscript.

Permission to use figures and tables from various texts and journal is gratefully acknowledged below:

Analytical Chemistry	Figures 12.1, 12.4, 12.7*a*, 12.8, 12.14, and Table 12.2
J. Appl. Polymer Sci.	Figure 12.1
J. Polymer Sci.	Figure 12.2
Thermal Analysis	Figures 12.2, 12.15, 12.16
Perkin-Elmer Corp.	Figures 12.2, 12.9, 12.10
Thermochim. Acta	Figures 12.6, 12.7*b*
Appl. Polymer Symp. No. 6	Figure 12.11
Techniques and Methods of Polymer Evaluation	Figure 12.12 and Table 12.1

REFERENCES

1. B. N. N. Achar, G. W. Brindley, and J. H. Sharp, *Proc. Int. Clay Conf. (Jerusalem)* **1,** 67 (1966).
2. H. C. Anderson, *Techniques and Methods of Polymer Evaluation,* Vol. I, P. E. Slade and L. T. Jenkins, eds., M. Dekker, New York, 1966 p. 87.
3. E. M. Barrall II and R. D. Diller, *Thermochim. Acta,* **1,509** (1970).
4. E. M. Barrall II, J. F. Gernert, R. S. Porter, and J. F. Johnson, *Anal. Chem.* **35,** 1837 (1963).
5. E. M. Barrall II and J. F. Johnson, *Appl. Polymer Symp. No. 8,* 191 (1969).
6. E. M. Barrall II and J. F. Johnson, *Techniques and Methods of Polymer Evaluation,* Vol. II, P. E. Slade and L. T. Jenkins, Eds., Dekker, New York, 1970, p. 9.
7. E. M. Barrall II, R. S. Porter, and J. F. Johnson, *Anal. Chem.* **37,** 1053 (1965).
8. E. M. Barrall II, R. S. Porter, and J. F. Johnson, *Anal. Chem.* **36,** 2172 (1964).
9. E. M. Barrall II, R. S. Porter, and J. F. Johnson, *J. Appl. Polymer Sci.* **9,** 3061 (1965).

10. E. M. Barrall II and L. B. Rogers, *Anal. Chem.* **34,** 1101 (1962).
11. R. A. Baxter, *Thermal Analysis,* Vol. 1, R. F. Schwenker, Jr. and P. D. Garn, Eds., Academic Press, New York, 1969, p. 75.
12. J. Block, *Anal. Chem.* **37,** 1414 (1965).
13. J. Block and A. P. Gray, *Inorg. Chem.* **4,** 304 (1965).
14. R. L. Bohon, *Anal Chem.* **35,** 1845 (1963).
15. A. Broido, *J. Polymer Sci. A-2* **7,** 1761 (1969).
16. J. Chiu, *Anal. Chem.* **36,** 2058 (1964).
17. A. W. Coats and J. P. Redfern, *Nature* **201,**, 68 (1964).
18. A. W. Coats and J. P. Redfern, *Analyst* **88,** 906 (1963).
19. C. Duval, *Inorganic Thermogravimetric Analysis,* R. E. Oesper, Trans., 2nd ed., Elsevier, New York, 1963.
20. C. D. Doyle, *Anal. Chem.* **33,** 77 (1961).
21. C. D. Doyle, *Techniques and Methods of Polymer Evaluation,* Vol. I, *op. cit.,* p. 113.
22. P. A. Einhorn, *Thermal Analysis, op. cit.,* p. 149.
23. F. T. Eggertsen, H. M. Joki, and F. H. Stross, *Thermal Analysis, op. cit.,* p. 341.
24. R. A. Fava, *Polymer* **9,** 137 (1968).
25. C. R. Foltz and P. V. McKinney, *Anal. Chem.* **41,** 687 (1969).
26. A. P. Gray, *Thermochim. Acta* in press 1971.
27. A. P. Gray and N. Brenner, *Amer. Chem. Soc. Div. Polymer Chem., Preprints* **6,** 956 (1965).
28. P. D. Garn, *Thermoanalytical Methods of Investigation,* Academic Press, New York, 1965.
29. P. D. Garn and J. E. Kessler, *Anal. Chem.* **32,** 1563 (1960).
30. K. Honda, *Sci. Rept. Tohoku Univ.* **4,** 97 (1915).
31. W. H. King, Jr., A. F. Findeis, and C. T. Camilli, *Analytical Calorimetry,* R. S.. Porter and J. F. Johnson, Eds., Plenum, New York, 1968, p. 261.
32. A. A. Krawetz and T. Trevorg, *Rev. Sci. Instr.* **33,** 1465 (1962).
33. H. G. Langer and T. P. Brady, *Thermal Analysis, op. cit.,* p. 295.
34. R. C. MacKenzie, *The Differential Thermal Investigation of Clays,* Mineralogical Society, London, 1957.
35. R. C. MacKenzie, Ed., *Differential Thermal Analysis,* Academic Press, New York, 1970.
36. S. L. Madorsky, *J. Polymer Sci.* **9,** 133 (1952).
37. J. H. Magill, *Polymer* **2,** 221 (1961).
38. C. Mazieres, *Anal. Chem.* **36,** 602 (1964).
39. I. C. McNeill and D. Neil, *Thermal Analysis,* Vol. 1, *op. cit.,* p. 353.
40. L. B. Morgan, *Progress in High Polymers,* Vol. 1, J. C. Robband and F. W. Peaker, Eds., Heywood, London, 1961, p. 272.
41. M. J. O'Neill, *Anal. Chem.* **36,** 1238 (1964).
42. W. C. Roberts-Austen, *Proc. Inst. Mech. Eng.* (*London*) **1,** 35 (1899).
43. R. F. Schwenker, Jr. and R. R. Zuccarello, *J. Polymer Sci.* **C6,** 1 (1964).

44. E. L. Simmons and W. W. Wendlandt, *Anal. Chim. Acta* **35,** 461 (1966).
45. W. J. Smothers and Y. Chiang, *Differential Thermal Analysis,* 2nd ed., Chemical Publishing Co., New York, 1966.
46. S. Strella, *J. Appl. Polymer Sci.* **7,** 569 (1963).
47. *Thermal Analysis Newsletter, No. 9,* Perkin-Elmer Co., Norwalk, Conn., 1970, p. 3
48. *Thermal Analysis Newsletter, No. 5, op. cit.,* pp. 4, 5.
49. *Thermal Analysis Newsletter, No. 7, op. cit.,* p. 4.
50. D. A. Vassalo and J. C. Harden, *Anal. Chem.* **34,** 132 (1962).
51. M. J. Vold, *Anal. Chem.* **21,** 683 (1949).
52. E. S. Watson, M. J. O'Neill, J. Justin, and N. Brenner, *Anal. Chem.* **36,** 1233 (1964).
53. W. W. Wendlandt, *Anal. Chim. Acta* **27,** 309 (1962).
54. W. W. Wendlandt, *Thermal Methods of Analysis,* Interscience, New York, 1964.
55. M. Zief, Ed., *Fractional Solidification,* Vol. 2, Dekker, New York, 1969.

Index